# Building Stone Decay: From Diagnosis to Conservation

## Geological Society books refereeing procedures

The Society makes every effort to ensure that the scientific and production quality of its books matches that of its journals. Since 1997, all book proposals have been refereed by specialist reviewers as well as by the Society's Books Editorial Committee. If the referees identify weaknesses in the proposal, these must be addressed before the proposal is accepted.

Once the book is accepted, the Society Book Editors ensure that the volume editors follow strict guidelines on refereeing and quality control. We insist that individual papers can only be accepted after satisfactory review by two independent referees. The questions on the review forms are similar to those for *Journal of the Geological Society*. The referees' forms and comments must be available to the Society's Book Editors on request.

Although many of the books result from meetings, the editors are expected to commission papers that were not presented at the meeting to ensure that the book provides a balanced coverage of the subject. Being accepted for presentation at the meeting does not guarantee inclusion in the book.

More information about submitting a proposal and producing a book for the Society can be found on its web site: www.geolsoc.org.uk.

It is recommended that reference to all or part of this book should be made in one of the following ways:

PŘIKRYL, R. & SMITH, B. J. (eds) 2007. *Building Stone Decay: From Diagnosis to Conservation*. Geological Society, London, Special Publications, **271**.

MCCABE, S., SMITH, B. J. & WARKE, P. A. 2007. An holistic approach to the assessment of stone decay: Bonamargy Friary, Northern Ireland. *In*: PŘIKRYL, R. & SMITH, B. J. (eds) *Building Stone Decay:From Diagnosis to Conservation*. Geological Society, London, Special Publications, **271**, 77–86.

GEOLOGICAL SOCIETY SPECIAL PUBLICATION NO. 271

# Building Stone Decay: From Diagnosis to Conservation

EDITED BY

R. PŘIKRYL
Charles University, Prague

and

B. J. SMITH
Queen's University, Belfast

2007
Published by
The Geological Society
London

# THE GEOLOGICAL SOCIETY

The Geological Society of London (GSL) was founded in 1807. It is the oldest national geological society in the world and the largest in Europe. It was incorporated under Royal Charter in 1825 and is Registered Charity 210161.

The Society is the UK national learned and professional society for geology with a worldwide Fellowship (FGS) of 9000. The Society has the power to confer Chartered status on suitably qualified Fellows, and about 2000 of the Fellowship carry the title (CGeol). Chartered Geologists may also obtain the equivalent European title, European Geologist (EurGeol). One fifth of the Society's fellowship resides outside the UK. To find out more about the Society, log on to www.geolsoc.org.uk.

**The Geological Society Publishing House** (Bath, UK) produces the Society's international journals and books, and acts as European distributor for selected publications of the American Association of Petroleum Geologists (AAPG), the American Geological Institute (AGI), the Indonesian Petroleum Association (IPA), the Geological Society of America (GSA), the Society for Sedimentary Geology (SEPM) and the Geologists' Association (GA). Joint marketing agreements ensure that GSL Fellows may purchase these societies' publications at a discount. The Society's online bookshop (accessible from www.geolsoc.org.uk) offers secure book purchasing with your credit or debit card.

To find out about joining the Society and benefiting from substantial discounts on publications of GSL and other societies worldwide, consult www.geolsoc.org.uk, or contact the Fellowship Department at: The Geological Society, Burlington House, Piccadilly, London W1J 0BG: Tel. +44 (0)20 7434 9944; Fax +44 (0)20 7439 8975; E-mail: enquiries@geolsoc.org.uk.

For information about the Society's meetings, consult *Events* on www.geolsoc.org.uk. To find out more about the Society's Corporate Affiliates Scheme, write to enquiries@geolsoc.org.uk.

Published by The Geological Society from:
The Geological Society Publishing House, Unit 7, Brassmill Enterprise Centre, Brassmill Lane, Bath BA1 3JN, UK

*Orders*: Tel. +44 (0)1225 445046, Fax +44 (0)1225 442836
Online bookshop: www.geolsoc.org.uk/bookshop



**British Library Cataloguing in Publication Data**

A catalogue record for this book is available from the British Library.

ISBN: 978-1-86239-218-2

Typeset by Techset Composition Ltd, Salisbury, UK
Printed by The Cromwell Press, Wiltshire, UK.

**Distributors**

***North America***
For trade and institutional orders:
The Geological Society, c/o AIDC, 82 Winter Sport Lane, Williston, VT 05495, USA
*Orders*: Tel +1 800-972-9892
Fax +1 802-864-7626
Email gsl.orders@aidcvt.com

For individual and corporate orders:
AAPG Bookstore, PO Box 979, Tulsa, OK 74101-0979, USA
*Orders*: Tel +1 918-584-2555
Fax +1 918-560-2652
Email bookstore@aapg.org
Website http://bookstore.aapg.org

***India***
Affiliated East-West Press Private Ltd, Marketing Division, G-1/16 Ansari Road, Darya Ganj, New Delhi 110 002, India
*Orders*: Tel. +91 11 2327-9113/2326-4180
Fax +91 11 2326-0538
E-mail affiliat@vsnl.com

# Contents

# Preface

Stone buildings and monuments form the cultural centres of many of the world's urban areas. Frequently these areas are also prone to high levels of atmospheric pollution that promote a variety of aggressive stone decay processes. Because of this, stone decay is now widely recognized as a severe and extremely costly threat to much of our cultural heritage. If this threat is to be successfully addressed it is essential that the symptoms of decay are clearly recognized, that appropriate stone properties are accurately characterized and that decay processes are precisely identified. For it is undoubtedly the case that successful conservation has to be underpinned by a comprehensive understanding of the causes of decay and the factors that control them. Parallel to the need for an understanding of decay processes is a requirement for the accurate specification of new and replacement stone linked to its performance, both as predicted from durability tests and as observed via its performance in use. To accomplish these demanding goals requires an interdisciplinary approach that, whilst underwritten by geological expertise, builds on co-operation between geologists, environmental scientists, chemists, materials scientists, civil engineers, restorers and architects.

In pursuit of this collaboration, this Special Publication aims to strengthen the knowledge base dealing with the causes, consequences, prevention and solution of stone decay problems. Most of the papers contained in this volume were presented during the European Geosciences Union General Assembly ('Volcanology, Geochemistry, Mineralogy 25' special session) held in Vienna (Austria) on 25–29 April 2005. In addition to these there are a number of invited contributions chosen to fill gaps in the coverage of the meeting's original aims. Preparation of this volume would not have possible without help from numerous colleagues who provided their reviews. Their in-time work highly improved the level of the papers. The following people were involved in the review process:

M. Auras
A. Bonazza
H.-G. Brokmeier
S. Brüggerhoff
A. Calia
J. Cassar
J. Curran
Delgado Rodriguez
W. Dubelaar
S. W. Faryad
E. Galan
M. Gardner
K. Germann
A. Goudie
E. Hyslop
R. Ketcham
W. Klemm
P. Marini
I. Maxová
J. Meneely
P. Mikula
P. W. Mirwald
D. Mottershead
D. Nicholson
T. Paradise
S. Pavia
C. Price
R. Přikryl
J. Přikrylová
D. Robinson
A. Ruffell
C. Saiz-Jimenez
R. Sandrone
B. J. Smith
R. Snethlage
Á. Török
E. K. Tschegg
A. Turkington
D. Urquhart
J. R. Vidal Ramoní
H. A. Viles
P. Warke
Z. Weishauptová
R. Williams
T. Yates
M. Young and F. Zezza

Finally, we would like acknowledge help from the Geological Society staff during production of this volume.

Richard Přikryl & Bernie Smith

# Diagnosing decay: the value of medical analogy in understanding the weathering of building stones

B. J. SMITH[1] & R. PŘIKRYL[2]

[1]*School of Geography, Archaeology and Palaeoecology, Queen's University Belfast BT7 1NN, UK (e-mail: b.smith@qub.ac.uk)*

[2] *Institute of Geochemistry, Mineralogy and Mineral Resources, Faculty of Science, Charles University in Prague, Albertov 6, Prague, CZ 128 43, Czech Republic*

**Abstract:** This paper represents the first element of the introduction to this volume, and as such investigates its principal underlying rationale; namely the importance of accurate diagnosis of stone decay in the formulation of effective conservation strategies. It does this by exploring ways in which perceived similarities between stone decay and human disease have influenced attitudes towards conservation, and how refinements within medical diagnostic strategies can inform future condition assessments of building stones. In doing so, it identifies the importance of looking beyond obvious symptoms to the isolation of the fundamental causes of decay and the factors that control them. These controls are strongly conditioned by accumulated stresses within the stonework. In many buildings these are the product of a complex history involving exposure to a variety of environmental conditions and successive human intervention. Only by understanding these memory effects is it possible to explain current decay phenomena, attempt any prediction of future behaviour or recommend appropriate intervention. The concept of appropriateness is further developed through an examination of the TNM (Tumours, Nodes and Metastases) Staging System for cancer diagnosis. This holistic scheme embodies a progressive approach to diagnosis that begins with a clinical assessment based on how the patient presents, and leads on to more detailed pathological investigations involving sampling, testing and analysis. The scheme also requires an assessment of the certainty of the diagnosis and proposed treatments must be viewed in terms of a cost benefit analysis. A modified version of this staging system has already been developed for use in the physical assessment of buildings. It is suggested that the next stage in its development, and that of any other condition assessment procedure that deals solely with the fabric of a building, is the inclusion of a value-based appraisal of its cultural significance.

The decay of building stones is often compared to the effects of an illness – most commonly a cancer – undermining the health of a building and eventually leading to its demise. This analogy has the value of all anthropomorphic comparisons, in that it allows the lay observer to place complex issues within a conceptual framework that relates to their own experience. It also carries with it an assumption that stone ages and has a lifespan that can be drastically shortened by illness. Obviously, there are dangers in pursuing this strategy too far and many risks in imbuing inanimate objects with the capability and desire to shape their own future. However, there remain potentially rewarding avenues along which the medical analogy can be followed that stop short of an invocation of human motivations behind the operation of stone decay systems. Most important is the opportunity it provides for exploring and exploiting underlying strategies developed for the characterization, classification and treatment of disease. The most obvious route is through the adaptation of medical technology (Vlassenbroeck *et al.* 2007), but we can also learn from mistakes associated with delayed intervention, the pursuit of quick fixes, the search for a universal panacea and misdiagnosis.

## The nature of the illness

One very useful area of analogy is the recognition that, just as with illnesses, stone decay can be diagnosed as chronic or acute. This includes the possibility that long experience of a chronic complaint can gradually undermine resistance and may eventually manifest itself in a rapid deterioration in the patient's condition. This eventual rapid decline may be a response to the original, underlying condition finally exploiting an enfeebled immune system. Alternatively, it may result from additional stress related to a new, superimposed illness to which the patient now has no effective resistance. Deterioration need not, however, be as complex as this. Sometimes, patients are simply

*From*: PŘIKRYL, R. & SMITH, B. J. (eds) *Building Stone Decay: From Diagnosis to Conservation.* Geological Society, London, Special Publications, **271**, 1–8.
0305-8719/07/$15.00 

laid low by a particularly severe or virulent illness that rapidly overwhelms the immune system and for which there is no effective cure.

In contrast to decline that is related to or driven by illness, some patients may remain disease free. This does not, however, shield them from the gradual deterioration that accompanies growing old, or mask the reality that we do not live forever. The clearest example of this pathway in the realm of stone decay is the gradual, karstic dissolution of limestone in response to the natural acidity of unpolluted rainfall. In contrast to this, the behaviour of many quartz sandstones, and granular limestones, can be complex, largely unpredictable and commonly characterized by episodic decay (Smith *et al.* 1994). In this context, sandstones may show little surface evidence of change for many years. However, during this time there may be a build-up of internal stress as natural and pollution-derived salts accumulate within the stone and/or surface layers become indurated. The latter could result from, for example, the outward migration and near-surface precipitation of iron cement that leaves the subsurface structurally weakened; or from the growth of a black gypsum crust that could act as a reservoir of potentially damaging salts that are gradually washed into the underlying stone. Eventually the apparent quiescence can be disrupted by, for example, the delamination of the surface and the falling away of a contour scale. This breakdown may result from chronic, fatigue effects generated by the slow build-up and repeated expansion and contraction of salts in intergranular pores (Ruedrich *et al.* 2007), or it may be triggered by an additional, acute stress such as a particularly severe frost or over-energetic cleaning (Svobodová *et al.* 2003). Once the outer layer is lost, the new surface may stabilize if, for example, pollution levels and gypsum deposition rates are high enough in relation to removal by surface wash for a new black crust to quickly develop (Smith *et al.* 2003; Török *et al.* 2007*a*). In many instances, however, positive feedbacks are generated, whereby the more humid environments within surface depressions created by localized scaling, together with reduced washout of deposited salts in areas now protected from rainfall, combine to accelerate retreat of the stone through flaking and disaggregation. In which case the stone/patient experiences a rapid and ultimately fatal deterioration (Rothert *et al.* 2007).

Underlying all of the above thinking is the realization that no stone lasts forever and that using it in construction, especially in a polluted urban setting (Winkler 1997; Schaffer 2004), will invariably shorten its lifespan. This acknowledgement amongst researchers contrasts with the apparent belief of many building owners that placing stone in a building somehow immunizes it from even natural decay and renders it immutable. A consequence of which is that, when decay does occur, it has to be the result of some kind of mistake, that somebody has to be to blame and that any damage can be readily cured. Building owners often find it difficult to accept that, as with all construction materials, stone has a design life. This may be curtailed by mistreatment, by exposure to a variety of hazards and by accident. Conversely, it may be prolonged by regular and appropriate maintenance or by an initial immunization – such as the artificial creation of a protective surface patina (Vazquez-Calvo *et al.* 2007) – but even then its lifespan can be cut short by catastrophic, extreme events. Included in these are natural catastrophes such as severe meteorological conditions and earthquakes, unnatural ones such as conflict damage, and some, such as fire, that can be either natural or human in origin (Sippel *et al.* 2007). Because of these preconceptions, it is rare that significant buildings are allowed to 'grow old gracefully'. Just as in the world of medicine the demand for facelifts and other plastic surgeries has continued to increase, so too has the desire amongst building owners for regular, often aggressive, cleaning, the removal and cosmetic replacement of non-life-threatening blemishes and the presentation of façades that are forever young. Outright opposition to radical, technology-driven intervention runs the risk of being portrayed as complacency. Whilst a proposal for an alternative, less drastic conservation strategy might be marginalized by the establishment as the equivalent of recommending an unproven and potentially dangerous form of fringe medicine.

## Treating causes not symptoms

Despite the aspersions that are cast on many alternative medicines, it should not be forgotten that, even though it may be difficult to prove their efficacy, many of them are commendably holistic and aim to treat the whole body. Thus, even though many treatments may turn out to be ineffective, the diagnostic approach taken has the virtue that it focuses on underlying causes, rather than a rush to treat symptoms. Treating symptoms may produce an initial, often short-lived, improvement in condition or appearance, but it is a strategy that allows potentially debilitating changes to continue while their worst effects are temporarily masked. Ultimately, even more severe symptoms will materialize, by which time either only drastic intervention will have any effect or the patient is beyond treatment. Accurate, holistic diagnosis is thus one of the keys to early, effective treatment that does not exacerbate any overall deterioration in condition.

From the world of stone decay, an example of this is provided by the study of a badly decayed sandstone church in the moist, polluted maritime environment of central Belfast reported by Smith *et al.* (2002, 2005). By the mid 1990s, the late 18th century church of St Matthew's in East Belfast was in an extremely poor state of external repair. Many of the Triassic sandstone blocks exhibited rapid, catastrophic salt weathering through contour scaling and granular disaggregation, which in turn was fuelled by a combination of pollution-derived gypsum and sodium chloride from marine aerosols. Funding for conservation was obtained from the UK's Heritage Lottery Fund, but on condition that, apart from replacement of the most damaged stone blocks, the only conservation permitted was a standard procedure involving the physical removal (dressing back) of the loose outer layers of stone that exhibited the most obvious symptoms of salt weathering. This came with the further proviso that the grant for restoration was not to be used to fund any research into the precise nature of the decay processes operating. Fortunately, the architects responsible for the project were sufficiently concerned to pay for their own, targeted research. This established that under the moist conditions experienced by the church, the salts had in fact penetrated throughout the outer stonework. A test wall also showed that within 4 months of the surface being dressed back, the newly exposed stone began to flake and scale as 'deep salts' were activated by surface wetting and drying. In the end, the Heritage Lottery Fund were persuaded to allow the use of a water repellent, selected through the use of test walls, that to-date has effectively 'switched off' subsequent salt weathering. What this example illustrates is the danger of basing diagnosis solely on a cursory examination of symptoms, the peril of conservation by formula and the value of detailed research – even on the most humble building – that identifies conservation intervention attuned to the specific requirements of stone and environment (Carò & Di Giulio 2007).

## Understanding the patient's background

In medicine, the first stage in any diagnosis is the taking of a patient's clinical history. This provides the opportunity to clarify symptoms and to explore any underlying causes or contributory factors that may help to pin down an illness. It also allows the identification of potentially adverse reactions to possible treatments based on previous allergic responses. The importance of establishing a case history applies equally to buildings, the stones from which they are constructed, and how these stones have been used and abused (see Calcaterra *et al.* 2007; Dionísio 2007; Figueiredo *et al.* 2007; Frangipane 2007; Šimunić Buršić *et al.* 2007). The value of this approach is embodied in the so-called 'memory effect'. This proposes that all building stones carry with them a stress history that reflects their origins, prior exposure to a range of environmental conditions, and treatment at the hands of quarrymen, builders and possibly conservators. The most obvious example of this in built environments is where stones have been loaded with pollutants under pollution regimes that no longer pertain. It is because of this that stones may continue to decay even after clean air legislation is enacted and when owners have convinced themselves that it is now safe to clean and renovate their buildings. The nature of stress inheritance, and the idea that stone behaviour is strongly conditioned by its past history, was explored by Warke (1996). Warke identified two categories of memory: pre- and post-emplacement. Pre-emplacement effects (Hoffmann & Siegesmund 2007) could include dilatation caused by pressure release as the stone was quarried, microfracturing induced by the quarrying process (especially if explosives are used), chemical and physical changes that occur as the stone 'cures' whilst awaiting transport from the quarry (Rothert *et al.* 2007), surface and near-surface changes conditioned by cutting and dressing, and the construction process itself. Included in the latter could, for example, be the loading of non-calcareous stone with calcium as mortar soaks into the bonded surfaces (Smith *et al.* 2001) or by chemicals used for cleaning and conservation (Přikryl *et al.* 2004). Post-emplacement effects are even more varied – principally because of the infinite variety of ways in which people modify and damage stonework. Included in these are the soiling and induration of stone surfaces in response to atmospheric pollution (Smith *et al.* 2007), salt accumulation from pollution, road de-icing and groundwater rise, stone cleaning, and conservation treatments that can range from surface consolidation to re-pointing with hard mortars. In addition, it should not be forgotten that stone used in buildings remains susceptible to natural, climate-driven weathering processes and will gradually accumulate a memory that may include exposure to freeze–thaw, chemical processes such as dissolution and hydrolysis, and a wide range of changes associated with biological colonisation (Krumbein 1983; Hoppert *et al.* 2002; Pohl & Schneider 2002).

The effects of stress inheritance are most clearly seen through changes in the physical and chemical characteristics of building stone – especially porosity (e.g. McKinley & Warke 2007). These either influence their susceptibility to decay mechanisms to which they are already susceptible or expose them to new processes to which they were previously immune. For example, freeze–thaw might promote the inward migration of a microfracture network that exposes a once impermeable stone to salt ingress. In this way one process acts as an essential precursor to another, whereas others may act in parallel and some may even work synergistically to accelerate decay beyond the sum of their individual effects. Understanding the role and importance of stress inheritance does not, therefore, lie simply in listing all the stresses to which a stone has been subject, but in establishing the sequence of events and understanding the superimposed interactions between the various stress factors (Andriani & Walsh 2007).

## Understanding complex stress histories

The consequences of a complex stress history for the lifetime behaviour of buildings have been recently investigated by McCabe *et al.* (2007*a*), through the study of a medieval sandstone church on the NE coast of Ireland just outside the town of Ballycastle. The ruins of Bonamargy Friary date from 1500, and exhibit complex and varied patterns of decay (see McCabe *et al.* 2007*b*) that are interpreted in part as the response to subtle variations in factors such as the porosity and iron content of the Carboniferous sandstones from which it was primarily constructed. Superimposed upon these lithological controls is, however, a range of post-emplacement factors. These include an early fire that destroyed the roof of the building, the lime rendering of the walls, followed by its removal as religious fashions changed, abandonment and re-use, exposure to a markedly different climatic regime during the Little Ice Age (*c.* 1590–1850), and, in more recent times, a number of conservation interventions. The latter includes extensive re-pointing with a comparatively hard mortar that has triggered the rapid retreat of less rigid sandstone blocks. This, and more gradual retreat elsewhere on the building, has been propelled over the years mainly by natural salt weathering in a moist maritime environment. In some of the more iron-rich sandstone, there is also evidence of the outward migration of iron to form a thin, indurated surface layer. Whilst this may temporarily stabilize the surface, there is also evidence that once this layer is eventually breached the weakened subsurface layer is prone to removal by salt weathering (see Smith *et al.* 2007). In addition to changes driven by specific processes, there is also the less specific, but nonetheless significant, gradual increase in compressive loading of individual stones constrained within any wall. This is the consequence of volume increases that are associated with virtually all weathering. Finally, on many stones there is now extensive surface and subsurface biological colonization by algae and lichen. To observe the complex spatial patterns of decay that are observed on the present-day building, as well as the complex decay histories of individual blocks, it is essential that its complex case history be established. This must be linked to identification of the roles played by individual factors and processes in controlling decay and an understanding of how they combine. Nowhere is this more obvious than in the need to examine the interactions between freeze–thaw and salt weathering. There is some history of investigation into the combined effects of these processes, but it has mainly involved the freezing of salt solutions within test blocks. In contrast, the benefit of an approach that emphasizes the importance of understanding the history of a building is that it also emphasizes the significance of interactions over time. Thus, the significance of freezing is not just a question of how cold it gets, but also one of the frequency with which freezing occurs, the number of intervening salt weathering cycles and how one process facilitates the effectiveness of others (see Thomachot & Matsuoka 2007; Warke & Smith 2007). To be successful, this analysis must also be based on the thorough, consistent and meaningful assessment of the building's present-day condition (see Frangipane 2007; Figueiredo *et al.* 2007; McCabe *et al.* 2007*b*).

## Formalizing condition assessment

The importance of condition assessment has been long understood within the medical profession, especially in the treatment of cancer, where 'patient diagnosis and assessment schemes are used as a means of conveying clinical information in an unambiguous way' (Warke *et al.* 2003, p. 1114). One of the most widely used medical classification schemes is the TNM (Tumours, Nodes and Metastases) Staging System for cancer (Hermanek & Sobin 1987), which, because of its holistic approach, recognizes that many factors influence the disease process and that these must be considered before arriving at a condition assessment (Warke *et al.* 2003). In their paper, Warke and her colleagues explored in detail the relevance of the internationally recognized TNM Staging System as a conceptual basis for the condition assessment of buildings and propose an equivalent scheme for buildings using Unit, Area and Spread. This is based on the need for a scheme that can provide a rapid initial condition assessment of a building as a whole and treats it as the product, rather than the sum, of its individual parts. This is in contrast to many established systems (e.g. Fitzner *et al.* 1992, 1995) that rely on the detailed mapping of individual blocks using a complex classification scheme. As Warke *et al.* (2003) point out, such mapping is especially useful on iconic monuments, where it could be argued that each stone has an intrinsic value (Fitzner & Heinrichs 2002; Fitzner *et al.* 2002), and also for noting detailed changes between surveys (Rothert *et al.* 2007), but it is time-consuming and rarely cost effective for more 'commonplace' built heritage. This is especially the case where it can be argued that it is the integrity of the structure as a whole that is important rather than the preservation of individual stones.

The argument in favour of the adaptation of the TNM system is further supported by the close analogy between the ways in which the two conditions (cancer and stone decay) attack their victims. For example, the original chemical and physical characteristics of a stone can determine susceptibility to decay in much the same way that a genetic predisposition can heighten the risk of developing a specific cancer (see papers by Marszalek 2007; Pereira *et al.* 2007; Scheffzük *et al.* 2007; Török *et al.* 2007*b*). Likewise, environmental factors such as long-term exposure to carcinogens can lead to cancer in the same way that exposure to atmospheric pollution can ultimately cause stone decay (Lefèvre *et al.* 2007). Moreover, as previously discussed, removing the source of pollution does not completely remove the risk that pollution-related decay may eventually develop, in the same way that stopping smoking still leaves a person with an elevated risk of developing a variety of cancers depending on how long and how many cigarettes they once smoked. However, the relevance of the TNM system for informing the assessment of stone decay goes far beyond detailed analogies between the pathologies of the two 'illnesses'. What is more important is the framework it establishes for organizing information and identification of the sequence of steps that must be taken before a diagnosis can be made and a likely prognosis arrived at. For example, the TNM system uses two levels of assessment: 'the first is the clinical assessment which relies on the patient's medical history and presenting symptoms. The second comprises a pathological classification based on results from the clinical assessment combined with data from biopsies, blood tests, scans etc.' (Warke *et al.* 2003, p. 1114). These assessments are then combined with a 'certainty factor' that reflects the extent and reliability of the diagnostic tools employed. Embedded within this rationale is a flexibility of response that entertains a range of options from radical intervention to palliative treatment in cases where the condition is beyond cure.

The TNM approach also focuses attention on the importance of the spread of cancer and whether localized removal of the tumour will suffice, whether a more radical removal of surrounding tissue is required, and/or the need for subsequent treatments such as radio- and chemotherapy. An example of this thinking applied to building stone decay is provided in a paper by Turkington & Smith (2004), in which they mapped decay and stone type for individual blocks on the previously mentioned St Matthew's Church in Belfast. For each decay type they then calculated its degree of connectivity by adding the number of adjacent blocks that exhibited the same type of decay. From this they were able, for example, to assess whether minor variations in sandstone lithology created a 'genetic predisposition' to particular types of decay. The connectivity data also showed that some forms of decay, such as contour scaling, tend to occur on isolated blocks and that their development is most probably influenced by intrinsic stone properties. In contrast, decay phenomena such as black crust development and biological colonisation showed a greater degree of connectivity. This in turn could imply the stronger influence of different environmental conditions across the mapped façade. In terms of possible conservation strategies, low connectivity suggests that it is safe to remove and replace individual affected blocks. Where connectivity is greater, it may be necessary to remove and replace a large area of wall surrounding the affected stones and treat the area with, for example, a biocide to prevent recurrence – in much the same way that doctors may recommend a course of chemotherapy following surgery.

A final benefit of the TNM approach is its recognition that in the real world choices are constrained not just by what is technically feasible, but also by what is economically and politically possible and by what is socially appropriate. Because of this, tough choices, literally between life and death, have to take into account the personal circumstances of the patient and in some way place a value on their treatment. In the equally constrained environment of building conservation it is inevitable that the values that society place on a building must influence the desire for and availability of funds to support restoration. Thus, although this brief introduction focuses specifically on the mechanics of stone decay, the next logical step is the development of evaluation procedures that combine physical assessment of condition with a value-based approach to the assessment of cultural heritage (see Grossi & Brimblecombe 2007). Only by application of this twin-track approach is it possible to focus limited resources on those structures that have meaning and resonance within society. It could be argued that such deliberations already play a part in the allocation of restoration funds through existing heritage agencies, government departments and wealthy charities. However, a true value-based approach must undertake to involve all stakeholders in the consultation and decision-making processes. Without this wider involvement, decisions over which elements of our built heritage survive into the future will continue to be taken by elite groups or, increasingly, decided for us by market forces. This is particularly the case for the 'commonplace' heritage that constitutes much of our urban fabric. Rarely do these buildings attract public attention or academic study, but for most of the population they constitute the backdrop to their everyday lives and as such are at the core of their cultural heritage. In which case it is only right that the values that ordinary people place on their conservation are factored into any decisions about their future.

## Conclusions

At the beginning of this introduction, the danger of attributing human motives and fallibilities to stones was highlighted. But this does not mean that building owners and those with a duty of care are themselves immune from emotion-led perceptions as to the nature and significance of stone decay. Through an examination of the medical analogy for stone decay, it has been possible to explore some of these perceptions. All too often it appears that decisions and actions regarding conservation intervention are driven by the view of stone decay as a disease that has to be fought by application of the latest technology and medicines. Understanding what drives this decision-making process is possibly the first step towards changing attitudes and introducing decision makers to solutions that are appropriate and not just technically feasible. One way of approaching this change is to build upon the analogy and use medical diagnostic strategies to show that treatment can take many forms, including that of allowing patients to grow old gracefully with only palliative care. A second step towards achieving appropriate conservation strategies is through a thorough diagnosis and understanding of the causes of change within stonework. Indeed, it was this need for accurate diagnosis that drove the initial proposal for a session on stone decay and conservation at the European Geosciences Union meeting from which this volume has stemmed. Within this first part of the introduction we have concentrated on the conceptual framework within which diagnoses can be made and, in particular, the early stages of assessment based primarily on visual appearance and the most overt symptoms of decay. As indicated in the discussion of the TNM Staging System for cancer diagnosis and treatment, such clinical assessments must be supported by pathological assessments that involve detailed sampling, testing and analysis. In the second part of the introduction, the investigation of building stone pathology is explored through an examination of the role that earth scientists can play in its study.

## References


ANDRIANI, G. F. & WALSH, N. 2007. The effects of wetting and drying, and marine salt crystallization on calcarenite rocks used as building material in historic monuments. *In*: PŘIKRYL, R. & SMITH, B. J. (eds) *Building Stone Decay: From Diagnosis to Conservation*. Geological Society, London, Special Publications, **271**, 179–188.

CALCATERRA, D., CAPPELLETTI, P., DE' GENNARO, M., DE GENNARO, R., DE SANCTIS, F., FLORA, A. & LANGELLA, A. 2007. The rediscovery of an ancient exploitation site of Piperno, a valuable historical stone from the Phlegraean Fields (Italy). *In*: PŘIKRYL, R. & SMITH, B. J. (eds) *Building Stone Decay: From Diagnosis to Conservation*. Geological Society, London, Special Publications, **271**, 23–32.

CARÒ, F. & DI GIULIO, A. 2007. Rock petrophysics v. performances of protective and consolidation treatments: the case of Mt Arzolo Sandstone. *In*: PŘIKRYL, R. & SMITH, B. J. (eds) *Building Stone Decay: From Diagnosis to Conservation*. Geological Society, London, Special Publications, **271**, 287–294.

DIONÍSIO, A. 2007. Stone decay induced by fire on historical buildings: the case of the cloister of Lisbon Cathedral (Portugal). *In*: PŘIKRYL, R. & SMITH, B. J. (eds) *Building Stone Decay: From Diagnosis to Conservation*. Geological Society, London, Special Publications, **271**, 87–98.

FIGUEIREDO, C. A. M., AIRES-BARROS, L., BASTO, M. J., GRAÇA, R. C. & MAURÍCIO, A. 2007. The weathering and weatherability of Basílica de Estrela stones, Lisbon, Portugal. *In*: PŘIKRYL, R. & SMITH, B. J. (eds) *Building Stone Decay: From Diagnosis to Conservation*. Geological Society, London, Special Publications, **271**, 99–108.

FITZNER, B. & HEINRICHS, K. 2002. Damage diagnosis at stone monuments – weathering forms. *In*: PŘIKRYL, R. & VILES, H. A. (eds) *Understanding and Managing Stone Decay*. Karolinum Press, Prague, 11–56.

FITZNER, B., HEINRICHS, K. & KOWNATZKI, R. 1992. Classification and mapping of weathering forms. *In*: RODRIGUES, D. J., HENRIQUES, F. & JEREMIAS, F. T. (eds) *Proceedings of the 7th International Congress on Deterioration and Conservation of Stone*, Lisbon, Vol. 2, Laboratorio Nacional de Engenharia, Lisbon, 15–18.

FITZNER, B., HEINRICHS, K. & KOWNATZKI, R. 1995. Weathering forms – classification and mapping. *In*: SNETHLAGE, R. (ed.) *Verwitterungsformen – Klassifizierung und Kartierung. Denkmalpflege und Naturwissenschaft, Natursteinkonservierung 1*. Ernst & Sohn, Berlin, 41–88.

FITZNER, B., HEINRICHS, K. & LA BOUCHARDIERE, D. 2002. Limestone weathering of historic monuments in Cairo, Egypt. *In*: SIEGESMUND, S., WEISS, T. & VOLLBRECHT, A. (eds) *Natural Stone, Weathering Phenomena, Conservation Strategies and Case Studies*. Geological Society, London, Special Publications, **205**, 217–239.

FRANGIPANE, A. 2007. Natural stone portals of the town of Udine (Italy): their design, construction and materials between the 15th and 20th centuries. *In*: PŘIKRYL, R. & SMITH, B. J. (eds) *Building Stone Decay: From Diagnosis to Conservation*. Geological Society, London, Special Publications, **271**, 33–42.

GROSSI, C. M. & BRIMBLECOMBE, P. 2007. Effect of long-term changes in air pollution and climate on the decay and blackening of European stone buildings. *In*: PŘIKRYL, R. & SMITH, B. J. (eds) *Building Stone Decay: From Diagnosis to Conservation*. Geological Society, London, Special Publications, **271**, 117–130.

HERMANEK, P. & SOBIN, L. H. 1987. *TNM Classification of Malignant Tumours*. Springer, Berlin.

HOFFMANN, A. & SIEGESMUND, S. 2007. The dimension stone potential of Thailand – overview and granite site investigations. *In*: PŘIKRYL, R. & SMITH, B. J. (eds) *Building Stone Decay: From Diagnosis to Conservation*. Geological Society, London, Special Publications, **271**, 43–54.

HOPPERT, M., BERKER, R., FLIES, C., KÄMPER, M., POHL, W., SCHNEIDER, J. & STRÖBEL, S. 2002. Biofilms and their extracellular environment on geomaterial: methods for investogation down to nanoscale. *In*: SIEGESMUND, S., WEISS, T. & VOLLBRECHT, A. (eds) *Natural Stone, Weathering Phenomena, Conservation Strategies and Case Studies*. Geological Society, London, Special Publications, **205**, 207–215.

KRUMBEIN, W. (ed.). 1993. *Microbial Geochemistry*. Blackwell, Oxford.

LEFÈVRE, R.-A., IONESCU, A., AUSSET, P., CHABAS, A., GIRARDET, F. & VINCE, F. 2007. Modelling of the calcareous stone sulphation in polluted atmosphere after exposure in the field. *In*: PŘIKRYL, R. & SMITH, B. J. (eds) *Building Stone Decay: From Diagnosis to Conservation*. Geological Society, London, Special Publications, **271**, 131–138.

MARSZAŁEK, M. 2007. The mineralogical and chemical methods in investigations of decay of the Devonian black 'marble' from Dębnik (southern Poland). *In*: PŘIKRYL, R. & SMITH, B. J. (eds) *Building Stone Decay: From Diagnosis to Conservation*. Geological Society, London, Special Publications, **271**, 109–116.

MCCABE, S., SMITH, B. J. & WARKE, P. A. 2007*a*. A legacy of mistreatment: Understanding the decay of medieval sandstones in NE Ireland. *Building and Environment*.

MCCABE, S., SMITH, B. J. & WARKE, P. A. 2007*b*. An holistic approach to the assessment of stone decay: Bonamargy Friary, Northern Ireland. *In*: PŘIKRYL, R. & SMITH, B. J. (eds) *Building Stone Decay: From Diagnosis to Conservation*. Geological Society, London, Special Publications, **271**, 77–86.

MCKINLEY, J. M. & WARKE, P. A. 2007. Controls on permeability: implications for stone weathering. *In*: PŘIKRYL, R. & SMITH, B. J. (eds) *Building Stone Decay: From Diagnosis to Conservation*. Geological Society, London, Special Publications, **271**, 225–236.

PEREIRA, D., YENES, M., BLANCO, J. A. & PEINADO, M. 2007. Characterization of serpentinites to define their appropriate use as dimension stone. *In*: PŘIKRYL, R. & SMITH, B. J. (eds) *Building Stone Decay: From Diagnosis to Conservation*. Geological Society, London, Special Publications, **271**, 55–62.

POHL, W. & SCHNEIDER, J. 2002. Impact of endolithic biofilms on carbonate rock surfaces. *In*: SIEGESMUND, S., WEISS, T. & VOLLBRECHT, A. (eds) *Natural Stone, Weathering Phenomena, Conservation Strategies and Case Studies*. Geological Society, London, Special Publications, **205**, 177–194.

PŘIKRYL, R., SVOBODOVÁ, J., ŽÁK, K. & HRADIL, D. 2004. Anthropogenic origin of salt efflorescences on sandstone sculptures (Charles bridge, Prague, Czech Republic) – mineralogical and stable isotope geochemistry evidence. *European Journal of Mineralogy*, **16**, 609–617.

ROTHERT, E., EGGERS, T., CASSAR, J., RUEDRICH, J., FITZNER, B. & SIEGESMUND, S. 2007. Stone properties and weathering induced by salt crystallization of Maltese Globigerina Limestone. *In*: PŘIKRYL, R. & SMITH, B. J. (eds) *Building Stone Decay: From Diagnosis to Conservation*. Geological Society, London, Special Publications, **271**, 189–198.

RUEDRICH, J., SEIDEL, M., ROTHERT, E. & SIEGESMUND, S. 2007. Length changes of sandstones caused by salt crystallization. *In*: PŘIKRYL, R. & SMITH, B. J. (eds) *Building Stone Decay: From Diagnosis to Conservation*. Geological Society, London, Special Publications, **271**, 199–210.

SCHAFFER, R. J. 2004. *The Weathering of Natural Building Stones*, 3rd reprinted edn. Donhead, London.

SCHEFFZÜK, C., SIEGESMUND, S., NIKOLAYEV, D. I. & HOFFMANN, A. 2007. Texture, spatial and orientation dependence of internal strains in marble: a key to understanding the bowing of marble panels? *In*: PŘIKRYL, R. & SMITH, B. J. (eds) *Building Stone Decay: From Diagnosis to Conservation*. Geological Society, London, Special Publications, **271**, 237–250.

ŠIMUNIĆ BURŠIĆ, M., ALJINOVIĆ, D. & CANCELLIERE, S. 2007. Kirmenjak *Pietra d'Istria*: a preliminary investigation of its use in Venetian architectural heritage. *In*: PŘIKRYL, R. & SMITH, B. J. (eds) *Building Stone Decay: From Diagnosis to Conservation*. Geological Society, London, Special Publications, **271**, 63–68.

SIPPEL, J., SIEGESMUND, S., WEISS, T., NITSCH, K.-H. & KORZEN, M. 2007. Decay of natural stones caused by fire damage. *In*: PŘIKRYL, R. & SMITH, B. J. (eds) *Building Stone Decay: From Diagnosis to Conservation*. Geological Society, London, Special Publications, **271**, 139–152.

SMITH, B. J. MAGEE, R. W. & WHALLEY, W. B. 1994. Breakdown patterns of quartz sandstone in a polluted urban environment: Belfast, N. Ireland. *In*: ROBINSON, D. A. & WILLIAMS, R. B. G. (eds) *Rock Weathering and Landform Evolution*. Wiley, Chichester, 131–150.

SMITH, B. J., MCALISTER, J. J., BAPTISTA NETO, J. A. & SILVA, M. A. M. 2007. Post-depositional modification of atmospheric dust on a granite building in central Rio de Janerio: implications for surface induration and subsequent stone decay. *In*: PŘIKRYL, R. & SMITH, B. J. (eds) *Building Stone Decay: From Diagnosis to Conservation*. Geological Society, London, Special Publications, **271**, 153–166.

SMITH, B. J., TURKINGTON, A. V. & CURRAN, J. M. 2001. Calcium loading of quartz sandstones during construction: implications for future decay. *Earth Surface Processes and Landforms*, **26**, 877–883.

SMITH, B. J., TURKINGTON, A. V. & CURRAN, J. M. 2005. Urban stone decay: the great weathering experiment. *In*: TURKINGTON, A. V. (ed.) *Stone Decay in the Architectural Environment*, Geological Society of America, Special Publications, **390**, 1–9.

SMITH, B. J., TÖRÖK, A., MCALISTER, J. J. & MEGARRY, Y. 2003. Observations on the factors influencing stability of building stones following contour scaling: a case study of oolitic limestones from Budapest, Hungary. *Building and Environment*, **38**, 1173–1183.

SMITH, B. J., TURKINGTON, A. V., WARKE, P. A., BASHEER, P. A. M., MCALISTER, J. J., MENEELY, J. & CURRAN, J. M. 2002. Modelling the rapid retreat of building sandstones. A case study from a polluted maritime environment. *In*: SIEGESMUND, S., WEISS, T. & VOLLBRECHT, A. (eds) *Natural Stone, Weathering Phenomena, Conservation Strategies and Case Studies*. Geological Society, London, Special Publications, **205**, 339–354.

SVOBODOVÁ, J., SLOVÁK, M., PŘIKRYL, R. & SIEGL, P. 2003. Effect of low and high fluence on experimentally laser-cleaned sandstone and marlstone tablets in dry and wet conditions. *Journal of Cultural Heritage*, **4**, (Suppl. 1), 45–49.

THOMACHOT, C. & MATSUOKA, N. 2007. Dilation of building materials submitted to frost action. *In*: PŘIKRYL, R. & SMITH, B. J. (eds) *Building Stone Decay: From Diagnosis to Conservation*. Geological Society, London, Special Publications, **271**, 167–178.

TÖRÖK, Á., SIEGESMUND, S., MÜLLER, C., HÜPERS, A., HOPPERT, M. & WEISS, T. 2007*a*. Differences in texture, physical properties and microbiology of weathering crust and host rock: a case study of the porous limestone of Budapest (Hungary). *In*: PŘIKRYL, R. & SMITH, B. J. (eds) *Building Stone Decay: From Diagnosis to Conservation*. Geological Society, London, Special Publications, **271**, 261–276.

TÖRÖK, Á., FORGÓ, L. Z., VOGT, T., LÖBENS, S., SIEGESMUND, S. & WEISS, T. 2007*b*. The influence of lithology and pore-size distribution on the durability of acid volcanic tuffs, Hungary.*In*: PŘIKRYL, R. & SMITH, B. J. (eds) *Building Stone Decay: From Diagnosis to Conservation*. Geological Society, London, Special Publications, **271**, 251–260.

TURKINGTON, A. V. & SMITH, B. J. 2004. Interpreting spatial complexity of decay features on a sandstone wall: St. Matthew's Church, Belfast. *In*: SMITH, B. J. & TURKINGTON, A. V. (eds) *Controls and Causes of Stone Decay*. Donhead, London, 149–166.

VAZQUEZ-CALVO, C., ALVAREZ DE BUERGO, M. & FORT, R. 2007. Overview of recent knowledge of patinas on stone monuments: the Spanish experience. *In*: PŘIKRYL, R. & SMITH, B. J. (eds) *Building Stone Decay: From Diagnosis to Conservation*. Geological Society, London, Special Publications, **271**, 295–308.

VLASSENBROECK, J., CNUDDE, V., MASSCHAELE, B., DIERICK, M., VAN HOOREBEKE, L. & JACOBS, P. 2007. A comparative and critical study of X-ray CT and neutron CT as non-destructive material evaluation techniques. *In*: PŘIKRYL, R. & SMITH, B. J. (eds) *Building Stone Decay: From Diagnosis to Conservation*. Geological Society, London, Special Publications, **271**, 277–286.

WARKE, P. A. 1996. Inheritance effects in building stone decay. *In*: SMITH, B. J. & WARKE, P. A. (eds) *Processes of Urban Stone Decay*. Donhead, London, 32–43.

WARKE, P. A. & SMITH, B. J. 2007. Complex weathering effects on durability characteristics of building stone. *In*: PŘIKRYL, R. & SMITH, B. J. (eds) *Building Stone Decay: From Diagnosis to Conservation*. Geological Society, London, Special Publications, **271**, 211–224.

WARKE, P. A., CURRAN, J. M., TURKINGTON, A. V. & SMITH, B. J. 2003. Condition assessment for building stone conservation: a staging system approach. *Building and Environment*, **38**, 1113–1123.

WINKLER, E. M. 1997. *Stone in Architecture*. 3rd rev. extended edn. Springer, Berlin.

# Understanding the Earth scientist's role in the pre-restoration research of monuments: an overview

R. PŘIKRYL

*Institute of Geochemistry, Mineralogy and Mineral Resources, Faculty of Science, Charles University in Prague, Albertov 6, Prague, CZ 128 43, Czech Republic (e-mail: prikryl@natur.cuni.cz)*

**Abstract:** To understand the role of the earth scientist in the pre-restoration research of stone monuments, it is necessary to summarize the tasks that he/she can fulfil. Pre-restoration research into building materials is generally conducted to provide information on types of material, their damage and repair. Although the technologist and restorer must manage the practical aspects of repair, the earth scientist can make a significant contribution in terms of material research. First, he or she can answer questions on the nature of the stone(s) used, their provenance (location of the quarry), and their weathering characteristics in terms of the deterioration of physical and mechanical properties and destruction of rock fabric. Second, the earth scientist can research the physical and mechanical properties of new stone proposed for as a replacement for decayed stonework, including recommendations for alternative materials where stone from the original quarry is no longer available.

Natural stone is a prominent material used on many monuments from the very beginning of the civilization (Shadmon 1996). Stone has been admired as a long-lasting or even immutable material. Unfortunately, this is not the case of natural rocks, and their performance and susceptibility to weathering is influenced by their genesis, composition and conditions of use (Winkler 1997). Pre-restoration research into materials that make up monuments has conventionally focused on the types of materials, their sources, decay forms, extent of damage and the possible prevention of decay. A restoration technologist who is responsible for decisions as to what approach and what types of materials should be used to reduce, for example, future stone decay, often manages this research. Prior to this analysis, the nature of the stone and its properties must be determined. Natural stone is not a simple, uniform material and its uniqueness requires the participation of an experienced specialist – a geologist or earth scientist – in the process of pre-restoration material research of monuments. This person is responsible for identifying:

- which types of natural stones and other materials have been used in a monument or building;
- where a natural stone comes from, or how an artificial material has been prepared;
- what types of decay are or have operated, their extent, impact;
- whether and where it is possible to find appropriate materials for replacement;
- whether existing materials and the fresh replacement material chosen (recently quarried natural stone or artificial replacement material) are compatible and will perform in a similar way?

To answer these questions the earth scientist has to utilize and be proficient in a range of research fields including: petrography and microscopy, geochemistry, mineralogy, rock mechanics, geophysics and the geology of mineral deposits. In light of these requirements, this paper considers the basic problems that an earth scientist participating in pre-restoration material research of stone monument may encounter and to which solutions are required. Owing to limited space it is not possible to list all aspects and methods in detail, and the paper focuses mainly on the methodological philosophy.

## Determination of stone type

### *Macroscopic examination*

The nature of a building stone can be studied by macroscopic observation (visual inspection) of individual pieces of stone (e.g. ashlars) *in situ*. Determination of stone type is based mainly on macrofabric characteristics, colour and minerals macroscopically visible on the exposed surface. All these parameters can be partly or completely obliterated by weathering, causing loss of information due to colour change, surface deposits, biological overgrowth, crust formation and/or dissolution of stone material. Because of this, it is advisable to conduct any macroscopic examination in two stages. The first, *preliminary phase* should precede restoration, but is open to error because

*From*: PŘIKRYL, R. & SMITH, B. J. (eds) *Building Stone Decay: From Diagnosis to Conservation*. Geological Society, London, Special Publications, **271**, 9–21.
0305-8719/07/$15.00 

of the factors listed above. The second stage is therefore a *verification (correction) phase* that can be completed during and/or after the restoration/repair, when cleaning of the stone surface and/or removal of mortar from joints often reveals the original appearance of the stone.

## *Sampling*

Sampling presents the second research element and should follow macroscopic observation and mapping of the monument. Sampling of rock material from monuments (Smith & McAlister 2000) raises several questions. These include: who should conduct the sampling, what is the purpose of the sampling, which sampling method should be used, and what is the required quality (size, state, orientation) and quantity (sampling strategy) of samples? In many cases, the earth scientist is not allowed to conduct the sampling but must rely on the restorer to collect material. This is especially the case for prestigious, iconic monuments with high levels of protective designation for which permission is required for sampling from national and international cultural heritage bodies such as UNESCO for World Heritage Sites.

The purpose of the sampling is twofold: (1) determination of the type of stone (petrographical analysis for the correct rock classification that can subsequently be used to determine provenance); and (2) its state (degree of weathering, physical properties, deviation from the intact state). The later goal is connected with understanding of decay process of the material itself (Smith & McAlister 2000) but also of external variables promoting weathering (Smith 1996).

In general, there are two sampling procedures for natural stone from monuments – manual mechanical sampling without water and machine-facilitated with/without water. The sampling procedure is dictated by the nature of the material, its state (intact coherent v. loose friable debris) and type of measurement to be performed (Smith & McAlister 2000). *Manual mechanical sampling* generally means sampling by hammer and chisel. *Machine-facilitated sampling* is conducted by diamond-core drilling, which may require water for cooling. Both approaches possess certain advantages and drawbacks. Manual sampling by hammer and chisel without water does not change the content of water-soluble salts and moisture, which are two important parameters often studied during pre-restoration research. Drilling with water cooling can, in contrast, remove all water-soluble salts (or may even introduce new anions and cations from the water used for cooling) and makes measurement of moisture content irrelevant. Drilling without water can cause rapid heating of surrounding stone inducing new fractures in the sampled material, leading to erroneous measurement of porosity, permeability or fracture density. The advantage of sampling by drilling, on the other hand, is that it often minimizes the impact on the monument compared to manual sampling by hammer and chisel. The other advantage is that drill cores can be used directly for measurement of certain, mainly mechanical properties. The selection of sampling method may thus depend on which type of analyses will be conducted on the sampled material.

The major challenge following sampling is the extrapolation of results from small samples to large objects (Gy 1992), a problem well known to exploration geologists (see Evans 1995 and references therein). In general, those responsible for the care of monuments do not like samples to be taken. If sampling is allowed, only small pieces (samples not exceeding a few $cm^3$) are permitted, which may be insufficient for correct measurement of many properties, in particular physical and mechanical properties.

Even if relatively large samples can be taken, orientation can significantly influence results (Delgado Rodrigues 1994; Strohmeyer & Siegesmund 2002). Anisotropy of rock fabric and of rock physical properties is generally related to the formation (genesis) of the rock, its later history in the rock mass (namely brittle deformation under regional stress resulting in uneven microcrack and fracture patterns, development of exfoliation joints and microcracks), and also with weathering that takes place after rock extraction and use in the monument.

## *Detailed petrographic study from microscopic observation*

Optical microscopy of rock in thin section presents a basic observational method that should be applied to any stone material sampled from monuments. The microscopy should not be restricted to the basic description of present rock-forming minerals and respective rock fabric. Modern techniques such as computer-assisted image measurement (sometimes called petrographic image analysis – see e.g. Ehrlich *et al.* 1984) can facilitate accurate analysis of microstructures (e.g. grain size, grain shape) and modal composition (proportion of individual phases in the rock).

Each rock represents, at least, a two-phase medium (e.g. Schön 2004) of which the solid part is composed of the rock-forming minerals and the pore space is occupied by air-filled voids. Both rock-forming minerals and pores (*sensu lato*) exhibit a geometry and are arranged spatially according to genetic factors and the later history of the rock mass. The spatial arrangement and geometric properties of rock-building constituents is referred to as the rock's fabric (Sander 1966). This, in turn, consists of the texture (crystallographic preferred orientation in polycrystalline aggregate: see, for example, Bunge 1997) and (micro)structure of rock-forming minerals (geometrical or morphological parameters of grains) (see, for example, Panozzo-Heilbronner 1994), and of the (micro)structure of the pore space (see for example, Lama & Vutukuri 1978; Walsh 1993) (Table 1). Rock fabric includes both scalar (directionless data such as grain-size distribution and description of shape of crystals or voids) and vector (shape and crystallographic or void preferred orientation) data (Pincus 1989).

Quantitative analysis of some rock fabric parameters, especially microstructures and pores, can be carried out using the image measurement system applied to a thin section (Siegesmund *et al.* 1994; Přikryl 2001). Although fully automated image analysis is available, semi-automated or manual image measurement software is preferred for petrographical examination owing to its greater accuracy.

Typical image measurement procedure consists of *image acquisition* (selection of measured area and preparation of 'map' – i.e. hand-drawn picture of mineral boundaries obtained from photomicrographs of individual grains and minerals in thin section), *digitizing* (conversion of the 'map' to a digital form), *measurement* and *data analysis* (Fig. 1). Parameters such as the area of individual grains or areas occupied by certain phases (from modal analysis), the size of individual grains (for grain-size distribution) and shape parameters of grains can be measured. The quantitative analysis of microstructures is advantageous for the precise petrographic classification of the rock, determining provenance and/or the interpretation of variations in rock mechanical properties (Přikryl 2001).

Microscopic analysis of rock fabric and weathering phenomena can be enhanced by various special treatments of samples before preparation of thin section (Taylor & Viles 2000). This consists mostly of pore and microcrack staining by various techniques. Saturation of the pore system by the epoxy resin – fluorescent dye mixture preserves not only pores but also microcracks connected with rock break-up during weathering. This method can also be used for the interpretation of other porosimetric studies (Weishauptová & Přikryl 2004) and as an additional technique for the description of crack/pore geometry. In the author's experience, a two-step resin penetration is more reliable than the precutting impregnation technique originally described by Nishiyama & Kusuda (1994), as it presumes connectivity of all pores and cracks in the rock. The procedure (Fig. 2) consists of: (I) penetration of the sample before cutting by diamond saw; (II) diamond-saw cutting of the plane that will be later glued to the glass plate; (III) fine grinding of the sawn surface; (IV) ultrasonic cleaning and removal of particles produced by cutting and grinding; (V) drying of the sample at about 40 °C; (V) second penetration (preferably under vacuum) on the cut plane to penetrate non-interconnected pores that were not accessible during the first penetration; (VI) gentle grinding of the excess resin – dye mixture from the surface of the sample; and (VII) preparation of the ordinary thin section (without cover glass). The thin section is then observed through a conventional optical microscope (e.g. Leica DMLP) equipped with the source of UV light.

## *Sourcing stone material and dimension stone lithotheques*

Determination of the source localities of stone used in monuments presents one of the most challenging tasks to the conservator and the earth scientist alike. If successful, it not only allows identification of the right replacement stone, but can also be important for dating and provenancing artefacts or for identifying copies of sculptures. Along with ordinary petrographic investigation and microscopic analysis (Lazzarini *et al.* 1980; Renzulli *et al.* 1999), and

**Table 1.** *Division of rock fabric elements observable by microscope (based on Lama & Vutukuri 1978; Panozzo-Heilbronner 1994 and original consideration of the author)*

| Elements of rock fabric | Aspect of material | Example |
|---|---|---|
| Texture | Preferred orientation of lattice of crystallites (related to the solid matrix) | Crystallographic preferred orientation, (e.g. quartz *c*-axes, etc.) |
| (Micro)structure | Geometry (morphology) of crystallites (related to the solid matrix) | Vector data (shape preferred orientation); scalar data (grain shape, grain size) |
| Voids (microcracks and pores) | Disturbance of crystallites (free-space in the solid matrix) | Different types of voids (microcracks, pores), their size, orientation and distribution |

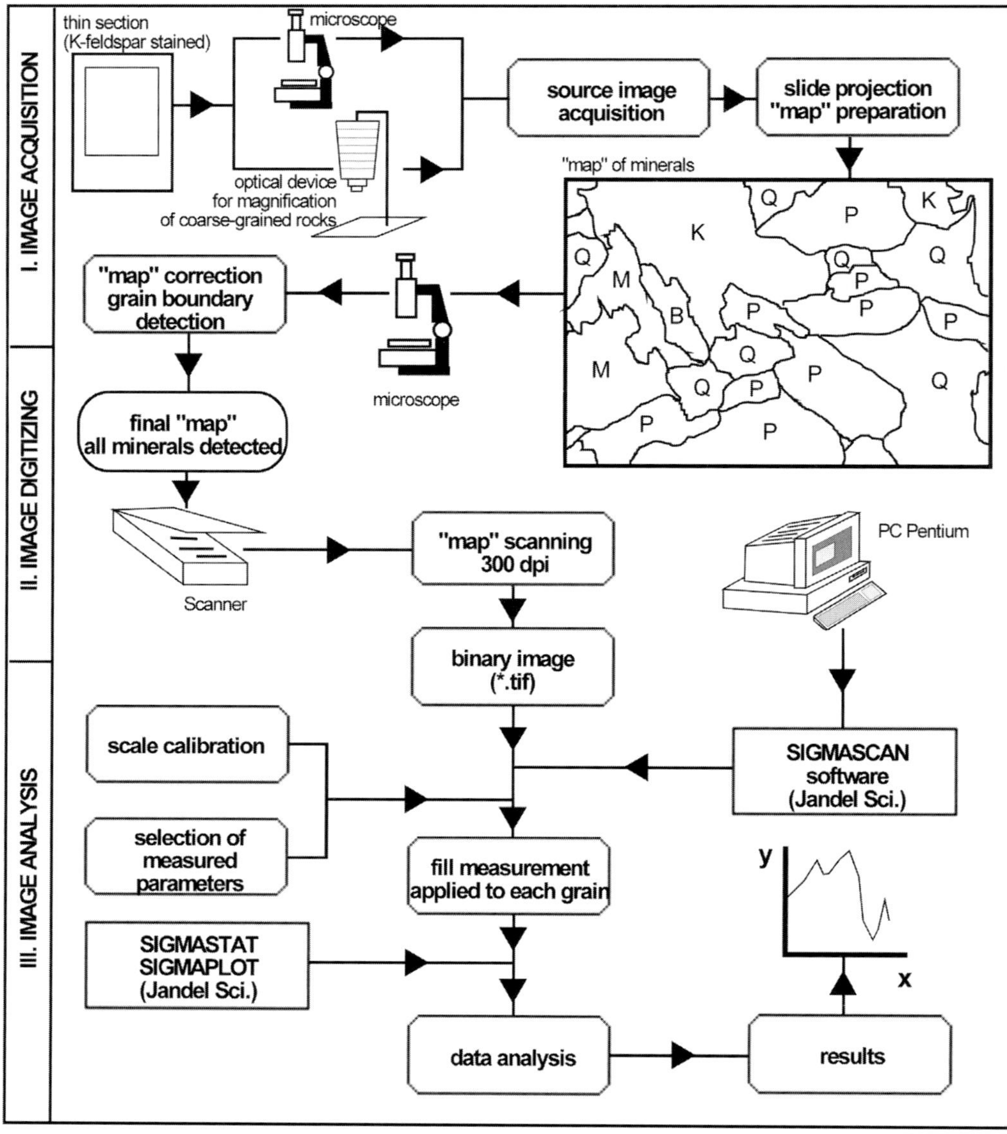


**Fig. 1.** One of the possible approaches for the image analysis of rock microstructures (i.e. geometrical aspects of mineral grains) from thin sections (adopted from Přikryl 2001).

quantitative fabric analysis (Schmid *et al.* 1999), numerous analytical techniques may be applied. These include identification of the geochemistry of major, minor and trace elements of the whole rock by X-ray fluorescence (XRF) (Rapp 1985), laser ablation microprobe with inductively coupled plasma-mass spectrometry (Mallory-Greenough *et al.* 1999*a*), electron microprobe (e.g. Mallory-Greenough *et al.* 1999*b*, 2000), electron paramagnetic resonance spectroscopy (Baïetto *et al.* 1999), gamma ray spectrometry (Williams-Thorpe *et al.* 2000), analysis of stable isotopes (especially C and O in carbonates) (Craig & Craig 1972; Herz 1985; Germann & Cramer 2005), and cathodoluminiscence of carbonates in marbles (Barbin *et al.* 1992) and quartz in sandstones (Matter & Ramseyer 1985; Michalski *et al.* 2002; Götze & Siedel 2004). The selection of appropriate method(s) depends largely on the mineralogical composition and rock fabric (grain size) of the tested rock, and on the amount of material available for analysis.

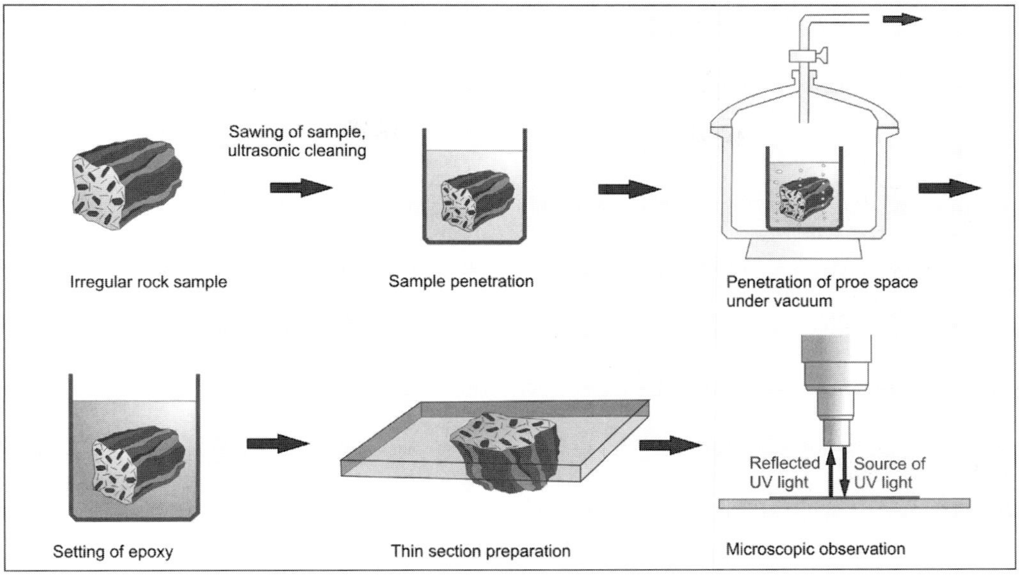


**Fig. 2.** Scheme of the penetration of microcracks and pores by the mixture of epoxy resin and fluorescent dye (technique adopted and modified from Nishiyama & Kusuda 1994).

## *Dimension stone lithotheques*

Correct sourcing of the stone in a monument requires not only application of the right analytical method(s) but also knowledge of the rock varieties that have been quarried in certain areas and their appearance and properties. This can be achieved by having a complete collection of stone types (lithotheque) and a database of mineralogical, geochemical, physical and mechanical properties of these rocks. Despite the existence of some valuable dimension stone collections deposited in museums and comprising, for example, antique stones (e.g. Cooke 2004), complete collections of stone types used from antiquity to the present day are generally missing in most European countries.

The Czech Republic is currently confronted with this situation and research is in progress to establish the Lithotheque of Czech historical dimension stones (Přikryl *et al.* 2001, 2002, 2004*a*). This research programme consists of an archive/literature study, fieldwork (location of historical quarries and stone sampling), laboratory study, lithoteque preparation (sawing of stone slabs) and compilation of an 'atlas' of stone types.

The historical resources for dimension stone can be categorized as follows:

- currently exploited deposits, often with geological exploration data and/or calculated reserves (these are conventionally recorded in the databases of national geological surveys);
- deposits not exploited at present but known from literature, archives, etc. (historical resources);
- newly discovered resources (not exploited in the past or at present) based on geological exploration;
- historical resources non-documented in literature, archives, resource inventories, etc.

The first three groups are easily located in the field because the quarry locations are known, whereas non-documented abandoned quarries are challenging to find. Such historical resources can be discovered either by detailed field study (regional reconnaissance, mapping of historical quarries) or by study of historical materials used in monuments in certain areas (found during restoration work on monuments and the sampling of original stone material) followed by a focused search for potential sites of historical quarrying. The project requires the co-operation of researchers in the field of earth science (geologists) and those involved in the history and restoration of monuments. Based on the author's experience, up to 5% of abandoned quarries and historical dimension stone varieties can be detected using this approach.

When an historical quarry is rediscovered, representative rocks are sampled and subjected to a range of common tests (petrographic examination, physical and mechanical properties, stone workability evaluation, etc. – see Table 2). The final output of the project will be:

- an archive of the stone types (sawn and/or polished slabs, thin sections);

**Table 2.** *Structure of stone inventory as proposed for the Czech Republic (modified and adopted from Přikryl* et al. *2001)*

| | |
|---|---|
| | **1. Basic data** |
| Locality | name of the locality |
| Location of the quarry | position of the quarry related to the important orientation points |
| Co-ordinates | |
| Name of the quarry | historical or current denomination of the quarry |
| State of the quarry | operating or abandoned, size |
| Map | on which map is the quarry visualized |
| | **2. Petrographic description** |
| Macroscopic description | rock type, grain size, macroscopically visible minerals, colour, macrofabric (i.e. macroscopic appearance of the stone) |
| Microscopic description | major and minor clastic minerals (in sedimentary rocks), accessories, opaque phases, organic remnants, matrix, pore or microcrack characteristics, microfabric (i.e. texture and structure) |
| | **3. Petrographic name** |
| | denomination of the stone according to the internationally accepted classifications |
| | **4. Analytical data** |
| Chemistry | major and minor oxides, trace elements (important for provenancing of stone, evaluation of stone susceptibility to weathering) |
| X-ray diffraction | mineralogy of extremely fine-grained rocks |
| Isotopic composition | O and C isotopes (mainly for provenancing of sediments) |
| | **5. Technical properties** |
| Physical properties | real and bulk density, porosity, adsorption, magnetic susceptibility |
| Mechanical properties | uniaxial compressive strength, tensile and bending strength, Young's modulus |
| Technological properties | abrasion, polishing, workability |
| Colour | measured by spectrophotometry |
| | **6. Deposit details** |
| Geological position | regional position in the context of the geology of the Bohemian Massif |
| Age | stratigraphic position or geochronological data |
| Previous geological exploration | evidence in the quarry directory, previous exploration |
| Overburden | for example, thickness |
| Joints | spacing and other characteristics of joints according to the ISRM |
| Block size | estimated or measured size in $m^3$ |
| Genesis of the deposit | i.e. sedimentary, igneous or metamorphic |
| Type of the deposit | i.e. complexity of its structure, thickness, etc. |
| Hydrogeology | quarrying above or below groundwater level |
| Possible current or future use | for which purpose (i.e. sawed, polished slabs, architectural, sculptural works of art, etc.) |
| | **7. Exploitation and historical use** |
| Period(s) of exploitation | |
| Documented historical use | for which purpose the stone has been used |
| Use on buildings | in which regions, buildings or monuments |
| Historical denomination | historical, commercial or scientific names |
| | **8. References** |

- a printed volume (*Atlas of Dimension Stones*) with all available data of each stone type;
- a collection (lithotheque) of the most important stones (in the form of blocks) that will be employed by sculptors and restorers for evaluation of stone properties.

Such a collection (lithotheque) can be employed during a search for stone provenance in a material research project that precedes monument restoration, and it is to be hoped that some of the new additions to the collection may have a potential for use in the restoration of historical monuments. Research on historical quarries also involves proposals for the protection of selected localities as potential resources of valuable dimension stone.

The study of historical resources of dimension stone provides important and often missing information on construction activity in certain regions, on the extent and period(s) of exploitation of the stone, and on the past trade and transport of building materials (local v. imported materials, export of domestic materials abroad, etc.).

### *Determination of the authenticity of the stone*

When no written documents exist on stone sources and replacement undertaken in the past it may be almost impossible to arrive at a solid conclusion as to which of the stones are original and which have been put in place during subsequent repairs. If several types of stone are present and no unambiguous evidence on stone authenticity exists it is advisable to use a rating matrix to facilitate decisions on which stone should be favoured for restoration. Such a rating matrix presents a mixture of factors related to the extent of the stone types used, length of service (authenticity or non-authenticity of the stone), durability (susceptibility of certain stone types to weathering, extent of weathering phenomena) and availability of the stone.

## Understanding weathering processes

### *General*

Weathering must be understood as a complex process of four major variables: material, environment, process and forms that develop both in time and space (Trudgill *et al.* 1991; Inkpen *et al.* 1994; Winkler 1997; Bland & Rolls 1998; Schaffer 2004). The material (a rock in our case) possesses inherited properties according to its composition and genesis (Warke 1996). These can be modified later by the exploitation method (quarrying), processing before emplacement in the monument, and possible intervention during the life of the monument as well as the application of cleaning and conservation methods during restoration (Přikryl *et al.* 2004*b*). The environment covers atmospheric (climatic) factors (Attewell & Taylor 1988), surrounding materials in the monument (mortars in joints, other stone, non-stone materials such as metal fixtures, etc.) and the indoor environment. Process (mechanical/physical or chemical weathering) represents the driving force behind weathering, resulting in macro- and microscopically visible decay forms.

### *Identification of weathering forms and processes*

Identification of weathering processes is essential for the proper maintenance (conservation) of a monument (Price 1996). The weathering process itself can, however, be correctly interpreted only from a sound understanding of visible or detectable weathering forms (Mottershead 2000). Along with qualitative visual assessment of various forms, analytical approaches like surface geometry measurement – i.e. retreat of surface due to weathering (Sharp *et al.* 1982; Cooke *et al.* 1995), colour changes of stone surfaces (Viles 1993), surface roughness measurement (Whalley & Rea 1994) or rock surface strength by Schmidt hammer (Day & Goudie 1977; Williams & Robinson 1983) – are performed. The intensity of stone deterioration, as well as effectiveness of conservation, is often evaluated by indirect methods such as ultrasound measurements (Chiesura *et al.* 1995 and references therein; Nicholson 2002) or stone permeability (Russel *et al.* 2002).

Detailed mapping of weathering features and quantitative assessment of the building and monument condition has been used extensively over recent decades (Emerick 1995; Benea 1996). Generally, two approaches are presently used: a very detailed one aiming to describe the state of each individual piece of the stone on monument, and a general (holistic) assessment approach focusing mainly on present decay processes (Smith *et al.* 1992).

The first approach is based on the detailed mapping of lithology and macroscopically visible weathering forms that are classified into many categories according to their nature and intensity (Fitzner *et al.* 1992, 1993, 1995, 1996; Snethlage 2005). This approach has been successfully adopted for prestigious internationally known monuments (Fitzner & Heinrichs 2002; Heinrichs 2005). Arguments against this approach concern mainly the high cost and time required, which restrict its application for the most valuable monuments.

An alternative approach presumes that for the less prestige monuments a cheap, quick and reliable system integrating weathering forms, intensity and distribution data (Smith *et al.* 1992) has to be used. This form of condition assessment of stonework and the staging of the severity of the damage (Warke *et al.* 2003; Smith & Přikryl 2007) is adapted from medicine where it is successfully applied to the staging tumours (Sobin & Wittekind 2002).

### *The analytical study of weathered stone*

The analysis of weathered stone involves petrographic description (as described in previous sections), the study of physical properties and the analysis of alien phases present in the rock. Salts can be considered as the most common destructive phases in stonework (Price 1996; Goudie & Viles 1997; Charola 2000). The presence of salts in stone monuments depends on a number of factors such as atmospheric pollution (Camuffo *et al.* 1983; Brimblecombe 1987; Whalley *et al.* 1992; Winkler 1997), binding materials (Smith *et al.* 2001) and/or restoration process (Přikryl *et al.* 2004*b*).

The analytical study of salts can be either qualitative or quantitative. The qualitative analysis concerns the phase analysis of salt efflorescence using X-ray diffraction of thermal analytical techniques (McAlister 1996). The optical microscopic study of thin sections or cross-sections is also possible (Arnold 1984; Bai *et al.* 2003). The phases can be interpreted from analytical techniques such as infrared spectroscopy (McAlister 1996), scanning electron microscope–electron dispersive X-ray spectroscopy (SEM/EDS) (Rao *et al.* 1996) or confocal microscopy (Rautureau *et al.* 1993). The amount of salt in stonework is generally measured by various chemical methods, among which ion exchange chromatography is favoured (McAlister 1996). This method is based on the water extraction of salts using deionized water and consequent analysis of the ionic content (Steiger *et al.* 1998).

### *The physical and mechanical properties of weathered stone*

Interpretation of physical and mechanical properties is one of the most frequently neglected roles of the earth scientist during research into stone monuments. Many studies have focused on chemical changes caused by weathering, interaction between stone and other materials (Smith *et al.* 2001) or atmospheric impact on degradation process (Lefèvre & Ausset 2002; Viles 2002). Relatively few studies, however, have focused on the changes of physical and/or mechanical properties of rock as a result of weathering (Dobereiner *et al.* 1993; Nicholson 2002).

The dynamics of physical and/or mechanical changes are rarely studied experimentally (Goudie 1999). Understanding the dynamics of changes (linear evolution, exponential, step like – compare, for example, Warke & Smith 2007) is crucial for understanding and interpretation of the current state of a specific stone in a monument.

The method of investigation used also significantly influences the results obtained. Discrepancies between methods suggested by the International Society for Rock Mechanics (ISRM) (see, for example, the summary of methods published by Brown 1981) and EN standards used in the stone industry are evident in the testing of, for example, uniaxial compressive strength. The cubic shape (or cylindrical with diameter to height ratio 1:1) of specimens (used in EN standards) is adopted from concrete testing. The ISRM suggested shape (Brown 1981) of specimens (height to diameter ratio 2–3:1) provides a more realistic view of real strength of the rock (Bieniawski 1968) and allows simultaneous measurement of deformation. Such knowledge is vital for the evaluation of the prefailure stress state of the stone in monuments.

## Finding replacement stone

The importance and difficulty of selecting the most suitable type of natural stone to act as a replacement are often underestimated when planning and undertaking monument repair/restoration, even though is clearly understood by some architects (Ashurst & Dimes 1998). Many European countries have faced the challenge of a decrease in supplies of traditional stone varieties over recent decades or longer. The Czech Republic can serve as a typical example. Over the last 10 centuries, more than 500 quarries have supplied about 800 stone varieties for use in construction (Hanisch & Schmid 1901; Rybařík 1994; Přikryl *et al.* 2001, 2002, 2004*a*). However, only a tiny number (about 15%) are now currently available (Přikryl 2004).

When the original stone used on a certain monument is no longer available (owing to the closure, renaturalization or recultivation of the quarry, and/or the mining out of reserves) three possibilities exist:

- use available stone that has properties that approach those of the original material;
- use stone that is currently available irrespective of its properties;
- use stone identical to the original stone.

The use of alternative stone can solve problems of stone availability, but it may not be a desirable solution. Properties that differ from those of the original stone (appearance, colour of weathered stone, mechanical properties, durability, etc.) may result in dissimilar weathering patterns that may not manifest themselves for many years. The use of any available stone type (irrespective of its properties) presents the most extreme case of erroneous care of a monument. The application of such stone can result in serious future problems due to increased susceptibility to weathering and the different visual character of the new stone once it has weathered. The third possibility usually involves the reopening of an abandoned quarry that supplied the original material, but it does represent the most acceptable option. If the original quarry is unknown or cannot be reopened an alternative new locality may be explored in the same geological formation. Such a task can be more easily solved if a complete database of historical stones from a particular area exists.

The question of the use of either the original type of stone or a replacement one is also a question of their relative durability. The evaluation of the resistance of the stone to decay factors should be based not only on conventional laboratory testing but also on the design methodology. It has been noted (Duffy & O'Brien 1996) that durability testing according to standards produces fragmented and outdated data sets that are of little help for predicting the dynamics of changes in stone properties (Warke & Smith 2007).

## Concluding remarks

The earth scientist's role in the material research of monuments is primarily limited to the diagnosis of the rock types used, their provenance, degree and type of decay, and their intrinsic properties. The earth scientist can, however, contribute when a new (alternative) stone material is needed for replacement. In such a case, she or he can assist in the search for a stone of similar properties, composition and weathering characteristics instead of a stone that is of the 'best' quality, but which may significantly differ from the original stone. It is important that the earth scientist does not set out to compete with restorers and other technologists to recommend, for example, restoration methods.

Petrographic study and the correct classification of rock type used on monuments is the essential first step of the pre-repair material research. Sources of natural stones can be correctly identified only if there is a good knowledge of the stone types quarried and used in the area, although historical documents may also assist in the identification of original stone types. A lithotheque of building stones from a certain area facilitates such determination.

Understanding the properties of natural stones (used on site or freshly quarried) requires a detailed knowledge of petrophysics and rock mechanics. The same is true for the study and interpretation of weathering processes. The application of non-destructive techniques for the measurement of physical and mechanical properties (water absorption technique and permeability techniques, geophysical methods such as microradar or P-wave velocity measurement, rock mechanical techniques such as rebound hardness using a Schmidt hammer) are advantageous for the assessment of weathering degree.

The study and interpretation of chemical weathering processes requires a detailed knowledge of geochemistry and mineralogy. This knowledge is particularly important for the study of salt efflorescence and determination of sources of salts.

Restorers often ask the earth scientist whether adequate natural stone is available for replacements. A knowledge of mineral deposits and building material resources is therefore a crucial qualification. The best solution to answering such questions is, as mentioned above, the creation of a lithotheque of available stone types. The earth scientist must, therefore, also be prepared to lead the evaluation (prospecting, exploration, assessment of drilling and testing and calculation of reserves) of potential deposits for use as dimension stone.

The publication of this paper would not be possible without financial support from the Ministry of the Education, Youth and Sports of the Czech Republic through research project MSM 520000001. The research on the porosity was partially supported by the project from the Grant Agency of the Academy of Sciences of the Czech Republic (project no. A 3046401). The part concerning rock mechanics benefited from research project of the Grant Agency of the Czech Republic (project no. 205/04/0088). The author highly acknowledged the critical and thorough reviews of A. Ruffel and E. Hyslop. Special thanks to B.J. Smith for valuable discussions and final checking of the manuscript, including the English.

## References

ARNOLD, A. 1984. Determination of mineral salts from monuments. *Studies in Conservation*, **29**(3), 129–138.

ASHURST, J. & DIMES, F. G. 1998. *Conservation of Building and Decorative Stone*. Elsevier, Amsterdam.

ATTEWELL, P. B. & TAYLOR, D. 1988. Time-dependent atmospheric degradation of building stone in a polluting environment. *In*: MARINOS, G. & KOUKIS, G. (eds) *Engineering Geology of Ancient Works, Monument and Historical Sites*. A. A. BALKEMA, Rotterdam, 739–753.

BAI, Y., THOMPSON, G. E., MARTINEZ-RAMIREZ, S. & BRÜGGERHOFF, S. 2003. Mineralogical study of salt crusts formed on historic monuments. *The Science of the Total Environment*, **302**, 247–251.

BAÏETTO, V., VILLENEUVE, G., SCHVOERER, M. & BECHTEL, F. 1999. Investigation of electron paramagnetic resonance peaks in some powdered Greek white marbles. *Archaeometry*, **41**, 253–265.

BARBIN, V., RAMSEYER, K., DECROUEZ, D., BURNS, S. J., CHAMAY, J. & MAIER, J. L. 1992. Cathodoluminiscence of white marbles: An overview. *Archaeometry*, **34**, 175–183.

BENEA, M. 1996. Representative stones and weathering forms at Histria Fortress, Romania. *In*: ZEZZA, F. (ed) *Origin, Mechanisms and Effects of Salts on Degradation of Monuments in Marine and Continental Environments*. European Commission Research Workshop, Bari, 207–216.

BIENIAWSKI, Z. T. 1968. The effect of specimen size on compressive strength of coal. *International Journal of Rock Mechanics and Mining Science & Geomechanical Abstracts*, **5**, 325–335.

BLAND, W. & ROLLS, D. 1998. *Weathering*. Arnold, London.

BRIMBLECOMBE, P. 1987. *The Big Smoke, A History of Air Pollution in London Since Medieval Times*. Methuen, London.

BROWN, E. T. 1981. *Rock Characterization, Testing, and Monitoring: ISRM Suggested Methods*. Pergamon Press, Oxford.

BUNGE, H.-J. 1997. X-ray texture analysis in materials and Earth sciences. *European Journal of Mineralogy*, **9**, 735–761.

CAMUFFO, D., DEL MONTE, M. & SABBIONI, C. 1983. Origin and growth mechanism of the sulphated crusts on urban limestone. *Water, Air and Soil Pollution*, **19**, 351–359.

CHAROLA, A. E. 2000. Salts in the deterioration of porous material: an overview. *Journal of the American Institute for Conservation*, **39**(3), 327–343.

CHIESURA, G., MECCHI, A. M. & ROTA ROSSI DORIA, P. 1995. La technique d'auscultation microsismique pour le diagnostic et l'evaluation des traitments sur materiaux pierreux. *In*: *Methods of Evaluation Products for the Conservation of Porous Building Materials in Monuments*. Preprints of the International Colloquium, Rome, 19–21 June 1995. ICCROM, Rome, 131–145.

COOKE, L. 2004. The shared passion for stone of William Spencer Cavendish, 6th Duke of Devonshire, and Faustino Corsi, a lawyer in Rome. *In*: PŘIKRYL, R. & SIEGL, P. (eds) *Architectural and Sculptural Stone in Cultural Landscape*. Karolinum Press, Prague, 41–51.

COOKE, R. U., INKPEN, R. J. & WIGGS, G. F. S. 1995. Using gravestones to assess changing rates of weathering in the United Kingdom. *Earth Surface Processes and Landforms*, **20**, 531–546.

CRAIG, H. & CRAIG, V. 1972. Greek marbles: Determination of provenance by isotopic analysis. *Science*, **176**, 401–403.

DAY, M. J. & GOUDIE, A. S. 1977. Field assessment of rock hardness using the Schmidt hammer. *British Geomorphological Research Group Technical Bulletin*, **18**, 19–29.

DELGADO RODRIGUES, J. 1994. Measurement and significance of physical properties on granitic rocks. *In*: *Stone Material in Monuments: Diagnosis and Conservation*. Second Course, Community of Mediterranean Universities, Heraklion, Crete, 24–30 May 1993, 71–81.

DOBEREINER, L., DURVILLE, J. L. & RESTITUITO, J. 1993. Weathering of the Massaic gneiss (Massif Central, France). *Bulletin of the International Association of Engineering Geologists*, **47**, 79–96.

DUFFY, A. P. & O'BRIEN, P. F. 1996. A basis for evaluating the durability of new building stone. *In*: SMITH, B. J. & WARKE, P. A. (eds) *Processes of Urban Stone Decay*. Donhead, London, 253–260.

EHRLICH, R., KENNEDY, S. K., CRABTREE, S. J. & CANNON, R. L. 1984. Petrographic image analysis. 1. Analysis of reservoir pore complexes. *Journal of Sedimentary Petrology*, **54**, 1365–1378.

EMERICK, K. 1995. The survey and recording of historic monuments. *Quarterly Journal of Engineering Geology*, **28**, 201–205.

EVANS, A. M. 1995. *Introduction to Mineral Exploration*. Blackwell Science, Oxford.

FITZNER, B. & HEINRICHS, K. 2002. Damage diagnosis on stone monuments – weathering forms, damage categories and damage indices. *In*: PŘIKRYL, R. & VILES, H. A. (eds) *Understanding and Managing Stone Decay*. Karolinum Press, Prague, 11–56.

FITZNER, B. & HEINRICHS, K. 2005. Kartierung und Bewertung von Verwitterungsschäden an Natursteinbauwerken. *Zeitschrift der Deutscheu Gesellschaft für Geowissenschaften*, **156**(1), 7–24.

FITZNER, B., HEINRICHS, K. & KOWNATZKI, R. 1992. Classification and mapping of weathering forms. *In*: RODRIGUES, D. J., HENRIQUES, F. & JEREMIAS, F. T. (eds) *Proceedings of the 7th International Congress on Deterioration and Conservation of Stone*. Laboratory National de Engenharia Civil, Lisbon, 957–968.

FITZNER, B., HEINRICHS, K. & KOWNATZKI, R. 1995. Weathering forms – classification and mapping. *In*: SNETHLAGE, R. (ed.) *Denkmalpflege und Naturwissenschaft. Natursteinkonservierung I*. Ernst & Sohn, Berlin, 41–88.

FITZNER, B., HEINRICHS, K. & VOLKER, M. 1996. Monument mapping – a contribution to monument preservation. *In*: ZEZZA, F. (ed.) *Origin, Mechanisms and Effects of Salts on Degradation of Monuments in Marine and Continental Environments*. European Commission Research Workshop, Bari, 348–355.

FITZNER, B., HEINRICHS, K., VOLKER, M. & CAMPOS PEREZ, R. 1993. Damage investigation on natural stone monuments in Minas Gerais. *In*: THIEL, M. J. (ed.) *Conservation of Stone and Other Materials*, Volume 1. Spon, London, 360–367.

GERMANN, K. & CRAMER, T. 2005. Methoden der Herkunftsbestimmung für Naturwerksteine – das Beispiel des Marmors. *Zeitschrift der Deutscheu Gesellschaft für Geowissenschaften*, **156**(1), 25–31.

GÖTZE, J. & SIEDEL, H. 2004. Microscopic scale characterization of ancient building sandstones from Saxony (Germany). *Materials Characterization*, **53**, 209–222.

GOUDIE, A. 1999. Experimental salt weathering of limestones in relation to rock properties. *Earth Surface Processes and Landforms*, **24**, 715–724.

GOUDIE, A. S. & VILES, H. 1997. *Salt Weathering Hazards*. Wiley, Chichester.

GY, M. 1992. *Sampling of Heterogeneous and Dynamic Material Systems*. Elsevier, New York.

HANISCH, A. & SCHMID, H. 1901. *Österreichs Steinbrüche*. von Carl Graeser, Wien.

HEINRICHS, K. 2005. Diagnose der Verwitterungsschäden an den Felsmonumenten der antiken Stadt

Petra/Jordanien. *Aachener Geowissenschaftliche Beiträge*, **41**, 1–244.

HERZ, N. 1985. Isotopic analysis of marble. *In*: RAPP, G., JR & GIFFORD, J. A. (eds) *Archaeological Geology*. Yale University Press, New Haven, CT, 331–351.

INKPEN, R. J., COOKE, R. U. & VILES, H. A. 1994. Processes and rates of urban limestone weathering. *In*: ROBINSON, D. A. & WILLIAM, R. B. G. (eds) *Rock Weathering and Landform Evolution*. British Geomorphological Research Group Symposia Series, **10**. Wiley, Chichester, 119–130.

LAMA, R. D. & VUTUKURI, V. S. 1978. *Handbook on Mechanical Properties of Rocks. Vol. 4 – Testing Techniques and Results*. Trans Tech Publications, Clausthal.

LAZZARINI, L., MOSCHINI, G. & STIEVANO, B. M. 1980. A contribution to the identification of the Italian, Greek and Anatolian marbles through a petrological study and the evaluation of Ca/Sr ratio. *Archaeometry*, **22**, 173–183.

LEFÈVRE, R. A. & AUSSET, P. 2002. Atmospheric pollution and building: stone and glass. *In*: SIEGESMUND, S., VOLLBRECHT, A. & WEISS, T. (eds) *Natural Stone, Weathering Phenomena, Conservation Strategies and Case Studies*. Geological Society, London, Special Publications, **205**, 329–345.

MALLORY-GREENOUGH, L. M., GREENOUGH, J. D., DOBOSI, G. & OWEN, J. V. 1999*a*. Fingerprinting ancient Egyptian quarries: preliminary results using laser ablation microprobe–inductively coupled plasma-mass spectrometry. *Archaeometry*, **41**, 227–238.

MALLORY-GREENOUGH, L. M., GREENOUGH, J. D. & OWEN, J. V. 1999*b*. The stone of predynastic basalt vessels: Mineralogical evidence for quarries in northern Egypt. *Journal of Archaeological Science*, **26**, 1261–1272.

MALLORY-GREENOUGH, L. M., GREENOUGH, J. D. & OWEN J. V. 2000. The origin and use of basalt in Old Kingdom funerary temples. *Geoarchaeology*, **15**, 315–330.

MATTER, A. & RAMSEYER, K. 1985. Cathodoluminesce microscopy as a tool for provenance studies of sandstones. *In*: ZUFFA, G. G. (ed.) *Provenance of Arenites*. Riedel, Dordrecht, 191–211.

MCALISTER, J. J. 1996. Analytical techniques for the examination of building stone. *In*: SMITH, B. J. & WARKE, P. A. (eds) *Processes of Urban Stone Decay*. Donhead, London, 173–194.

MICHALSKI, S., GÖTZE, J., SIEDEL, H., MAGNUS, M. & HEIMANN, R. B. 2002. Investigation into provenance and properties of ancient building sandstones of the Zittau/Görlitz region (Upper Lusatia, Eastern Saxony, Germany). *In*: SIEGESMUND, S., VOLLBRECHT, A. & WEISS, T. (eds) *Natural Stone, Weathering Phenomena, Conservation Strategies and Case Studies*. Geological Society, London, Special Publications, **205**, 283–297.

MOTTERSHEAD, D. N. 2000. Identification and mapping of rock weathering surface forms and features. *Zeitschrift für Geomorphologie, Supplementbände*, **120**, 5–22.

NICHOLSON, D. T. 2002. Quantification of rock breakdown for experimental weathering studies. *In*: PŘIKRYL, R. & VILES, H. A. (eds) *Understanding and Managing Stone Decay*. Karolinum Press, Prague, 59–74.

NISHIYAMA, T. & KUSUDA, H. 1994. Identification of pore spaces and microcracks using fluorescent resins. *International Journal of Rock Mechanics and Mining Science & Geomechanical Abstracts*, **31**, 369–375.

PANOZZO-HEILBRONNER, R. 1994. Orientation and misorientation imaging: integration of microstructural and textural analysis. *In*: BUNGE, H. J., SIEGESMUND, S., SKROTZKI, W. & WEBER, K. (eds) *Textures of Geological Materials*. Deutsche Gesellschaft für Materialkunde-Informationsgesellschaft-Verlag, Oberursel, 147–164.

PINCUS, H. J. 1989. Constitution of rocks. *In*: TOULONKIA, Y. S., JUDD, W. R. & ROY, R. F. (eds) *Physical Properties of Rocks and Minerals*. CINDAS Data Series on Material Properties, **II-2**. Hemisphere, New York, 1–19.

PRICE, C. A. 1996. *Stone Conservation*. The Getty Conservation Institute, Santa Monica, CA.

PŘIKRYL, R. 2001. Some microstructural aspects of strength variation in rocks. *International Journal of Rock Mechanics and Mining Science & Geomechanical Abstracts*, **38**, 671–682.

PŘIKRYL, R. 2004. The natural stone industry in the Czech Republic. *Roc Maquina*, **53**, 32–35.

PŘIKRYL, R., SVOBODOVÁ, J. & SIEGL, P. 2001. Search for historical resources of dimension stone in the Czech Republic. *In*: SANDRONE, R. (ed.) *Proceedings of the International Workshop 'Dimension Stones of the European Mountains'*, 10–12 June 2001, Luserna san Giovanni – Torre Pellice (Italy), 307–309.

PŘIKRYL, R., SVOBODOVÁ, J. & SIEGL, P. 2002. Provenancing of dimension stones and 'Atlas of monumental stones of the Czech Republic'. *In*: PŘIKRYL, R. & PERTOLD, Z. (eds) *Czech Geology at the Beginning of Third Millenium*, 20 May 2002. Charles University, Prague, 105–110.

PŘIKRYL, R., SVOBODOVÁ, J. & SIEGL, P. 2004*a*. Historical dimension stone resources the Czech Republic. *Roc Maquina*, **53**, 28–31.

PŘIKRYL, R., SVOBODOVÁ, J., ŽÁK, K. & HRADIL, D. 2004*b*. Anthropogenic origin of salt efflorescences on sandstone sculptures (Charles bridge, Prague, Czech Republic) – mineralogical and stable isotope geochemistry evidence. *European Journal of Mineralogy*, **16**, 609–617.

RAO, S. M., BRINKER, C. J. & ROSS, T. J. 1996. Environmental microscopy in stone conservation. *Scanning*, **18**, 508–514.

RAPP, G., JR. 1985. The provenance of artifactual raw materials. *In*: RAPP, G., JR. & GIFFORD, J. A. (eds) *Archaeological Geology*. Yale University Press, New Haven, CT, 353–375.

RAUTUREAU, M., COOKE, R. U. & BOYDE, A. 1993. The application of confocal microscopy to the study of stone weathering. *Earth Surface Processes and Landforms*, **18**, 769–775.

RENZULLI, A., ANTONELLI, F., SANTI, P., BUSDRAGHI, P. & LUNI, M. 1999. Provenance determination of lava flagstones from the Roman 'Via Consolare Flaminia' pavement (central Italy) using petrological investigations. *Archaeometry*, **41**, 209–226.

RUSSEL, M. I., HARMON, N. G., CURRAN, J. M., MUHAMMED BASHEER, P. A. & SMITH, B. J. 2002. Permeation properties of building stone: the Autoclam Permeability System. *In*: PŘIKRYL, R. & VILES, H. A. (eds) *Understanding and Managing Stone Decay*. Karolinum Press, Prague, 75–84.

RYBAŘÍK, V. 1994. *Decorative Building and Sculptural Stones of the Czech Republic*. Nadace Střední průmyslová školy kamenické a sochařské v Hořicích v Podkrkonoší, Hořice v Podkrkonoší [in Czech].

SANDER, B. 1966. *An Introduction to the Study of Fabrics of Geological Bodies* (translated from German). Pergamon Press, Oxford.

SCHAFFER, R. J. 2004. *The Weathering of Natural Building Stones*, 3rd reprinted edn. Donhead, London.

SCHMID, J., AMBÜHL, M., DECROUEZ, D., MÜLLER, S. & RAMSEYER, K. 1999. A quantitative analysis approach to the discrimination of white marbles. *Archaeometry*, **41**, 239–252.

SCHÖN, J. H. 2004. *Physical Properties of Rocks. Fundamentals and Principles of Petrophysics*. Handbook of Geophysical Exploration, Seismic Exploration, **18**, Elsevier, Amsterdam.

SHADMON, A. 1996. *Stone – An Introduction*, 2nd edn. Intermediate Technology Publications, London.

SHARP, A. D., TRUDGILL, S. T., COKKE, R. U., PRICE, C. A., CRABTREE, R. W., PICKLES, A. M. & SMITH, D. I. 1982. Weathering of the balustrade on St. Paul's Cathedral, London. *Earth Surface Processes and Landforms*, **7**, 387–389.

SIEGESMUND, S., HELMING, K., & KRUSE, R. 1994. Complete texture analysis of a deformed amphibolite – comparison between neutron diffraction and U-stage data. *Journal of Structural Geology*, **16**, 131–142.

SMITH, B. J. 1996. Scale problems in the interpretation of stone decay. *In*: SMITH, B. J. & WARKE, P. A. (eds) *Processes of Urban Stone Decay*. Donhead, London, 3–18.

SMITH, B. J. & MCALISTER, J. J. 2000. Sampling and pre-treatment strategies for the chemical and mineralogical analysis of weathered rocks. *Zeitschrift für Geomorphologie, Supplementbände*, **120**, 159–173.

SMITH, B. J. & PŘIKRYL, R. 2007. Diagnosing decay: the value of medical analogy in understanding the weathering of building stones. *In*: PŘIKRYL, R. & SMITH, B. J. (eds) *Building Stone Decay: From Diagnosis to Conservation*. Geological Society, London, Special Publications, **271**, 1–8.

SMITH, B. J., WHALLEY, B. & MAGEE, R. 1992. Assessment of building stone decay: a geomorphological approach. *In*: WEBSTER, R. G. M. (ed.) *Stone Cleaning and the Nature, Soiling and Decay Mechanisms of Stone*. Donhead, London, 249–257.

SMITH, B. J., TURKINGTON, A. V. & CURRAN, J. M. 2001. Calcium loading of quartz sandstones during construction: Implications for future decay. *Earth Surface Processes and Landforms*, **26**, 877–883.

SNETHLAGE, R. 2005. *Leitfaden Steinkonservierung*. Fraunhofer IRB, Stuttgart.

SOBIN, L. H. & WITTEKIND, C. (eds). 2002. *TNM Classification of Malignant Tumours*, 6th edn. Wiley-Liss, New York.

STEIGER, M., NEUMANN, H.-H., GRODTEN, T., WITTENBURG, C., RICHTER, U. & DANNECKER, W. 1998. Salze in Natursteinwerk – Probenahme, Messung und Interpretation. *In*: SNETHLAGE, R. (ed.) *Denkmalpflege und Naturwissenschaft. Natursteinkonservierung II*. Ernst & Sohn, Berlin, 61–91.

STROHMEYER, D. & SIEGESMUND, S. 2002. Anisotropic technical properties of building stones and their development due to fabric changes. *In*: SIEGESMUND, S., VOLLBRECHT, A. & WEISS, T. (eds) *Natural Stone, Weathering Phenomena, Conservation Strategies and Case Studies*. Geological Society, London, Special Publications, **205**, 115–135.

TAYLOR, M. P. & VILES, H. A. 2000. Improving the use of microscopy in the study of weathering: Sampling issues. *Zeitschrift für Geomorphologie, Supplementbände*, **120**, 145–158.

TRUDGILL, S. T., VILES, H. A., COOKE, R. U., INKPEN, R. J., HEATHWAITE, A. L. & HOUSTON, J. 1991. Trends in stone weathering and atmospheric pollution at St Paul's Cathedral, London, 1980–1990. *Atmospheric Environment*, **25A**, 2851–2853.

VILES, H. A. 1993. Observations and explanations of stone decay in Oxford, U.K. *In*: THIEL, M. J. (ed.) *Conservation of Stone and Other Materials*, Volume 1. Spon, London, 115–120.

VILES, H. A. 2002. Implications of future climate change for stone deterioration. *In*: SIEGESMUND, S., VOLLBRECHT, A. & WEISS, T. (eds) *Natural Stone, Weathering Phenomena, Conservation Strategies and Case Studies*. Geological Society, London, Special Publications, **205**, 407–418.

WALSH, J. B. 1993. The influence of microstructure on rock deformation. *In*: HUDSON, J. A. (ed.) *Comprehensive Rocks Engineering. Principles, Practice & Projects*, Volume 1, 1st edn. Pergamon Press, Oxford, 243–254.

WARKE, P. A. 1996. Inheritance effects in building stone decay. *In*: SMITH, B. J. & WARKE, P. A. (eds) *Processes of Urban Stone Decay*. Donhead, London, 32–43.

WARKE, P. A. & SMITH, B. J. 2007. Complex weathering effects on durability characteristics of building stone. *In*: PŘIKRYL, R. & SMITH, B. J. (eds) *Building Stone Decay: From Diagnosis to Conservation*. Geological Society, London, Special Publications, **271**, 211–224.

WARKE, P. A., CURRAN, J. M., TURKINGTON, A. V. & SMITH, B. J. 2003. Condition assessment for building stone conservation: a staging system approach. *Building and Environment*, **38**, 1113–1123.

WEISHAUPTOVÁ, Z. & PŘIKRYL, R. 2004. Porosimetric studies in rocks: methods and application for weathered building stone. *In*: PŘIKRYL, R. (ed.) *Dimension Stone. New Perspectives for a Traditional Building Material.* A. A. BALKEMA, Leiden, 237–241.

WHALLEY, W. B. & REA, B. R. 1994. A digital surface roughness meter. *Earth Surface Processes and Landforms*, **19**(7), 9–14.

WHALLEY, B., SMITH, B. J. & MAGEE, R. 1992. Effects of particulate air pollutants on materials: investigation of surface crust formation. *In*: WEBSTER, R. G. M. (ed.) *Stone Cleaning and the Nature, Soiling and Decay Mechanisms of Stone*. Donhead, London, 227–234.

WILLIAMS, R. B. G. & ROBINSON, D. A. 1983. The effect of surface texture on the determination of the surface hardness of rock using the Schmidt hammer. *Earth Surface Processes and Landforms*, **8**, 877–883.

WILLIAMS-THORPE, O., WEBB, P. C. & THORPE, R. S. 2000. Non-destructive portable gamma ray spectrometry used in provenancing Roman granitoid columns from Leptis Magna, North Africa. *Archaeometry*, **42**, 77–99.

WINKLER, E. M. 1997. *Stone in Architecture. Properties, Durability*, 3rd rev. edn. Springer, Berlin.

# The rediscovery of an ancient exploitation site of Piperno, a valuable historical stone from the Phlegraean Fields (Italy)

D. CALCATERRA[1], P. CAPPELLETTI[2], M. DE' GENNARO[2], R. DE GENNARO[2], F. DE SANCTIS[1], A. FLORA[1] & A. LANGELLA[3]

[1]*Dipartimento di Ingegneria Geotecnica, Università Federico II, Piazzale V. Tecchio 80, Naples, Italy (e-mail: domcalca@unina.it)*

[2]*Dipartimento di Scienze della Terra, Università Federico II, Via Mezzocannone 8, Naples, Italy*

[3]*Dipartimento di Studi Geologici ed Ambientali, Università del Sannio, Via Port'Arsa 11, Benevento, Italy*

**Abstract:** This paper reports the research results over several years on Piperno, the most important ornamental architectural stone of Naples. Particular attention is paid to the rediscovery of the old exploitation sites of this rock and to the survey of the last underground quarry site, still accessible, at the base of the Camaldoli Hill (western Naples) at Pianura. The conservation state was assessed by means of specific surveys in view of possible future utilization. At present, the re-opening of abandoned quarries is not possible owing to unsafe site conditions. The cultural relevance of the Pianura quarry site could suggest its possible restoration as a museum of mining and a centre for teaching the working of ornamental stone within the Campania Region.

Piperno represents the most widely used stone in the historical architecture of Naples, Campania region, Italy. In addition, its use was also recorded in many minor centres (Calcaterra *et al.* 2003) and even outside the region, including historical buildings in the town of Gallipoli, Puglia Region (Calcaterra pers. comm.). Notwithstanding the limited extent of Piperno's occurrence and its difficulty of exploitation (mainly extracted from underground), it has been used since Greek–Roman times and intensively from the 18th century until after World War II, mainly in Naples and its province. This study of Piperno is part of a wider multidisciplinary project of the Earth Science and Geotechnical Engineering departments of the 'Federico II' University of Naples, supported by the Campania regional government. It aims to provide a detailed petrophysical characterization of the numerous ornamental stones used as part of the important architectural heritage of the Campania Region. Within this framework, Piperno plays a significant role as it represents the most used natural stone in Naples architecture and its surroundings. The unusual quarrying procedures, at least in southern Italy, conditioned by its peculiar outcrop pattern, also means that the abandoned underground quarries are worthy of study in their own right and require detailed survey to assess their heritage value.

## Historical notes on the exploitation of Piperno

The geological formation of Piperno is only clearly exposed at the foot of the Camaldoli Hill, within the urban area of Naples, even though some authors (Di Girolamo 1968) have reported further outcrops at different sites (Fig. 1). The first traces of its use as a building stone date back to the Greek period (Cardone & Papa 1993). For example, at the archaeological site of Cuma, Piperno was used to produce the drums of the columns adorning the temples of the acropolis (8th century BC) and also to partially pave some roads (Cardone & Papa 1993).

Historical sources (Cardone & Papa 1993) testify to quarrying in the rural village of Pianura (nowadays an urban district of Naples) since the 13th century. At that time, under the Angevin kings, Piperno, along with the Neapolitan Yellow Tuff, represented the most used building stone for some of the most outstanding monuments that are still today a marker in the urban setting of Naples. These include Santa Chiara Church, San Domenico Maggiore Church and the San Pietro a Maiella Church. Further proof of the importance that this quarrying gained with time is given by the name of Soccavo (in Latin, *sub cava* = near the quarry), another village located at the foot of the Camaldoli hill. Under the Aragonese domination

*From*: PŘIKRYL, R. & SMITH, B. J. (eds) *Building Stone Decay: From Diagnosis to Conservation.* Geological Society, London, Special Publications, **271**, 23–31.

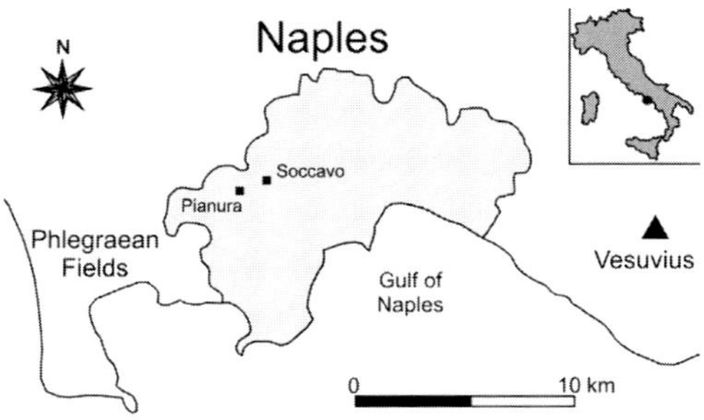


**Fig. 1.** Sketch map showing the location of Piperno underground site at the foot of Camaldoli Hill.

(15th century) the demand for Piperno greatly increased, as a consequence of its use in the main buildings of that time (e.g. the renovation of Maschio Angioino Castle, the Royal Palace, the Sanseverino Palace and, currently, the Gesù Nuovo Church). The importance of the stone also led to the creation of a specific guild of workers (*pipernieri*), which increased in importance from the 15th to the 18th century.

The exploitation of Piperno continued mainly through the exploitation of underground quarries at Pianura, Soccavo and Verdolino. The environmental conditions were, however, very dangerous and, on 22 October 1739, 11 miners died as a consequence of a vault collapse while working in one of the underground quarries (Cardone & Papa 1993).

From the 18th century onwards, Piperno was progressively replaced by less expensive materials, such as lavas of the Phlegraean and Vesuvian districts that are now seen in many buildings of that period in Naples and other Campanian towns. However, the Piperno quarries of Pianura remained active until the first decades of the 20th century. Today, the textural imprinting of Piperno, when used as a dimension stone in modern buildings, is improperly replaced with a similar volcanoclastic rock coming from the Viterbo area (Lazio region), known as *peperino*.

## The Piperno formation within the geology of the Phlegraean Fields

Piperno is the product of volcanic activity that developed about 39 ka BP in the Phlegraean Fields (De Vivo *et al.* 2001). The few outcrops are all located at the foot of the Camaldoli Hill, on its western and southern side. The maximum exposed thickness never exceeds 20 m and the base of the formation is not exposed. The age of Piperno is not so different from another important volcanic formation of the Phlegraean Fields, the Campanian Ignimbrite, and some authors (Rosi *et al.* 1983; Rosi & Sbrana 1987) have interpreted the Piperno and some breccia deposits (Breccia Museo) of Camaldoli as the proximal deposits of the Campanian Ignimbrite. This could in part be explained by the high level of stratigraphic variability of this formation, as remarked by Maggiore (1934) and evidenced by successive layers that differ in terms of their scoriae dimension and frequency. In view of this, Maggiore (1934) identified six layers and, based on field observations carried out on the few outcrops and the walls of the underground quarry in Pianura, a reconstruction of the stratigraphical succession and the main petrophysical parameters of the most exploited layers (2, 3/5H and 5L) was created (Fig. 2).

## Mineralogical and petrographical features of Piperno

Piperno is characterized by an eutaxitic fabric with black flattened scoriae (*fiamme*) set in a hard and light grey matrix. At a macroscopic scale (Fig. 3) Piperno shows centimetre- to decimetre-sized *fiamme* with a maximum length of 30–40 cm and an average flattening ratio of 1:10. Similar fabrics can be seen at a microscopic scale with tiny shards flattened and moulded one over another (Fig. 4). The main phases are sanidine ($Or_{68-43}$), subordinate plagioclase ($An_{86-28}$), clinopyroxene ranging from diopside to salite ($Mg_{85-47}$), biotite, amphibole ($Mg_{62-56}$), magnetite ($Ulv_{40-37}$) and sodalite. These phases are set in a totally recrystallized matrix, where alkali feldspar ($Or_{53-34}$) represents the neoformed phase (Calcaterra *et al.* 2000). *Fiamme* are also recrystallized by tiny new crystals of alkali feldspar with the same composition as those of the matrix.

Chemically (Calcaterra *et al.* 2000), no substantial differences have been noted between Piperno sampled in different localities (Soccavo and Pianura). The composition ranges from trachyte to trachyphonolite ($SiO_2$, 60.9–63.5 wt%, $K_2O$, 6.8–7.3 wt%, on a dry basis). In some cases it shows a peralkaline character (A.I., agpaitic index, up to 1.14). Minor elements show a restricted range in concentration; Nb and Zr exhibit their incompatible characteristics and concentrations ranging from 90 to 121 ppm and 581 to 713 ppm, respectively; Sr and Ba exhibit low concentrations (from 22 to 40 ppm and from 18 to 54 ppm, respectively) (Calcaterra *et al.* 2000). All these geochemical features are typical of the differentiated rocks occurring in the Phlegrean Fields and account for the residual character of the magma that produced the Piperno deposit.

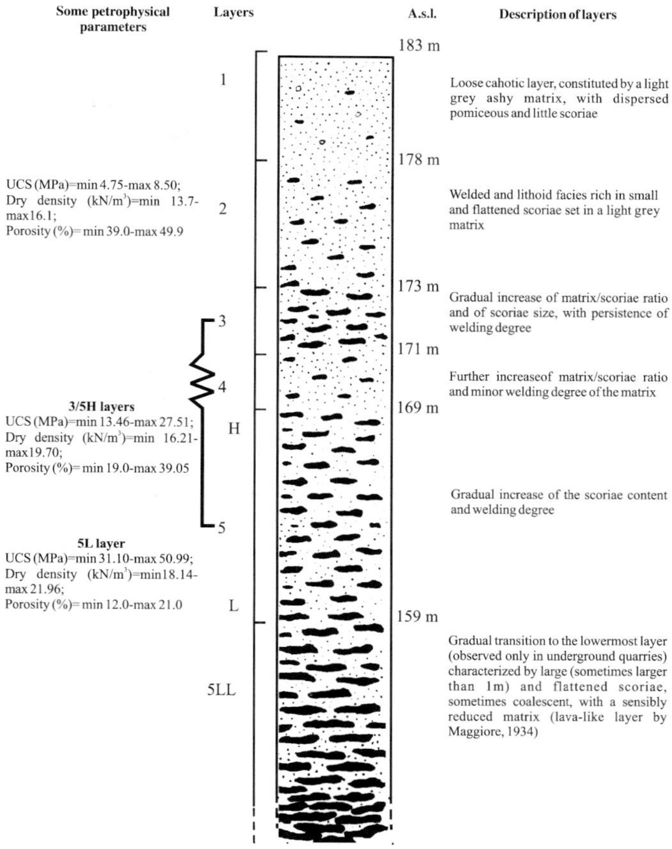


**Fig. 2.** Reconstruction of the stratigraphical succession of the Piperno Formation and main petrophysical parameters of the most exploited layers (2, 3/5H and 5L). A short description of the features of each layer is also reported (modified after Calcaterra *et al.* 2005).

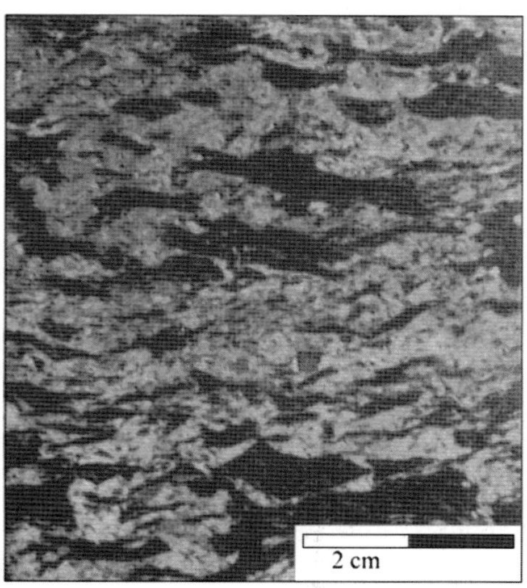


**Fig. 3.** Eutaxitic fabric of Piperno characterized by collapsed black scoriae in a grey ashy matrix.

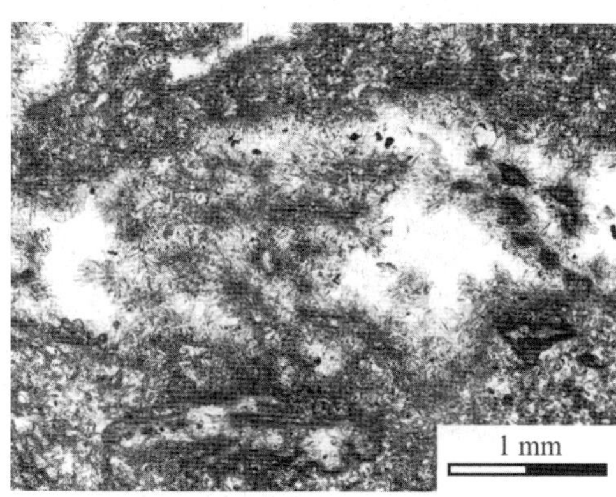


**Fig. 4.** Plane polar micrograph (×40) of a flattened scoria totally recrystallized by tiny acicular sanidine.

**Table 1.** *Mineralogical composition of Piperno (Calcaterra* et al. *2000)*

| | Total feldspars | Sodalite | Magnetite | Biotite | Amphibole | Amorphous |
|---|---|---|---|---|---|---|
| Pianura | 95.4 | 3.5 | 0.5 | – | tr. | 0.8 |
| Soccavo | 89.3 | 3.9 | 1.5 | tr. | – | 5.4 |

tr., trace.

Table 1 shows the results of a quantitative mineralogical evaluation of representative samples of Piperno from Pianura and Soccavo. For both groups of samples the prevailing phase is sanidine, ranging between 89 and 95% (Calcaterra *et al.* 2000, p. 421, table II). Subordinate amounts of sodalite and magnetite were also recognized. The only Pianura sample shows a residual fraction of unreacted glass (about 5.5%). Only a very limited portion of feldspar can be ascribed to a primary genesis, most of it derives from a devitrification process (vapour phase crystallization) that involved the glassy fraction in both the matrix and scoriae. These processes led to significant lithological changes. The large glassy scoriae, as well as the matrix, lose their primary features thus becoming hard and compact as a consequence of welding and/or feldspar crystallization that also reduces the available pore space.

The products of vapour-phase crystallization in Piperno are alkali feldspars with a narrow range in chemical composition ($Or_{53-34}$) (Calcaterra *et al.* 2000). Vapour-phase crystallization results from hot gases passing up through the body of the deposit. Some fluids may be of juvenile origin, exsolved from pumice and vitric particles, and some may be from heated groundwater (Calcaterra *et al.* 2000). These authigenic feldspars are observed in *fiamme* as well as in the matrix, and their composition is distinguishable from the few phenocrysts present in the rock ($Or_{60-53}$). The minerogenetic process seems to be confirmed by many gas-escape pipes present in the upper breccia (e.g. at Verdolino); these vertical channels testify to the wide degassing of the underlying Piperno unit. Electron microscopy observations (SEM) confirmed the above considerations and demonstrated the presence of feldspar crystals, with a typical tabular shape, growing on the glassy matrix (Fig. 5).

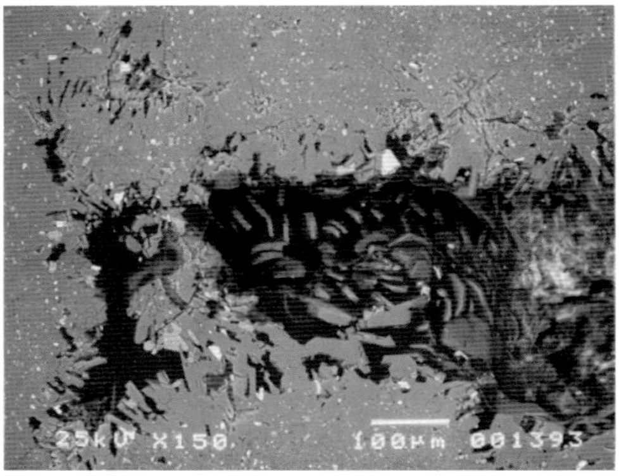


**Fig. 5.** Backscattering scanning electron micrograph of a thin section of Piperno.

## The Pianura underground quarry

One of the main aims of this research is to rediscover the former exploitation sites of Piperno, at the foot of the Camaldoli Hill. A preliminary investigation showed that, among the main historical underground quarries, the one located in Pianura (Masseria del Monte), and the object of the present investigation, was the only one accessible for study. The entrance of another important site on the Soccavo side of the hill was totally obliterated by dumped materials, whereas the Verdolino underground quarry, also located on the Soccavo side of the hill, was described in an old survey as having an extremely limited exploitation area (Cardone & Papa 1993).

The above considerations led the study to focus on the underground quarry located in Pianura at Masseria del Monte. The study of this site started with a topographical survey carried out following the standard techniques adopted for spelaeological investigations. The instruments used were a Leica laser stadia, a fibreglass metric tape, and a Suunto spelaeological inclinometer and compass.

The survey consisted of the measurement of the parameters (distance, orientation and dip) necessary for the construction of a traverse representing the framework on which the main structure of the hypogeum was based. Radial or closed traverses, as well as triangulations, were carried out as a function of the dimension, the morphological complexity and access difficulties over of the investigated sites. Data were processed by means of Microsoft Excel, followed by a plano-altimetric rendering of the hypogeum on an Autocad platform. The final report of the survey enabled the editing of a 1:200 scale map, and a relevant number of longitudinal and transversal sections. Finally, the main joints, including their dip direction, persistence, the width and possible filling materials, were surveyed.

The main entrance of the underground site can be accessed by following a trench, about 20 m long and gently inclined from the initial ground surface, until the Piperno layers are intersected. The trench leads to a wide yard that most probably represents a former quarry front exhibiting a tectonic contact with a loose whitish pyroclastic material. In this area, an abrupt deepening of the Piperno Formation is recorded, as a consequence of a caldera collapse following the huge eruption that emplaced the Neapolitan Yellow Tuff (Orsi *et al.* 1996). The evidence of the caldera collapse was also confirmed by a 22 m-deep borehole drilled almost above the tectonic contact, which did not reach the top of the Piperno Formation and only encountered loose pyroclastic deposits mixed with Piperno blocks of different size. The presence of this pyroclastic deposit, resting over Piperno, required the quarrymen to reinforce the entrance with masonry structures. From this area, two underground quarries were opened. The one opening northwards is the object of the present investigation, whereas the one facing southwards is now almost totally obstructed by debris and waste, and is impossible to explore.

The surveyed quarry covers an area of about 5000 m$^2$. As a whole, its development does not show any predefined exploitation scheme or any preferential direction (Fig. 6). The initial cross-section of the quarry is trapezium shaped, 3 m wide and 2 m high. This section continues for about 30 m in a NNW direction. In the final portion of this initial track the continuity of the

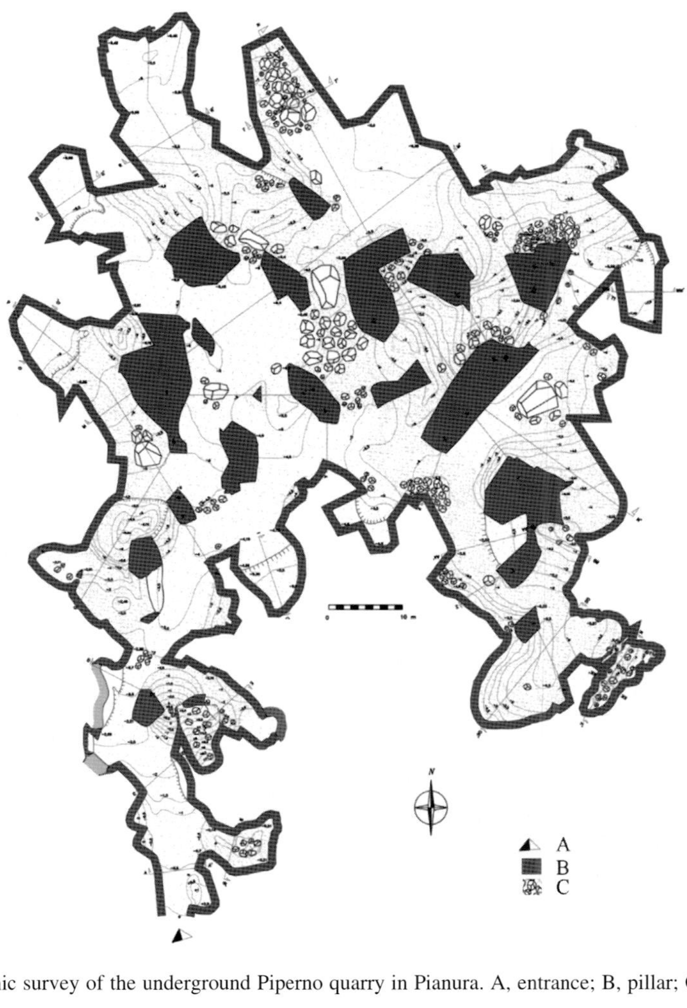


**Fig. 6.** Topographic survey of the underground Piperno quarry in Pianura. A, entrance; B, pillar; C, debris cone.

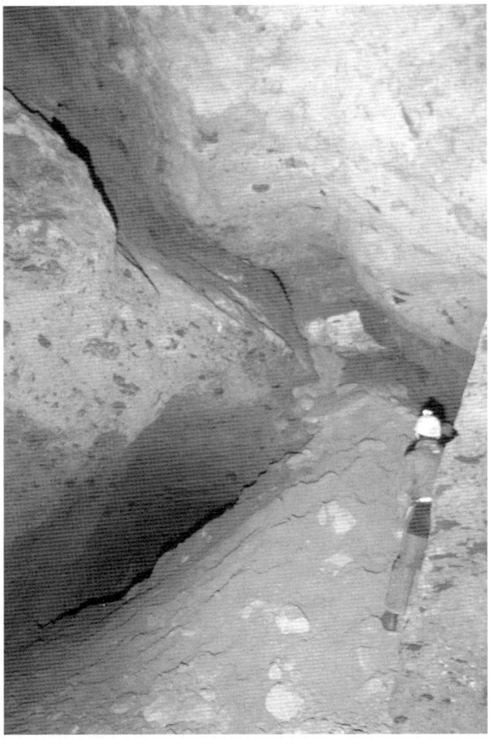

Fig. 7. A major joint observed in the SE branch of the cavity.

Fig. 8. A typical debris cone, constituted by heterometric blocks.

Piperno layers is interrupted by the presence of pyroclastic deposits of a whitish pumice in an ashy matrix that most probably filled pre-existing trenches in the Piperno deposit and preserved their primary attitude. The cavity then branches off in a SE direction for about 20 m following a persisting joint (N230°/45°) at a higher elevation (Fig. 7). From this point onwards three sectors can be schematically identified: a NW one, a central one and a SE one. The development of the NW sector is partly conditioned by the previously cited joint. The final portion of this sector shows a collapsed vault that produced a debris deposit of variable grain size at its base (Fig. 8). The SE sector is controlled by a persisting vertical joint (N40°), about 1 m wide, partly filled by Piperno blocks, and characterized by a continuous air flow, most probably from an external conduit. The central portion of the underground site is definitely the most chaotic area of the hypogeum. Pillars are scattered over the area without any logical distribution, showing irregular and different shapes. Evident indication of a static fatigue also determines fracture systems (Fig. 9) that cut off rock prisms that in turn toppled to the floor. Some chimneys and connected heterometric deposits can be related to block detachment from the vaults (Fig. 10). The natural stratigraphic sections exposed along these chimneys show the Piperno–Breccia Museo transition.

The planimetric development of the hypogeum suggests that exploitation did not guarantee the stability of the site and, consequently, the safety of quarrymen. The exploitation conforms to a so-called 'abandoned pillars' geometry with dimensions defined on the basis of the skill of the individual quarryman. Thus, pillar distribution and shape are irregular in every part of the hypogeum, and most probably reflect the variable strength of the

Fig. 9. Stress-related open fracture in a pillar.

**Fig. 10.** Collapsed chimney in the vault of the cavity.

rock mass that led workers to follow the main joints. The total surface of these pillars does not exceed 14% of the underground area and is indicative of the low safety margins that characterized the site's exploitation. Indeed, it is clear that the surfaces of many pillars, as well as the perimeter walls, are connected to the joints along which it was easier to quarry the rock blocks. This pattern of exploitation created static stresses on the pillars and some walls that continue to the present day.

In the quarry some evidence of past activity is found, such as old quarrying tools or electrical wires and traces left by the miners' tools that indicate procedures common to the quarrying of other volcaniclastic products, such as the Neapolitan Yellow Tuff. For example, blocks were roughly shaped and reduced to requested dimensions on site, presumably to minimize additional costs such as transportation. However, the final size and shape of the stone was given by the *pipernieri* during the construction of the building.

## Evaluation of the static conditions of the underground site

In order to evaluate the static conditions of the underground site, the stress state induced by the excavation has been evaluated and analysed taking into account the mechanical properties of Piperno. Uniaxial compression tests were carried out on Piperno specimens from different sites, some of which were taken directly from the Pianura quarry. The uniaxial compression strength shows a large scatter, being included in a wide range (4.75–67.5 MPa), depending essentially on the welding degree, the textural features that characterize each layer and the void ratio.

The safety factors (SF) were evaluated for critical sections of the cavity using analytical methods. The static conditions of pillars and roofs were separately analysed using standard procedures (e.g. Evangelista *et al.* 2000, 2002). Owing to the lack of information about rock thickness in the roof ($t$), four different values ($t = 1.5, 2, 3$ and $4$ m) were considered.

The vertical stress in the pillars was evaluated by assuming that each pillar sustains the shared overburden weight with adjacent pillars. In this preliminary analysis of safety conditions, discontinuities and the irregular geometry of the pillars were not considered. It must be stressed that this simplification is not conservative, and its relevance will be carefully analysed in the near future. The safety factor of the pillar is $SF_{pil} = \sigma_{lim}/\sigma_{pil}$, where $\sigma_{lim}$ is the average uniaxial compressive strength of Piperno from Pianura quarry (12 MPa) and $\sigma_{pil}$ is the vertical stress. Only one pillar, out of six, was characterized by unsafe conditions using this method (Fig. 11).

Simple tools were used to estimate the stability conditions of the roofs. A general failure mechanism is considered in which, starting from tension zones, a crack may develop in the roof mid-span and at the two edges. By imposing the equilibrium to rotation of a half beam, the critical length $L_{critical}$, which gives rise to this 'arch mechanism', is (Fig. 12a):

$$L_{critical} = 1225 \times t \times (\sigma_{c\ lim}/\sigma_v)^{0.5}$$

where $t$ and $\sigma_c$ are the thickness and the uniaxial compressive strength of the rock mass of the roof, respectively, and $\sigma_v$ is the vertical stress at the roof depth, prior to cavity digging. The latter is due to the weight of both the rock beam and the layers above. The safety factor of the roof is $SF_{roof} = L_{critical}/L$. As expected, the thicker the roof, the better the general safety conditions (Fig. 12b). The effect of the excavations was

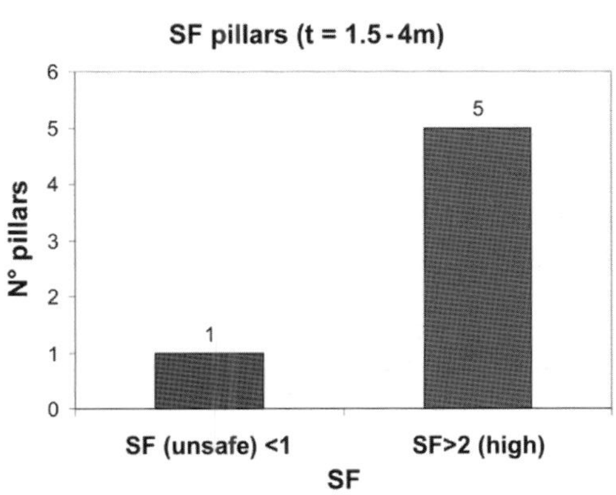


**Fig. 11.** Results of the stress simulation carried out on pillars.

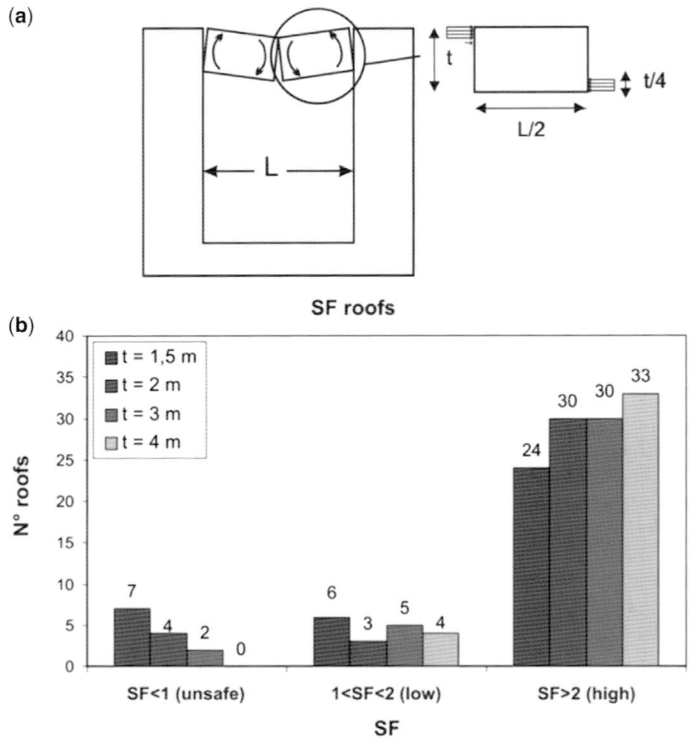


**Fig. 12.** (**a**) Description of the 'arch mechanism' (see text for details); and (**b**) results of the simulations.

determined by developing a two-dimensional (2D) parametrical analysis with FLAC2D (Evangelista *et al.* 2000), which is a widely used DEM computer code. The mechanical features were taken into account by varying the stiffness and the strength of the undisturbed rock; the role of pre-existing fractures was also considered, with both a discontinuum and an equivalent continuum approach. The constitutive models used for the numerical simulations were calibrated considering the laboratory results.

## Discussion

The need for further knowledge of Piperno, in terms of the identification of its exploitation sites, derives from the fact that it represents, along with the Neapolitan Yellow Tuff, the most used architectural and structural building material in the local historical architecture. On the one hand, information on mineralogical and petrographical features may provide a useful contribution to the correct interpretation of the weathering processes affecting the stone, and, on the other, the rediscovery of old exploitation sites allows evaluation of material still available for any replacement restoration. It may also encourage the preservation of exploitation sites that can help to demonstrate the traditional activities of their local populations.

Mineralogical and petrographical data show that, following deposition, this volcanoclastite was subject to welding and secondary minerogenetic processes that led to an almost complete feldspathization of the former glassy matrix. As a consequence of juvenile gas exsolved from pumice and vitric particles, a profound change in mineralogical characteristics occurred with the complete lithification of the deposit. The feldspathization process, as testified by the lack or very low amount of residual glass, has not produced an homogeneous degree of welding throughout the deposit, as indicated by the variability in UCS values (Calcaterra *et al.* 2005). The eutaxitic fabric provides the rock with its typical pattern as a function of the cut direction (*contro*, *verso*, *secondo*). *Contro* or *secondo* directions have always been used in local architecture to produce the drums of columns that serve a structural purpose, and also for the production of coating slabs. In Neapolitan architecture, however, many examples of slabs cut along the *verso* direction can also be found, mainly for staircases or even coating slabs.

At the Pianura quarry site, the planimetric scheme of the cavity, the hypothesized thickness above the Piperno vaults and observed collapses appear to indicate that most of the productive layers of Piperno have already been exploited. Confirmation of this hypothesis will only be possible after geognostic and geophysical prospecting to be carried out in the near future by the research group. However, collapsed material from the vaults, sometimes exceeding 1 $m^3$ in volume, could represent a resource for architectural replacement. Although it cannot be discounted that these blocks were originally dispensed with because they belonged to layers with poor petrophysical properties.

## Conclusions

This study is part of a wider research programme aimed at rediscovering and evaluating important stone exploitation sites of the Campania region (southern Italy) that provided valuable materials for the most important monuments of this area. This research topic is apposite given the numerous initiatives, funded by the regional government (Centro Regionale di Competenza), aimed at safeguarding their most important cultural and scientific geosites.

The most significant finding arising from the research is the discovery and characterization of one of the old quarry sites at Pianura that is still within a reasonable state of conservation. If this site is to be restored and used it is strongly recommended that rapid steps be taken to ensure its safety and stability. In terms of its future use, the Pianura quarry site provides a possible location for the construction of a museum of mining and working of ornamental stone within the Campania Region. This could in turn be linked to a school aimed at promoting the crafts associated with the working of stone. Such a development would be highly valuable in an area with a very high density of ancient cultural heritage that is in continuous need of restoration. There is also the possibility that the hypogeum site could function as a tourist attraction in its own right, with the specific function of explaining the geological background to visitors.

The authors wish to thank the reviewers (A. Calia, W. Dubelaar and an anonymous referee) who substantially improved the paper. The authors are also grateful to L. Melluso and V. Morra (CISAG/Dipartimento Scienze della Terra – Università FEDERICO II) for useful discussions. Work carried out with the financial support of Campania Region Legge no. 5 – Annualità 2002 (P. Cappelletti) and MIUR COFIN 2003 (M. de' Gennaro), and within the scientific activities of the Centro di Competenza 'INNOVA' – Dimostratore Campi Flegrei.

## References

Calcaterra, D., Cappelletti, P., Colella, A., de' Gennaro, M., de Gennaro, R. & Langella, A. 2003. Le pietre dell'architettura storica della Campania. *Arkos – Scienza e Restauro dell'Architettura*, **2**, 40–46.

Calcaterra, D., Cappelletti, P., Langella, A., Morra, V., de Gennaro, R. & Colella, A. 2000. The building stones of the ancient centre of Naples (Italy): the Piperno from Phlegrean Fields. Contributions to the knowledge of features of a long-time used stone. *Journal of Cultural Heritage*, **1**, 415–427.

Calcaterra, D., Langella, A., de Gennaro, R., de' Gennaro, M. & Cappelletti, P. 2005. Piperno from Campi Flegrei: a relevant stone in the historical and monumental heritage of Naples (Italy). *Environmental Geology*, **47**, 341–352.

Cardone, V. & Papa, L. 1993. *L'identità dei Campi Flegrei*. CUEN, Napoli.

De Vivo, B., Rolandi, G., Gans, P. B., Calvert, A., Bohrson, W. A., Spera, F. J. & Belkin, H. E. 2001. New constraints on the pyroclastic eruptive history of the Campanian volcanic Plain (Italy). *Mineralogy and Petrology*, **73**, 47–65.

Di Girolamo, P. 1968. Petrografia dei tufi campani: il processo di pipernizzazione (tufo > tufo pipernoide > piperno). *Rendiconti Accademia Scienze Fisiche e Matematiche, Napoli, IV*, **35**, 5–70.

Evangelista, A., Feola, A., Flora, A., Lirer, S. & Maiorano, R. M. S. 2000. *Numerical analysis of roof failure mechanisms in soft rocks*. GeoEng 2000, Melbourne. Technomic, Lancaster.

Evangelista, A., Flora, A., Lirer, S., De Sanctis, F. & Lombardi, G. 2002. *Studi ed interventi per la tutela di un patrimonio sotterraneo: l'esempio delle cavità di Napoli*. L'Aquila, XXI Convegno Nazionale di Geotecnica, Patron Ed., Bologna.

Maggiore, L. 1934. Notizie sui materiali vulcanici della Campania utilizzati nelle costruzioni. *Estratt Relazione Servizio Minerario Statistiene Industria Estrattiva, Rome*, **45**, 60.

Orsi, G., De Vita, S. & Di Vito, M. 1996. The restless, resurgent Campi Flegrei nested caldera (Italy): constraints on its evolution and configuration. *Journal of Volcanology and Geothermal Research*, **74**, 179–214.

Rosi, M. & Sbrana, A. 1987. Phlegrean Fields. *In*: *Quaderni de 'La Ricerca Scientifica'*, CNR, Progetto Finalizzato Geodinamica, **114**(9).

Rosi, M., Sbrana, A. & Principe, C. 1983. The Phlegrean Fields: structural evolution, volcanic history and eruptive mechanism. *Journal of Volcanology and Geothermal Research*, **17**, 273–288.

# Natural stone portals of the town of Udine (Italy): their design, construction and materials between the 15th and 20th centuries

ANNA FRANGIPANE

*Dipartimento di Ingegneria civile, Università degli studi di Udine, via delle Scienze 206, 33100 Udine, Italy (e-mail: anna.frangipane@uniud.it)*

**Abstract:** The research focuses on the features of 250 natural stone portals of the civil buildings of the town of Udine (NE Italy), dating between the 15th and 20th centuries. In order to clearly define the number, characteristics and uniqueness of these architectural elements, three strategies were implemented: (i) a concise database of all the portals; (ii) a concise reference database of more than 100 portals of five significant nearby towns; and (iii) a detailed inventory, consisting of data and photographs of about 200 portals selected for relevance or because they represent a recurrent type. The analysis of the data collected, supported by reference studies of quarry location, stone-cutter activity, the work of architects, cultural relationships with immediate and distant influences permitted the definition of an interdisciplinary framework describing the main features of portal production, as related to formal evidence, stone materials, historical building and carving techniques. The rational organization of the huge set of data collected represents an effective working tool, interconnecting different aspects of the portals' realization, which was indispensable for the research, but will also be useful for further research on the role of stone material in the historic buildings of Udine.

The great variety of natural stone portals in Italy has attracted relatively few investigations, those which have taken place have mainly focused on the characteristics and relevance of selected or local samples. Some researches have addressed the analysis of a single field of interest. For example, McGraw (1929), Romano (1992) and Sardella (1998) considered, respectively, the formal evidence of important Italian, Sicilian and Neapolitan portals. Whereas the work by Biraghi (1992) was concerned with understanding the philosophical and cultural meanings in portals designed, drawn and described throughout history.

Recent research, mainly in the field of restoration and engineering construction, has focused on the portal as a part of wider research topics. These can involve the history of construction, the identification of stone materials and weathering features, the understanding of design and construction phases, and characterization of different elements. In this context the research by Grandesso (1988) involved the study of medieval stone portals in Venice in terms of their history, form and materials. While Fianchino & Sciuto (1999) referred to natural stone portals in a wider study, concerning Sicilian building interventions that followed the reconstruction of the 1693 earthquake and focusing on building materials, intervention techniques and costs, Cervellini & Ippoliti (2000) included the analysis of formal and technical aspects of natural stone portals within a wider study of the Ascoli Piceno historical centre. Sansone (2002) implemented a detailed catalogue of civil architecture portals of the ancient centre of Naples, focusing attention on form, materials and weathering features.

Building on these approaches, the aim of the present research is to define the characteriztic features of the natural stone portals of civil architecture, built between the 15th and 20th centuries in Udine (the regional capital of Friuli, in NE Italy), in order to analyse in detail their form, materials and construction techniques in the context of the history of building in the town. The study of the portals permitted a parallel study of the capabilities of the artists and the craftsmen in the area, of the provenance of the stone material employed, and of carving and building techniques. It was also the starting point for further research into the construction history of Udine.

## Historical notes

The production of portals in the town of Udine is strictly related to historical circumstances, mainly owing to the fact that the town experienced, in little more than 1000 years of documented history, a variety of different political and cultural influences. Until the early Middle Ages, Udine was a minor settlement. It was named for the first time in an official document in 983, in which the jurisdiction of the town was allocated by the German

*From*: PŘIKRYL, R. & SMITH, B. J. (eds) *Building Stone Decay: From Diagnosis to Conservation.* Geological Society, London, Special Publications, **271**, 33–42.
0305-8719/07/$15.00 

Emperor to the Patriarch of Aquileia. In 1230 and 1248 the Patriarch of Aquileia bestowed free market and town rights. It was the beginning of a commercial development that put Udine at the centre of European trade routes, as demonstrated by the presence of Lombard, Tuscan and Venetian bankers in the town. In art and architecture attention was mainly paid to religious buildings, a habit that, following the Florentine and Roman examples, changed during the 15th and 16th centuries, leading to the construction of important palaces. In 1420 the Venetian army occupied the town and the surrounding area, thus ending a five century long Patriarchal government. From then on, politics and culture were strongly influenced by the Venetians. In 1593 the construction of a Venetian fortress at Palmanova, 20 km south of Udine, began. The influence of such an important military building site on portal building is revealed by this work. Defeat in the Candia War (1645–1669) lead the Venetians to make important investments in inland areas. Extensive cultivable estates were acquired and country villas were built. Contemporary town palaces showed a peculiar similarity in portal style because of the presence of shared skilled workers on the building sites. The action of Austrian Emperor Charles VI, in bestowing the free market status to the Adriatic harbours of Trieste and Fiume in 1719, damaged the commercial importance of Udine. It was also a sign of a decrease in the leading role of Venice, marked in 1797 by the victory of the French army over the Republic. Udine experienced two French governments between 1797–1801 and 1805–1813, interspersed with a brief Austrian presence. A second Austrian government (1813–1866) brought a substantial halt to artistic and architectural activities. The annexation to Italy in 1866 put Udine, for the first time, directly in the sphere of influence of the peninsula, where it played a secondary role. This minor cultural position was overcome by only a few leading figures at national level, such as Raimondo D'Aronco, an Art Nouveau Italian master and the architect of the last natural stone portal of the town.

Thus, depending on historical circumstances, Udine played alternately a primary and a secondary role in the artistic and architectural fields. After a leading cultural role, dating back to the Patriarchate of Bertrand of Saint-Geniès (1334–1350), the presence of important architects marks two significant moments experienced in the first half of the 16th and 18th centuries. In 1527 Giovanni da Udine, a Raphael disciple and the re-inventor of stucco, escaped from Rome and brought new ideas on architecture to the town. His work in Udine marks the transition from the Gothic to the Renaissance style. The work of Andrea Palladio, present in the town in 1556 and in 1563, influenced the production of portals in a rustic ashlar style throughout the 16th–18th centuries, which used the Bollani Arch, the main access to the Castle (Puppi 1999), as a model. Survey visits to the Palmanova fortress by Michele Sammicheli (1532) and Vincenzo Scamozzi (1593 in Udine) are documented. Later constructions were probably influenced by their advice. Udine was at the centre of the architectural scene again in the first part of the 18th century owing to the local activity of the Venetian architects Domenico Rossi (who was working on the renovation of the Cathedral interior, the construction of the Manin Chapel and the enlargement of the Patriarch Palace (1708–1735)), and Giorgio Massari (who was working on the façade of St Antony's Church (1732–1735)). These important buildings influenced taste and development for many decades.

## Research methods

Repeated surveys within the town centre lead to the identification of 250 natural stone portals of relevance, dating from the 15th to the 20th century. A total of 172 portals were selected for their representative features, 100 of which have an evident artistic value. Out of this number, 21 portals were surveyed in detail.

In order to investigate in depth the form, material and technical aspects of these architectural elements, three strategies were adopted, involving the construction of: (i) a concise database of all the 250 portals; (ii) a concise reference database of 124 portals of five significant nearby towns; and (iii) a detailed inventory of the 172 selected portals. The complete description of these resouces was recently presented (Frangipane 2004*b*). The html version of the work is in implementation and it is expected to be published in 2006.

### *The object and reference databases*

The first tool, the concise object database, which comprises the 250 portals, was constructed using Microsoft Access©. It is composed of 250 cards, summarizing the main information and features of each portal. It is organized in 16 fields: a common heading; the building number (which follows an old ordering, dating back to the French government); the address; the building's official name (Palace ...); when it came into use; the stone employed; the documented or presumed date of construction; an indication of a traditional or valuable character; the presence of a single v. a multiple architectural order; an indication of the kind of portal beam (an architrave, a plat band, an archivolt or an arch); the kind of arch, if present; the beam component elements and their finish; the presence of upper or lateral openings; the function of the portal (carriage v. pedestrian gateway); principal references; notes; and an image of the portal.

The second tool is the concise reference object database of 124 portals of five significant towns in the area: Palmanova, Cividale del Friuli, Tolmezzo, Gradisca d'Isonzo and Trieste. These were chosen for their specific characteristics: Palmanova as an important military reference settlement; Cividale del Friuli – of Roman origin and once the capital of the Friuli Lombard Duchy – as an important town close to the quarries of PIasentina Stone, the most common material employed; Tolmezzo as an important town close to the mountains, the widest possible source of building material; Gradisca d'Isonzo for its mixed Venetian and Austrian artistic influences; and Trieste for a development restricted to a short and precise period, which crossed the 18th and 19th centuries, permitting a clear comparison in contemporary matching of styles. The basic card is almost the same as the object database card of Udine portals.

## The inventory

The third tool is the inventory of 172 portals, selected out of 250, in which a precise description of each portal feature is provided. The inventory contains, for each portal, a long checklist and 10 photographs of the portal and of its architectural components, organized in different sections. In this case it was not possible to adopt 'statically implemented' database software, as used for the catalogues, due to the enormous size of the photographic files. The choice was, therefore, to construct the inventory using drawing software (Corel Draw©) and to overcome the absence of a quarrying tool by using a parallel Microsoft Excel© spreadsheet, containing the checklist input. The construction of 'dynamically implemented' database software, allowing both the visualization of photographs and the quarrying facility, is ongoing.

Part I of the inventory, referred to as 'References and relationships', concerns portal information and relational aspects (Fig. 1a). Its first section contains the data regarding the building, matching the first

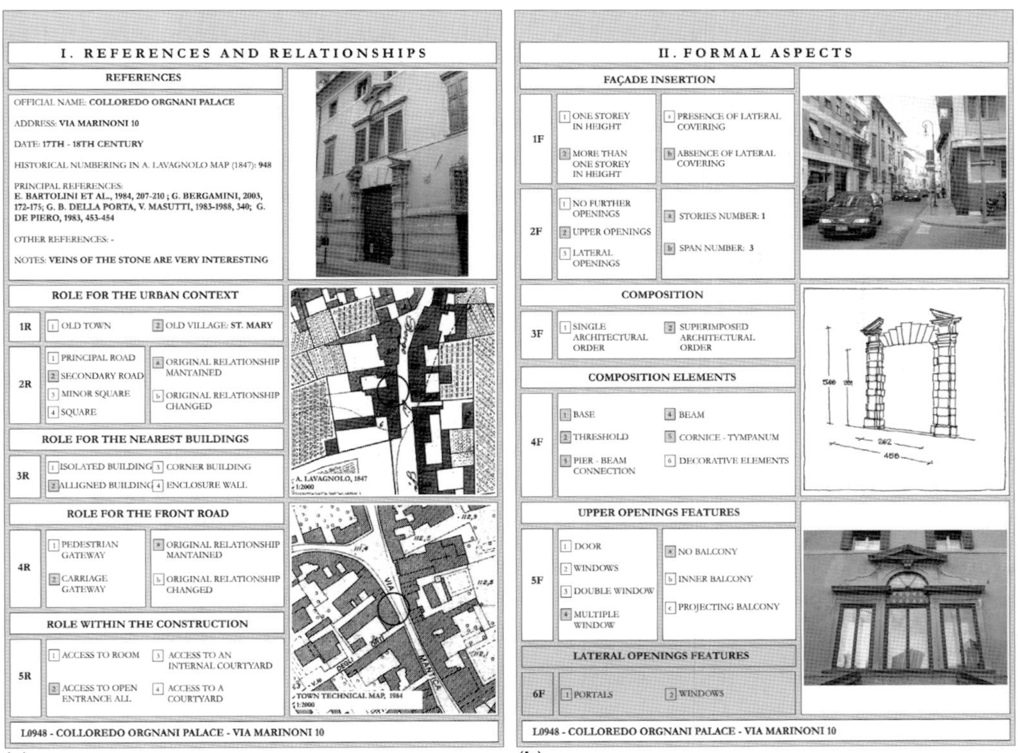


(a) (b)

**Fig. 1.** (**a**) Part I of the inventory 'References and relationships' contains general information on the portal and shows the relationships existing with the town, the closest buildings, with the road in front and with the building itself. (**b**) Part II of the inventory, 'Formal aspects', defines the characterizing formal features of the portal (object: Palazzo Colloredo – Orgnani).

part of the above object database, and a portal photograph. In its second section the role of the construction is documented according to: (i) the urban context (building overlooking a square, a minor square, a principal or secondary road); (ii) the nearest buildings (isolated, aligned or corner building or enclosure wall); and (iii) the front access (pedestrian or carriage gateway). Furthermore, the role of the portal within the construction (access to a room, to an open entrance hall, to an internal courtyard or to a courtyard) is described with reference to both the original and the present condition. This documentation is supported by two map cuttings (1:2000) indicating the urban structure of the area in 1847 and in 1984, and showing recent demolitions and additional constructions that changed the role of the building and, implicitly, of the portal.

Part II of the inventory, entitled 'Formal aspects', defines the formal features of the portal (Fig. 1b). A first section describes the relationship between the façade and the portal size (one or more than one storey in height) and the presence of additional lateral or upper openings if part of the design. A view of the façade is provided. A second section investigates aspects of the composition of the portals themselves, such as the presence of a single or superimposed architectural order and the presence of composition elements defining its formal structure (base, threshold, pier, beam connection, beam, trabeation, tympanum and decorative elements). A sketch of the portal, indicating its main dimensions, completes the section. The third section refers to the characteristics of upper openings (doors, single, double or multiple windows) and of lateral ones (doors or windows), with, if possible, a photograph.

Part III, refers to 'Technical and material aspects', and describes in detail the construction aspects regarding the base, threshold, piers, beam connection, beam, trabeation, tympanum and decoration elements (Fig. 2). Technical descriptions are given for each element, together with the kind of stone of which they are made. For piers, beam connection and beam the stone finish is described. The section of each element section is provided with a corresponding image.

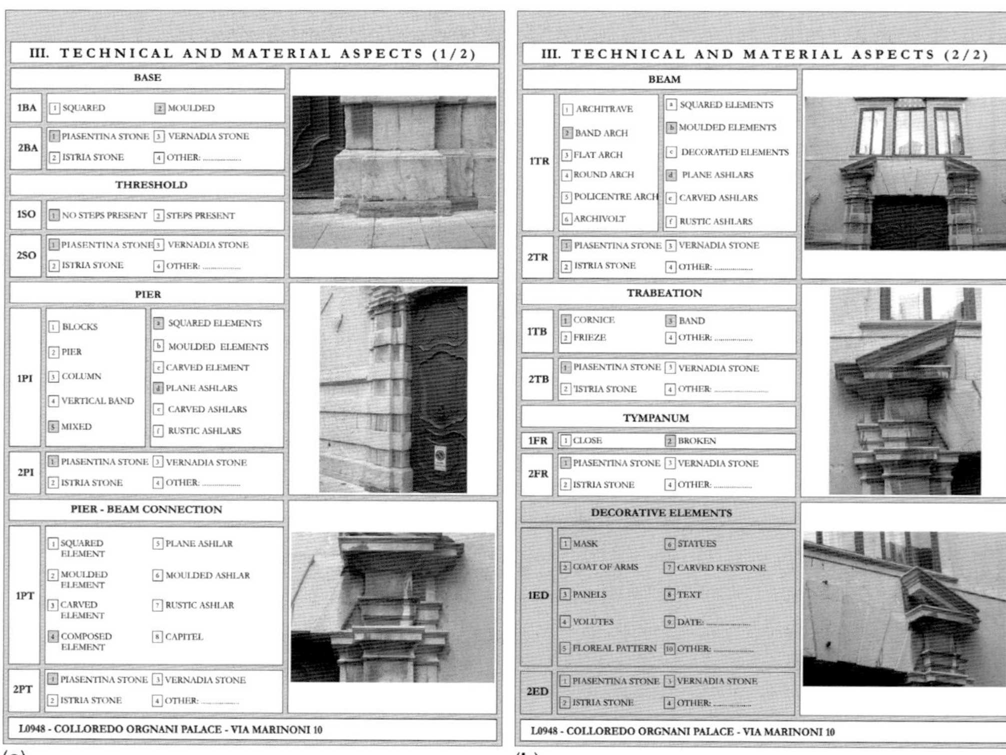


**Fig. 2.** Part III of the inventory 'Technical and material aspects' describes in detail the constructive aspects regarding: (**a**) base, threshold, piers, beam connection; and (**b**) beam, trabeation, tympanum and decoration elements (object: Palazzo Colloredo – Orgnani).

For 21 of the 172 portals, chosen for their significance, detailed surveys were carried out in order to analyse both the geometry and the finishing of the stone elements. The data acquired for these 21 portals were summarized in three drawings, comprising three additional parts of the inventory card.

Part IV provides a 'Geometrical survey (centimetres)'. It is the basic tool for indicating portal dimensions. However, this survey does not help to reveal the ideas leading to the project, as it does not match the original unit system on which the design of the portals was based. This mismatch runs the risk of hiding the true meaning of the original project.

Part V, the 'Geometrical survey (feet and inches)' seeks to overcome this problem (Fig. 3a). The comparison of all surveys, based on feet and inches (where 1 Udine building ft $= 34.048$ cm and 1 Udine inch $= 1/12$ Udine building ft $= 2.84$ cm), showed the presence of recurrent measures, as, for example, 6 or 7 ft for the width of the opening, 1 ft or 1 ft and 6 inches for ashlar width, and 11 inches for their height, and so on. The proof of a clear rationality in the design of many architectural elements was one of the surprising results of the research.

Part VI, entitled 'Block survey and finishing features', is concerned with block shapes and the visible traces of carving tools (Fig. 3b). This revealed that the shape of the stone blocks does not fit the visible lines of the portal. Their shape and disposition fulfil static and constructive requirements that are proved by this table. The survey of block shapes is completed by detailed images showing the finish of stones, related to the use of certain carving tools.

## Results

Analysis of the data collected, of the images and the surveys have brought to light some previously unknown aspects of the portals of Udine with regard to their construction forms, stone materials, and their weathering and historical techniques.

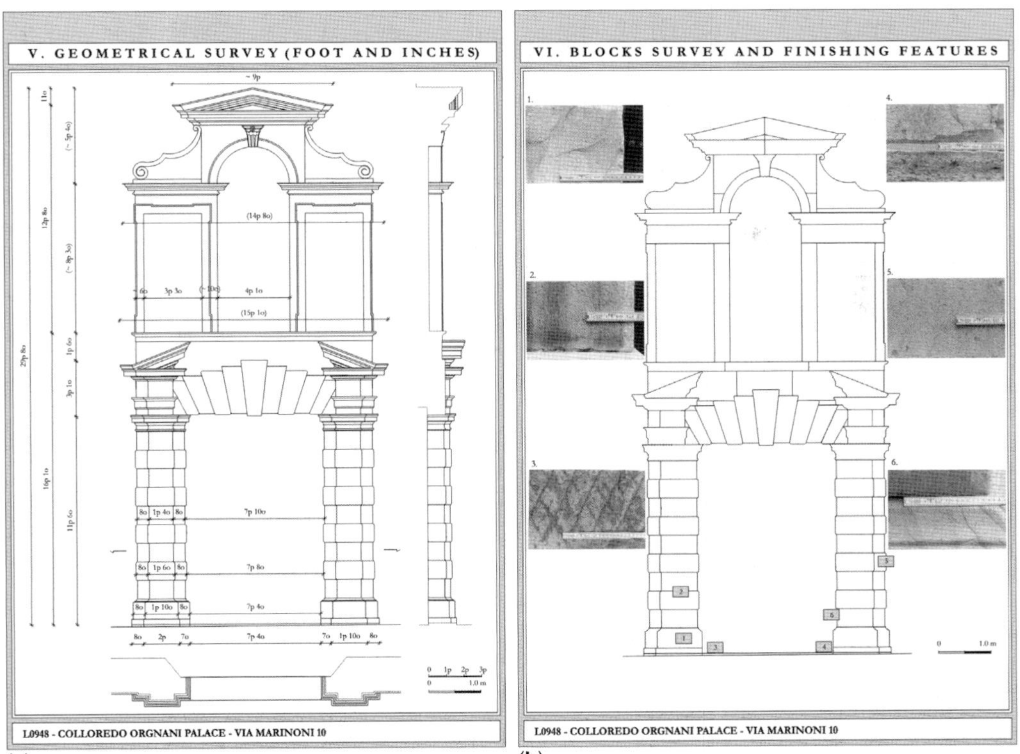


(a) (b)

**Fig. 3.** (**a**) Part V of the inventory 'Geometrical survey (feet and inches)' presents the geometrical survey, based on the town of Udine ancient units. (**b**) Part VI of the inventory 'Blocks survey and finishing features' shows the composing blocks shapes and highlights the visible traces of carving tools (object: Palazzo Colloredo – Orgnani).

## *Construction forms*

The change in formal features, as observed for the different historical periods (Frangipane 2004*a*), can be outlined as follows.

The 15th century was almost entirely characterized by a style that simplified the Gothic lines of the rich holy portals. This trend ceased in the first part of the 16th century, when the style of the portals of Venetian churches by Mauro Codussi was imported. In the same period classical portals, composed of a simple moulded architrave with an upper cornice, were introduced by Giovanni da Udine, following Roman examples (Fig. 4a). The present study offers evidence that the work of Andrea Palladio, who understood the effectiveness of the rough carving of the local Piasentina Stone, brought a new input in the second half of the century. He was probably the first to apply a finish of this type to portals, which was close to the roughness of Roman aqueduct arcades. The Bollani Arch (Fig. 4b) by Palladio was, in the following century, an example that influenced most palace carriage entrances. The 17th century experienced the influence of Venetian military architecture in the severe definition of palaces' main portals (Fig. 4c), as well as the reference to 'ready to use models', diffused by the architectural treatises of the period (Fig. 4d). The 17th and the 18th centuries were characterized by the almost total absence of Baroque elements, and pursuit of a sort of continuing classicism, a feature common to contemporary Venetian architecture. Forms of the 17th century were repeated with austerity, the presence of upper windows in continuity with the portal being the only characteristic frequent feature (Fig. 4e). In this period links are evident with the architecture of contemporary country estates. The last years of the century experienced the introduction of neoclassical lines (Fig. 4f). The production of portals during the 18th century is, however, limited, when compared to that of the two preceding centuries. Owing to political and economic factors, the 19th century shows a substantial lack of important architectural activity. The secondary role assumed by the town is also reflected

**Fig. 4.** Typical forms of Udine portals. Sixteenth century: (**a**) castle entrance, attributed to Giovanni da Udine; (**b**) Bollani Arch by Andrea Palladio. Seventeenth century: (**c**) Deciani Daneluzzi Braida Palace; (**d**) Torriani Palace. Eighteenth century: (**e**) Zignoni Margreth House; (**f**) Pavona Asquini Palace. Nineteenth century: (**g**) Morelli de Rossi House. Twentieth century: (**h**) Moisesso Liruti Biasutti House, by Raimondo D'Aronco.

in the small number of portals built. Attention is mostly paid to the refurbishment of buildings and portals of minor importance (Fig. 4g), following the framework provided by a rigorous Town Administration. The époque of portals is concluded by the work of the master Raimondo D'Aronco, the designer of the only important portal of the 20th century (Fig. 4h).

## *Stone materials and their weathering*

The use of different stone materials and associated weathering features were identified by macroscopic and phenomenological analysis, supported by historical reference and data matching. No petrophysical analyses were carried out.

Four important types of natural stone were identified: (i) a first local sandstone, the Piasentina Stone; (ii) a second local sandstone, the Vernadia Stone; (iii) a compact limestone coming from Istria; and (iv) two kinds of local fossil limestone, the Travesio and Aurisina stones. Other stones of minor importance were identified and, although rarely used, sometimes played an important role in the history of the town buildings.

The Piasentina Stone (Fig. 5a) is an Eocene calcareous breccia with a calcareous–marly cement. Its colour is grey, with a weak pale brown tonality; white quartzite veins are present in the blocks. The different size of the grains defines coarse, medium and fine qualities. It has been used in nearly all the portals classified. It comes from the eastern hilly area of the region, in the neighbourhoods of Tarcento, Cividale del Friuli and Gorizia. The use of Piasentina Stone in rustic ashlar portals is documented, as mentioned, in the Bollani Arch by Andrea Palladio, who was the first to introduce this rough finish for important buildings. Macroscopic analysis shows severe weathering of the Piasentina Stone, despite early opinions regarding its strength and resistance (Pitacco 1884). This is mainly due to water absorption and temperature changes, which generate severe stress within the composite structure of the stone, leading to slow deterioration of the marly cement. This behaviour was reported since the first geological studies of the portals as long ago as the late 19th century (Marinoni 1881). Archive images show how disaggregation has increased as a result of air pollution. Elements dating back several centuries appear to be intact in pictures taken less than a century ago, while they are dramatically weathered today.

The Vernadia Stone (Fig. 5b) is an Eocene calcareous-quartzose sandstone, quite micaceous, once quarried in the same sites as the Piasentina Stone (Marinoni 1881). Its colour is grey, with blue reflections, sometimes with an ochre tonality due to its iron content. It is very common in the simplest portals and window frames, and is a material of evidently inferior resistance compared to the Piasentina Stone. No petrophysical property studies of Vernadia Stone are known, owing to

**Fig. 5.** Natural stones employed in Udine portals and their characteristic weathering features: (**a**) Piasentina Stone; (**b**) Vernadia Stone; (**c**) Istria Stone; (**d**) supposed Travesio Stone; (**e**) dolomite limestone; and (**f**) 'Red Ammonite' Stone.

the absence of real interest in this poor building material. Its use is, however, well documented in archives. The evident visible features of its weathering consist of significant flaking parallel to bedding planes, producing flakes several centimetres in size.

The Istria Stone (Fig. 5c) is the limestone widely used for decorative elements of Venetian buildings. It is a sedimentary, fine-grained limestone, dating to the lower Cretaceous period (D'Ambrosi & Sonzogno 1962). It is a very compact material, sometimes crossed by dark narrow veins that do not affect its resistance. However, reddish veins, indicating the presence of clay materials, are the origin of fractures (Dalla Costa & Feiffer 1981). The provenance is the Istria peninsula. The location of the quarries and their exploitation are well documented in the 17th century Scamozzi treatise (Scamozzi 1615). Owing to difficulties in transport, it was rarely employed in Udine, and only in small dimension blocks for window and door frames. Nevertheless, some of the most important portals dating from the 16th–18th centuries are built in Istria Stone. The stone is very resistant to weathering and for this reason it was widely used in extreme conditions such as coastal areas. Petrophysical properties and the weathering of Istria Stone employed in Udine were analysed in detail by Biscontin *et al.* (1990), with reference to the Manin Chapel, the 18th century architectural jewel by Domenico Rossi. Macroscopic analysis shows a weak weathering of the Istria Stone, corresponding to visible veins.

The Travesio and Aurisina stones are fossiliferous limestones employed in the town in different periods, depending on different historical conditions. They are both Cretaceous–Eocene limestones with fossils present in varying sizes, almost always visible. They both have a tonality between white and grey, passing through a pale brown. Their certain identification would require laboratory analysis. The Travesio Stone was quarried in the foothill area west of Udine during the 15th and 16th centuries, and was employed for the most important Renaissance religious portals in the region. They are masterpieces of the so-called 'Lombard School', named after the provenance of the sculptors in the areas of Como and Ticino. Petrophysical property studies of the stone are presented in Carulli & Onofri (1966), with reference to the Clauzetto Stone. The Travesio Stone is often confused with the similar Aurisina limestone, even if archive documents (Bergamini & Goi 1982; Goi 1998) clearly state its provenance. In Udine it was probably used only for a few portals of the early 16th century. The use of Aurisina Stone (Carulli & Onofri 1969) is, on the other hand, documented in the mid 17th century for the enlargement of the Town Hall building (Joppi & Occioni-Bonaffons 1877; Spadea *et al.* 2000). It comes from the Karst, the arid hilly area surrounding the town of Trieste. Changes in customs rates and in political alliances could have justified such a distant provenance. The fossiliferous limestone used in the Torriani Palace (Fig. 4d) is supposed to come from that area, possibly from the quarries of the Torriani family described by Scamozzi in his treatise. Petrophysical property data are presented in Cucchi & Gerdol (1985). Travesio and Aurisina stones do not exhibit evident signs of weathering, with the exception of a white patina.

Other stones of different origin were rarely found in the portals of the town. Two of them deserve attention, on the grounds of historical circumstances. The first one is a fine-grained yellowish dolomitic limestone (Fig. 5e), only present in a small portal of minor importance belonging to Pignat House. This stone had an important role in the 14th century, as demonstrated by its use for the building of the main portals of the cathedrals of Udine (Spadea *et al.* 1996) and Spilimbergo. Petrophysical property data are presented in Spadea (1995) for some samples from Udine Cathedral. Weathering takes place in terms of small sized chipping. A second stone of interest is an ammonite stone of uncertain provenance, which is pale red in colour. Even if 'Ammonitico rosso' stones are commonly believed to come from Verona, their availability close to the mountain area, near the town of Gemona, and in the western mountain area, close to the villages of Erto and Casso, is reported in old references to active quarries (Marinoni 1881; Pitacco 1884). The matter of the provenance deserves, therefore, further investigation and implies an interesting consequence in the history of the constructions of the area. Weathering features are those common to 'Ammonitico rosso' materials, mostly consisting of alteration along veins crossing the blocks and foliation damage.

## *Historical construction techniques*

The research considered different aspects of historical construction techniques, taking advantage of the repeated detailed surveys carried out and of archive data analysis.

The first aspect investigated concerns the relationship existing between hand-worked features and tools. Recurrent patterns in surface finish were identified, and the use of a limited number of successive tools was recognized. In terms of their increasing accuracy for the definition of the surface, these are: the roughing chisel, the rough pointed chisel, the point chisel, the rough tool axe, the fine tool axe and the flat chisel. Each tool left a characteristic mark on the surface, closely related to the quality of the stone. This is, perhaps, the reason for the preference shown by Palladio for the Piasentina Stone. The use of the roughing chisel to obtain rough hewn blocks was, in fact, particularly effective, due to the way that the stone is broken obtaining a very specific stone finish not possible in other stone varieties.

Normally invisible features of blocks were also observed. The importance of parts commonly hidden from view indicated the role of unfinished surfaces for mortar and plaster adhesion. Hidden iron elements, contributing to portal stability, were sometimes discovered. The parallel direct observation of dismantled portals was very helpful and instructive.

Building site characteristics and block-laying techniques were then looked for, by referring mainly to iconography and archive documents, which enabled the understanding of where and how stone carving and finishing were carried out.

Detailed surveys were the way to truly comprehend the function of minor elements, such as mouldings. It became clear that the design of mouldings actually answered to real needs, such as the sheltering of surfaces from rain action, and that great care was taken in defining the size of component elements, not only from a formal point of view but also from a functional one.

All the data and analyses presented contributed to providing an effective idea of the complex framework within which the portals were built.

## Conclusions

The aim of this research was the definition of the main formal, material and technical aspects of the natural stone portals within the civil architecture of Udine (NE Italy), as related to the history of its buildings.

A concise database, including 250 portals, a concise database of 124 portals of relevance in nearby towns and a detailed inventory of 172 portals, integrated by the survey of 21 of them, were the tools that underpinned the analysis. The systematic study of the huge amount of data and images helped to bring together the salient and noteworthy aspects identified above. Archive and reference studies placed local production within the wider frame of Italian historical architecture.

The influence of Roman architecture, as imported by the work of Giovanni da Udine and Andrea Palladio, is, for instance, the leading element of portals constructed across the 16th–18th centuries. In that period the Piasentina Stone played a major role, both for rustic ashlars and for classical framed portals. On the other hand, the influence of Venetian architectural culture became evident at the beginning of the 17th century. The architecture of Venetian palaces and military constructions conditioned the drawing and the choice of material of the most important portals incorporating Istria Stone. Similarities with the country estate buildings of the Venetian aristocracy in the area support this evidence. The 'golden age' of Udine portals declines, in a sense, with the power of Venice and no substantial contribution was provided by the Austrian and Italian cultures that followed. Working techniques remained almost unchanged during all the period considered, as did building techniques. Observation of these techniques nonetheless helps to provide a better understanding of the complex framework within which the portals were realized.

The picture that emerges from the integration of all these different components and from comparison with similar studies elsewhere in Italy (Grandesso 1988; Fianchino & Sciuto 1999; Cervellini & Ippoliti 2000; Sansone 2002) is one of provincial production, depending for its construction forms on a few important external trends. The stones employed were basically quarried locally, with the exception of the Istria Stone, and even the architects and craftsmen involved were mainly local.

This work not only offers a detailed study of these architectural elements, but also provides, owing to the systematic organization of the data collected, a framework of analysis of materials, weathering and finishing techniques that could constitute the starting point for further studies regarding the use of natural stones in historical and traditional architecture in Udine. Specifically, it provides the availability of an operational tool for conservation activities, multidisciplinary studies on building history and an appreciation of local cultural heritage.

The research was carried out by the author during the PhD studies at the Faculty of Engineering of Naples University Federico II. The helpful guidance of R. Iovino, director of the research, is grateful acknowledged. My sincere thanks to the late I. Bulfone, stone master, whose help was essential in stone identification and evaluation of the quality of stone work. Important input to the investigation came from the works carried out in Udine University with P. Spadea on the use of natural stone in monuments. A grateful thanks to the reviewers, whose work was essential in improving to the paper quality.

## References

BERGAMINI, G. & GOI, P. 1982. Bernardino da Bissone a Tricesimo. *In*: CICERI, A. & MIOTTI, T. (eds) *Tresésin*. Societât Filologjiche Furlane, Udine, 351–362.

BIRAGHI, M. 1992. *Porta multifrons: forma, immagine, simbolo*. Sellerio, Palermo.

BISCONTIN, G., LONGEGA, G., PAGANI, T., PERUSINI DE PACE, T. & SPADEA, P. 1990. *In Restauro nel Friuli Venezia Giulia*. Memorie del Centro regionale di restauro. Centro Regionale di Catalogazione e restauro dei Beni Culturali. Regione Autonoma Friuli Venezia Giulia, Trieste, 65–156.

Carulli, G. B. & Onofri, R. 1966. *Il Friuli: i marmi*. Camera di commercio, industria e artigianato, Udine.

Carulli, G. B. & Onofri, R. 1969. *I marmi del Carso*. Regione Autonoma Friuli-Venezia Giulia, Assessorato Industria e Commercio, Trieste.

Cervellini, F. & Ippoliti, E. 2000. *Per un atlante architettonico e urbano di Ascoli Piceno: portali*. Gangemi, Rome.

Cucchi, F. & Gerdol, S. (eds). 1985. *I marmi del Carso triestino*. Camera di commercio industria artigianato e agricoltura, Trieste.

D'Ambrosi, C. & Sonzogno, G. 1962. *La cava romana*. Marmi e pietre del Carso e dell'Istria. Cava Romana, Aurisina, Trieste.

Dalla Costa, M. & Feiffer, C. 1981. *Le pietre nell'architettura veneta e di Venezia*. La stamperia di Venezia editrice, Venice.

Fianchino, C. & Sciuto, G. 1999. *Materiali, procedimenti e costi della ricostruzione nel '700 in Sicilia*. Gangemi, Rome.

Frangipane, A. 2004*a*. The use of natural and artificial stone in the portals of the town of Udine (Italy). *In*: Přikryl, R. & Siegl, P. (eds) *Architectural and Sculptural Stone in Cultural Landscape*. Karolinum Press, Prague, 73–90.

Frangipane, A. 2004*b*. *I portali lapidei nell'edilizia civile della città di Udine: aspetti formali, materici e tecnologici*. PhD Thesis, Naples University Federico II, Naples.

Goi, P. 1998. Lapicidi Lombardi a Tolmezzo: verifiche e considerazioni. *In*: Ferigo, G. & Zanier, L. (eds) *Tumieç*. Societât Filologjiche Furlane, Udine, 595–611.

Grandesso, E. 1988. *I portali medievali di Venezia*. Helvetia, Venice.

Joppi, V. & Occioni-Bonaffons, G. 1877. *Cenni storici sulla loggia comunale di Udine con 48 documenti inediti, per cura dell'Accademia e a spese del Comune di Udine*. Tipografia di Giuseppe Seitz, Udine, 14.

Marinoni, C. 1881. *Sui minerali del Friuli*. Annuario Statistico della Provincia di Udine. Tipografia Seitz, Udine.

McGraw, C. B. 1929. *Italian Doorways: Measured Drawings and Photographs*. Hansen, Cleeveland.

Pitacco, L. 1884. *Descrizione delle pietre e dei marmi naturali che si impiegano nelle costruzioni in provincia di Udine*. Tipografia di G. B. Doretti e soci, Udine.

Puppi, L. 1999. *Palladio*. Electa, Milan, 305–306.

Romano, M. 1992. *Il portale barocco di Siracusa: con itinerario dei portali e itinerario monumentale*. Emanuele Romeo, Siracusa.

Sansone, C. 2002. *I portali lapidei dei palazzi nel centro antico di Napoli: lettura tipologica e analisi del degrado*. PhD Thesis, Naples University Federico II, Naples.

Sardella, F. M. (ed.). 1998. *Fra le mura; dai portali al verde nascosto*. Soprintendenza per i Beni ambientali e architettonici di Napoli e provincia & Comune di Napoli, Elio de Rosa Editore, Naples.

Scamozzi, V. 1615. *L'Idea dell'Architettura Universale*. Anastatic reprint, 1997. Centro Internazionale di Studi Andrea Palladio, Vicenza, 206.

Spadea, P. 1995. *Studio mineralogico e petrografico dei materiali lapidei e delle malte*. In Duomo di Udine. Ricerca per il restauro del portale della Redenzione. Centro Regionale di Catalogazione e restauro dei Beni Culturali. Regione Autonoma Friuli Venezia Giulia, Trieste, 95–116.

Spadea, P., Perusini, T. & Frangipane, A. 1996. Dolostones used in Middle Age in Friuli (Ne Italy). *In*: *Proceedings of the 8th International Congress on Deterioration and Conservation of Stone, Berlin*, Vol 1, Möller Druck und Verlag GmbH, Berlin, 155–157.

Spadea, P., Perusini, T., Frangipane, A. & Maddaleni, P. 2000. The Loggia del Lionello of Udine (15th century): weathering of the stone facing. *In*: *Proceedings of the 5th International Symposium of the Conservation of Monuments in the Mediterranean Basin, Seville*, Departamento de Cristalografía, Mineralogía y Química Agrícola, Facultad de Química, Universitad de Sevilla, 146–147.

# The dimension stone potential of Thailand – overview and granite site investigations

A. HOFFMANN & S. SIEGESMUND

*Geoscience Centre, University of Göttingen, Department of Structural Geology & Geodynamics, Goldschmidtstrasse 3, 37077 Göttingen, Germany (e-mail: ahoffma2@gwdg.de)*

**Abstract:** The production of dimension stones is well established in Thailand and the country has considerable processing capacities in the region, second only to China. The geological background of Thailand provides a huge potential of dimension stones, including magmatic, metamorphic and sedimentary rocks. The NE part of the country is made up by the Khorat Plateau with the main sandstone resources at its western margin. Metamorphic carbonate rocks are predominantly distributed along the border of a basin area in central Thailand. The western part of the country is characterized by magmatic belts that comprise the resources of igneous rocks. Large quantities of the dimension stone potential were used in the first part of the 1990s, when the domestic economy underwent a considerable upturn. The most important region for the production of granitoid rocks is the Tak batholith in northern Thailand. Therefore, the Tak granitoids are discussed as a case study with respect to petrophysical and depositional characteristics.

Over the last few years the Asian continent has shown important changes in the production of dimension stones. With an impressive expanding rate, producers from Asia have grown remarkably strong and gained ground on the international dimension stone market by the supply of raw materials and finished products. Certainly, the overwhelming quantity of rocks coming from Asian producers, particularly from China, is one of the most impressive phenomena in the recent history of the dimension stone sector.

Apart from China and other major producers in the region such as India, the production of dimension stones is also well established in Thailand. The country holds considerable processing capacity, but, in contrast to China and India, products from Thailand are mainly distributed on the domestic market. The major part of the products from Thailand were used during times of economic wealth in the first part of the 1990s, when intensive construction took place especially in the Bangkok Metropolitan Area. Production declined in 1997, when Asian countries and in particular Thailand were seriously affected by an economic crisis. As a consequence, many companies were forced to produce at a very low levels for years.

In times of prosperity, the import of dimension stones from China, Brazil, Vietnam or Norway was significant in terms of quantity and value. Because of governmental restrictions as a result of the economic crisis, the import of dimension stones has been relatively low during recent years. Since an economic recovery in 2001, the domestic building stone industry in Thailand has grown at an accelerating rate and its prospects are still promising today. Some of the restrictions do not exist anymore, so that since March 2003 unprocessed marble blocks are allowed to be imported (Duerrast *et al.* 2003). In terms of export trade, only limited quantities of rock material have been sent to Japan, Taiwan and Korea, and also to the USA and Australia. As the export and import of dimension stones are limited, the country can be considered as a relatively closed market for dimension stones.

Dimension stone quarrying has become an integral part of the mineral industry in Thailand, which is generally well developed and growing. In the year 2004 the output value of the mining and quarrying sector contributed 2.2% to the gross domestic product (Wu 2004). Apart from considerable operations for mineral fuels (lignite, natural gas) and metallic minerals (iron ore, zinc ore), the continued growth of the mining sector is also due to the increasing production of industrial minerals such as barite, dolomite, feldspar, gypsum and limestone among others. In fact, Thailand was one of the world's top producers of feldspar and gypsum in the year 2004, and took one of the world's leading positions regarding the export of cement, feldspar and gypsum (Wu 2004).

## Geological setting

Thailand comprises three principal units with respect to accretional events from the Late Palaeozoic to Early Mesozoic. Those units are the two relatively stable blocks Shan Thai and Indochina, which occupy the western and eastern parts of the

*From*: PŘIKRYL, R. & SMITH, B. J. (eds) *Building Stone Decay: From Diagnosis to Conservation.* Geological Society, London, Special Publications, **271**, 43–54.

country, respectively, and a mobile belt in between (Fig. 1). The Shan Thai terrane covers western Thailand and Myanmar, and extends northwards into China and southwards through the peninsular Thailand into Malaysia. The Indochina block extends over NE Thailand and the territories of Cambodia, Laos and parts of Vietnam. The mobile belt, is usually referred to as the Yunnan Malay mobile belt, and covers the eastern part of northern Thailand and the western part of NE Thailand. In northern and NE Thailand, the belt can be subdivided into the western Sukhothai Foldbelt and the eastern Loei Foldbelt (Hahn *et al.* 1986). According to Bunopas & Vella (1978) and Bunopas (1981), the Sukhothai Foldbelt belongs to the Shan Thai block, while the Loei Foldbelt is part of the Indochina block (Fig. 1).

Since the Upper Palaeozoic, these units have been intensively affected by the collision of microterrains or island arcs. From the Upper Permian to Upper Triassic an extensive tectonic regime in SE Asia led to an extension of the continental crust and, as a consequence, to the formation of half-graben structures in the northern and NE parts of the country (Helmcke 1983; Gabel *et al.* 1993). In northern Thailand half-grabens developed between Lampang and Phrae (Gabel 1991), while in NE Thailand such structures underlie the sediments of the Khorat Plateau (Fig. 1). The Triassic formation of half-grabens in northern and NE Thailand is displayed as an initial, rift-like stage of a subsequent long-lasting thermal subsidence (Drumm *et al.* 1993). The events resulted in the formation of the Khorat Basin that demonstrates a wide deposition area for clastic sediments in the northern and NE parts of the country. The lithostratigraphic units of the Khorat Basin in NE Thailand were combined by Ward & Bunnang (1964) under the term 'Khorat Group'. Between the Upper Triassic and Palaeogene, the Khorat Basin was filled with 4500 m of the mainly continental series of the Khorat Group (Heggemann 1994).

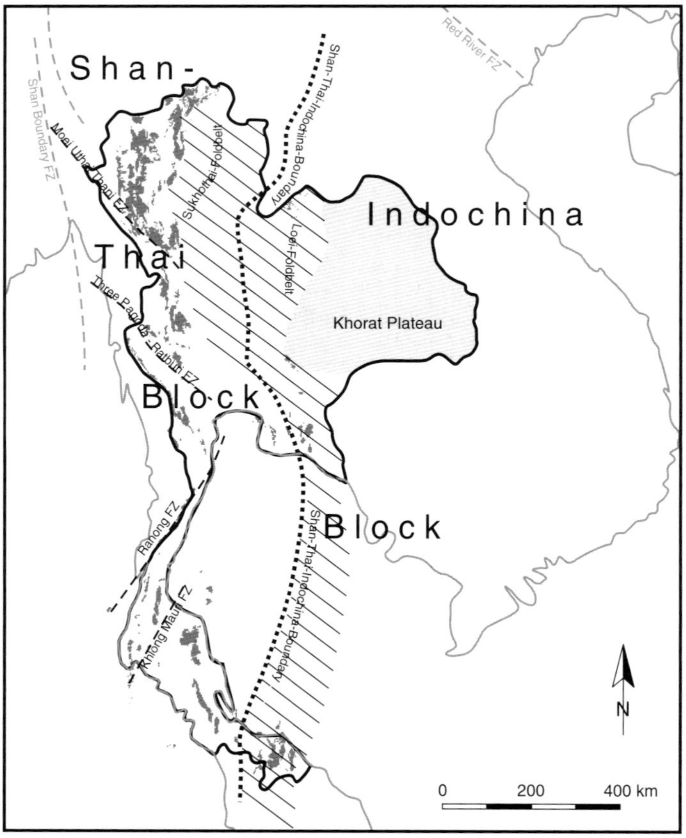


**Fig. 1.** Tectonic framework of Thailand. The Yunnan Malay mobile belt (parallel lines) is N–S-trending along the boundary between the Shan–Thai and Indochina Block (pointed line). The light grey area in the east represents the extension of the Khorat Plateau in Thailand. The dark grey areas in the west represent the SE Asian batholithic intrusions in Thailand. Modified after Hahn *et al.* (1986) and Bunopas & Vella (1992).

The tectonic events in SE Asia were accompanied by magmatic activity, represented by the SE Asian batholithic intrusions that extend from Indonesia to the provinces of south China, covering the Thai–Malay Peninsular, eastern Myanmar, NE Thailand and western Laos (Fig. 1). According to Nakapadungrat & Putthapiban (1992), the emplacement of these granitoid rocks occurred during four periods of magmatism in the Early Triassic, Late Triassic, Early Cretaceous and Late Cretaceous. Based on field geology, petrography and geochemistry, the granites can be broadly divided into three belts: the Eastern, Central and Western belts (Mitchell 1977). Granites of the Western belt feature mixed or equigranular hornblende–biotite granites or porphyritic two mica megacrystic K-feldspar granites. The Central belt comprises mainly porphyritic biotite–muscovite monzogranites and granites, while the Eastern belt is characterized by equigranular hornblende–biotite granodiorites and minor bodies of hornblende diorites. A foliation is well developed in the Central belt.

## Regionalization of the dimension stone potential in Thailand

The geological evolution of Thailand has provided facies conditions and depositional environments for magmatic, metamorphic and sedimentary rocks (Fig. 2). A regionalization of those rocks used as dimension stones is broadly reflected by three components that characterize the overall geology of the country: namely, the magmatic belts, the Khorat Plateau and the Yunnan Malay mobile belt.

The magmatic belts of Thailand represent the geological setting for granitoid rocks in the country. The distribution of the magmatic dimension stone deposits allows a separation into two distinct axes, both reflecting the N–S-trending orientation of the magmatic belts. The first of those axes is defined by four mining provinces that reach from the Thai–Myanmar border in the north to the beginning of the peninsular in the south. The operations known so far focus on the provinces Chiang Rai, Tak, Ratchaburi and Prachuap Khiri Khan (Fig. 2). The second axis of granitoid dimension stones in Thailand is located on the eastern side of the central basin area, flanking the Khorat Plateau at its western margin. While the majority of rocks from the aforementioned provinces demonstrate similarities in terms of texture, the dimension stones on the second axis differ significantly from each other in texture, colour or mineral content. The rocks are quarried from north to south in the provinces Loei, Nakhon Sawan, Nakhon Ratchasima and Chachoengsao (Fig. 2).

The presently known sedimentary resources of Thailand involve carbonate and clastic sediments (Fig. 2). Mining operations are developed in Sra Kaeo Province in SE Thailand, where Permian reddish and grey limestone is quarried. Occurrences of Permian carbonate rocks are quarried in central and NE Thailand, such as black limestone in Saraburi Province and black and grey limestone in Nakhon Ratchasima Province, respectively. Investigations reveal that limestone as a dimension stone is also obtained in Tak Province in northern Thailand and travertine resources occur in Lop Buri Province in central Thailand. Unfortunately, no information on the exact position of the two sites is available. However, limestone resources in Tak would be the most western of the presently known deposits for sedimentary rocks in Thailand. Clastic sediments are predominantly found in NE Thailand, where the influence of a continental facies enabled the formation of considerable sandstone resources. But although sandstones occur in the entire NE region of the country, the presently known activities for sandstone mining focus on an area at the western margin of the Khorat Plateau. In relatively close geographic areas, sandstones with different colours are quarried from three stratigraphic units of the Khorat Group: namely, the Phu Kradung Formation (Middle Jurassic; green sandstone), the Phra Wihan Formation (Middle–Late Jurassic; white, yellow, brown sandstone) and the Khok Kruat Formation (Aptian–Albian; red sandstone).

The mobile zone between the granite belts and the Khorat Plateau, as well as the eastern part of northern Thailand, can be considered as a region in which deformed and metamorphosed lithologies were uplifted and dissected. The mobile zone and its neighbouring areas comprise important metamorphic dimension stone resources in Thailand, which are mostly located in the vicinity of the central basin. These resources include metamorphic carbonate rocks with varying grades of calcite recrystallization. Marble is developed in Permian strata of the Uttaradit and Nakhon Ratchasima Provinces. Both marble locations define a N–S-trending axis on the eastern side of the central basin. On the prolongation of this axis further to the south of Thailand, marble is also quarried in Permian sequences of Yala Province. Along the western margin of the central basin, limestones with tendencies to marble occur in the provinces of Kamphaeng Phet and Sukhothai (Fig. 2).

## Case study: granite deposits in the Tak batholith, northern Thailand

During a field campaign on the investigation and economic assessment of dimension stone deposits in Thailand, different rock samples were taken from the Tak batholith in northern Thailand (Fig. 3). The Tak batholith is part of the Eastern

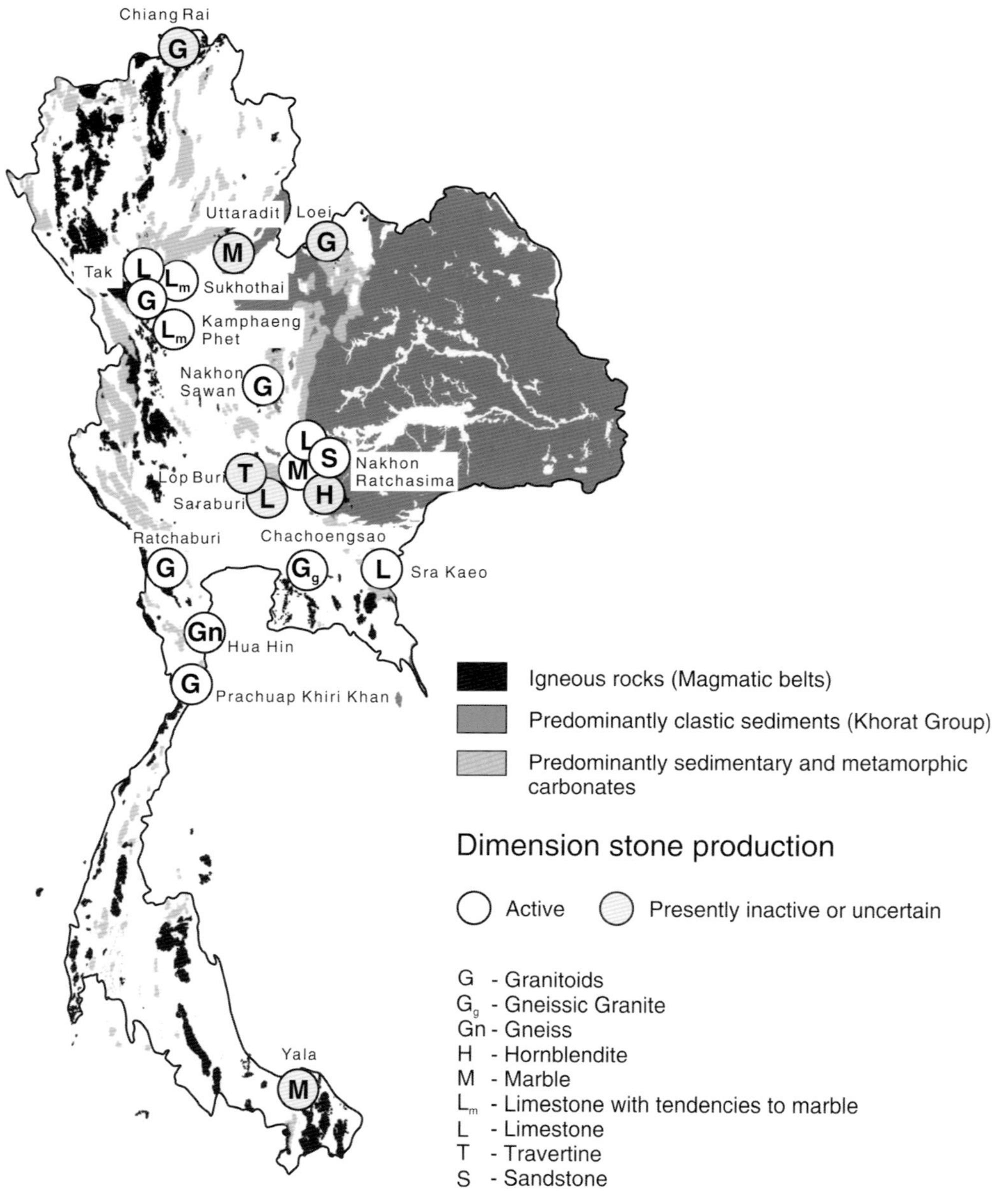


**Fig. 2.** Presently known regions for dimension stone production in Thailand.

granite belt, which usually occurs as small plutons (Nakapadungrat & Putthapiban 1992). However, the Tak granites are exposed as a large body with a N–S-trending axis measuring approximately 80 km and an E–W-trending axis of approximately 40 km (Cobbing & Pitfield 1986). The rocks crop out over an area of at least 3000 km$^2$ between the district centres of Tak and Thoen. Tak granites were intensively studied by Mahawat (1982), Mahawat *et al.* (1990) and Atherton *et al.* (1992), who classified the rocks into four composite plutons. In chronological order of emplacement, these are the Eastern pluton, the Western pluton, the Mae Salit pluton and the Tak pluton (Fig. 3). Similar to other rocks of the Eastern belt, the Tak granites have a wide range of composition with granite, granodiorite and quartz-diorite–tonalite in the Eastern pluton, and quartz-monzonites or

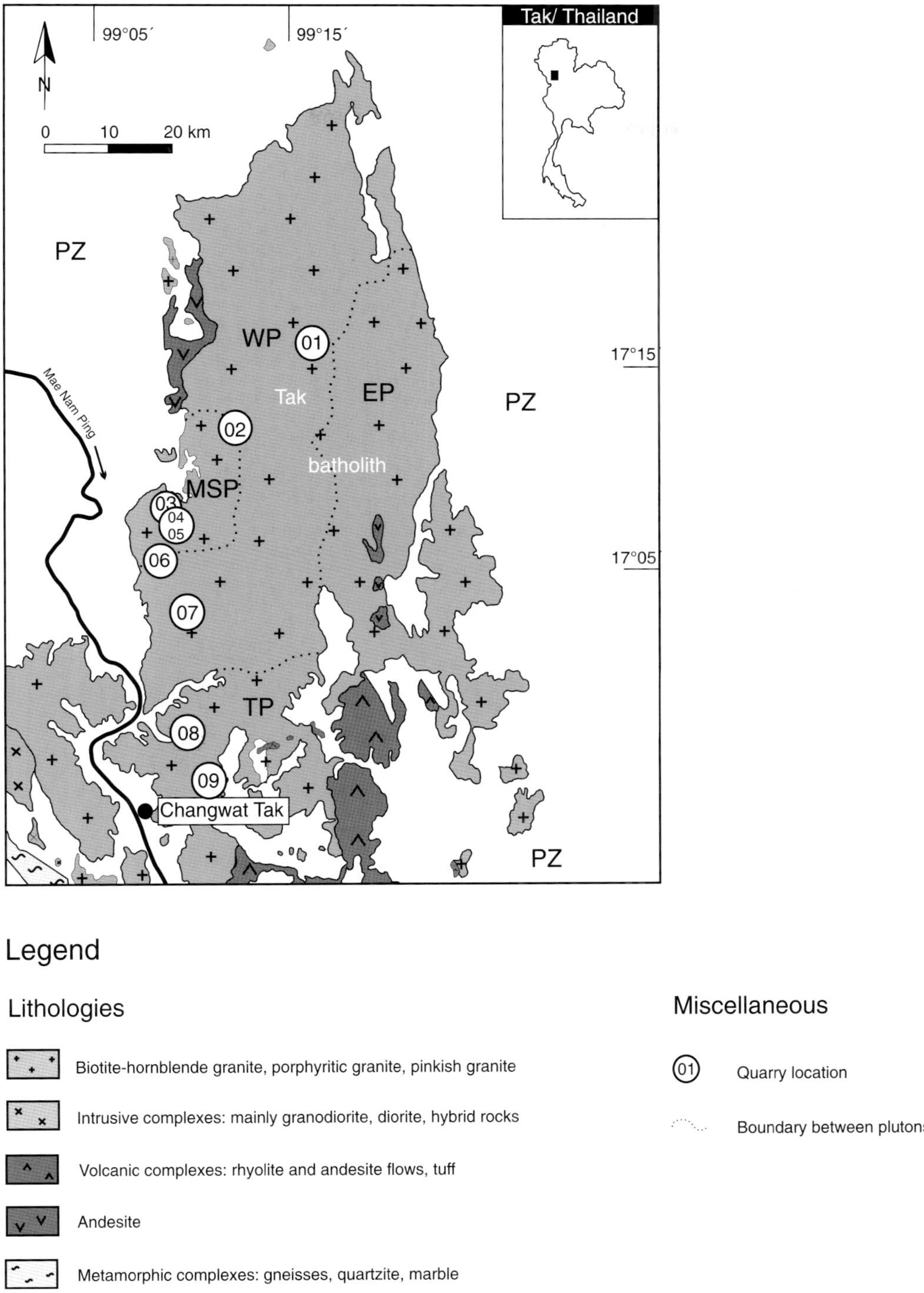


**Fig. 3.** Site locations in the Tak batholith. WP, Western pluton; EP, Eastern pluton; MSP, Mae Salit pluton; TP, Tak pluton; PZ, Palaeozoic.

monzogranite–syenogranite in the Western, Mae Salit and Tak plutons (Nakapadungrat & Putthapiban 1992).

## *Economic aspects of the granite production in the Tak batholith*

Compared to other granite-producing provinces, like Nakhon Sawan, Prachuap Khiri Khan or Loei, the Tak Province ranked first over the years from 1997 to 2001 with respect to the annual production of granite raw material in Thailand. The share of the region in total granite production was about 77% in 1997 and 69% in 2001. Although production rates decreased from 18.460 $m^3$ in 1997 to 4.585 $m^3$ in 2001 (United Nations 2002) due to the economic crisis, the Tak Province held its first rank until the beginning of recovery in 2001. There are no updated data available; however, it is assumed that the province still holds this position.

The deposits are large with good conditions for quarrying. Some of the investigated sites have been active for 15–20 years and reach a production rate of 50–700 $m^3$ per month. However, all production sites in the Tak area are still characterized by near-surface mining, although the opening of deeper quarrying levels would be possible in many cases. The material is obtained from boulders and walls in flat areas or from the slopes and tops of mountains by carefully applied explosives.

Factories for processing are usually located close to the quarry area and require a relatively short distance for the transport of blocks. The sizes of investigated blocks range from 1 to 13 $m^3$. Only in one case do blocks have to be carried to processing plants outside of Tak Province. The companies produce blocks or tiles, the latter by using gang-saws, block-saws or other kinds of cutting equipment. The technical inventory is preferably from Italy, which allows the companies to reach high standards in cutting and polishing. Less frequently, but also in use, are machines of Asian manufacture from China and South Korea. In a final stage of the processing, tiles are sorted by colour and structure, which define different grades of a product group. In sum, a relatively stable quality can be expected.

## *Site location and product range*

Products from the Tak batholith can be separated on the basis of colour and mineralogy (Fig. 4). Two varieties that occur in the south of the batholith are defined by a high quartz content. Both rocks are medium grained, and distinguishable by soft orange colours and intensive orange colours (sites 08 and 09) (Fig. 4a, b). Approximately 15 km to the north of the sites, granite with a poorly developed texture, medium grain size and flesh-like colour of feldspar is mined in the Western pluton (site 07). The mining area is situated at a lower elevation along the eastern side of a NNE–SSW-trending mountain range. The elevation demonstrates a distinct morphological feature in the region, separating the deposit from other mining activities to the NW and NNW. One further operation inside this pluton concentrates on a medium-grained, porphyritic granite that occurs isolated in the northern parts of the batholith (site 01). Those rocks quarried on the western side of the mountain range belong to the Mae Salit pluton and its bordering area to the Western pluton. The products can be defined as fine- to coarse-grained dark quartz diorites (site 04) (Fig. 4c), medium-grained granites with slightly blueish shade of feldspar (sites 05 and 06) (Fig. 4d) and fine- to medium-grained granites with light–grey colours (site 03). It is assumed that the material of sites 05 and 06 are identical, as their outward appearance and fabric is the same. Both rock types are arranged on a NE–SW-trending axis over a distance of approximately 3 km. From field observations it may be guessed that their alignment probably follows a shear zone or graben structure, since the positions are in a depression and bordered by the ranges and isolated elevations in the east and west, respectively. The position of site 02 coincides with this lineament, but here a slight increase of the topographic level is recorded towards the NNE. The quarrying activities cover every unit of the batholith except the Eastern pluton.

## *Evaluation of orange granite sites in the Tak batholith*

The two orange granites are located on a NNW–SSE-trending line with approximately 7 km distance to each other (sites 08 and 09) (Fig. 3). The stones are predominantly distinguished by their different shades of orange feldspar that vary from soft orange to intensive orange in the quarrying region. The orange tint is moreover irregularly disseminated throughout the quarrying region, which results in an additional grey variety, where the orange tone is absent. The grey variety is also exploited in each of the two mining areas.

### *Petrography*

The northern quarry (site 08) is arranged along the foot of a mountain with a lateral dimension of more than 300 m. The mountain is one of several isolated elevations in the Tak pluton, with an altitude difference of approximately 100 m to the surrounding plain. The massive rock has an almost equigranular

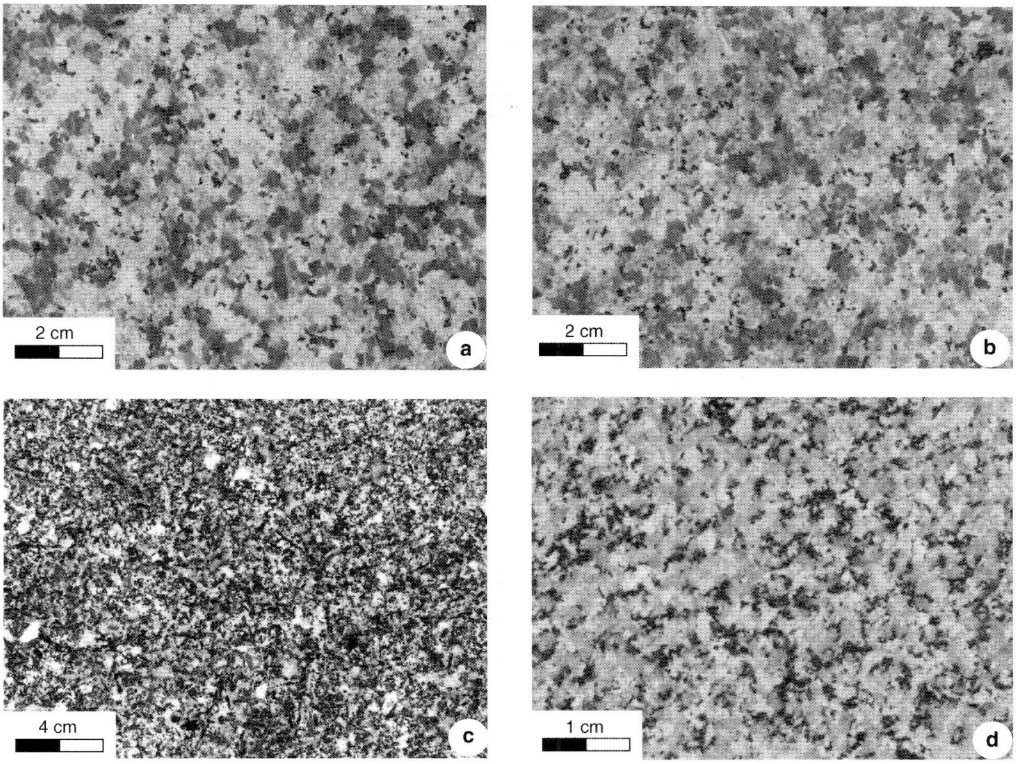


**Fig. 4.** Selected dimension stones from the Tak batholith. (**a**) Quartz-rich granitoids with a soft orange colour (site 08). (**b**) Quartz-rich granitoids with intensive orange colour (site 09). (**c**) Fine-grained quartz diorites (site 4). (**d**) Medium-grained granitoids with partially blueish shade of feldspar (site 05).

structure and is generally characterized by a large amount of cloudy quartz (5 mm) and by a smooth orange and white colour, resulting from K-feldspar and plagioclase (5–10 mm) (Fig. 4a). Thin sections of the rock reveal that feldspar crystals are coated with brownish patches. Some plagioclase crystals display a brittle overprint that is recognized by fracturing of grains. In almost every case these microfractures are healed by a later permeation of quartz. Other quartz aggregates show undulose extinction, pointing to a certain strain that affected the rock. Often, intra- and transgranular cracks can be recognized in quartz crystals, which are filled by finer grained epidote. Felsic minerals are medium sized (5 mm), while mafic components are partly smaller and generally scarce. Mafic minerals occur preferentially along the margins of feldspar grains.

The other orange granite deposit (site 09) is also bound to one of those isolated hills that characterize the morphology in the area. The rock is comparable with that previously mentioned, as it is likewise characterized by a medium-grained, equigranular structure, and a mineral composition of predominantly orange K-feldspar (5 mm) and grey, cloudy quartz (5 mm) (Fig. 4b). However, stones from this location demonstrate a more intensive orange tone compared to those from site 08. Plagioclase (2–3 mm) is white–slightly greenish and less visible in the macroscopic appearance. Feldspar is more frequently covered by microscopically observable brownish stains in this variety. Similar to site 08, the rock is characterized by epidote that occurs as both fracture fillings in quartz and single, isolated minerals. The latter form is predominantly bound to scarce biotite. Deformation features are also apparent by undulose extinction in quartz.

Ratios in the mineral constitution can change in both sites, so in some parts of the deposits a relatively higher amount of quartz with respect to feldspar is possible. The low quantity of mafic minerals can even decrease within some areas of the deposits. Mafic xenoliths are generally scarce and the few forms observed outline similar characteristics. Common features are their small size (<5 cm), their fine-grained mineral composition and a rounded–spherical shape with sharp contact to the host rock.

Although there is generally no preferred orientation of the minerals, the rocks partially display a very poor WNW–ESE-trending foliation at site 08 and an ENE–WSW-trending foliation at site 09. Both directions are barely recognizable and only roughly estimated by scarce sections of the granites. The foliation of rocks at site 08 coincides with WNW–ESE-trending joints in the deposit. Other directions of joints are ENE–WSW and NNW–SSE. Almost the same strike of fractures was recorded at site 09. While many of the investigated joints can be attributed to pressure release during uplift of the material, others (in particular subhorizontal, ENE–WSW-trending joints at site 08 and steep dipping, WNW–ESE-trending joints at site 09) mirror the tectonic overprint of the mining area by the lineation of fibrous epidote and chlorite. The lineation indicates movement on the joint surfaces. Further minerals such as hematite and limonite cover joints to a minor extent. However, the occurrence of these minerals could indicate that the microscopic brownish stains on many feldspar crystals result from hematite or subsequent hydration of hematite to limonite.

## *Geochemistry and physical–technical properties*

The geochemical compositions of the materials demonstrate almost similar contents of $SiO_2$ and $K_2O$, with minor variations in CaO and $Na_2O$ (Table 1). Testing of basic physical parameters reveals that the rocks are also comparable in terms of density (2.61–2.63 g $cm^{-3}$) and porosity (0.35–0.39 vol.%) (Table 2). Minor differences are given by the strength parameters, since here, the rocks from site 08 show slightly higher values with around 9 MPa for the tensile strength and around 17 MPa for the flexural strength (Table 2). Apparent is the difference in the compressive strength. In this case, samples from site 08 reach almost 185 MPa, while rocks from site 09 attain only 160 MPa. However, the compressive strength of both rocks, whilst low–medium, are common values for granitoid rocks. The flexural strength of around 15–17 MPa (Table 2) ranges in the upper half among comparable lithologies (Mueller 1996). All data for strength analyses in this context should be regarded as mean values from measurements, which were conducted with respect to three different directions for each sample.

## *Controlling parameters of the deposits*

One fundamental aspect regarding the evaluation of the deposits is the distribution of the orange feldspar colour. Locally, the orange colour is related to the presence of chlorite and epidote minerals that occur as straight thin veins in massive areas of rock, as thin reticular bandings in the adjacent areas of faults or as coverings on joint surfaces (Fig. 5a–d). At site 08, the estimation of this influence is made on the basis that the concentration of orange feldspars is abnormally high in areas located directly along epidote veins and epidote-coated joint surfaces. If the granite from this site is represented by material that occurs directly next to epidote mineralization, its appearance is identical to the intensive orange granite from site 09. The colour intensity at site 08 decreases with greater distance from veins and joints, and at about 1 m distance from those elements the granite becomes distinguishable from that at site 09. Although epidote veins were also identified at site 09, this quarry is further affected by chlorite veins. In contrast to the orange discoloration along epidote veins, the adjoining parts of chlorite veins are characterized by a considerable amount of violet feldspar and remarkably large quartz crystals. Similar to the changes in orange feldspar colour along epidote veins, the violet feldspar colour decreases with increasing distance from chlorite veins.

**Table 1.** *Geochemical composition of rock material from selected sites of the Tak batholith (%)*

| | $SiO_2$ | $TiO_2$ | $Al_2O_3$ | $Fe_2O_3$ | MnO | MgO | CaO | $Na_2O$ | $K_2O$ | $P_2O_5$ |
|---|---|---|---|---|---|---|---|---|---|---|
| Site 01 | 72.6 | 0.26 | 13.90 | 2.63 | 0.06 | 0.63 | 1.78 | 2.84 | 4.14 | 0.10 |
| Site 02 | 69.30 | 0.31 | 15.10 | 2.22 | 0.10 | 0.54 | 1.91 | 3.91 | 5.20 | 0.10 |
| Site 03 | 69.70 | 0.31 | 14.80 | 3.12 | 0.08 | 0.37 | 1.88 | 3.61 | 4.89 | 0.07 |
| Site 04 | 53.00 | 0.96 | 15.70 | 8.07 | 0.13 | 6.82 | 7.59 | 2.64 | 2.68 | 0.35 |
| Site 05 | 66.40 | 0.33 | 16.30 | 3.61 | 0.09 | 0.66 | 1.84 | 3.83 | 6.07 | 0.11 |
| Site 06 | 66.40 | 0.39 | 16.00 | 3.29 | 0.08 | 0.76 | 1.95 | 3.51 | 6.11 | 0.12 |
| Site 07 | 69.70 | 0.34 | 14.80 | 2.48 | 0.10 | 0.74 | 2.06 | 3.33 | 5.12 | 0.11 |
| Site 08 | 73.90 | 0.16 | 13.40 | 1.58 | 0.05 | 0.22 | 1.00 | 3.42 | 5.12 | 0.04 |
| Site 09 | 74.70 | 0.10 | 13.20 | 1.22 | 0.06 | 0.14 | 0.76 | 3.71 | 4.97 | 0.03 |

**Table 2.** *Some technical properties of granitoids from the Tak batholith*

| | Site 01 | Site 02 | Site 03 | Site 04 | Site 05 | Site 06 | Site 07 | Site 08 | Site 09 |
|---|---|---|---|---|---|---|---|---|---|
| Density (g cm$^{-3}$) | – | 2.64 | 2.66 | 2.88 | 2.65 | – | – | 2.62 | 2.60 |
| Porosity (vol.%) | – | 0.41 | 0.32 | 0.35 | 0.51 | – | – | 0.39 | 0.35 |
| Compressive strength (MPa) | – | 153.14 | – | 174.49 | 183.95 | – | – | 184.86 | 150.90 |
| Tensile strength (MPa) | – | 8.61 | 10.58 | 13.84 | 8.49 | – | – | 8.90 | 8.58 |
| Flexural strength (MPa) | – | 14.61 | 19.72 | 28.32 | 14.76 | – | – | 16.91 | 15.35 |

The occurrence of veins and the associated changes in orange colour intensity might originate from infiltrating fluids. These fluids probably carried iron oxides that coated feldspar grains and caused epidote and chlorite mineralization. As the intensive orange rocks always occur in the vicinity of epidote and chlorite veins, it is deduced that fluids related to this mineralization have a major

**Fig. 5.** (**a**) Steep-dipping fault, site 09. (**b**) Epidote–chlorite mineralizations (arrows) in the adjacent area of the fault in site 09. (**c**) Major fault with dominating epidote mineralizations in site 08. (**d**) Close-up of an epidote vein in site 08.

impact on the orange feldspar colouring in the Tak batholith.

Field observations reveal that chlorite veins at site 09 are developed as a dense network that parallels a steep dipping WNW–ESE-trending fault (Fig. 5a, b). A relationship between tectonics and vein mineralization is also demonstrated at site 08, where epidote veins were activated by major thrusts (Fig. 5c). On the one hand, it could be possible that after the formation of the rock, tectonic strain acted on pre-existing veins and thereby initiated the fracturing of the rocks along these discontinuities. In this case the infiltration of fluids could be attributed to a late-stage magmatic event. On the other hand, the infiltration might be related to a syn- to post-tectonic hydrothermal circulation. This scenario would imply that epidote and chlorite veins were affected by minor tectonics, since faults and fractures had already been arranged prior to the injection of fluids. As a consequence, post-tectonic veins could be healed to such an extent that they do not represent potential planes of weakness for the rock.

The near-surface position of the quarrying sites could be another aspect associated with the origin of colours. As the geographical area is characterized by intensive surface weathering, alteration of the stones should start preferably along their discontinuities. However, it is uncertain if the specific colour can be attributed to surface processes, since general indications for weathering are scarce. Although some weathering is locally apparent in the form of biotite alteration in the vicinity of fractures, the stones are characterized by the fresh condition of their constituent minerals.

## *Consequences for mining*

The dependence of orange colour on the presence of certain mineral veins is disadvantageous for mining. First, the orange stones in the Tak pluton can only be quarried in limited areas, which are affected by specific mineral veins. Second, the colour can change gradually in the quarries, which complicates the supply of consistent material. In case of a coloration related to surface weathering, the continuation of the orange tone towards deeper quarrying levels would become more questionable. Third, the frequent occurrence of veins near faults suggests that orange colours in the Tak pluton seem to be related to tectonic overprints, no matter if tectonic events occurred before or after the activity of fluids. This conclusion implies another disadvantage for the mining operations, since faults and fractures as a consequence of tectonic processes significantly influence the volume and shape of the raw blocks.

Both volume and shape are fundamental quality criteria for the deposits, as the blocks obtained must meet the requirements of the further processing and should not fall below unprofitable dimensions. To quantify the influence of fractures on the mining as a quality factor, the spacing of dominant or subordinate fracture systems was examined. While site 08 reveals a maximum vertical joint spacing of 20.0 m, site 09 only attains 3.3 m. Horizontal fractures in the deposits are spaced up to 5.0 and 2.8 m apart, respectively (Table 3). The data are representative for the joint spacing in the Tak granite sites that reaches up to 20.7 m for vertical and 14.4 m for horizontal discontinuities (both maxima at site 02). In almost every quarry in the Tak batholith, the horizontal spacing is less than the vertical spacing (Table 3).

Taking into account that each of the deposits should yield raw blocks with measurements of between 1.5 m and 2.5 m length, the block productivity reaches more than 75% in active quarries of the Tak batholith (Table 3). Such maximum values for the block production were also recorded from site 08, but these data are, however, only valid for a section of the quarry that measures about 60–20 m for the floor space and 10 m for the height. Generally, the mining operations are complex here as the ENE–WSW-trending tectonic joints are large-scale thrusts that affect the quarried mountain in its middle parts. Because of a high fracture density in the hanging wall of the faults, the

**Table 3.** *Joint spacing and block productivity of selected sites of the Tak batholith**

| | Site 01 | Site 02 | Site 03 | Site 04 | Site 05 | Site 06 | Site 07 | Site 08 | Site 09 |
|---|---|---|---|---|---|---|---|---|---|
| Max. fracture spacing (m), vertical | 8.9 | 20.7 | 12.6 | 7.2 | 2.4 | 12.0 | 7.5 | 20.0 | 3.3 |
| Max. fracture spacing (m), horizontal | 6.0 | 14.4 | 2.7 | 4.0 | 1.6 | 4.0 | 8.0 | 5.0 | 2.8 |
| Block productivity (%) | 50–75 | 50–75 | >75 | 50–75 | <10 | 25–50 | 50–75 | >75 | <10 |

*Calculations for the block productivity are valid for blocks with dimensions 1.5–2.5 m length, width, height. The data must be evaluated as a guiding level, since the calculation was made under the assumption of an orthogonal joint system (after Mueller 2004), although such conditions were not given in every case. Moreover, the calculation refers to selected parts of a quarry, which implies that different terms might prevail in other areas of the sites.

portion towards the top of the mountain must be considered as unsuitable material for the dimension stone processing. In fact, the overlying material interferes with the mining in the lower parts of the deposit, as removed blocks from the bottom induce unstable conditions in the overlying sequences. The risks of rock fall in these parts are considerable and complicate the mining procedures. In addition, the removal of such large amounts of overburden requires enormous technical effort. As a consequence, questions remain as to how long the deposit might be viable and what investment is necessary to maintain the mining operation.

As site 09 holds a similar rock type, the mining situation is similarly as complex as at site 08. But in contrast to site 08, where mining is still feasible, site 09 was actually closed in the year 2005 due to a low block productivity of approximately 3%. High investment costs are also said to have been a factor that caused the decline and closure of the site. Alternatives, such as a move of the operations from tectonic zones to other locations, were not practicable as they would not have maintained the decorative value and properties of the rock. As a consequence, it can be seen that quarrying has been ultimately controlled by the dense fracture spacing during the late stage of the operation, which resulted in small, uneconomic block sizes.

## Summary and concluding remarks

The regional distribution of dimension stone deposits in Thailand demonstrates that the country offers a broad spectrum of different stones, from which the majority known so far are located in northern Thailand (e.g. Tak, Kamphaeng Phet, Sukhothai and Uttaradit) and NE Thailand (Nakhon Ratchasima). Personal communications with quarry owners revealed that a few of these sites were studied prior to our investigations by governmental institutions, companies or private geologists. But as there are usually no ongoing geological investigations of structural elements or geological conditions in the quarries, many dimension stones are mined without any account for the geological environment and without any elaborate strategy.

However, special care must be exercised where dimension stones are going to be mined over long periods. To establish a dimension stone on the market, the colour and the final appearance of the material must be consistent during further exploitation, since discoloration critically influences the commercial value of the material. As demonstrated, the orange character of the two granites from Tak is influenced by fluid infiltrations and vein mineralization. A uniform picture of both rocks seems to be only guaranteed by mining operations in areas affected by tectonic overprints. In turn, tectonic overprints are often accompanied by discontinuities that define the block sizes of extractable raw material and limit the quantities of high-quality material. Knowledge of the depositional and structural environment of dimension stones is thus of prime importance for determining the viability of a quarry and the prospects for long-term mining.

Since mining in the Tak batholith is still characterized by near-surface operations and most of the boulders demonstrate wide fracture spacing, even more favourable conditions might prevail at deeper levels. But, whilst mining conditions might improve with depth, the decor of the product might also change. Finally, such a change could be disadvantageous for the marketing of the existing product, but might also lead to the discovery of new products. However, if major issues such as exploitation strategies can be solved, the Tak granites could be regarded as having a promising potential for further development.

## References


Atherton, M., Brotherton, M. & Mahawat, C. 1992. Integrated chemistry, textures, phase relations and modelling of a composite granodioritic–monzonitic batholith, Tak, Thailand. *Journal of Southeast Asian Earth Sciences*, **7**, 89–112.

Bunopas, S. 1981. *Palaeogeographic history of western Thailand and adjacent parts of SE Asia – a plate tectonics interpretation*. PhD thesis, Victoria University of Wellington, New Zealand.

Bunopas, S. & Vella, P. 1978. Late Palaeozoic and Mesozoic structural evolution of northern Thailand, a plate tectonics model. *In*: Nutalaya, P. (ed.) *Proceedings of the Third Regional Conference on Geology and Mineral Resources of SE Asia, Bangkok*, 133–140.

Bunopas, S. & Vella, P. 1992. Geotectonics and geologic evolution of Thailand. *In*: *Proceedings of the National Conference on 'Geologic Resources of Thailand: Potential for Future Development', 17–24 November 1992*. Department of Mineral Resources, Bangkok, 209–228.

Cobbing, E. J. & Pitfield, P. E. J. 1986. *South-East Asia Granite Project: Field Report for Thailand 1985*. British Geological Survey, Keyworth, Overseas Directorate, **86/16/R**.

Duerrast, H., Siegesmund, S. & Stein, J. 2003. Natursteine in Thailand. *Naturstein, Ulm*, **9**, 68–74.

Drumm, A., Heggmann, H. & Helmcke, D. 1993. Contribution to the sedimentology and sedimentary petrology of the non-marine Mesozoic sediments in Northern Thailand (Phrae and Nan provinces). *In*: Thanasuthipitak, T. (ed.) *Biostratigraphy of Mainland Southeast Asia: Facies and Palaeontology, Proceedings International Symposium, Chiang Mai*, **2**, 299–318.

Gabel, J. 1991. *Geodynamische Entwicklung Thailands im höheren Palaeozoikum und*

*Mesozoikum unter besonderer Berücksichtigung der intramontanen Becken*. PhD thesis, University of Goettingen.

GABEL, J., TRZEBSKI, R., ZWINGMANN, H., CHONGLAKMANI, C., HELMCKE, D. & MEISCHNER, D. 1993. Triassic extension basins in northern Thailand. (Abstract.) *In*: *Late Orogenic Extension in Mountain Belts*. International Symposium, Montpellier.

HAHN, L., KOCH, K. E. & WITTEKINDT, H. 1986. *Outline of the Geology and the Mineral Potential of Thailand*. Bundesanstalt für Geowissenschaften und Rohstoffe, Hannover, Geologisches Jahrbuch, **59**.

HEGGEMANN, H. 1994. *Sedimentaere Entwicklung der Khorat-Gruppe (Ober-Trias bis Palaeogen) in NE- und N-Thailand*. PhD thesis, University of Goettingen.

HELMCKE, D. 1983. On the variscan evolution of central mainland SE Asia. *Earth Evolution Science*, **4**, 309–319.

MAHAWAT, C. 1982. *The petrology and geochemistry of the granitic rocks of the Tak batholith, Thailand*. PhD thesis, University of Liverpool, Liverpool.

MAHAWAT, C., ATHERTON, M. P. & BROTHERTON, M. S. 1990. The Tak Batholith, Thailand: the evolution of contrasting granite types and implications for tectonic setting. *Journal of Southeast Asian Earth Sciences*, **4**, 11–27.

MITCHELL, A. H. G. 1977. Tectonic settings for emplacement of Southeast Asian tin granites. *Bulletin of the Geological Society of Malaysia, Kuala Lumpur*, **9**, 123–140.

MUELLER, F. 1996. *Gesteinskunde*. 5th edn, Ebner Verlag, Ulm.

MUELLER, S. 2004. *Lagerstättenkundliche Untersuchungen zu den Bankkalken des Deutschen Jurakalkes der Formation Malm delta zu naturwerksteinkundlichen Sortendefinition anhand der geochemischen Zusammensetzung und der gesteinstechnischen Eigenschaften sowie der Rohblockhöffigkeit an einem definierten Abbaublock nach unterschiedlichen Methoden*. Unpublished thesis, University of Freiberg, Freiberg.

NAKAPADUNGRAT, S. & PUTTHAPIBAN, P. 1992. Granites and associated mineralisation in Thailand. *In*: *Proceedings of the National Conference on 'Geologic Resources of Thailand: Potential for Future Development', 17–24 November 1992*. Department of Mineral Resources, Bangkok, 153–171.

UNITED NATIONS. 2002. *Atlas of Mineral Resources of the ESCAP Region: Mineral Resources of Thailand*, Volume 16. United Nations, New York.

WARD, D. E. & BUNNANG, D. 1964. *Stratigraphy of the Mesozoic Khorat-Group in Northeastern Thailand*. Department of Mineral Resources, Report of Investigation, Bangkok.

WU, J. C. 2004. The mineral industry of Thailand. *In*: *US, Geological Survey Minerals Yearbook 2004*, Volume 3.

# Characterization of serpentinites to define their appropriate use as dimension stone

D. PEREIRA, M. YENES, J. A. BLANCO & M. PEINADO

*Depto. de Geología, Universidad de Salamanca, 37008 Salamanca, Spain (e-mail: mdp@usal.es)*

**Abstract:** Many questions arise when considering the appropriate use of building stones such as serpentinites. The commercial names of these rocks, collectively grouped as 'Green marbles', have no correspondence to their actual mineralogy, geochemistry and/or physical characteristics. Serpentinite being the hydrated product of an ultramafic parent rock and not a metamorphic product of limestone as implied by the term 'marble'. However, the serpentinites most widely used for ornamental purposes come from India (e.g. Rajasthan Green, Emerald Green) and in these the original mineralogy has been almost completely converted into carbonates. By contrast, serpentinites from elsewhere (e.g. Vermont Verde Antique from the USA and Verde Pirineos from Spain) do preserve some of their original mineralogy. The different physical and chemical behaviour of carbonates and serpentine minerals can result in significantly different behaviour of commercial building stones. Thus, carbonates are resistant to weathering but suffer from acidic cleaning agents in interior use; whereas serpentinites, with a high content of talc, used on external faces undergo an increase in volume and a consequent rapid degradation.

Accurate and precise characterization of serpentinites, including information on their mineralogy and geochemistry (including major, trace and volatile elements together with oxygen isotopes), in conjunction with their physical properties, would enable architects to select the appropriate interior or exterior use of these handsome building stones.

Serpentinites are formed from the transformation of ultramafic rocks (Moody 1976; O'Hanley 1996), with pre-existing anhydrous minerals, olivine, pyroxene, and other Mg-rich silicates and carbonates replaced by assemblages such as calcite/dolomite–tremolite, calcite/dolomite–diopside–quartz, calcite/dolomite–tremolite–talc, calcite/dolomite–olivine–diopside–serpentine–brucite–magnetite, etc. The rocks can be partly or totally serpentinized, and therefore exhibit different varieties of textures. To understand the behaviour of serpentinites as ornamental stones, it is necessary to characterize the mineralogy, geochemistry and mechanical properties of these rocks. Serpentinites have been widely used in monuments (e.g. Greek and Italian heritage), and they are very popular in civil construction nowadays (Meierding 2005).

We have studied the serpentinization process of a harzburgite in Cabo Ortegal, NW Spain (Fig. 1, see Díaz García *et al.* 1999 for a description of the complex). This is one of the ultramafic complexes that is exposed in Spain (the other important one being in the SE of Spain, where the Ronda peridotite crops out: Pereira *et al.* 2003 and references therein). The fresh rock is mined for refractory material to use in thermic plants, and the serpentinized rock was once traded as dimension stone, with the commercial name 'Verde Pirineos'. The serpentinization process starts by conducting fluids through shears, producing local weathering (Fig. 2). Fluids involved can be sea water, meteoric water, magmatic water and/or surface-derived hydrothermal fluids (Pereira *et al.* 2003).

Serpentinization obliterates the primary mineralogy of the rock, and only a few remnants can be identified. Different serpentine phases occur, depending on the transformation conditions (i.e. pressure, temperature, fluid origin). Some rocks show carbonate alteration: the serpentinite has been converted to talc-carbonate paragenesis, and there is no evidence of the previous mineralogy (Fig. 3a–c). These serpentinites are known commercially as 'Green marbles'. However, if some of the mineral precursors are still in the rock, weathering can affect the rock selectively.

Serpentinites are affected by late shearing, which produced veins filled by calcite. These veins act with different responses to weathering, being weaker than the host rock. The rock can break through these parts once it has been installed as facing stones in a building, thus leading to degradation of the whole slab (Fig. 4).

Serpentinites exhibit a wide spectrum of colours (light–dark green to almost black) and patterns that result from alteration of rock types of diverse bulk-rock compositions and structures. The colour of serpentinite also varies with the extent of hydration of the protolith and with the extent of deformation. Once in place, serpentinites can evolve in different ways, depending on their composition.

*From*: Přikryl, R. & Smith, B. J. (eds) *Building Stone Decay: From Diagnosis to Conservation.* Geological Society, London, Special Publications, **271**, 55–62.
0305-8719/07/$15.00 

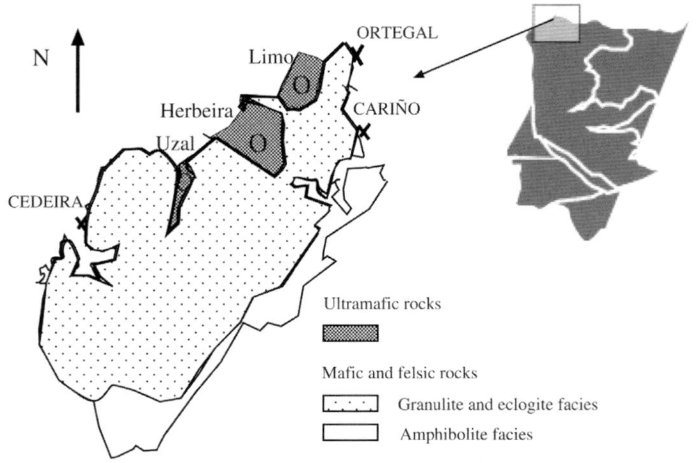


**Fig. 1.** Location of the Cabo Ortegal complex and the three different massifs studied in this work.

To characterize the serpentinites properly for their use as building stones, their study (and dissemination of results) should include geochemistry, mineralogy and mechanical properties, because all of these parameters strongly depend on the rock, serpentinites being different in different locations (e.g. Verde Macael and Verde Pirineos, from Spain; Verde Alpi and Verde Prato, from Italy; Rajasthan Green, from India). Knowledge of serpentinite behaviour is important not only in monument conservation (Malesani *et al.* 2003; Marino *et al.* 2004), but in prevention of rock decay, as this dimension stone is widely used these days in tiling in commercial buildings.

This work includes an example of the characterization of serpentinite from Cabo Ortegal, which explains the behaviour of this rock as building stone.

**Fig. 2.** Serpentinization of ultramafic rocks in Cabo Ortegal. The process starts via shearing, which acts as conduits for fluids.

## Mineralogy

Although a very detailed petrographic study was carried out on thin sections of samples from Cabo Ortegal, it is extremely difficult to distinguish different varieties of serpentine minerals. When serpentinization affects ultramafic rocks severely, only a few remnants of the original mineralogy can be distinguished. Studied samples are formed mainly by several serpentine phases, with relicts of olivine, orthopyroxene presenting many exsolutions, clinopyroxene and chromiferous spinel in different proportions. As accessories green spinel and amphibole were observed. Some phases (e.g. clinopyroxene, amphibole) could have crystallized during metamorphism. The main secondary minerals, besides serpentine, are chlorite, talc and dolomite. From the normative anhydrous composition, most of the samples were determined as harzburgites (except one dunite and one lherzolite).

There should be a correlation between the original lithology and the degree of alteration: dunite (olivine >90%) leads to 100% altered rock; harzburgite, made of orthopyroxene, leads to less altered rocks (D'Antonio & Kristensen 2004). Fine-grained rock allows fluids to penetrate and serpentinize. If large orthopyroxene grains are present, then the degree of serpentinization is lower. Microfabric is a very important feature when considering the possible use of serpentinite, because the evolution of any weathering will depend on it.

Carbonated serpentinites present different phases for carbonate substitution. Total carbonate replacement was not found in Cabo Ortegal, this substitution being the most prominent feature in serpentinites from Macael. In these carbonated rocks it is possible to distinguish between a

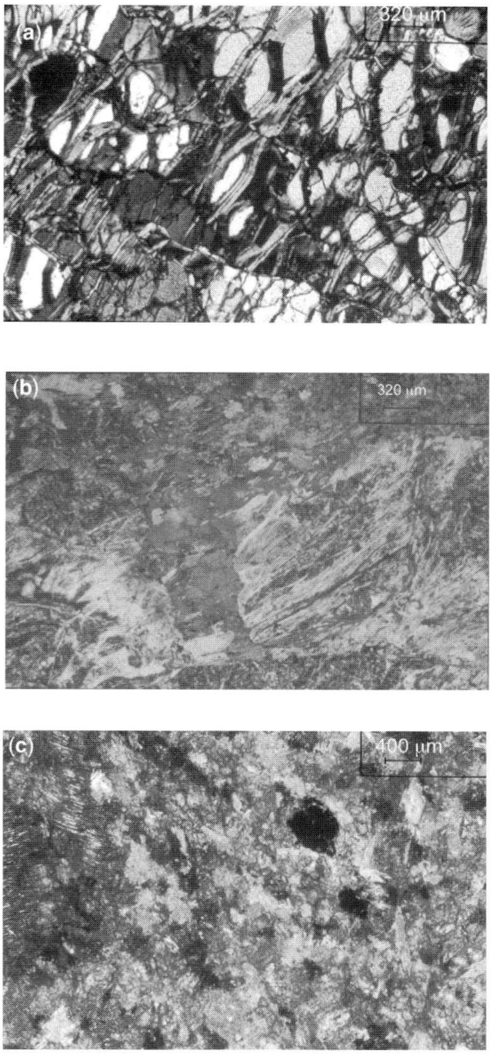


**Fig. 3.** Changes in mineralogy during a serpentinization process: (**a**) from Fe–Mg silicates; (**b**) to serpentine phases; (**c**) to carbonates.

microsparite matrix and the large crystals of sparite filling the fractures. Fabric evidence in these rocks suggests that, although veining and fracture filling predominate, serpentine minerals are extensively replaced by $CaCO_3$. This process, which is probably related to exposure to a circulating fluid, has to be studied in detail, together with its consequences, as it could be the key property in the behaviour of a serpentinite as ornamental stone.

X-ray diffraction was used in the study of some of the samples to distinguish the main components of serpentinization, taking into account that some accessory minerals are present in a quantity below 2% and

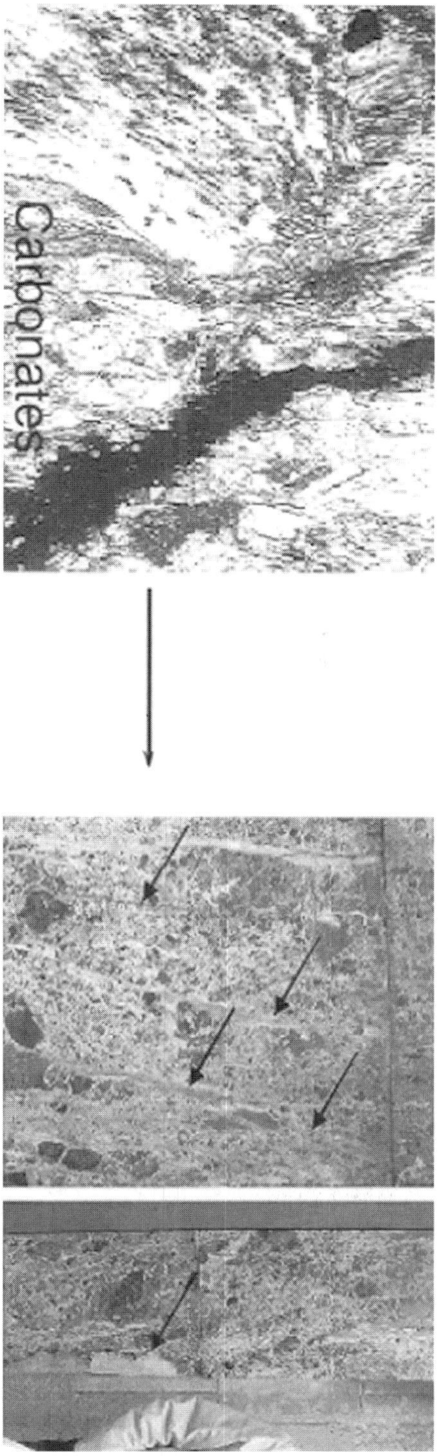


**Fig. 4.** Shear affecting a serpentinite building stone. Carbonates filling the shear develop a weakness that ends by cracking the tile.

might not be detected by diffraction. The proper characterization of the secondary minerals is very important for our purpose. In very fresh rocks forsterite, enstatite and tremolite were found as the main mineral phases. The weathered samples contain lizardite and in some of them dolomite has been observed (Fig. 5a). A polishing treatment will be different in samples containing only one mineral phase (or phases from the same group) to those made up of several phases: serpentine, talc, carbonates and spinel. The evolution of the ornamental stone once it has been emplaced in a building will be different as well. X-ray diffractograms can help to find minerals other than serpentine (Fig. 5b). The weathering of different phases can result in the disintegration of the rock. Knowing the presence of such phases can facilitate prior treatment of the rock or the decision to use the ornamental rock for a different purpose subject to less aggressive weathering.

## Geochemistry

The chemistry of serpentinization depends on the origin of the fluid causing the weathering. Although some authors have not found a major variation in the serpentinization process (Mével 2003 and references therein), an opposite conclusion was arrived at here by comparing fresh to weathered samples in the same massif (Tables 1 and 2). Serpentinization introduced changes in the geochemistry of the rocks, both in major and trace elements (including rare earth elements and volatiles such as B and Cl). Stable isotopes are changed as well, and these are very good tracers for the nature of the weathering fluid. At Cabo Ortegal values indicating high-temperature interaction with surface-derived fluids were found, together with values suggesting the evolution of hydrothermal fluids with a more complex history that could involve a mixture with igneous origin fluids. Although most of the serpentinization processes found in the literature call for sea water as the main fluid involved in the process, it is not thought to be the case for Cabo Ortegal serpentinization, where there is no clear enrichment in Sr, which is the key element pointing to a sea-water interaction.

It is apparent that some of the changes are dependent on the nature of the precursor, and knowing the original rock gives a clearer picture of the evolution of weathering. Once the rock has been emplaced as tiles or in any other function as ornamental stone, if it is under different conditions such as very humid conditions, or acid rain or an otherwise contaminated atmosphere, the weathering process can continue on the rock, the same as it would as if it were in outcrop. Therefore, precautions must be taken to avoid the disintegration of the rock. However, if the damage is severe and there is a need for replacement, knowing the geochemistry of the serpentinite would help us to find out where it comes from, so

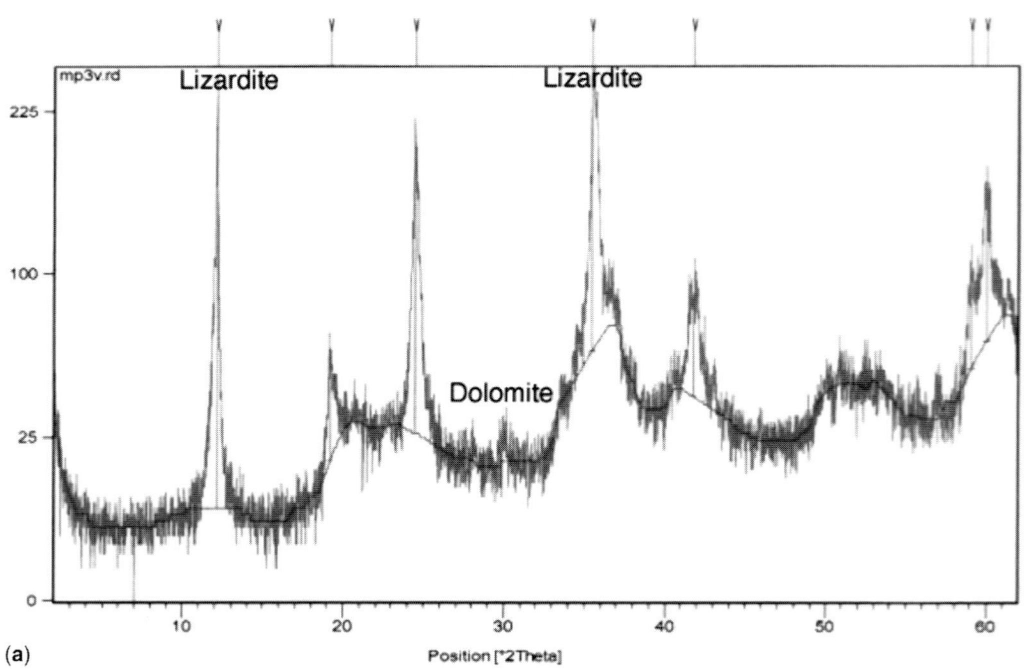


**Fig. 5.** X-ray diffraction differentiates the different serpentine phases (**a**) and relicts of the original mineralogy (**b**), which makes it easier to recognize the possible source of the dimension stone.

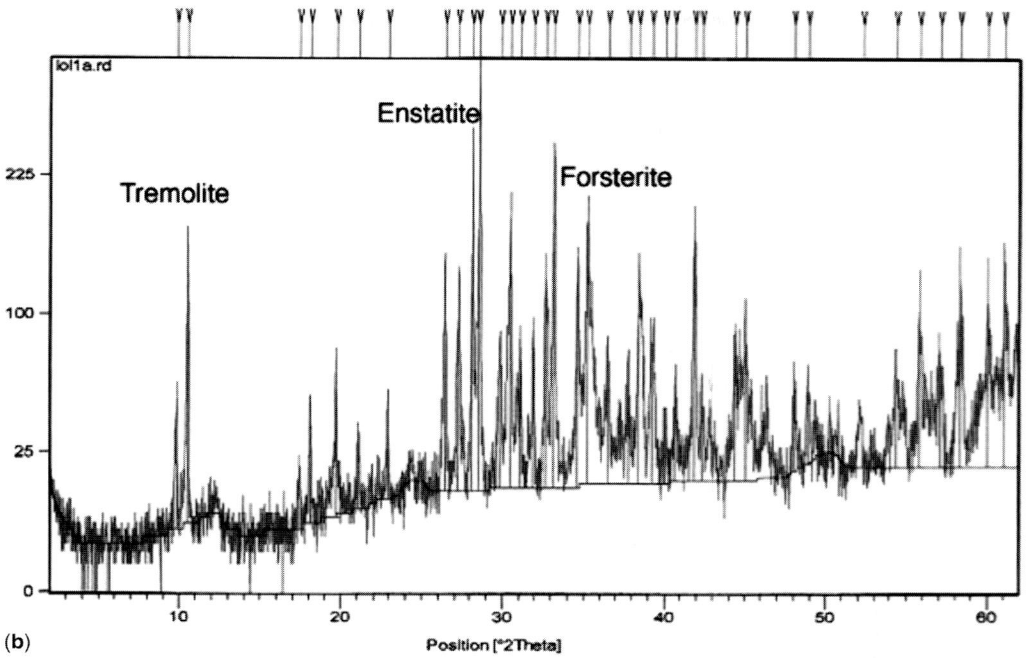


(b)

**Fig. 5.** *Continued.*

that a similar rock could be used to minimize any changes in the general aspect of the building or structure affected by the weathering (Malesani *et al.* 2003; Marino *et al.* 2004). For example, serpentinites coming from Macael (Verde Macael, SE Spain) have undergone a complete transformation to carbonates, and that is reflected in their geochemistry (Table 3). Serpentinites coming from Moeche (Verde Pirineos, NW Spain) are totally transformed to serpentine phases, and only a few mineral precursors are preserved. The geochemistry of a totally serpentinized rock differs from that of a talc-carbonate converted serpentinite. It may also have a different evolution during the *in situ* meteoric weathering, related to mineral and volume changes.

Stable isotopes can be good tracers for the origin of serpentinites. Wenner & Taylor (1973) came to the conclusion that, depending on the mineralogy of serpentinites, these should be described either as 'continental' or 'oceanic'; however, Tzen-Fu *et al.* (1990), taking as a basis stable isotopes, suggested that the domains proposed by the former authors should be extended. In ornamental

**Table 1.** *Major element composition of samples from Cabo Ortegal*

| Ref. | $SiO_2$ | $TiO_2$ | $Al_2O_3$ | FeO | MnO | MgO | CaO | $Na_2O$ | $K_2O$ | LOI |
|---|---|---|---|---|---|---|---|---|---|---|
| Ort-A | 41.73 | 0.04 | 2.48 | 8.41 | 0.10 | 38.31 | 2.05 | 0.06 | 0.00 | 8.44 |
| Ort-B | 41.33 | 0.09 | 2.46 | 8.12 | 0.09 | 37.50 | 2.51 | 0.06 | 0.00 | 7.35 |
| Ort-C | 38.99 | 0.00 | 1.06 | 8.06 | 0.05 | 38.09 | 0.11 | 0.06 | 0.00 | 13.46 |
| Ort-D | 40.35 | 0.04 | 2.19 | 8.63 | 0.08 | 39.03 | 1.04 | 0.07 | 0.00 | 8.89 |
| Ort-F | 38.59 | 0.05 | 1.67 | 8.68 | 0.12 | 37.92 | 1.34 | 0.06 | 0.00 | 11.61 |
| Ort-S | 40.63 | 0.03 | 1.11 | 9.37 | 0.11 | 41.38 | 0.69 | 0.07 | 0.01 | 7.16 |
| Ort-25 | 40.54 | 0.07 | 2.52 | 8.70 | 0.12 | 37.49 | 1.97 | 0.06 | 0.00 | 7.80 |
| Ort-26 | 42.99 | 0.13 | 3.57 | 7.96 | 0.08 | 34.23 | 5.01 | 0.04 | 0.06 | 4.74 |
| Ort-29 | 42.78 | 0.07 | 3.21 | 8.22 | 0.10 | 37.26 | 2.93 | 0.06 | 0.00 | 4.74 |
| O-718 | 34.83 | 0.01 | 0.19 | 11.37 | 0.18 | 40.09 | 0.02 | 0.07 | 0.01 | 14.46 |
| O-724 | 43.71 | 0.08 | 2.67 | 8.27 | 0.10 | 35.60 | 2.23 | 0.04 | 0.07 | 7.00 |
| O-730 | 43.75 | 0.08 | 3.44 | 7.78 | 0.09 | 33.83 | 3.37 | 0.05 | 0.03 | 7.64 |
| Limo-1 | 42.02 | 0.07 | 3.07 | 8.15 | 0.10 | 34.90 | 2.49 | 0.05 | 0.02 | 9.10 |

LOI, loss on ignition.

**Table 2.** *Trace element composition of the samples*

| Sample ID | Li | Rb | Cs | Be | Sr | Ba | Sc | V | Cr | Co | Ni | Cu | Zn | Ga | Y | Nb | Ta | Zr | Hf | Mo | Sn | Tl | Pb | U | Th |
|---|---|---|---|---|---|---|---|---|---|---|---|---|---|---|---|---|---|---|---|---|---|---|---|---|---|
| ORT-A | 1.36 | 0.07 | 0.05 | 0.51 | 11.50 | 7.05 | 15.39 | 58.50 | 2458 | 102 | 2398 | 14.49 | 21.69 | 3.64 | 1.58 | 0.23 | 0.03 | 9.00 | 0.05 | 0.27 | b.d.l. | 0.01 | b.d.l. | 0.03 | b.d.l. |
| ORT-B | 1.36 | 0.27 | 0.05 | 0.58 | 12.22 | 9.47 | 17.05 | 64.20 | 2428 | 106 | 2433 | 18.50 | 19.80 | 3.52 | 1.88 | 0.28 | 0.06 | 7.50 | 0.09 | 0.37 | b.d.l. | 0.01 | 0.26 | 0.03 | 0.01 |
| ORT-C | 0.44 | b.d.l. | 0.02 | 0.57 | 2.41 | 7.69 | 3.32 | 25.76 | 3272 | 119 | 2909 | 5.10 | 50.56 | 3.87 | 0.15 | 0.16 | 0.01 | 7.00 | 0.01 | 0.52 | b.d.l. | 0.01 | b.d.l. | 0.02 | b.d.l. |
| ORT-D | 1.03 | 0.06 | 0.05 | 0.42 | 17.49 | 9.18 | 11.09 | 54.62 | 3646 | 106 | 2421 | 7.12 | 43.23 | 4.09 | 1.04 | 0.16 | 0.05 | 7.50 | 0.02 | 0.28 | 44.90 | 0.01 | 0.06 | 0.06 | 0.04 |
| ORT-S | 1.48 | 0.09 | 0.04 | 0.47 | 16.85 | 11.04 | 7.19 | 35.75 | 1804 | 132 | 2760 | 13.31 | 38.56 | 3.50 | 0.61 | 0.65 | 0.22 | 9.30 | 0.04 | 3.54 | 53.95 | 0.01 | b.d.l. | 0.07 | 0.10 |
| O-25 | 1.64 | 0.17 | 0.04 | 0.41 | 10.17 | 14.26 | 15.33 | 62.87 | 1844 | 105 | 2431 | 23.49 | 22.27 | 3.69 | 2.01 | 0.26 | 0.04 | 8.03 | 0.05 | 0.35 | b.d.l. | 0.01 | b.d.l. | 0.05 | 0.01 |
| O-26 | 1.65 | 0.26 | 0.02 | 0.55 | 30.13 | 24.55 | 25.82 | 100.45 | 2910 | 102 | 2207 | 132.96 | 72.05 | 4.60 | 3.57 | 0.87 | 0.07 | 10.03 | 0.12 | 0.45 | 61.00 | 0.02 | 8.85 | 0.11 | 0.29 |
| O-29 | 1.51 | 0.22 | 0.01 | 0.38 | 19.66 | 20.93 | 18.33 | 77.89 | 2438 | 103 | 2378 | 23.01 | 22.42 | 3.93 | 2.36 | 0.36 | 0.06 | 8.39 | 0.04 | 0.22 | 0.05 | 0.01 | b.d.l. | 0.10 | 0.25 |
| O-724 | 4.20 | 0.35 | 0.11 | 1.00 | 103.56 | 28.89 | 17.36 | 67.30 | 2300 | 100 | 2220 | b.d.l. | 34.06 | 4.68 | 3.30 | 1.28 | 0.08 | 8.30 | 0.18 | 0.30 | b.d.l. | 0.03 | 1.03 | 0.12 | 0.22 |
| O-730 | 2.33 | 0.56 | 0.15 | 0.49 | 30.24 | 43.81 | 21.50 | 91.93 | 2707 | 88 | 2030 | 2.49 | 18.73 | 4.53 | 2.82 | 0.81 | 0.21 | 12.67 | 0.09 | 0.34 | b.d.l. | 0.02 | b.d.l. | 0.14 | 0.25 |
| O-718 | 0.48 | b.d.l. | 0.01 | 0.69 | 1.97 | 1.30 | 3.97 | 8.23 | 234 | 151 | 1820 | 0.49 | 36.23 | 2.16 | 0.23 | 0.27 | 0.03 | 9.58 | 0.04 | 0.25 | b.d.l. | 0.02 | b.d.l. | 0.03 | b.d.l. |
| Limo-1 | 2.93 | 0.66 | 0.18 | 0.39 | 47.23 | 29.20 | 17.42 | 8b.d.l. | 2301 | 94 | 2241 | 31.40 | 27.52 | 4.16 | 2.38 | 0.43 | 0.06 | 7.39 | 0.07 | 0.29 | b.d.l. | 0.02 | b.d.l. | 0.10 | 0.09 |

b.d.l. below detection limit.

**Table 3.** *Major elements in Spanish rocks: a totally serpentinized rock from Moeche (Galicia) and a carbonated serpentinite from Macael (Andalucía)*

| | Verde Macael* | Verde Pirineos[†] |
|---|---|---|
| $SiO_2$ | 0.17 | 38.99 |
| $TiO_2$ | 0.01 | 0.00 |
| $Al_2O_3$ | 0.00 | 1.06 |
| FeO | 0.28 | 8.06 |
| MgO | 0.07 | 38.09 |
| MnO | 0.15 | 0.05 |
| CaO | 49.64 | 0.11 |
| $Na_2O$ | 0.06 | 0.05 |

*Source: http://www.dipalme.org.
[†]This work.

serpentinite, the real stable isotopic signature probably includes not only the primary values of the rock, but the addition of these plus the changes introduced by meteoric water and the isotopic values of environment and anthropogenic influence. Experiments on changes in isotopic composition in serpentinites from different origins can give useful results in terms of source identification.

## Technical properties

Mechanical strength depends, amongst other factors, on the available anisotropic surfaces. This is why any mechanical characterization has to take into account the orientation of these surfaces regarding the direction of applied efforts. Cabo Ortegal serpentinites show local anisotropies that can change their mechanical behaviour. In this work, selected physical properties of non-oriented samples were determined as a first evaluation of their usefulness as ornamental rock. Future studies will have to deal with oriented samples, to study the changes in properties related to anisotropy.

Owing to the lack of a European specific standard for serpentinites, the physical properties of some of these rocks used as dimension stone were compared with the ASTM standards (C-1526-02, ASTM 2002). This standard specifies the minimum requirements for a serpentinite to be used as ornamental stone, either for interior or exterior use. Parameters analysed in Cabo Ortegal are shown in Tables 4 and 5, together with some other serpentinites from Spain and elsewhere. Absorption according to ASTM C-97 (ASTM 2002), compressive strength according to ASTM C-170 (ASTM 2002) and flexural strength according to ASTM C-880 (ASTM 2002) values are essential for understanding the poor behaviour of the Spanish Verde Pirineos. The serpentinization degree of the rock is generally correlated to the

**Table 4.** *Physical properties of commercial serpentinites compared to the ASTM requirements*

| Physical properties | ASTM requirements | Vermont Verde Antique* | Verde Alpi† | Verde Macael‡ | Verde Pirineos‡ |
|---|---|---|---|---|---|
| Absorption (%) external/internal | 0.20 max./0.60 max. | 0.15 | 0.56 | 0.10 | 0.93 |
| Density (kg $m^{-3}$) | 2.560 (min) | 2.863 | 2.712 | 2.730 | 2.700 |
| Compressive strength (MPa) | 69 (min) | 180.29 | 195.6 | 149.4 | 34.46 |
| Flexural strength (MPa) | 6.9 (min) | 33.37 | 22.8 | 17.4 | 5.92 |

*Source: http://www.tnmarble.com.
†Source: http://www.chooseby.com.
‡Source: http://www.piedra.com.

**Table 5.** *Physical properties of Cabo Ortegal serpentinites studied in this work and compared to the ASTM requirements*

| Physical properties | ASTM requirements | Cabo Ortegal 2 | Cabo Ortegal 3a | Cabo Ortegal 3b | Cabo Ortegal 1 | Cabo Ortegal 4 |
|---|---|---|---|---|---|---|
| Absorption (%) external/internal | 0.20 max./0.60 max. | 13.06 | 0.25 | 0.13 | 0.09 | 0.09 |
| Density (kg $m^{-3}$) | 2.560 (min) | 1.690 | 2.920 | 2.94 | 2.94 | 2.500 |
| Compressive strength (MPa) | 69 (min) | 1.91 | 49.4 | 67.1 | 78.22 | 84.1 |

absorption value. Sample 2 presents a highly weathered aspect in hand specimen, and shows the highest absorption values, clearly outside the requirement values. However, sample 3a, which did not show a highly weathered aspect in the field, gave an absorption value of 0.25%, which does not favour its use for exterior tiling. Samples 1b and 4 have very low absorption values, but they cannot be classified as serpentinites because the hydration is not enough in these rocks, and they preserve most (80–90%) of the original mineralogy (olivine, enstatite and spinel) besides serpentine and some tremolite.

There is also a negative correlation between absorption and compressive strength (taking into account that the samples were tested without orientation). Sample 2 presents a very low value for compressive strength and very high absorption value. Sample 3a has a low absorption value and a compressive strength value that does not recommend its use as dimension stone. The rest of the samples reach the minimum physical requirements, where higher compressive strength values are associated with lowered absorption values. This correlation is significant because a first approach to the possibilities of using a serpentinite for ornamental purposes could be tested using the absorption study, which is easy, quick and cheaper, before testing other properties.

## Discussion and conclusions

Serpentinites can suffer a range of changes when exposed to atmospheric conditions (Meierding 2005). To avoid unexpected behaviour of serpentinites used as building stones, a complete characterization of the rock, with data regarding their mineralogy, geochemistry and mechanical properties, should be available. The mining of serpentinites from NW Spain has stopped very recently (Marmolera Gallega pers. comm.). The dimension stone coming from that location, in Moeche, was known as 'Verde Pirineos'. The main reason for stopping the mining appears to have been the poor behaviour of a rock that crumbles very easily, through the lack of any component (e.g. carbonated cement) to hold all the other phases together. In Macael, carbonates hold the serpentine minerals in place and they do not break so easily. The weathering of different phases can lead to the disintegration of the rock. If the presence of such phases is known, the rock can be treated or the decision can be made to use the ornamental rock for a different purpose with less aggressive weathering.

The geochemistry of serpentinites can help to characterize the rock properly, and can be used to create an 'identification card' for the rock. Major and trace elements vary substantially, depending on the precursor of the rock. By knowing the chemistry of the serpentinite, it is possible to obtain a petrogenetical picture of the outcrop from which it could come. This would help in the conservation process of the rock, including replacement. Analyses of stable isotopes can also serve as a possible experiment to see how the atmospheric changes and the meteoric water (and other fluids) to which the rock may be subjected will affect the evolution of the serpentinite once placed as tiling.

X-ray diffractograms can help to find minerals other than serpentine. This technical approach can aid in identifying the possible source of an ornamental serpentinite, as some of these rocks are mined from massifs with different petrogenesis, and therefore different mineralogy and different mineralogical evolution.

Physical properties for these rocks are very relevant, as they may fulfil the requirements for interior use, but not for exterior use. Some countries do not have a specific standard values table for serpentinites, and they follow the requirements for marbles. It must be insisted that in origin they are not marbles, and, therefore, mineral precursors can be found in them that change some properties, especially absorption, abrasion resistance and even flexural strength. It is recommended that ASTM requirements for serpentinites (ASTM 2002) should be followed. This decision will help to avoid any unexpected behaviour of the stone when used in construction.

It would also be desirable to include the proper petrological name in any description or classification. Including serpentinites within marbles ('green marble') can lead to their incorrect use as building stones. Although some companies suggest the same applications for marbles and limestones, they do not recommend their use as polished surfaces. But the latter purpose is the most common one for this rock, because it shows off all the beauty of the colours (as can be seen in monuments from Greece and Italy). The problems with polishing can be overcome by knowing the correct mineralogy of the rock.

It is proposed that commercial companies should submit a complete characterization of the rock, with data regarding its mineralogy, geochemistry and physical properties. It is also suggested that large commercial companies maintain a 'Geology' section that would advise on the proper use of the rock, together with the possible consequences regarding the behaviour of the rock if the recommendations are not followed. Although this could be expensive for the company, it would be compensated for by offsetting any economic punishment attached to the wrong characterization of the stone when exporting to countries with severe restrictions (US Customs 2001). In addition, it would be a desirable tool for studying building façade deterioration, distress, normal aging, and the possibility for innovation in the maintenance, repair and replacement of degraded serpentinite used as dimension stone.

This work has been funded by Projects BTE 2003-04812 and CGL2005-03048/BTE from the Spanish government. Comments by R. Přikryl, R. Sandrone and an anonymous reviewer helped to improve the manuscript.

## References

ASTM. 2002. *Standard Specification for Serpentine Dimension Stone. C 1526-02*. ASTM International, West Conshohocken, PA.

D'Antonio, M. & Kristensen, M. B. 2004. Serpentine and brucite of ultramafic clasts from the South Chamorro Seamount (Ocean Drilling Program Leg 195, Site 1200): inferences for the serpentinization of the Mariana forearc mantle. *Mineralogical Magazine*, **68**, 887–904.

Díaz García, F., Arenas, R., Martínez Catalán, J. R., González del Tánago, J. & Dunning, G. R. 1999. Tectonic evolution of the Careón ophiolite (Northwest Spain): A remnant of oceanic lithosphere in the Variscan belt. *Journal of Geology*, **107**, 587–605.

Malesani, P., Pecchioni, E., Cantisani, E. & Fratini, F. 2003. Geolithology and provenance of materials of some historical buildings and monuments in the centre of Florence (Italy). *Episodes*, **26**, 250–255.

Marino, L., Corti, M., Coli, M., Tanini, C. & Nenzia, C. 2004. The 'Verde di Prato' stones of cathedral and baptistery of Florence. (Abstract.) *In*: *32nd IGC Florence*, T16.03.

Meierding, T. C. 2005 Weathering of serpentine stone buildings in the Philadelphia, Pennsylvania, region: A geographic approach related to acidic deposition. *In*: *Stone Decay in the Architectural Environment*. Geological Society of America, Special Paper, **390**, 17–25.

Mével, C. 2003. Serpentinization of abyssal peridotites at mid-ocean ridges. *Comptes Rendus Geosciences*, **335**, 825–852.

Moody, J. B. 1976. Serpentinization: a review. *Lithos*, **9**, 125–138.

O'Hanley, D. 1996. *Serpentinites*. Oxford University Press, New York.

Pereira, D., Shaw, D. M. & Acosta, A. 2003. Mobile trace elements and fluid-dominated processes in the Ronda peridotite, Southern Spain. *Canadian Mineralogist*, **41**, 617–625.

Tzen-Fu, Y., Hsueh-Wen, Y. & Chihming Wang, L. 1990. A stable isotope study of serpentinization in the Fengtien ophiolite, Taiwan. *Geochimica et Cosmochimica Acta*, **54**, 1417–1426.

US Customs. 2001. *Classification of Marble. What Every Member of the Trade Community Should Know About*. Informed Compliance Publication, US Customs Service, Washington, DC.

Wenner, D. B. & Taylor, H. P. 1973. Oxygen and hydrogen isotope studies of the serpentinization of ultramafic rocks in oceanic environments and ophiolite complexes. *American Journal of Sciences*, **273**, 207–239.

# Kirmenjak–*Pietra d'Istria*: a preliminary investigation of its use in Venetian architectural heritage

M. ŠIMUNIĆ BURŠIĆ[1], D. ALJINOVIĆ[2] & S. CANCELLIERE[3]

[1]*University of Zagreb, Faculty of Architecture, Kačićeva 26, HR 10000 Zagreb, Croatia (e-mail: marinasimunic@yahoo.com)*

[2]*University of Zagreb, Faculty of Mining, Geology and Petroleum Engineering, Pierottijeva 6, HR 10000 Zagreb, Croatia (e-mail: daljin@rgn.hr)*

[3]*Università IUAV-DSA-LAMA, Pal. Badoer, S. Polo 2468, 30125 Venezia, Italy (e-mail: stefanoc@iuav.it)*

**Abstract:** Kirmenjak – white limestone from the quarries near the village of Kirmenjak in Istria (Croatia), in the past known as *Pietra d'Istria* – has been regularly used in the construction of the basal zone of Venetian buildings since the 14th century. Its characteristics – durability, extremely low water absorption and high compressive strength – made it an ideal material for the lowest parts of Venetian buildings in the zone between foundation (wooden piles) and brick walls. In this zone, exposed to tidal flooding and low-tide drying, materials deteriorate very quickly, but Kirmenjak has proved to be durable even in this aggressive saline environment. Moreover, this dense micritic or pelmicritic stylolitized limestone from the Upper Jurassic (Tithonian) was used as an efficient barrier to rising damp. Preliminary *in situ* investigation of how Kirmenjak blocks were laid shows that the prevalent stylolite orientation is horizontal in the basal parts of buildings, while in other structural elements this orientation varies. This inspired the hypothesis that the Venetian constructors took advantage of horizontally laid stylolite discontinuities (partially filled with clay) as a multilayer humidity barrier.

Venice – *'La Serenissima'* – the city that dominated the Eastern Mediterranean for centuries, was built in an unfavourable environment, hidden in a shallow lagoon of the northern Adriatic. Its isolated, swampy site, with adverse conditions for construction (small sandbars, partly submerged, poor load-bearing strength of soil, limited area for building) was deliberately chosen by the first inhabitants to escape from the barbarians after the fall of ancient Rome (Howard 1989). Because of this, builders were faced with problems that did not exist elsewhere – or at least not to such a degree. Building on a muddy, unstable and submerged soil was a problem that the Venetians solved by driving wooden piles into the ground to form a dense network that improved the load-bearing capacity of the soil. Oak was preferred because it is more durable in water than most other timber. However, it is not durable enough for the tidal zone, where damage by periodic tidal wetting with sea water and drying is more aggressive and rapid than constant submergence. This is aggravated by the mechanical effect of waves and creates very harsh conditions for any building material. Indeed, most kinds of stone also deteriorate relatively quickly in this aggressive saline environment at the boundary between air and sea water. These include expensive and exotic stones, such as marbles, porphyries, granites, serpentines and basalts (Lazzarini 2004), brought mostly from Constantinople after its fall in 1204, during the 4th Crusade (Howard 1989) and re-used in Venice for sumptuous palaces and churches. Behind Venetian façades, richly adorned with these rare decorative stones, there are load-bearing structures of brickwork walls. Brickwork, when exposed to rising humidity, soon experiences serious damage as a result of the action of soluble salts (cracking and crumbling). Only the so-called *Pietra d'Istria* (Istrian stone) proved resistant and durable in this tidal zone.

## Kirmenjak (Pietra d'Istria) in Venice – historical notes

*Pietra d'Istria* – white Istrian limestone nowadays called Kirmenjak after the quarries near the village of Kirmenjak in Istria (Croatia) – was much appreciated by the Venetian builders who were aware of its high resistance to weathering (Zbîrnea 2000). According to the 17th century chronicler Prospero Petronio, they believed that Kirmenjak, when exposed to air and rain, improved in quality, becoming stronger and more resistant

*From*: PŘIKRYL, R. & SMITH, B. J. (eds) *Building Stone Decay: From Diagnosis to Conservation*. Geological Society, London, Special Publications, **271**, 63–68.

(Bertoša 1992). Moreover, Kirmenjak has a rare quality, very important for structures that 'rise from the sea', in that it has an extremely low water absorption. The builders of Venice were well aware of this precious quality and they widely used Kirmenjak as a damp-proof course. A glance at the basal zone of Venetian churches, palaces and houses shows that throughout Venice the lower parts of buildings are built consistently of this white stone (Fig. 1).

The lower part of the protective layer of Kirmenjak was permanently or from time to time under water, while the top of the Kirmenjak course was designed to be always above sea level – even during extremely high tides (*acqua alta*). Thus, the height of the Kirmenjak base is strictly defined by the highest expected level of high tides. Although recently, the *acqua alta* has been known to rise over the Kirmenjak course more frequently than before because of the relative sinking of Venice.

In this way the basal zone built of Kirmenjak, with its low water absorption and porosity, protected the luxurious stone facings and brick structures of Venetian buildings from rising damp. Kirmenjak has never been plastered or protected in any way. In the rare cases where Kirmenjak was painted (the traces of which can be seen, for example, on reliefs of the Scuola Grande di San Marco), it was only for decorative reasons. Thus, Venetian master-builders and their clients were aware of the excellent properties of Kirmenjak and made good use of it. Kirmenjak was used not only as a damp-proof course, but very often also for structural or pseudo-structural elements of buildings (e.g. portals, window frames, cornices). References to the use of Kirmenjak for many important Venetian buildings are found in a number of documents. For example, according to a contract with the Senate of Venice in the 15th century, Zuanne and Bartolomeo Bon was required to use stone from the quarry of Rovinj (Istrian town near Vrsar–Orsera) for carved parts of the Porta della Carta, the main entrance to the Ducal Palace. In 1484 Antonio Rizzo, 'proto' ('first master', i.e. architect) of the Ducal Palace, stipulated in a contract with the brothers Zuanne and Simon from Rovinj that they should deliver stone that was used in the Cortile d'Onore and in the Scala dei Giganti of the Ducal palace (Bertoša 1992). The use of Kirmenjak for the other parts of the Ducal Palace, as well as for many other Venetian edifices, is documented in numerous historical sources (Betoša 1992).

**Fig. 1.** Kirmenjak blocks used in the basal zone of Venetian buildings, laid with horizontally oriented stylolitic discontinuities.

It cannot be claimed with certainty when the builders of Venice first discovered the important characteristic of Kirmenjak and began to use it as a damp-proof course. However, it might be supposed that the use of Kirmenjak in Venice is connected to the political and social history of the Republic of St Mark's. In fact, Istria was a part of the Venetian Republic from 1279, but the first evidence of the use of Kirmenjak in Venice stems from the beginning of the 14th century (Lazzarini 1986). Nevertheless, as early as 1307 the Statutes of the Venetian craft-guild of stonecutters prescribed severe penalties for those who palmed off '*pietre d'aspetto simile e di ben diverso pregio*' (stone of a similar aspect but different i.e. lower value), trying either to sell it or use it instead of Kirmenjak (Zbîrnea 2000; Cotman 2004). This citation proves that Venetian masters of that time were well aware of the special value (*pregio*) of Kirmenjak. This special quality was obviously not its aspect, so it may be assumed that even at the beginning of the 14th century Venetian master-builders appreciated the excellent characteristics of Kirmenjak, which made it an irreplaceable building material for the harsh conditions of the Lagoon.

In important Gothic palaces (e.g. Ca' d'Oro constructed 1424–1437: Arslan 1986) Kirmenjak was consistently used in the basal layer of structures as a damp barrier (preliminary survey by the authors), and in the 16th–17th centuries the whole façades of important buildings (e.g. churches S. Giorgio Maggiore and Redentore designed by the famous architect Andrea Palladio) and important palaces (e.g. Ca' Rezzonico, Palazzo Grimani) were entirely covered by Kirmenjak (Lazzarini 1986).

An additional advantage is that the deposits of Kirmenjak are located not far from Venice on the western coast of Istria – Croatian peninsula in the northern Adriatic. For the sea-oriented Venetian merchants it was very important for quarries to be located very close to the coast, so that blocks could be loaded directly on to Venetian boats and transport costs kept very low (Lazzarini 1986).

The Venetian government, aware of the importance of Kirmenjak for the construction of Venetian buildings, established a monopoly for extracting this kind of stone as early as in the 14th century. This monopoly lasted for centuries, and even in the middle 18th century Kirmenjak for the reconstruction of the famous sanctuary of Santa Casa (Holy House) of Loreto (which at the time belonged to the Papal State) had to be literally smuggled because the export of such stone was strictly forbidden. It was only in July 1753 that the Government of Venice withdrew the embargo on the export of Kirmenjak (Bertoša 1992).

## Geological and petrographical characteristics of Kirmenjak

Although often referred to as 'the brilliant white marble' by past writers and scholars (and even by the *Encyclopaedia Britannica* 2002), Kirmenjak is not a marble, but a dense micritic or pelmicritic stylolitized limestone from the peninsula of Istria (Croatia). It has a beautiful ivory colour, and in subaerial conditions the stone obtains a snow-white patina. Only when exposed to gentle abrasion does the original ivory colour appear on its smoothed surface.

Kirmenjak limestone was deposited at the beginning of the second large-scale depositional sequence in Istria lasting from the Upper Tithonian to the Lower–Upper Aptian (Vlahović *et al.* 2003). Chronostratigraphically it corresponds to the Tithonian. The depositional environment can be envisaged as part of the shallow Adriatic carbonate platform. The limestones used as architectural-building stones are typically dense mudstones. Occasionally the oldest dinosaur tracks in Istria can be found in these mudstone layers (Mezga *et al.* 2003). Kirmenjak deposits represent the beginning of the oscillatory transgression over emerged relief, i.e. the beginning of deposition of the informal Kirmenjak Unit (Vlahović *et al.* 2003). In the ancient quarry on the Zlatni rt cape an approximately 35 m-thick sequence of the Kirmenjak Unit crops out, characterized by typical shallowing-upwards cycles. According to Vlahović *et al.* (2003) generally shallowing-upwards cycles consist of three members:

(1) a thin, laterally variable bed of black-pebble breccia with a carbonate, clayey or marly matrix. This member was formed by the redeposition of material originating from marsh deposits enriched in organic matter that was eroded and transported during a relative sea-level rise;

(2) a thick (100–200 cm) stylolitized mudstone with rare bioclasts of *Clypeina jurassica* Favre, *Salpingoporella annulata* Carozzi and *Campbelliella striata* (Carozzi). In some cycles its upper part is characterized by vertical bioturbation, fenestral fabric, desiccation cracks and erosion surfaces. This member was deposited in a low-energy subtidal environment;

(3) the upper part of cycles are characterized by variation in thickness, lithology and structural fabric, and are predominantly represented by vadose fabric, pisoids and, in places, by stromatolites. The upper bedding surfaces are sharp, irregular with desiccation cracks and/or erosion features. This member was formed in an intertidal and/or vadose zone.

The entire sequence of the Kirmenjak Unit is characterized by a gradual change in the cycle's composition and thickness. Generally, the succession shows a decrease in the thickness of the subtidal members (second member), and an increase in the thickness of the intertidal, supratidal and vadose members (third member). The first member, black-pebble breccia, is present only in cycles of the lowermost part of the sequence (Vlahović *et al.* 2003).

For architectural stone utilization of the second, subtidal dense micritic member is of the greatest importance. Kirmenjak has very low porosity (1.1 vol.%) and an extremely low water absorption rate (0.24 wt%). A comparative study has been performed to test the most important architectural stone from Croatia (Crnković & Šarić 1992) using standard methods (ISRM 1977) (Table 1). This

**Table 1.** *Values of water absorption and porosity for some natural stones in Croatia*

| Type of stone | Water absorption (wt%) | Porosity (vol.%) |
|---|---|---|
| Biosparite 'Rasotica' | 0.19 | 1.50 |
| Micrite 'Kirmenjak – *Pietra d'Istria'* | 0.24 | 1.10 |
| Oncolith 'Kanfanar' | 0.64 | 3.00 |
| Biosparite 'Kupinovo Unito' | 0.95 | 5.20 |
| Biosparite 'Veselje fiorito' | 1.06 | 7.50 |
| Biosparite–biomicrite 'Lucija' | 1.51 | 5.60 |
| Biosparite 'Visočani Unito' | 1.72 | 6.48 |
| Dolomitic limestone 'Sivac' | 2.32 | 7.70 |
| Dolomitic limestone 'San Giorgio' | 2.48 | 8.80 |
| Coquina 'Vinkuran Unito' | 4.00 | 12.10 |

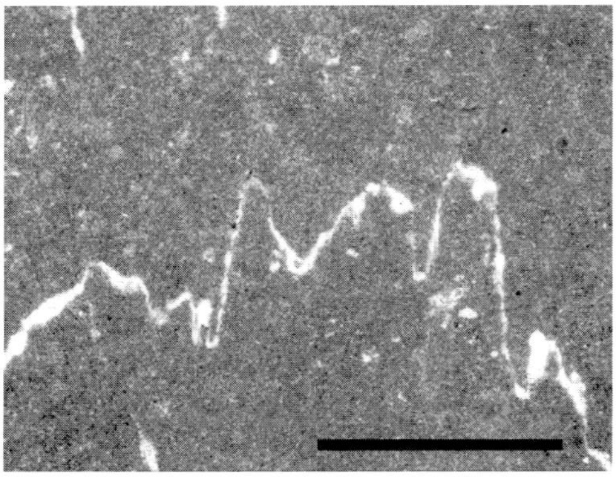

**Fig. 2.** Photomicrograph of the dense micritic structure of Kirmenjak with a stylolite in the middle. Plane polarized light. Scale bar is 1 mm.

shows that, of the 10 tested stones, Kirmenjak has the lowest values for both porosity and water absorption. This can be explained by its extremely dense micritic structure (Fig. 2). Nevertheless, Kirmenjak is visually and petrographically characterized by horizontal stylolitic discontinuities (Fig. 2). Stylolites represent an anisotropic feature of the stone, which is a possible disadvantage. Therefore, it was decided to examine how this could influence geotechnical properties. To do this, the direction of stylolitic planes built into the basal zone and in other structural and pseudo-structural elements of Venetian buildings are examined. The aim is to establish whether stylolitic discontinuities represent an advantage or a disadvantage in construction.

## Kirmenjak in the basal course of Venetian buildings

As Kirmenjak displays different patterns of stylolitic planes on its surface, depending on how it is cut and which plane is visible (perpendicular to the stylolites or parallel to them), it was possible to make a non-destructive visual *in situ* survey of how Kirmenjak was laid in the different elements of Venetian structures. The first step of the preliminary visual *in situ* survey was carried out on 30 important Venetian buildings, among them Ca' d'Oro (the most famous Gothic palace of Venice: Arslan 1986), the Early Renaissance palace of Vendramin-Calergi (architects Mauro Codussi and Tullio Lombardo: McAndrew 1980), the Gothic church of S. Maria Gloriosa dei Frari and the world-famous bridge Ponte di Rialto.

It is known that Kirmenjak blocks were usually extracted from the quarries with bedding planes (and consequently stylolitic discontinuities) parallel to the longest dimension of blocks. Historical sources also mention the standard size of a block (4 ft long – approximately 136 cm), which influenced typical dimensions of some Venetian building elements – for example, 4 ft was the typical width of Venetian decoratively carved window frames (Zbîrnea 2000). Stonecutters carved structural and decorative elements of Kirmenjak following the direction of the 'cut' of a block from the quarry, in order to make use of the length of blocks. Therefore, in the elements of portals, columns and window frames made of Kirmenjak, stylolites are parallel to the longer dimension, no matter how they are oriented within the structure (i.e. horizontal, vertical, oblique).

The preliminary non-destructive *in situ* survey showed that in structural, pseudo-structural and decorative elements of the buildings investigated (e.g. in the portals of the church of S. Maria Gloriosa dei Frari), stylolite orientation follows the longest dimension of the construction element. So that in vertical building elements the stylolites within Kirmenjak blocks are also vertical. The *balaustre* (vertical elements of the balustraded parapet) of Ponte Rialto are also carved from blocks cut parallel to the bedding planes, so that their stylolite planes are parallel to the longest dimension of the architectural elements, but the builders obviously did not care whether they were oriented parallel to the bridge span or in any other vertical plane. The orientation of stylolites in built-in stone blocks was merely a function of the technology involved in the extraction of stone blocks in quarries.

However, in the basal zones of buildings Kirmenjak blocks are consistently laid with horizontal stylolite planes (e.g. Fig. 1). Preliminary non-destructive research on 30 Venetian buildings from the Gothic, Renaissance and Baroque periods shows that it was a common and consistent practice. It can be concluded, therefore, that from centuries' long experience, Venetian builders knew that in the protective basal zone of buildings Kirmenjak had to be laid with its stylolite planes horizontal. It can also be assumed that they used the stylolite discontinuities as an efficient, multi-layer damp-proof course. This is because the stylolites are partially filled with clay. Results of X-ray diffraction analyses of the material from stylolite fills show that there are expanding 14 Å minerals (vermiculite or/and smectite) that minimize the already very low water absorption of the stone (Fig. 3). Scanning electron micrographs also show that filosilicates are oriented with (001) planes covering the surface of the stylolites (Fig. 4). This hypothesis needs to be verified by more samples, since the present conclusions are based only on a preliminary study.

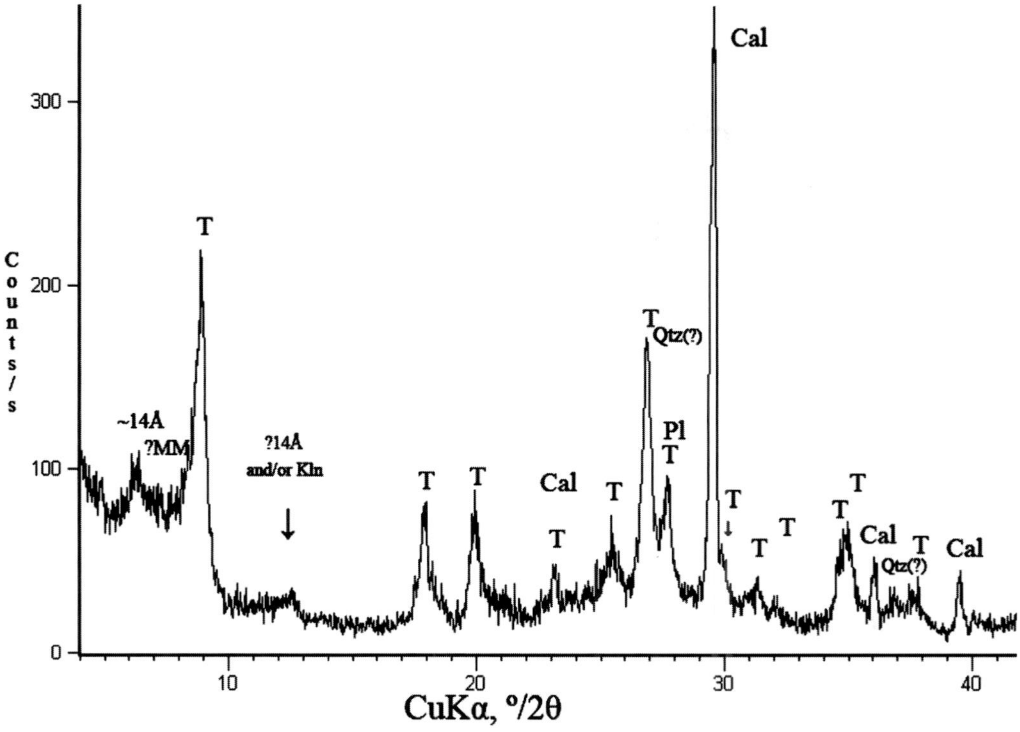


**Fig. 3.** X-ray diffraction analyses of the material from the stylolitic fill of Kirmenjak: MM, interstratified filosilicates; 14 Å, *c.* 14 Å filosilicates (smectite, vermiculite, chlorite); Kln, kaolinite?; T, illitic material (illite, ±muscovite, ±illite–smectite with high amount of illite layers); Qtz, quartz; Cal, calcite; Pl, plagioclase.

## *Acqua alta* and the importance of Kirmenjak

The sinking of Venice has resulted in a relative rise in sea level. Nowadays the *acqua alta* rises over the protective layer of Kirmenjak more frequently than ever before. In such cases the basal course of Kirmenjak cannot play its protective role. The saline water wets the superstructure and humidity persists even after the sea has withdrawn. In some cases rising damp reaches a height of 3 m. Humidity causes immense damage to the mainly brick structural parts of buildings, as well as to decorative stone façades. This damage shows how important Kirmenjak has been for Venetian architecture, and how it acted as an efficient rising damp barrier, in times when modern damp-proof layers did not exist.

The problem of very high tides in Venice is serious for conservators and restorers. It is a new threat that endangers old buildings. The basal layer of Kirmenjak, which in the past was a perfect protection against saline water, sunk together with the whole of Venice. With these changes in the environment Venetian architectural heritage is increasingly endangered.

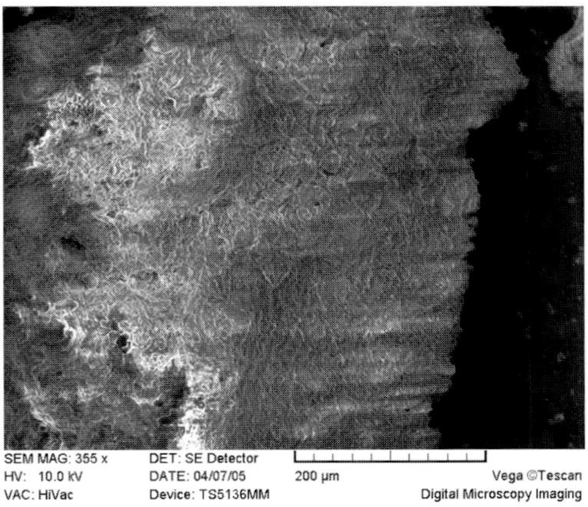


**Fig. 4.** Scanning electron micrograph of clay minerals covering the surface of a stylolite.

## Conclusions

Kirmenjak is the most important building stone of Venice. Its buildings typically have colourful marble, porphyric, granitic and other luxurious stone facings, but are nevertheless underpinned by the white Kirmenjak.

It is not known how and when the builders of Venice discovered the excellent qualities of Kirmenjak, especially its extremely low water absorption and impermeability, but it is well known that they made good use of it and used Kirmenjak as a damp-proof course. In Venice the basal zone of buildings, exposed to tidal wetting and drying, is built consistently of this dense micritic limestone. Even in these highly aggressive conditions, the basal zone has never been plastered or protected in any way.

The importance of Kirmenjak was shown during the great flood in the second half of the 20th century, when the *acqua alta* rose above the Kirmenjak base and damaged the brickwork superstructure (walls) and the expensive stone facing of the world famous Venetian façades.

In order to investigate what makes Kirmenjak so appropriate for the basal damp barrier, preliminary *in situ* and laboratory analyses were carried out. This preliminary research on a limited number of important Venetian buildings, constructed from the 15th to 18th centuries, showed that in the basal zones of buildings Kirmenjak was laid strictly and consistently with horizontal stylolites. When used for other purposes, the orientation of stylolites in stone blocks is quite random.

These observations suggest that the original builders of Venice also took advantage of horizontal stylolite discontinuities, filled with 14 Å clay minerals, as a damp-proof course. The assumption is that this fill of stylolite discontinuities acts as a multilayer humidity barrier, which minimizes already very low water absorption. This hypothesis needs to be verified with a more comprehensive study of stone use and characteristics.

The authors would like to acknowledge Mr I. Cotman, MSc from 'Kamen-Pazin' who provided us with useful information and helpful literature references.

## References

ARSLAN, E. 1986. *Venezia gotica – L'architettura civile*. Edizioni Electa, Milano.

BERTOŠA, M. 1992. L'avorio istriano per Donatello. *Jurina i Franina, rivista di varia cultura istriana*, **51**, 38–41.

COTMAN, I. 2004. Kamenolomi – jučer, danas, sutra. *In*: BRATULIĆ, J. & COTMAN, I. (eds) *Kamen-Pazin 1954–2004*. Kamen d.d., Pazin, 102–149.

CRNKOVIĆ, B. & ŠARIĆ, L. J. 1992. *Građenje prirodnim kamenom*. University of Zagreb, Zagreb.

*Encyclopaedia Britannica*. 2002. *Deluxe Edition CD*. britannica.co.uk. Ltd, Bristol.

HOWARD, D. 1989. *The Architectural History of Venice*. B.T. Batsford, London.

ISRM. 1977. Suggested Method for Petrographic Description of Rocks. Committee on Laboratory Tests, Document no. 6. *International Journal of Rock Mechanics & Mining Science & Geomechanics Abstracts*, **15**(2), 41–45.

LAZZARINI, L. 1986. I materiali lapidei dell'edilizia storica veneziana. *Restauro & Città*, **3/4**, 84–100.

LAZZARINI, L. 2004. *Pietre e marmi antichi*. Casa editrice dott. Antonio Milani, Padova.

MEZGA, A., BAJRAKTAREVIĆ, Z., CVETKO TEŠOVIĆ, B. & GUŠIĆ, I. 2003. Dinosaur tracks as an evidence for the terrestriality in the Late Jurassic sediments of Istria, Croatia. *In*: 22nd *IAS Meeting of Sedimentology, Abstracts Book*, Institute of Geology – Zagreb, Zagreb, 126.

MCANDREW, J. 1980. *Venetian Architecture of the Early Renaissance*. MIT Press, Cambridge, MA.

VLAHOVIĆ, I., TIŠLJAR, J., VELIĆ, I., MATIČEC, D., SKELTON, P. W., KORBAR, T. & FUČEK, L. 2003. Main events recorded in the sedimentary succession of the Adriatic Carbonate Platform from the Oxfordian to the Upper Santonian in Istria (Croatia). *In*: VLAHOVIĆ, I. & TIŠLJAR, J. (eds) *Evolution of Depositional Environments from the Palaeozoic to the Quaternary in the Karst Dinarides and the Pannonian Basin*. Field Trip Guidebook of the 22nd IAS Meeting of Sedimentology, Institute of Geology – Zagreb, Zagreb, 19–71.

ZBÎRNEA, I. M. 2000. Rassegna veneziana. I tagliapietra e la loro arte. *Annuario. Istituto Romeno di Cultura e Ricerca Umanistica, Venice*, **2**, 51–120.

# Photo-based decay mapping of replaced stone blocks on the boundary wall of Worcester College, Oxford

M. J. THORNBUSH & H. A. VILES

*School of Geography, Oxford University Centre for the Environment, University of Oxford, South Parks Road, Oxford OX1 3QY, UK (e-mail: mary.thornbush@ouce.ox.ac.uk)*

**Abstract:** In this study on the boundary wall of Worcester College, Oxford, the decay mapping in Adobe Photoshop (DMAP) approach is introduced to test the use of simple daylight photographs in the long-term monitoring of stone decay. This is conducted primarily through measured changes in surface brightness and roughness based on close-up photographic images of walls. The Magic Wand Tool was applied to greyscale images in Lab Color Mode to select proportions of pixels with a lightness ($L$) value of 77%. This paper shows the effectiveness of the calibration procedure used to validate lightness between surveys so that cross-temporal comparisons have a greater validity. It also outlines and discusses errors associated with the method as well as its limitations. The DMAP approach proves to be particularly useful when applied to long-term monitoring exceeding 5 years of survey.

For years researchers within stone decay and heritage conservation have attempted to develop an efficient approach to map and classify stone deterioration in the field in order to describe and explain patterns of decay (e.g. Fitzner *et al.* 1992, 1996). Decay mapping has been, for instance, attempted recently using photogrammetry (e.g. Dixon *et al.* 1998). Inkpen *et al.* (2001) used GIS as a system for mapping decay also within a photogrammetric approach. De' Gennaro *et al.* (2000) likewise devised a GIS-like system based on a database linked to thematic maps of the historical architecture of Naples. They found the predominance of three main lithotypes: Piperno sensu strictu, Neapolitan Yellow Tuff and lavas of which roughly 80% were affected by moderate–severe weathering such as alveolization, patinas (from atmospheric pollution), encrusted dust, fissuring, disaggregation, spalling and exfoliation.

The use of photogrammetry is a successful approach to remotely mapped decay. However, it suffers from several complications, including the fact that it is a specialist approach requiring expensive technology for data analysis. GIS is also beneficial as a means of introducing a spatial element to raw data linking several layers containing different types of information. Although GIS can handle and store much data, it is another specialist approach requiring familiarity with the software that most non-geographers, such as photographers, may not be able to operate. The approach introduced in this paper provides those workers unfamiliar with specialized technology with the means of quickly and easily acquiring visually-based quantitative data.

Decay mapping has also been conducted manually in field surveys. Viles (1993), for instance, established a database of stone decay for 15 buildings in central Oxford, which included 12 colleges as well as the Bodleian Library, the Clarendon Building and the Ashmolean Museum. Details were taken at each site for each of the following: stone type, repair history, percentage and intensity of surface discolouration, decay features (i.e. blisters, blowouts, pits, etched bedding planes and standing fossils), biological cover (i.e. creepers, lichens, algae and mosses), aspect and road traffic density. Blistering and blowouts were the most common of all decay features. Over 25% of surfaces were discoloured (light–medium grey). Ground-floor ashlar was most affected by decay and discolouration, suggesting a low-level causation. She suggested winter road de-icing salt as one possibility and the other was traffic – for which she found no clear variation with decay features (e.g. blistering and blowouts appeared on walls where there was no traffic). She discovered more severe blistering on S-facing walls.

Likewise, Török (2002) found that orientation of stonework had an impact on sheltering v. exposure of the Citadella fortress ooidal limestone of Budapest, Hungary. He also used photography from 1920–1930 as part of his research to investigate the rate of stone decay and weathering mechanisms. His relevant findings are that on exposed walls most material loss occurred around edges and near ground level (from wetting rather than capillary rise, as the height of the water table was some tens of metres below the foundation) with frequent granular disintegration, and that black crusts appeared in sheltered areas as they require steady

*From*: PŘIKRYL, R. & SMITH, B. J. (eds) *Building Stone Decay: From Diagnosis to Conservation.* Geological Society, London, Special Publications, **271**, 69–75.
0305-8719/07/$15.00 

microenvironmental conditions (e.g. rainwash protection). Exposed sections of walls consisted of case-hardened (calcite-rich) crusts that resulted from exposure to wind and rain.

Antill & Viles (1998) found an abundance of hard thin grey crusts (38% coverage) and unconsolidated dusty deposits (30% coverage), as well as black thick ropey crusts (23% coverage) and hard thin black crusts (22%), on the wall section at Worcester College along Walton Street. Appearing less frequently were grey thick ropey crusts (15%), blisters and blowouts (13%), organisms such as fungi, algae and moss (10%) and brown crust (5%). They found that 39% of 700 blocks had 76–100% blackening; another 29% had 26–50% cover of blackening. These workers noticed that black crusts appear thick and ropey in sheltered areas and develop in thin layers on vertical sections. They also observed a 'zone of maximum decay' located where there was extensive blistering and blowouts of the wall, where sulphate concentrations were high (especially within blisters) and gypsum formation was evident. Because they examined only the E-facing section of the wall along Walton Street, their results do not encompass the greater spatial extent of this study such as the S-facing (unsheltered) portion presented here.

The present study examines the boundary wall of Worcester College, Oxford as an exemplar historical structure with recently replaced limestone blocks. Through long-term monitoring and assessment of these blocks, it is hoped that weathering of the stone surface can be captured with the 'decay mapping in Adobe Photoshop' (DMAP) approach – an analytical technique for processing close-up images of simple daylight photographs. The objectives of this study are: (1) to delineate a method using image analysis (namely, the DMAP approach) to assess weathering, particularly encrustation and surface roughness (affected by e.g. blistering, blowouts), of the replaced blocks; (2) to examine the usefulness/limitations of photographic surveying using simple daylight photographs; and (3) to test the calibration and comparability of annual photographic surveys.

## Method

According to Antill & Viles (1998), restoration work was performed at Worcester College at the beginning of the 20th century and then again in the 1960s and in 1983 before the most recent repairs planned for the perimeter wall. Their fieldwork was conducted in 1997, whereas the new blocks from 1999–2000 were remnants from other building work, probably Bath limestones with some Clipsham stone. Because of their mixed lithology, these replacement blocks had a varied appearance (e.g. colouration, texture, bedding orientation) even before any noticeable change associated with decay.

Four close-up photographic surveys were taken from ground level representing *c.* 50 × 50 cm of the wall surface within approximately 1.5 m of the pavement. These were taken subsequent to replacement in 2000, 2003, 2004 and 2005. Each survey was taken between 09:00 and 12:00 in April–July, representing spring–summer months of daylight. Although a film-based camera was used in the initial survey in 2000, subsequent surveys were taken with a tripod-mounted Nikon Coolpix 950 digital camera. The digital camera was set to automatic mode with the flash off. Digital photographs were originally 1600 (width) × 1200 (height) pixels with a resolution of 300 pixels/inch. Photographs were scanned to acquire a digital image only for the first survey of 2000. Thirteen sites are included in the study spanning the SE boundary of Worcester College along Hythe Bridge Street and into Worcester Street, but not into Walton Street, which had already examined by Antill & Viles (1998). Sites were photographed each time with the inclusion of a greyscale without any plastic covering. A total of 52 photographs were taken at the locations of replaced stonework within 93 and 206 cm of the wall surface and parallel to it.

The image analysis methodology was modified from Thornbush & Viles (2004*a*), using CIE $L^*a^*b^*$ lightness ($L$) for a selected range of pixels. Images were cropped so that they represented the same area of coverage across the wall. A decline in $L$ is known to occur in polluted environments (cf. Pio *et al.* 1998), which is representative of surface blackening or soiling. Images were saved in Adobe Photoshop (as .PSD files) and then converted from Lab Color Mode into greyscale images. Calibration was then conducted using both white (100%) and black (0%) areas of the greyscale bar used to distinguish/differentiate between greyscale images. The white scalebar was used in brightness adjustments and the black scalebar to adjust the contrast so that each was at least 90% brightness and 10% contrast, respectively. The greyscale bar was then selected (including any of its cast shadows) and the image was reverse selected, effectively excluding the greyscale bar area from analysis.

DMAP was used to select 'bright' pixels (approximately the original colour of the limestone) to track the areal coverage of the decay (through encrustation and blackening) of blocks. The methodology involved the selection of each image at a 50 tolerance (20%), which selects for a broader range of pixel colours than any lower value. The highest tolerance value possible is 255 (the broadest possible range of pixel colours). A tolerance of 100 would be the maximum recommended range in this approach allowing for suitable edge detection. The 'anti-aliased' option was not selected to avoid smoothing edges in pixel counts. 'Contiguous' was also not selected so that non-adjacent areas with the same colouration could be included. All layers were used including, for example, the background layer. The colour selected using the Magic Wand Tool was set at an $L$ value of 77% denoting the brightness of unexposed control samples (see, for example, Thornbush & Viles 2004*b*) based on image analysis of stone sensors photographed indoors with the same digital camera.

## Results and discussion

### *Deciphering 'bright' pixels*

Photographs from the year 2000 had the greatest proportion of pixels (76.23%) in the images, with an $L$ value of 77% (Fig. 1). This greater brightness of samples in 2000 is especially evident at sites 1 (91.38% of pixels) and 4 (90.20% of pixels). The survey in 2000 could be brighter either due to scanning, because of a greater outdoor brightness on the day in which this set of photographs was taken, or because of the S-facing orientation of this section of the wall. Subsequent surveys in 2003 and 2004 showed a generally reduced proportion of light pixels (65.52% and 62.02%, respectively). The drop between 2000 and 2003 is significant at most sites. There was an overall slight recovery in improved lightness in 2005 (67.09%).

In Figure 1 there appears to be a spatial effect of the proportion of selected (bright) pixels representative of 'clean' stone surfaces. The general trend is an initial reduction of bright pixels at sites 1 and 2, an increased proportion at sites 3 and 4, followed by a reduction at sites 5–8. Sites 8–10 have the least amount of bright pixels. Sites 11–13 are comparable to sites 5–8, although slightly reduced. These results could be a product of location along the wall. Sites 1–5 are located along Hythe Bridge Street; whereas, sites 6–13 are around the bend on Worcester Street. Whereas sites on Hythe Bridge Street are S-facing, those on Worcester Street are E-facing. Sites 3 and 4 consist of mostly bright pixels, perhaps because of their exposure to wind and wind-driven rain from the SW (cf. Oxfordshire County Council 1995), possibly leading to white-washing of the surface and dissolution exposing clean surfaces stripped of any crusts.

Lighting conditions are also another likely reason. More sunlight during the survey period could be hitting this S-facing part of the wall, quite possibly leading to a greater brightness. This is possible because soft box flash light was not used, as by professional photographers, to exclude the influence of variable daylight. Consistent lighting conditions are very difficult to obtain in outdoor photography.

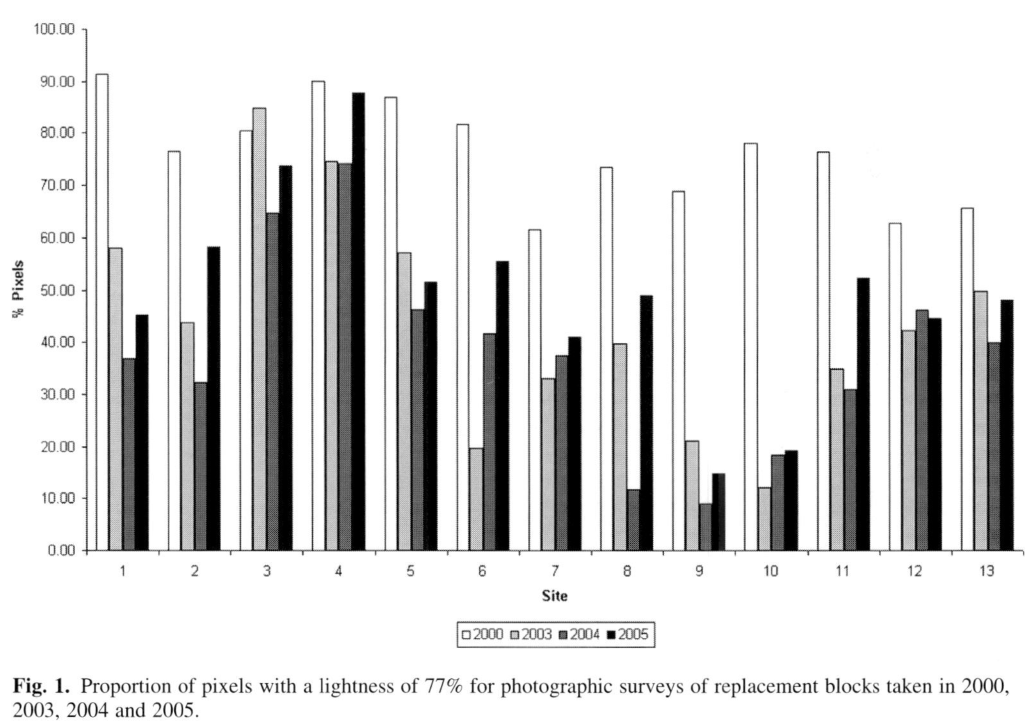


**Fig. 1.** Proportion of pixels with a lightness of 77% for photographic surveys of replacement blocks taken in 2000, 2003, 2004 and 2005.

Even if photographic surveys are taken with the same digital camera, at the same time of the day, and in the same season or even month, lighting conditions are likely to vary. This is due to cloud cover as well as climatic conditions and season. It may well be impossible, particularly in a maritime environment with unpredictable cloud cover, to obtain exactly comparable weather conditions in the uncontrolled out-of-doors environment.

Although the use of a digital camera clears up the problem of film quality as identified, for instance, by Inkpen *et al.* (2001), camera lenses and the vital issue of differences in lighting may be irresolvable in outdoor repeat photographic surveys. The former can produce distortions and, for this reason, zoom lenses are also to be avoided in good rectified photography (Swallow *et al.* 2004). The inclusion of a light meter in the study would have helped to at least track differences in lighting conditions. Using the same digital camera, however, is crucial as it can prevent other issues in photographic surveys. Coe *et al.* (1992), for instance, conducted a close-range photogrammetric study, which included rods for the alignment of stereoscopic pairs. Although their findings were precise in the measurement of long-term surface recession, they experienced problems with the resolution of an earlier set of benchmark photographs from 1987 disabling any viable short-term erosion rates. Similarly, the fact that a different type of camera was used in the 2000 survey is a limitation of the comparability of results in this study.

It is interesting to note, however, that there is variation in brightness/contrast adjustments even when the same camera is used. This may reflect differences in the lighting conditions associated with outdoor photography. That the calibration procedure actually detects differences can be considered to show that it works to produce more comparable lighting in the photographic images. Brightness adjustments for 2003 and 2005 are similar (Fig. 2), and this is probably the result of the successful selection of similar lighting conditions for the surveys. Likewise, contrast adjustments for 2004 and 2005 are also similar (Fig. 3), this time probably portraying a similarity in the time of day.

As aforementioned, the survey in 2000 was taken using photographic film and then scanned to obtain digital images. This appears to have affected its brightness and contrast adjustments, so that they have been reduced in both cases. In comparison, the photographs taken with a digital camera in 2003, 2004 and 2005 have a positive and wider range of brightness adjustment, and only their contrast was increased in the calibration procedure. The implication of this is that the results acquired with a digital camera are indeed different from the scanned survey taken in 2000.

It is difficult to discern whether weathering features such as dark encrustation or blowouts are expanding without the intrusion of shadows introducing error in any estimation. One of the problems with this approach is that, as it was applied here, it does not discriminate between areas of darkness with an $L$ value of less than 23% (e.g. black v. grey crusts; blowouts v. shadows). Shadows are cast by blowouts, making it very difficult to disentangle them (both are often less than 10% in lightness); however, it may be possible to further refine this method so that it can differentiate between grey ($L = 20$–$30\%$) v. black ($L = 10$–$20\%$) crusts. Surface roughness is an important factor in colour changes associated with stone decay (Benavente *et al.* 2003). Although shadows cast by the greyscale bar were removed, it was not possible to remove all shadows. They could be

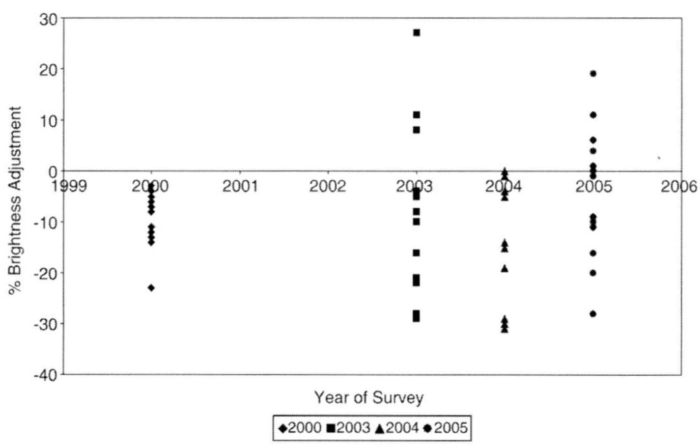


**Fig. 2.** Brightness adjustments for each of the photographic surveys at all sites.

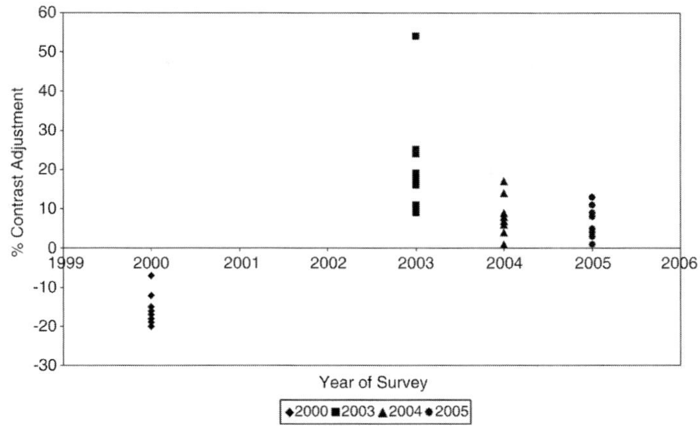


**Fig. 3.** Contrast adjustments for each of the photographic surveys at all sites.

indicative of surface roughness; however, the angle of incident sunlight could distort shadows (e.g. E-facing sites), contributing to error in using them to detect roughness changes.

There are other sources of error, however, such as vegetation – which has been effectively excluded amongst bright pixels at Site 5 in 2005 (Fig. 4). Interpretation remains a drawback of such methods, even within photogrammetric analysis (cf. Inkpen *et al.* 2000) because, even though the images are quantifiable, observer input is necessary to understand the nature of the change – this can be a problem generally experienced in the use of repeat photography (e.g. the issue of 'hidden time' presented by Bass 2004).

Qualitative assessment is necessary to supplement the quantitative results obtained through the use of this technique. Observations of the calibrated images suggest that direct v. indirect lighting causes images to appear different in their lightness. The survey taken in 2000 is notably brighter than the one in 2004, which is generally dark and without many shadows or apparent surface roughness. Another qualitative observation is that there is not much change in crust coverage on the replacement blocks (e.g. sites 1, 4, 6 and 12). Blowouts do not seem to get larger either (e.g. at site 11). There is only one visible instance of possible dissolution (at Site 8 in a sheltered section of the wall), which is especially notable in the 2005 survey and may be a product of lighting.

### *Technique development*

The DMAP approach can be modified to accommodate other research needs. Cossu & Chiappini (2004) presented a similar method using colour images for the segmentation of decay features such as holes and cavities. The DMAP approach can be altered (e.g. using colour images) to provide similar results with edge detection and histogram output. This can be performed, specifically, by changing the value used for selection by the Magic Wand Tool and ascertaining the *a* and *b* values. Edge detection can be controlled through the tolerance level. This study however, incorporated images in greyscale because of the limitation imposed by the original photographic survey in 2000.

Another suggestion is to take multiple photographs of a site, one including a greyscale bar and another without one. This way any error associated with shadows cast by the greyscale are eliminated as calibration is performed on the image with the greyscale bar and DMAP can then be applied on calibrated image without the greyscale bar. DMAP is a user-friendly approach that is easy to use in Adobe Photoshop, which is an accessible and versatile software.

## Conclusions

The DMAP approach is able to measure cross-temporal change in stone decay using close-up simple daylight photography. Through the selection of bright image pixels, it is possible to quantify the soiling of historical walls. It is evident, however, that the temporal scale should be greater than 5 years. Errors associated with wall orientation, shadows and sunlight, as well as those introduced through image cropping, the use of different cameras and the calibration procedure, are examples of what can go wrong with this approach. Simple daylight photographs are not easy to interpret because of variations in outdoor lighting, which can only be partly remedied by the use of

| SITE 5 | | | |
|---|---|---|---|
| **Year** | **Calibrated Image** | **Selected Pixels (L = 77%)** | **Pixel Proportion** |
| 2000 |  |  | 86.91% |
| 2003 |  | 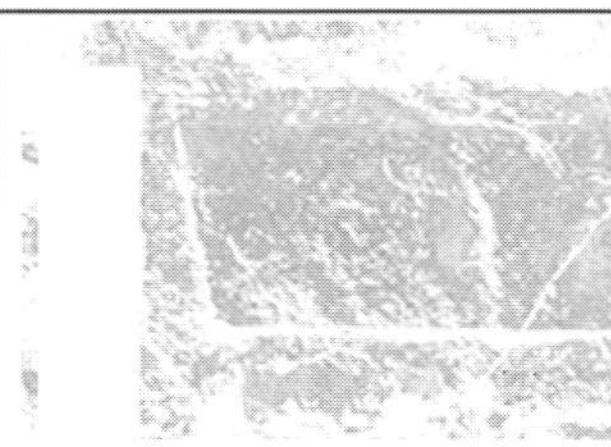 | 57.31% |
| 2004 |  |  | 46.29% |
| 2005 |  | 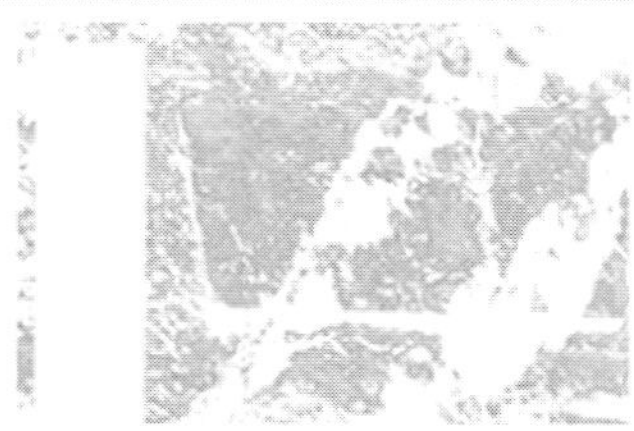 | 51.58% |

**Fig. 4.** Site 5 blocks showing the exclusion of vegetation amongst bright pixels in 2005.

greyscale bar calibration or even flash light. It is clear that though it is possible to calibrate images so that lighting is similar across surveys, the results are still laden with problems associated with lighting that can blur short temporal interpretations.

This study has also succeeded in capturing adjustments associated with lightness calibration for comparisons between surveys. However, it is strongly recommended that light-meter readings be taken in the field to further verify calibration. It may also be helpful to further verify calibration by using a portable colorimeter. Digital cameras are recommended for use in the DMAP approach as they portray a more consistent brightness/contrast calibration. Here, the 2000 survey, which consisted of scanned photographic prints, required negative adjustments of both brightness and contrast. It is questionable, therefore, whether the surfaces as they appear in the 2000 survey were truly as bright and rough.

Because of these errors, and the relatively short temporal span of these surveys, the results should be taken with caution. The focus here has been to introduce the DMAP approach, outlining it in some detail, and apply it to historical building stone in order to monitor change in newly replaced stone blocks. It is noteworthy that this technique does not work well with short photographic surveys as the error associated with the approach itself may be greater than any measurable change. Instead of using annual surveys, perhaps it is better to conduct photographic surveys at longer temporal intervals – even between 5 and 10 years. The temporal interval for photographic surveys should be derived from appraisals of the level of change specific to different conditions (e.g. stone type, environment). Future follow-up studies will incorporate the suggestions presented here for technique development and refinement, and will further test lightness calibration possibly through the use of a portable spectrophotometer.

Thanks to S. E. Thornbush and V. Robin for field assistance.

## References


ANTILL, S. J. & VILES, H. A. 1998. Deciphering the impacts of traffic on stone decay in Oxford: some preliminary observations from old limestone walls. *In*: JONES, M. S. & WAKEFIELD, R. D. (eds) *Stone Weathering and Atmospheric Pollution Network 1997: Aspects of Stone Weathering, Decay and Conservation*. Imperial College Press, Aberdeen, 28–42.

BASS, J. O. JR. 2004. More trees in the tropics. *Area*, **36**, 19–32.

BENAVENTE, D., MARTÍNEZ-VERDÚ, F. ET AL. 2003. Influence of surface roughness on color changes in building stones. *Color Research and Application*, **28**, 343–351.

COE, J. A., SHERWOOD, S. I., MESSERICH, J. A., PILLMORE, C. L., ANDERSEN, A. & MOSSOTTI, V. G. 1992. Measuring stone decay with close range photogrammetry. *In*: RODRIGUES, J. D., HENRIQUES, F. & JEREMIAS, F. T. (eds) *Proceedings of the 7th International Congress on Deterioration and Conservation of Stone, 15–18 June, Lisbon*. Laboratório Nacional de Engenharia Civil, Lisbon, 917–926.

COSSU, R. & CHIAPPINI, L. 2004. A color image segmentation method as used in the study of ancient monument decay. *Journal of Cultural Heritage*, **5**, 385–391.

DE' GENNARO, M., CALCATERRA, D., CAPPELLETTI, P., LANGELLA, A. & MORRA, V. 2000. Building stone and related weathering in the architecture of the ancient city of Naples. *Journal of Cultural Heritage*, **1**, 399–414.

DIXON, L. F. J., BARKER, R. ET AL. 1998. Analytical photogrammetry for geomorphological research. *In*: LANE, S. N., RICHARDS, K. S. & CHANDLER, J. H. (eds) *Landform Monitoring, Modelling and Analysis*. Wiley, Chichester, 63–94.

FITZNER, B., HEINRICHS, K. & KOWNATZKI, R. 1992. Classification of mapping of weathering forms. *In*: RODRIGUES, J. D., HENRIQUES, F. & JEREMIAS, F. T. (eds) *Proceedings of the 7th International Congress on Deterioration and Conservation of Stone, 15–18 June, Lisbon*. Laboratório Nacional de Engenharia Civil, Lisbon, 957–968.

FITZNER, B., HEINRICHS, K. & VOLKER, M. 1996. Monument mapping – a contribution to monument preservation. *In*: ZEZZA, F. (ed.) *Origin, Mechanisms and Effects of Salts on Degradation of Monuments in Marine and Continental Environments, Proceedings of the European Commission Research Workshop, 25–27 March 1996, Bari, Italy*. Protection and Conservation of the European Cultural Heritage Research Report, 4, 347–355.

INKPEN, R. J., COLLIER, P. & FONTANA, D. 2000. Close-range photogrammetric analysis of rock surfaces. *Zeitschrift för Geomorphologie (Supplementbände)*, **120**, 67–81.

INKPEN, R., FONTANA, D. & COLLIER, P. 2001. Mapping decay: integrating scales of weathering within a GIS. *Earth Surface Processes and Landforms*, **26**, 885–900.

OXFORDSHIRE COUNTY COUNCIL. 1995. *Oxfordshire's Environment: Environmental Monitoring Report, 1995*. Department of Planning and Property Services, Oxford.

PIO, C. A., RAMOS, M. M. & DURATE, A. C. 1998. Atmospheric aerosol and soiling of external surfaces in an urban environment. *Atmospheric Environment*, **32**, 1979–1989.

SWALLOW, P., DALLAS, R., JACKSON, S. & WATT, D. 2004. *Measurement and Recording of Historic Buildings*, 2nd edn. Donhead, Shaftesbury.

THORNBUSH, M. & VILES, H. 2004*a*. Integrated digital photography and image processing for the quantification of colouration on soiled surfaces in Oxford, England. *Journal of Cultural Heritage*, **5**, 185–190.

THORNBUSH, M. J. & VILES, H. A. 2004*b*. Surface soiling pattern detected by integrated digital photography and image processing of exposed limestone in Oxford, England. *In*: SAIZ-JIMENEZ, C. (ed.) *Air Pollution and Cultural Heritage*. A. A. Balkema, London, 221–224.

TÖRÖK, Á. 2002. The influence of wall orientation and lithology on the weathering of ooidal limestone in Budapest, Hungary. *In*: PŘIKRYL, R. & VILES, H. A. (eds) *Understanding and Managing Stone Decay*. Karolinum Press, Prague, 197–208.

VILES, H. A. 1993. Observations and explanations of stone decay in Oxford, UK. *In*: THIEL, M.-J. (ed.) *Conservation of Stone and Other Materials*. E. & F.N. Spon, London, 115–120.

# An holistic approach to the assessment of stone decay: Bonamargy Friary, Northern Ireland

S. McCABE, B. J. SMITH & P. A. WARKE

*School of Geography, Archaeology and Palaeoecology, Queen's University Belfast, Belfast BT7 1NN, Northern Ireland, UK (e-mail: stephen.mccabe@qub.ac.uk)*

**Abstract:** Decay mapping and condition assessment have proved to be useful tools in understanding stone decay and identification of remedial action. In this paper an holistic strategy is taken to the study of façade decay at the medieval Bonamargy Friary, on the north Antrim coast, Northern Ireland. After lithology and decay forms are mapped, interrelationships between decay form, stone type and environment are identified and quantified. This is accomplished through analysis of the spatial association of decay forms, and is used to inform our understanding of decay processes and environmental and lithological controls on those processes. This approach is combined with the application of the UAS ('Unit', 'Area', 'Spread') staging system developed by Warke *et al.* that is based upon a 'whole-building' approach to the assessment of stone condition, the spread to decay and a staged approach to conservation intervention. The case study demonstrates how the combination of these approaches improves our understanding of the factors that control stone decay whilst providing a clearer understanding of the cumulative impact of combined decay mechanisms.

Decay mapping and condition assessment have proved to be useful tools in understanding stone decay and identifying remedial action (Ball & Young 2004; Young *et al.* 2004). This paper combines decay mapping with techniques of mapping analysis to provide a clearer understanding of the factors that control stone decay and the cumulative impact of combined decay mechanisms. This is achieved by combining two approaches to understand the pathology of building stone decay and the resulting condition of the stone. Spatial association analysis (Turkington & Smith 2004) helps to quantify the spatial patterns and connectivity of decay features across a façade as determined by lithology, environment and time. The 'Unit', 'Area', 'Spread' (UAS) condition assessment scheme (Warke *et al.* 2003) gives a meaningful rating of the state of conservation of the subject monument, based on the extent of deterioration across a façade. It is a system for diagnosing stone decay on buildings and monuments, based on one of the most widely used cancer assessment schemes in medicine – the TNM (Tumour Node Metastases) staging system (Hermanek & Sobin 1987). The UAS staging system is based on a 'whole-building' approach to the assessment of stone decay and, rather than the time-consuming task of rating each stone making up a façade, a holistic approach is taken. An overall UAS rating is given, which correlates to different stages of decay that suggest if and when intervention is required. This paper aims to enhance understanding of the links between decay processes and features, to refine the process of diagnosis and condition assessment, and ultimately to inform any future conservation intervention through a case study of Bonmargy Friary, a ruined medieval sandstone structure in the NE of Ireland.

## Monument details – Bonamargy Friary, Northern Ireland

Bonamargy Friary is a ruined medieval ecclesiastical monument near the town of Ballycastle, on the north Antrim coast. The Friary was founded in 1500 and used by the Franciscan monks. As the tops of ruined walls are unsheltered, moisture is unhindered in flowing into a façade from the top. Bonamargy has undergone a complex decay history, experiencing several 'exceptional' factors (as well as background environmental factors – for example, marine salts, temperature and moisture cycles) that are likely to have influenced the deterioration of the sandstone through the centuries and its performance in the present day – a concept known as inheritance (Warke 1996). These 'exceptional' (McCabe *et al.* in press) factors, here defined as extreme events that have the potential to cause a step change in the equilibrium of the stone system, include the possible calcium loading of stone as a result of the lime render on the monument (Smith *et al.* 2001), fire (experienced in 1584), climate change (for example, the Little Ice Age, which may have had the effect of increasing the frequency and intensity of frost events) and conservation

*From*: PŘIKRYL, R. & SMITH, B. J. (eds) *Building Stone Decay: From Diagnosis to Conservation.* Geological Society, London, Special Publications, **271**, 77–86.

**Table 1.** *Sandstones at Bonamargy Friary and their characteristics*

| Name | Age | Colour | Mineralogy | Grain size | Water absorption (wt%) |
|---|---|---|---|---|---|
| Fair Head A | Carboniferous | Light brownish grey (Munsell 2.5Y 6/2) | Quartz, iron cement | 0.5–2 mm, coarse-grained, poorly sorted | 9.3 |
| Fair Head B | Carboniferous | Yellow (Munsell 2.5Y 7/6) | Quartz | 0.2–0.5 mm, fine-grained, well sorted | 8.0 |
| Fair Head C | Carboniferous | Brown–dark brown (Munsell 7.5 Y 4/4) | Quartz | 0.2 mm, fine-grained, very well sorted | 4.8 |

intervention. The present-day climate of the north Antrim coast is temperate maritime, and, because of the coastal location, is relatively mild. Average July temperatures are 16 °C (maximum) and 9 °C (minimum), with 72 mm of rainfall. Average January temperatures are 6 °C (maximum) and 1 °C (minimum), with 114 mm rainfall. Average annual rainfall is 1076 mm (climatic data was taken from the 1961–1990 base period). Three Carboniferous sandstone types have been identified at the site, and are described in Table 1. The natural outcrop of these stones is Fair Head, a large and varied Carboniferous succession approximately 2 km to the NE of Bonamargy Friary (see map, Fig. 1).

## Principal decay features

A brief discussion of the prominent weathering forms present at Bonamargy Friary, and the processes and mechanisms that are likely to have caused them, follows.

### *Alveolar weathering*

Alveolar weathering is common at the Friary, and is seen to affect various discrete portions of the monument (pictured, Fig. 2a). Bonamargy is an exposed coastal site, and salt carried inland from the sea is the likely cause of the alveoli, although there is still uncertainty about the processes involved in the development of cavernous weathering (Turkington & Paradise 2005). Alveolar weathering can also be seen at Fair Head, where the stone is likely to have been quarried. In McGreevy's 1984 study of the honeycomb weathering on Fair Head, gypsum was identified within the hollows, and it was suggested that salt weathering was the dominant process responsible for their formation. Other

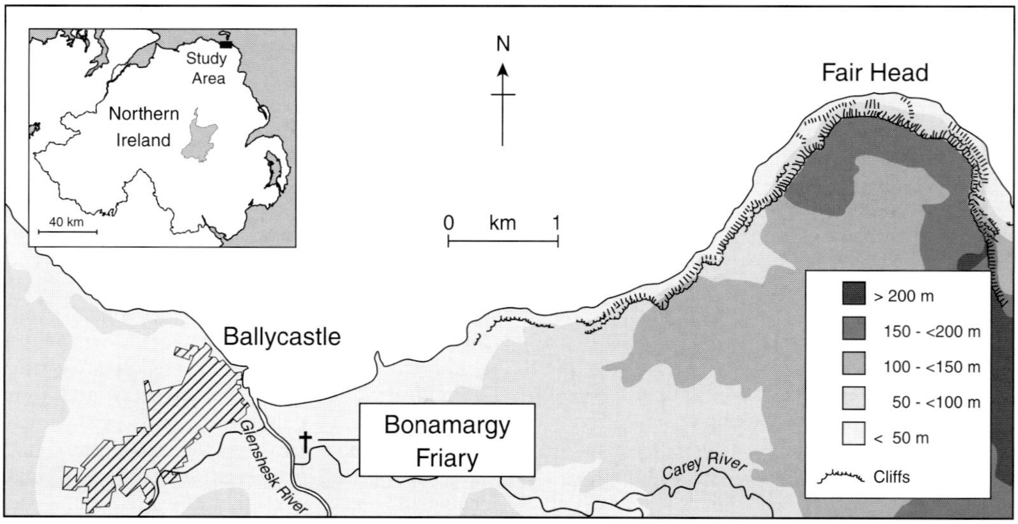


**Fig. 1.** Map of the Ballycastle area, showing Bonamargy Friary and Fair Head.

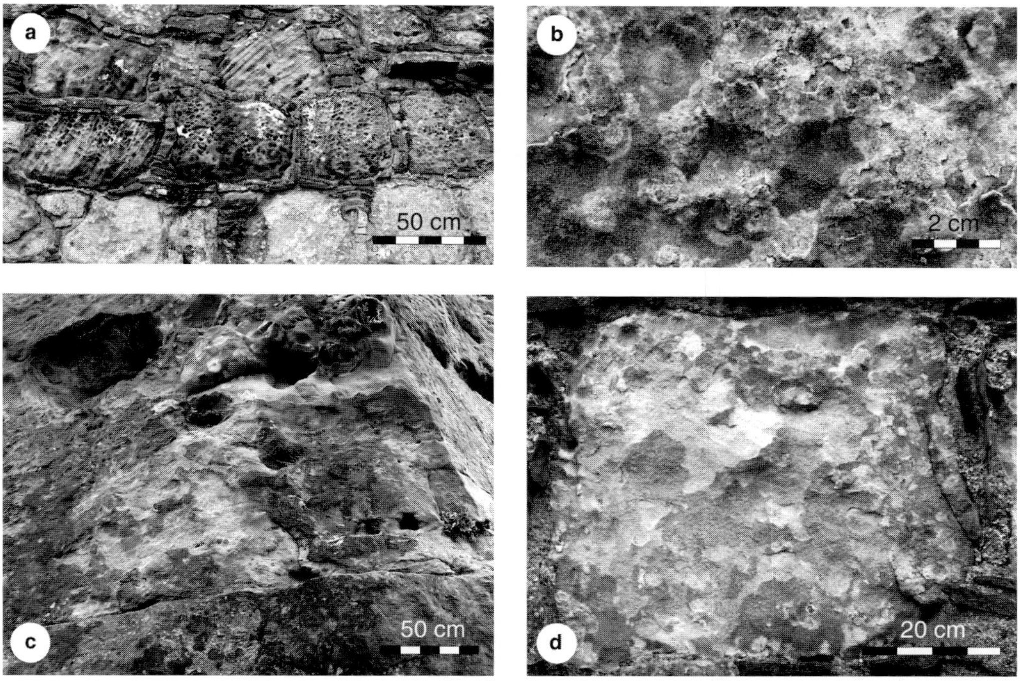


**Fig. 2.** Photographs of decay features common at Bonamargy Friary. (**a**) Alveolar weathering. (**b**) Blistering associated with lichen detachment and salt. (**c**) Iron crusts at Fair Head, the natural outcrop of the stone used in the construction of Bonamargy. (**d**) Scars of flaking – where recent, fresh stone is visible. Some fresh surfaces have been covered by the regrowth of secondary and tertiary crusts that vary in tone depending on age.

marine salts (e.g. sodium chloride and magnesium sulphate), more soluble than gypsum, are likely to have been removed by rainwash. Fair Head may be seen as analogous to Bonamargy Friary, and so it is likely that the alveoli seen at Bonamargy were formed by the deposition, ingress and precipitation of marine salts.

## *Biological colonization and associated decay*

Biological colonization and associated decay is widely seen at Bonamargy Friary – lichen, algae, mosses and higher plants are all visible on the surface of the stonework (and endolithic algae is also present in the substrate). Lichen is the most widespread of the organisms. It is possible to see lichen that has become integrated into the stone surface detaching from the stone and thus bringing surface grains of sand with it (pictured, Fig. 2b). This is the most obvious form of biological deterioration of stonework on the Friary. Lichen often works in combination with salts to cause material detachment in the form of granular disaggregation and blistering of the stone surface – salts may in fact be formed from the chemical action of the lichen (Chen *et al.* 2000). A study by Papida *et al.* (2000) also proposed that biological action occurring in conjunction with salt caused a greater degree of damage to stone than either would have acting in isolation. This integration of mineral fragments into the lichen thallus is an important aspect of the biophysical decay caused by the organism (Bjelland & Thorseth 2002). Moisture plays an important role in the biological decay of the Friary. Algae can be seen as a dry crust on parts of the monument in the summer months, and as a moist, spongy layer in the winter. In areas that are perennially damp, greening of the stone caused by algae occurs. Once greening occurs on a surface a positive feedback loop can be established – the algal growth encourages the retention of moisture and hence further greening. With projected climate change forecasted to bring milder, wetter winters in Northern Ireland, it is possible that greening of stone may increase at the site. Smith *et al.* (2004) have suggested that the diurnal wetting and drying cycles common in the present climate could shift to become more seasonal, and that stone could remain damp for several months at a time. This increased 'time-of-wetness' for stonework could result not only in further greening of the stone, but also in a greater depth of salt penetration into blocks (Smith *et al.* 2004).

### *Iron migration and the formation of indurated surface crusts*

Iron migration and the formation of indurated surface crusts can have a negative impact on both the aesthetics of the sandstone and ultimately on its durability. It is a process that has tended to be neglected by researchers into building stone decay, when compared with efforts in areas such as salt weathering. Surface iron staining can be evidence of iron migration from the substrate to the surface. Iron can be mobilized within sandstone and precipitated on the surface, often resulting in uneven discoloration – a widespread feature at Bonamargy, largely due to the use of Fair Head A, and highlighting the strong lithological controls on decay seen at the Friary. In the most extreme cases, a hard iron crust can form on the surface of the stone (there is some evidence for this at Bonamargy, but it is common at the natural outcrop, Fair Head – Fig. 2c). The problem is more than simply aesthetic, however. While strengthening the surface of the stone, the leaching of iron from the substrate can weaken the interior (McAlister *et al.* 2003), especially if the iron is drawn from the material cementing the sandstone. Thus, when the hard crust is breached by a high-magnitude stress event (or by the gradual convergence of decreasing surface strength with accumulating background stress effects) rapid retreat of the unstable substrate can ensue.

Ultimately, most decay processes are manifested in the *detachment of surface and near-surface material*. Especially interesting, at Bonamargy Friary, are scars of flaking, visible on each of the facades (pictured, Fig. 2d). These are reminiscent of the primary, secondary and tertiary gypsum crusts that form, detach and regrow on the Matthais Church, Budapest, noted by Smith *et al.* (2003). The crusts on the Friary, however, seem to be formed from the iron (and possibly manganese) in the stone cement. Where flaking is recent, fresh stone is visible (and often has loose grains at the surface). It is noteworthy, however, that some fresh surfaces have been covered by the regrowth of these secondary and tertiary crusts that vary in tone depending on age.

## Mapping of decay

An example of the decay mapping that has been carried out at Bonamargy Friary is shown in Figure 3. The mapped façade shown is west facing. Lithology was classified first (Fair Head A, B and C) and then overlain by maps of prominent decay features affecting each block. Classification of decay features into a limited number of significant categories – alveolar weathering, biological colonization, iron staining and detachment of material in the form of flaking and scaling (relating back to the discussion on principal decay features) – was based on field observation. This combined mapping approach is a valuable visual tool that provides help with condition assessment. After mapping has been carried out, it is possible to see how much of a façade, or a particular stone type, is affected by certain weathering forms, and a better understanding of the processes and controls resulting in those decay forms is gained. In this way mapping provides a basis from which an informed assessment of stone condition, and decisions concerning possible remedial action for the sandstones, can be made.

The following five steps were taken in the analysis of mapped decay features at Bonamargy Friary.

### *1. The amount of each stone type used was quantified*

Stone use is an important factor in the decay of the monument – Fair Head A is used more frequently than the other stone types (63% of the blocks on this particular façade section), and hence the performance of this stone is particularly important in the assessment of the decay of the monument. Fair Head B and C are deployed less on the mapped façade, making up 10 and 27% of blocks, respectively.

### *2. The percentage of overall decay on the facade was determined*

Taking the façade section as a whole, lichen affects almost 30% of the stone blocks, while alveolar weathering affects 14%, iron staining 7% and detachment in the form of flaking 3% (granular disaggregation is much more widespread and often associated with alveolar weathering).

### *3. The percentage of each stone type affected by decay was determined*

By analysing decay by stone type the picture becomes clearer. Over 90% of the alveolar weathering and iron staining on the façade occur on Fair Head A, with 80% of detachment and 60% of lichen growth also occurring on this stone type. This suggests that Fair Head A is susceptible to decay and exhibits a more transient form than the other stone types (the widespread decay is

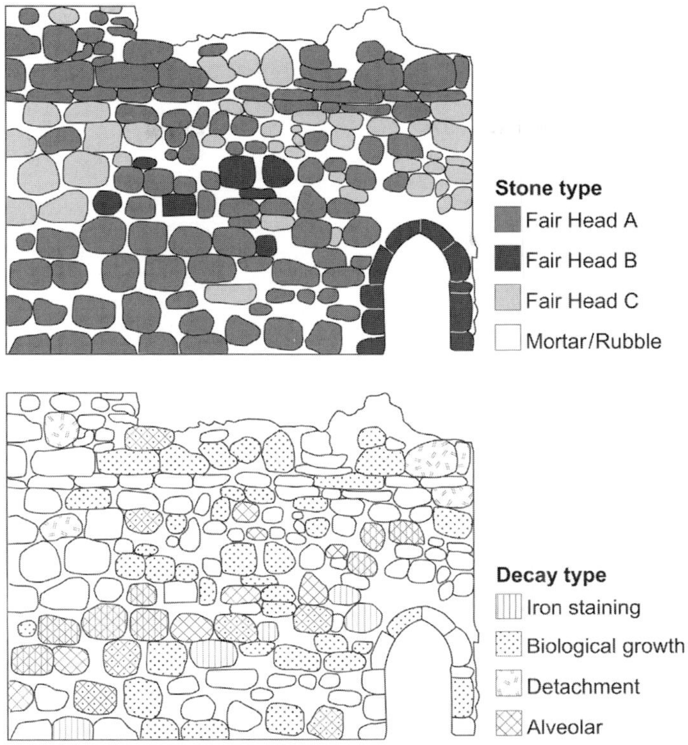


**Fig. 3.** Decay mapping of Bonamargy Friary west dormitory façade.

linked to lithology), but is also likely to be related to the frequency of use of Fair Head A as compared to Fair Head B and C.

On Fair Head A blocks 27% of blocks exhibit lichen growth, 21% alveolar weathering, 12% iron staining and only 4% detachment. The other stones show less decay in all categories (again, this could be attributed to their more limited use), with lichen growth and iron staining being the dominant decay forms on Fair Head B, while only lichen growth is common on Fair Head C. A summary of the percentage of each stone type exhibiting the different identified decay features can be seen in Figure 4.

This is clearly a somewhat simplistic way of viewing the decay at Bonamargy Friary – what is being observed here is simply a snapshot in time of the decay. Fair Head C does appear to be the more stable, durable stone type. However, when set in the context of the episodic nature of stone decay, it could be argued that decay may simply not have been triggered yet on Fair Head C blocks. That said, the stability of the stone is widespread throughout Bonamargy Friary and the well-sorted, tightly pack grain structure of the stone (see Table 1) seems to suggest that it is considerably more durable than the other stone types.

### *4. Spatial association/connectivity of decay features was determined*

A rating of the spatial association of decay can be derived from the number of sides of each block in one decay class adjacent to a block classified as exhibiting the same decay feature. This allows the spatial relationships between decay form, stone type and environment to be identified and analysed. Turkington & Smith (2004) suggest three ways of explaining the problem of the spatial variation of decay.

- Clustering of decay forms can imply that environmental controls are dominant in determining decay. This can be especially important for biological colonization.
- Isolation of decay forms can imply a dominance of rock properties in controlling decay – the susceptibility of particular blocks is a result of specific combinations of material properties.
- Spatial variability as a product of temporal variability – 'temporal variations in stone response to decay processes begets spatial variability in decay features' (Turkington & Smith 2004, p. 163).

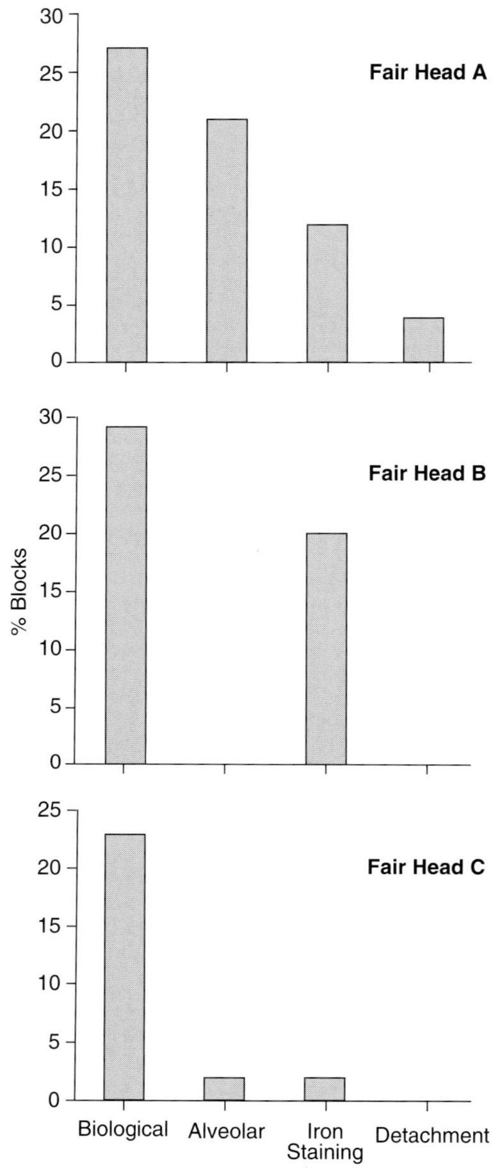


**Fig. 4.** Graphs showing the percentage of blocks of each stone type exhibiting decay features.

The graphs in Figure 5 show pronounced clustering of alveolar weathering and a tendency towards clustering in iron staining and lichen growth, while blocks exhibiting scars of detachment in the form of flaking are relatively isolated (on this particular façade). The use and placement of stone type in the monument has had a significant impact on the spatial patterns of decay. Fair Head A seems to be more susceptible to alveolar weathering, iron staining and detachment than the other stone types, so that whenever Fair Head A blocks are placed beside each other in a façade, clustering of those decay features would be expected. However, the roles of the local weathering environment and time should not be overlooked in determining the spatial associations of decay features. Each of the decay features explored in this paper relies on environmental factors as well as lithology, and patterns of decay should be viewed with the temporal response of the stone in question – blocks exhibiting the decay features described may alternatively be seen as simply further along the temporal decay sequence than those that appear stable (Turkington & Smith 2004).

Interestingly, alveolar weathering and iron staining are often seen occurring on the same blocks – in fact, on this façade, alveolar weathering only occurs on blocks exhibiting remnant areas of iron staining. It is likely that the alveolar weathering and iron staining evident on blocks are representative of the physical and chemical actions of salt weathering, respectively.

## *5. The application of UAS*

Condition assessment should provide a meaningful rating of the state of conservation of a façade and point to any remedial action that may be required. Assessment should be based on an understanding of the decay processes acting on the stone and the links between environment, stone type, decay process and decay form. This understanding can be enhanced through mapping, quantifying façade decay and carrying out spatial analysis of decay forms (steps 1–4, above).

The lack of standardized language and methods is a problem in condition assessment of monuments and buildings – 'project managers and/or building owners tend to rely on the assessment expertise of a particular contractor/employee, a tendency that has not encouraged the development of a common assessment method and descriptive language/terminology between different "experts" ' (Warke *et al.* 2003, p. 1113). The most detailed methods of recording and rating monumental decay have been developed and an overview provided by Fitzner & Heinrichs (2002). They classify weathering forms on the basis of four broad groups – loss of stone material; discoloration/deposits; detachment; and fissures/deformation. Within these categories are more specific levels, arranged hierarchically. Classification of individual weathering forms (of which there are 75 in all) correlates to damage categories from which possible interventions are determined. The methods employed by Fitzner & Heinrichs are costly, both in terms of time and money, and have been used to assess the state of

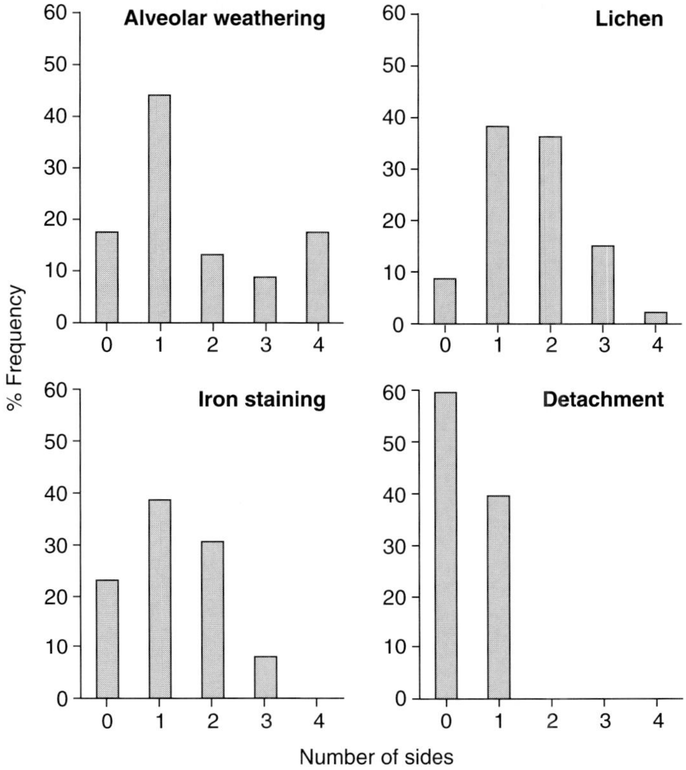


**Fig. 5.** Graphs showing the spatial association/connectivity of decay features on the mapped façade (see Fig. 3).

conservation of high profile world heritage sites, for example, Petra, Jordan (Fitzner & Heinrichs 2002). However, the Venice Charter 1964 (ICOMOS 1964. The Venice charter – international charter for the conservation and restoration of monuments and sites. http://www.icomos.org/venice_charter.html) has stated that the term 'historic monument' applies 'not only to great works of art but also to more modest works of the past which have acquired cultural significance with the passage of time'. There is a lack of standardized condition assessment methods for these smaller, less high profile monuments – a gap that should be filled if these worthy, culturally significant, monuments are to be successfully and efficiently conserved.

Warke *et al.* (2003) proposed a system for assessing and diagnosing stone decay on buildings and monuments, based on one of the most widely used cancer assessment schemes in medicine – the TNM (Tumour Node Metastases) staging system (Hermanek & Sobin 1987). The system describes the spread of a tumour based on unambiguous categories that then correlate to a 'Stage' classification. A likely prognosis can then be ascertained. Warke *et al.* (2003) have adapted this medical procedure to provide a rapid condition assessment method for building stone. The UAS staging system is based on a 'whole-building' approach to the assessment of stone decay, and rather than the time-consuming task of rating each stone making up a façade, a holistic approach is taken. 'Unit' refers to individual stone blocks, 'Area' to adjoining blocks and 'Spread' to the extent of deterioration across an entire façade. The UAS staging system model is shown in Figure 6 – this illustrates how 'U', 'A' and 'S' classifications are made. An overall UAS rating is given, which correlates to different stages of decay (see Table 2) that suggest if and when intervention is required. The UAS staging system is designed to be a relatively rapid and easy-to-use method of assessing the condition of a structure.

The importance of the certainty factor in the UAS staging system scheme should be highlighted, as it becomes especially important in the context of historically and culturally significant monuments. There are three levels of certainty in the system. C1 is a visual assessment only, with no sampling of the stone. As scientists become legally and morally constrained in heritage conservation, sampling is often restricted or denied, making accurate visual assessment vital. C2 comprises a visual

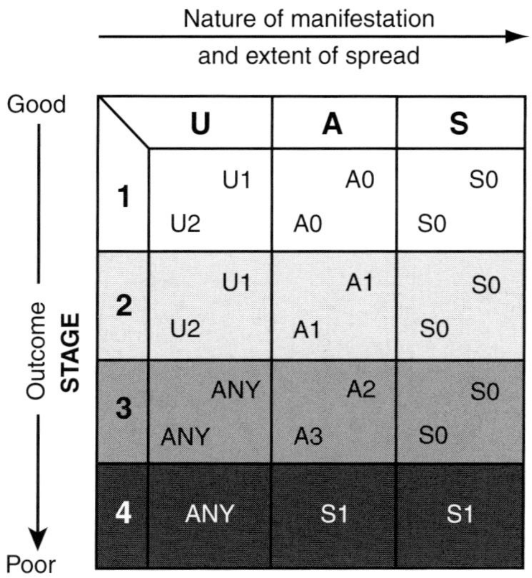


**UNIT**

**U0** No deterioration detectable.

**U1** Some surface alteration with minimal breakdown affecting only parts of individual blocks.

**U2** Well-developed surface alteration and/or obvious breakdown involving whole blocks.

**U3** Well-established surface breakdown with loss of original surfaces affecting approximately 10% of the façade.

**AREA**

**A0** No detectable involvement of surrounding blocks.

**A1** Positive involvement of adjoining blocks affecting less than 10% of the façade.

**A2** Positive involvement of adjoining blocks affecting 10–20% of the façade.

**A3** Extensive localised involvement of adjoining blocks affecting more than 20% of the façade.

**SPREAD**

**S0** Deterioration restricted to the specific sections of the façade.

**S1** Deterioration affecting distant unconnected portions of the façade involving more than 50% of the total surface area.

**Fig. 6.** Model of the UAS condition assessment scheme, reprinted from Warke *et al.* (2003), with permission from Elsevier.

assessment, ion chromatography, atomic absorption spectroscopy and X-ray diffraction. The highest level of certainty, C3, involves all of the above as well as scanning electron microscopy and coring of the stonework (Warke *et al.* 2003, p. 1116). As investigation becomes more destructive, a more conclusive diagnosis of the condition of the monument can be given.

The connectivity analysis of decay carried out in step 4 facilitates a UAS assessment, accurately rating the extent of spread of decay across a façade – the basis of the UAS condition assessment scheme. 'U', 'A' and 'S' values are determined by matching the criteria for decay ratings given in Figure 6 with what is observed on a façade in the field, or with the results of decay mapping. Using decay mapping to reach a UAS rating is more likely to provide a more accurate result than simply carrying out the work in the field – decay mapping and UAS condition assessment have a synergistic relationship. It can be seen from the mapped façade (Fig. 3) that well-developed surface alteration is present on the façade (biological growth, alveolar weather and iron staining are all common features) and that there is breakdown involving whole blocks of sandstone. The Unit rating is, therefore, U2. There is positive involvement of adjoining blocks affecting 10–20% of the façade (an issue already explored in the connectivity analysis), although deterioration is restricted to specific sections of the façade – there are significant areas of the façade that appear unaffected by decay (due to lithological controls). This gives Area and Spread ratings of A2 and S0, respectively. Ultimately the aim of this project is to inform and direct future conservation actions including future intervention. In this context UAS has allowed the classification of the west dorm façade (Fig. 2) as Stage 3 – significant intervention required (U2, A2, S0, certainty level 1 – visual assessment only). Up to 50% of the mapped façade (Fig. 3) shows evidence of deterioration by the decay processes and forms discussed above, although the UAS scheme suggests that appropriate conservation treatment should prolong the life of the façade.

## Discussion

Simple mapping, connectivity analysis of decay and UAS have all been important steps in gaining an understanding of the deterioration of Bonamargy Friary and in condition assessment of the mapped façade (Fig. 3). A study of the connectivity of mapped decay features was combined successfully with (and helped to facilitate) the UAS assessment scheme to give a meaningful rating of the monument condition, based on an understanding of decay processes and controls. Having studied the connectivity of decay, a more informed rating of the 'Unit', 'Area' and 'Spread' decay categories could be made.

The methods used in this paper are an effective way of diagnosing and assessing sandstone monument decay. Two low-intensity stages of analysis (mapping with decay connectivity analysis and UAS) have proved as effective and more efficient than one minutely detailed examination. While time-consuming and labour-intensive techniques

**Table 2.** *Summary guidelines outlining the extent of conservation treatment indicated for each of the four condition stages in the UAS scheme (Warke* et al. *2003)*

| Stage | Extent of intervention required |
|---|---|
| Stage 1 | A façade in this condition would require only localized remedial treatment concentrating on individual stone blocks. A staging classification of one may also indicate that no active intervention is required with only periodic reassessment of the façade advised |
| Stage 2 | Section-specific remedial action would be required in this case, but the extent of intervention should be relatively limited because of the lack of distant involvement within the façade boundaries |
| Stage 3 | Significant intervention will be required with up to 50% of the total façade surface showing evidence of deterioration. Although the extent of deterioration is severe, appropriate conservation treatment should prolong the life expectancy of the structure |
| Stage 4 | Serious deterioration affecting more than 50% of the total façade surface with stone decay detected on unconnected, distant portions of the façade. On a stage 4 category of façade, considerable intervention will be required to restore the stonework. If the structure is of limited historical and/or architectural merit then consideration should be given to the provision of palliative rather than restorative treatment |

such as those employed by Fitzner & Heinrichs (2002) may be applicable to key monuments, the techniques deployed in this paper are entirely appropriate for understanding and diagnosing decay on the smaller monuments so common in our landscape. While Fitzner & Heinrichs (2002) use 75 individual weathering form classifications for high-profile monuments, when studying less high-profile monuments it is suitable that less time-consuming techniques are used and a less detailed knowledge of processes is needed to effectively diagnose and assess decay. Thus, in the case of lower priority monuments, classification of weathering forms prominent at a site can be tailored to each individual case, based on field observation.

When concerned with minute detail and individual blocks there is a tendency to 'not see the building for the stones', so to speak. The UAS element of the study allows us to take a step back and look at the façade as a whole – looking from this point of view allows us to see the bigger picture and to pick up on decay patterns that might be missed in a more minutely detailed study. A clearer understanding of the interactions between environment, stone type and the resulting decay forms has been gained through mapping and analysis of the spatial association of decay. This understanding has underpinned the UAS condition assessment of the west dormitory façade at Bonamargy Friary, enabling an informed and meaningful rating of the state of conservation to be made, and a decay 'stage' to be assigned. Built on this understanding, steps can be made towards identifying event sequences at Bonamargy – sequences of events that produce a recognizable and possibly predictable response in the stone (Brunsden 2001).

This paper also seeks to inform those who have responsibility to carry out targeted intervention at the site (and it is hoped that the techniques used will have a wider application). In the case of 'isolated' decay this may include the stabilization of individual blocks, whereas decay controlled by environmental factors may be best approached through modification of the environment at the stone surface – for example, through inhibiting rainwash, frost incidence or moisture ingress from above and/or below. The integrated approach chosen also allows the identification and recognizes the importance of the exposure history of the building and the inappropriateness of certain past interventions. Results will feed into the Northern Ireland Natural Stone Database and directly into the Environment and Heritage Service. Hopefully this will aid in the selection of conservation measures 'fitted' to the specific stress histories of a façade and its individual blocks, and reduce the possibility that future intervention will exacerbate rather than inhibit decay.

## Conclusions

- A clearer understanding of the interactions between environment, stone type and the resulting decay forms at Bonamargy has been gained through mapping and analysis of the spatial association/connectivity of decay.
- Combining decay mapping and the accompanying spatial association/connectivity analysis with UAS has allowed a meaningful and accurate rating of the state of conservation of Bonamargy Friary.
- The techniques used are efficient and effective, and are appropriate for the decay diagnosis and condition assessment of less high-profile monuments.

The Department of Employment and Learning (DEL) are thanked for their funding. The Environment and Heritage Service (EHS) are thanked for site access. Climate data were made available by the British Atmospheric Data Centre (BADC). Thanks to G. Alexander for preparing figures. British Geomorphological Research Group are thanked for their conference funding. Table 2 was reprinted from Warke *et al.* (2003), with permission from Elsevier.

## References

BALL, J. & YOUNG, M. E. 2004. Comparative assessment of decay and soiling of masonry: methodology and analysis of surveyor variability. *In*: MICHELL, D. J. & SEARLE, D. E. (eds) *Stone Deterioration in Polluted Urban Environments*. Science Publishers, Plymouth, 167–190.

BJELLAND, T. & THORSETH, I. H. 2002. Comparative studies of the lichen-rock interface of four lichens in Vingen, western Norway. *Chemical Geology*, **192**, 81–98.

BRUNSDEN, D. 2001. A critical assessment of the sensitivity concept in geomorphology. *Catena*, **42**, 99–123.

CHEN, J., BLUME, H. & BEYER, L. 2000. Weathering of rocks by lichen colonization – a review. *Catena*, **39**, 121–146.

FITZNER, B. & HEINRICHS, K. 2002. Damage diagnosis on stone monuments – weathering forms, damage categories and damage indices. *In*: PŘIKRYL, R. & VILES, H. A. (eds) *Understanding and Managing Stone Decay*. Karolinum Press, Prague, 11–56.

HERMANEK, P. & SOBIN, L. H. 1987. *TNM Classification of Malignant Tumours*. Springer, Berlin.

MCALISTER, J. J., SMITH, B. J. & CURRAN, J. A. 2003. The use of sequential extraction to examine iron and trace metal mobilisation and the case-hardening of building sandstone: a preliminary investigation. *Microchemical Journal*, **74**, 5–18.

MCCABE, S., SMITH, B. J. & WARKE, P. A. in press. A legacy of mistreatment: understanding the decay of medieval sandstones in NE Ireland. *Building and Environment*.

MCGREEVY, J. P. 1984. A preliminary scanning electron microscope study of honeycomb weathering of sandstone in a coastal environment. *Earth Surface Processes and Landforms*, **10**, 509–518.

PAPIDA, S., MURPHY, W. & MAY, E. 2000. Enhancement of physical weathering of building stones by microbial populations. *International Biodeterioration and Biodegradation*, **46**, 305–317.

SMITH, B. J., TÖRÖK, Á., MCALISTER, J. J. & MEGARRY, Y. 2003. Observations on the factors influencing stability of building stones following contour scaling: a case study of oolitic limestones from Budapest, Hungary. *Building and Environment*, **38**, 1173–1183.

SMITH, B. J., TURKINGTON, A. V. & CURRAN, J. M. 2001. Calcium loading of quartz sandstones during construction: implications for future decay. *Earth Surface Processes and Landforms*, **26**, 877–883.

SMITH, B. J., WARKE, P. A. & CURRAN, J. M. 2004. Implications of climate change and increased 'time-of-wetness' for the soiling and decay of sandstone structures in Belfast, Northern Ireland. *In*: PŘIKRYL, R. (ed.) *Dimension Stone*. Taylor & Francis, London, 9–14.

TURKINGTON, A. V. & PARADISE, T. R. 2005. Sandstone weathering: a century of research and innovation. *Geomorphology*, **67**, 229–253.

TURKINGTON, A. V. & SMITH, B. J. 2004. Interpreting spatial complexity of decay features on a sandstone wall: St. Matthew's Church, Belfast. *In*: SMITH, B. J. & TURKINGTON, A. V. (eds) *Stone Decay – Its Causes and Controls*. Donhead, London, 149–166.

WARKE, P. A. 1996. Inheritance effects in building stone decay. *In*: SMITH, B. J. & WARKE, P. A. (eds) *Processes of Urban Stone Decay*. Donhead, London, 32–43.

WARKE, P. A., CURRAN, J. M., TURKINGTON, A. V. & SMITH, B. J. 2003. Condition assessment for building stone conservation: a staging system approach. *Building and Environment*, **38**, 1113–1123.

YOUNG, M. E., BALL, J. & LAING, R. A. 2004. Quantification of the decay rates of cleaned and soiled building sandstones. *In*: SMITH, B. J. & TURKINGTON, A. V. (eds) *Stone Decay – Its Causes and Controls*. Donhead, London, 13–32.

# Stone decay induced by fire on historic buildings: the case of the cloister of Lisbon Cathedral (Portugal)

A. DIONÍSIO

*Laboratory of Mineralogy and Petrology, Mining and Georesources Department, Centro de Petrologia e Geoquímica, Instituto Superior Técnico, Av. Rovisco Pais, 1049-001 Lisbon, Portugal (e-mail: amelia.dionisio@mail.ist.utl.pt)*

**Abstract:** Lisbon Cathedral, built in Late Romanesque style, is one of the most ancient Portuguese monuments. Its cloister was severely damaged by a fire that occurred in 1755 right after an earthquake. The aim of this investigation is to study stone thermal damage through the application of *in situ* and laboratory techniques. With this study it is possible to identify and characterize (chemically and mineralogically) the main thermal decay forms. Special attention is given to colour modification and granular disintegration. Through the application of an indirect ultrasound method it is verified that only a small number of stone blocks are relatively sound (11%). In terms of chromatic alteration, two factors are considered to explain heat-induced colour modification: the transformation of goethite into hematite and an increase in hematite single crystalline domains. It is also established that the most probable high-temperature range to which the cloister stones were subjected during the fire was 300–350 °C.

The current condition of many historical buildings, as well as sculptures, constructed using natural stone clearly reveals that they are not immune to the deleterious effects of weathering or accidents such as a fire. Fire can severely damage historic buildings and it is therefore important to establish the effects of fire on a building's structure and material in terms of any reduction in strength and change in appearance. Moreover, fire continues to present a serious threat to cultural heritage in all countries. In this connection a key problem in assessing the severity of fire damage is the need for a reliable method of assessing the temperature regime and duration of the fire (Hajpál 2002).

Stones that are most affected stones by fire include granites, sandstones, limestones, dolostones and marbles (Winkler 1997). Visible changes in stone appearance after exposure to fire have been observed and mentioned by several authors: for example the Cathedral of Nidaros in Throndheim (Dahlin 1988), the Cathedral of Mainz, the Cathedral of St Stephen's in Vienna (Kieslinger 1954), St Michael Church in Budapest, the Heidelberg Castle and Lobenfeld Monastery in Germany (Hajpál 1999), S. Domingos Church in Lisbon (Canas 1997), Lisbon Cathedral (Dionísio *et al.* 2005) and Windsor Castle in the UK (Chakrabarti *et al.* 1996). However, little if any attempt appears to have been made to study the detailed nature or causes of the observed damage. Damage phenomena produced by thermal weathering depend in particular on the thermal conductivity of the stone, the expansion coefficients and heat absorption capacities of different minerals, and intrinsic properties of the stone such as porosity, grain size, grain-boundary geometry, shape, preferred orientation, lattice preferred orientation and pre-existing microcracks (Siegesmund *et al.* 2000*a*). The damage is then driven by the heating circumstances: for example highest temperature reached, the rapidity of heating, the temperature distribution through the stone (is the heating one-sided or homogeneous) and the length of time to which the stone is exposed to heat (Galán 1991; Winkler 1997).

According to Hapjál (2002) fires can generally be grouped into two types: (1) small and localized fires that generally do not generate much heat (temperatures are in general lower than 800 °C) and their damaging effects are limited to surface effects and surface staining by smoke; and (2) large and widespread fires that generate higher temperatures (maximum temperatures of 1200 °C) and as a result have a significant effect on the physical–chemical properties of stone. The burning temperature can also be much higher if significant amounts of inflammable materials are present.

Building stone damage by fires is a theme that has mainly been studied in terms of physical phenomena. Several research groups have been working on stone thermal weathering. Marble has received much attention, probably because of its extensive use in cultural heritage. However, other stone materials, including sandstones, limestones, dolerites, gabbros, chalks and serpentinites, have also received attention. Studies that use laboratory simulations to measure the effects of fire/thermal stress on stone weathering and stone durability

*From*: PŘIKRYL, R. & SMITH, B. J. (eds) *Building Stone Decay: From Diagnosis to Conservation.* Geological Society, London, Special Publications, **271**, 87–98.

have been reported by several authors such as Goudie *et al.* (1992), Allison & Goudie (1994), Allison & Bristow (1999), Hapjál (1999, 2002), Royer-Carfagni (1999), Ehling & Köhler (2000), Siegesmund *et al.* (2000*a*, *b*), Malaga-Starzec *et al.* (2002), Weiss *et al.* (2002), Zeisig *et al.* (2002), Hapjál & Török (2004), Dionísio *et al.* (2005) and Török & Hajpál (2005).

The main recorded or interpreted effects of thermal decay induced by fire include spalling, contour scaling and flaking, granular disintegration, microcracks and even fractures. In addition to these effects, in certain types of stones, such as some varieties of limestone, dolostone and sandstone, heating results in a colour change. Our knowledge of colour modification conditions and of the phenomena involved is, however, sparse (Hajpál 2002; Dionísio *et al.* 2005). This is despite the fact that colour modification induced by fire is effectively non-reversible and, together with other physical and chemical damage, can eventually promote highly significant material loss, the possible rupture of stone and destruction of value. This is not to say, however, that society at large is not increasingly aware of a wide range of damage to stone buildings and monuments and the dangers of an irretrievable loss of cultural heritage. Consequently, great efforts (public and political) have been focused in recent years on monument conservation and preservation. Central to these efforts is the recognition of the need for precise damage diagnosis, based on comprehensive characterization, interpretation, rating and prediction of stone damage, if conservation intervention is to be sustainable. Such intervention, according to the Krakow Charter of 2000, take many forms (such as environmental control, maintenance, repair, restoration, renovation and rehabilitation), but must be based evaluation of a range of appropriate technical options and prepared via a cognitive process of knowledge gathering knowledge and a thorough understanding of the building and its materials (Riviera Blanco 2001). These actions have to be carried out in conjunction with systematic research, inspection, control, monitoring and testing. Possible future decay should be foreseen and reported on, and appropriate preventive measures taken.

In the spirit of these protocols for sustainable intervention, the present study focuses on fire-induced stone damage that occurred at one of the most ancient Portuguese monuments – Lisbon Cathedral, specifically its cloister. It is based on the application of *in situ* methods and laboratory analyses carried out on samples collected at the monument and at outcrops located near the cathedral. The ultimate aim of this work is the estimation of burning temperature.

## Lisbon Cathedral

Lisbon Cathedral, built in the Late Romanesque style, is one of the most ancient Portuguese monuments. It was built after the city was retaken from the Moors in 1147 by the first king of Portugal. The cathedral has suffered several modifications as well as restoration works during its history due to the modification of styles and from the occurrence of natural disasters (mainly earthquakes). The cathedral was severely damaged by several earthquakes, particularly by those in 1344, 1531 and 1755. It was reconstructed and repaired frequently, which led to a pleasant coexistence of architectural styles still visible today.

Since antiquity, Lisbon had experienced numerous earthquakes, but never with the intensity of the one in 1755. On 1 November, 1755, All Saints Day, at exactly 9:40 am, when most of the population was in church, a series of increasingly strong earthquakes, the effects of which were felt throughout Europe, surprised Lisbon. In Lisbon its intensity was IX–X (Mercalli scale) in the SE zones (downtown and central hills) and VIII (Mercalli scale) in the other zones of the town. The damage was severe and an area that had been built on top of unstable terrains reclaimed from an ancient branch of Tagus River was destroyed. After the earthquake a tsunami occurred, increasing the damage on the riverside zones, and a fire burned for 5–6 days in the central part of the town, destroying the majority of monuments that had withstood the tremors.

As for most of the buildings from the western part of the city, Lisbon Cathedral was affected by the earthquake, but not severely damaged. The structure remained intact apart from the South Tower and the Lantern Tower, which were partially destroyed. At the cloister the vaults located in the southern corner collapsed. According to several authors, the monument was much more damaged by the fire that occurred after the earthquake rather than by the telluric effect. The fire mainly affected the cloister, where large wooden elements promoted the spread of fire to other areas. This fire affected the stone materials in the vaults of the central gallery, mainly in SE area of the cathedral.

## Experimental studies

### Materials

According to historical information and a geological survey performed by Dionísio (2002), the main stone material used at the cloister of Lisbon Cathedral corresponds to a Miocene limestone of

Burdigalian age, 'calcário gresoso', exploited in the vicinity of the building. 'Calcário gresoso' is a yellow compact limestone (with chromatic co-ordinates $L^* = 72.20$, $a^* = 5.69$, $b^* = 23.82$), with an important detrital and fossiliferous component of low dimension (<0.1 mm). Under polarized microscope, this limestone shows a micritic matrix and a detritic fraction, mostly composed of quartz grains, feldspars, glauconite and quartzite. On the basis of the average Ca/Mg ratio 'calcário gresoso' is classified as a calcitic limestone (Chilingar *et al.* 1967). This carbonate rock does not have large amounts of iron oxides as confirmed by the Fe content in chemical analyses, with an average iron content of less than 1%. Iron in samples is only present as accessory minerals, principally iron oxide/hydroxides or in clay minerals.

## Methodology

In order to evaluate and characterize stone decay phenomena at the cloister of Lisbon Cathedral three sets of parallel studies were conducted:

- *In situ* studies, involving identification of the main decay forms and the use of indirect, non-destructive ultrasound methods to investigate the degree of stone weathering.
- Laboratory studies, involving the chemical and mineralogical characterization of decay products observed at the cloister.
- Laboratory studies to explain the chromatic alteration induced by fire. To achieve this goal a set of samples collected from outcrops near the monument were also studied. This stone material shows similar petrographical features to that used in the monument and was subjected to artificial heating. The results were compared with those obtained from samples collected in the monument.

Ultrasound measurements were conducted using a portable device (STEINKAMP-model BP-7). The method was applied in five of the 19 vaults of the cloister and the values were compared with those obtained from similar sound stone in vaults not affected by the fire. The transducers were positioned at distances that were a multiple of 5 cm, and the measurements obtained (times of passage of ultrasonic pulses) enabled construction of time–distance diagrams and, consequently, the establishment of changes in the condition of stones and depth of weathered layers.

The mineralogical, chemical and elemental composition of the samples taken from the monument was established by X-ray powder diffraction analysis (XRD), Fourier transform infrared spectroscopy (FTIR) and X-ray fluorescence spectrometry with wavelength dispersive system (XRFS/WDS). Heating tests were carried out on cylindrical test samples heated in an Ehret muffle furnace ($t_{max}$: 1100 °C) in an oxidizing environment, at different temperatures (150, 200, 225, 250, 300, 400, 500 and 600 °C) for 24 h. The warming took 30 min and after heat-exposure they were immediately allowed to cool to room temperature. For each temperature step, three stone samples were used. In order to evaluate possible heat-induced colour modification, several analyses were performed before and after heating, including: colour characterization, and chemical and mineralogical analyses, including clay mineralogy and $^{57}Fe$ Mössbauer spectroscopy. Colour characterization tests were carried out with a Minolta portable spectrophotometer (model CM508i) with integrating sphere (diffuse illumination/8° viewing angle), featuring an 8 mm-diameter area of measurement with diffuse illumination by means of xenon flash arc lamp and 10 nm diffuse bandwidth. In order to quantify colour, CIELAB values ($L^*$, $a^*$, $b^*$) for D65 average daylight illuminant including ultraviolet radiation and CIE 2° Standard Observer, according to the ASTM-D2244-79/-D2244-85 standard method, were used. The $L^*$ values refer to the luminosity, which varies from 0 black to 100 white; while $a^*$ and $b^*$ are the chromaticity co-ordinates: $+a^*$ is red, $-a^*$ is green, $+b^*$ is yellow and $-b^*$ is blue.

Colour differences can be determined as follows:

$$\Delta L^* = L_1^* - L_0^*;\ \Delta a^* = a_1^* - a_0^*;\ \Delta b^* = b_1^* - b_0^*$$

where $L_1^*$, $a_1^*$ and $b_1^*$ are the final values, and $L_0^*$, $a_0^*$ and $b_0^*$ are the original ones.

The total colour difference is determined as follows:

$$\Delta E^* = (\Delta L^{*2} + \Delta a^{*2} + \Delta b^{*2})^{1/2}.$$

Mineralogical studies of these samples were carried out by optical microscopy and XRD. The <63 μm fraction of the samples was submitted to an acid attack (20% $HNO_3$) for the study of the clay minerals and associated minerals concentrated in the <2 μm fraction.

The XRD analyses were obtained for the bulk rock and the <2 μm fraction. Oriented aggregates of the <2 μm fraction were obtained by sedimentation from an aqueous suspension onto glass slides and subjected to the following treatments: air-dried, ethylene glycol-solvated and heated. The XRD patterns were obtained using a Philips PW1710 diffractometer with CuK$\alpha$ radiation at 40 kV and 30 mA.

In limestones, as in many other environmental samples, Fe is distributed among a variety of different species, namely Fe oxide/hydroxides, of mainly very small particle size that escape detection by XRD or microscope techniques. Combined with XRD, Mössbauer spectra supply the best solution to the problem of identification. Fe Mössbauer spectra of the natural Fe species are generally unique and characteristic for each individual mineral form. The parameters that allow the identification of the different Fe oxides arise, however, from their magnetic properties. When the particle size of these oxides is very small these properties are only observed in the Mössbauer spectra at very low temperatures. $^{57}$Fe Mössbauer spectroscopy was therefore performed in the transmission mode using a constant acceleration spectrometer and a 25 mCi $^{57}$Co source in Rh matrix, at 295K and 10K. The powder samples of the clays were pressed together with lucite powder into Perspex holders, in order to obtain homogeneous and isotropic Mössbauer absorbers containing *c.* 5 mg/cm$^{-2}$ of natural iron. The velocity scale was calibrated using an $\alpha$-Fe foil at room temperature. Low-temperature spectra were collected using a liquid-helium flow cryostat. The spectra were fitted to Lorentzian lines using a non-linear least-squares computer method (Dionísio *et al.* 2005).

## Results

### *Decay forms*

At Lisbon Cathedral cloister different decay forms, mainly related to fire exposure, can be observed. Among these the most significant is stone colour modification (Fig. 1a). In these blocks a colour

**Fig. 1.** (**a**) Heat-induced stone colour modification in stone blocks from Lisbon Cathedral cloister associated with granular disintegration. (**b**) Colour gradation from dark orange to yellow in some of stone blocks of cloister's vaults. (**c**) Development of chromatic alteration within some stone blocks, reaching in some cases a depth greater than 3 cm. It is also possible to observe important material loss.

**Fig. 1.** *Continued.*

gradation from dark orange to yellow is observed (Fig. 1b). Colour measurements have shown, as revealed by simple visual inspection, that the total colour change ($\Delta E^*$) is significant, with an average value close to 20 (Table 1). These chromatic modifications lead to the orange and dark appearance of stone blocks affected by fire. In this specific case, the chromatic alteration is not only

**Table 1.** *Comparison of chromatic parameters of blocks affected and not affected by fire at Lisbon Cathedral cloister*

| | Stone affected by fire | Stone (apparently) not affected by fire |
|---|---|---|
| $L^*$ (luminosity) | 47.62 ± 9.33 | 60.14 ± 8.75 |
| $a^*$ (red hue) | 13.67 ± 6.01 | 4.55 ± 0.83 |
| $b^*$ (yellow hue) | 24.53 ± 10.23 | 29.03 ± 3.86 |
| ($\Delta E^*$) | 19.78 ± 9.76 | |

superficial but it continues below the surface, in some cases to depths greater than 3 cm (Fig. 1c).

In the Cathedral cloister chromatic alteration is usually associated with granular disintegration leading to significant material losses (Fig. 1a). The loss of surface material is, in some cases, greater than 5 cm in depth (Fig. 1c).

Along the boundary of the heated and unheated area it is also possible to observe spalling. Contour scaling, flaking, microcracks and fractures also occur. Other pathologies, not strictly related to fire but probably induced by air pollution and water infiltration, can also be observed, namely black and brown crusts and salt efflorescences in the vicinity of mortars.

## *Mineralogical composition of decay forms*

The mineralogical composition from some the above-described pathologies is summarized in Table 2. Calcite ($CaCO_3$) and gypsum ($CaSO_4 \cdot 2H_2O$) accompany weathering forms (spalling and granular disintegration) that are directly but not strictly related to fire action. Guanine ($C_5H_5N_5O$) was also identified in the spalling samples. The mineralogy of the crusts (black or brown) is mainly gypsum and they are characteristic of limestones located in sheltered areas where pollutants are retained. The salt efflorescences are mainly composed of syngenite ($K_2Ca(SO_4)_2 \cdot 7H_2O$), gypsum, calcite and nitrate.

## *Results from ultrasound tests*

Through analysis of travel time curves, i.e. through the alignment of experimental points, it should be possible to determine certain characteristics of the *in situ* fired stones related to their degree of weathering, in particular the ultrasound velocity and hence thickness of any subsurface layers corresponding to differential weathering/alteration. In most cases, however, and taking into account the method limitations, no interface was detected and thus stone blocks can be considered as composed of a material that does not change its characteristics with depth. The majority of these stone blocks (90%) presented ultrasound values bellow 2000 m s$^{-1}$. In contrast, when the same method was applied to stone blocks in vaults not affected by fire, it produced ultrasound values of 2551 ± 182 m s$^{-1}$. Figure 2 is an example of a studied vault with an indication of all the blocks tested and their corresponding ultrasound velocity values.

For a small number of studied blocks (5%), also subjected to fire, the alignments of experimental points revealed the presence of two different slopes, i.e. an interface. In these situations, through the analysis of direct-reverse time-travel curves and through the use in each curve of the crossing distance–intercept time, it is possible to verify that this interface is horizontal and to calculate weathering layer thickness: a superficial layer

**Table 2.** *Mineralogical composition of decay products sampled from the cloister of Lisbon Cathedral*

| | Stone decay forms | | | | | |
|---|---|---|---|---|---|---|
| | Saline efflorescences | Black crusts | Brown crusts | Spalling | Granular disintegration with light orange colour | Granular disintegration with dark orange colour |
| Calcite | A | VA | VA | VA | VA | VA |
| Gypsum | P | A | VA | VA | | |
| Singenite | VA | | | | | |
| Quartz | | | | | P | T |
| Feldspar | | | | | T | T |
| Guanine | | | | A | | |
| Hematite | | | | | T | T |
| Goethite | | | | | | T |
| Nitrate | P | P | T | | | |
| Silicate* | | T | T | | | |
| Ca oxalate* | P | P | | | | |

*Only detected by FTIR.

VA, very abundant; A, abundant; P, present; T, traces.

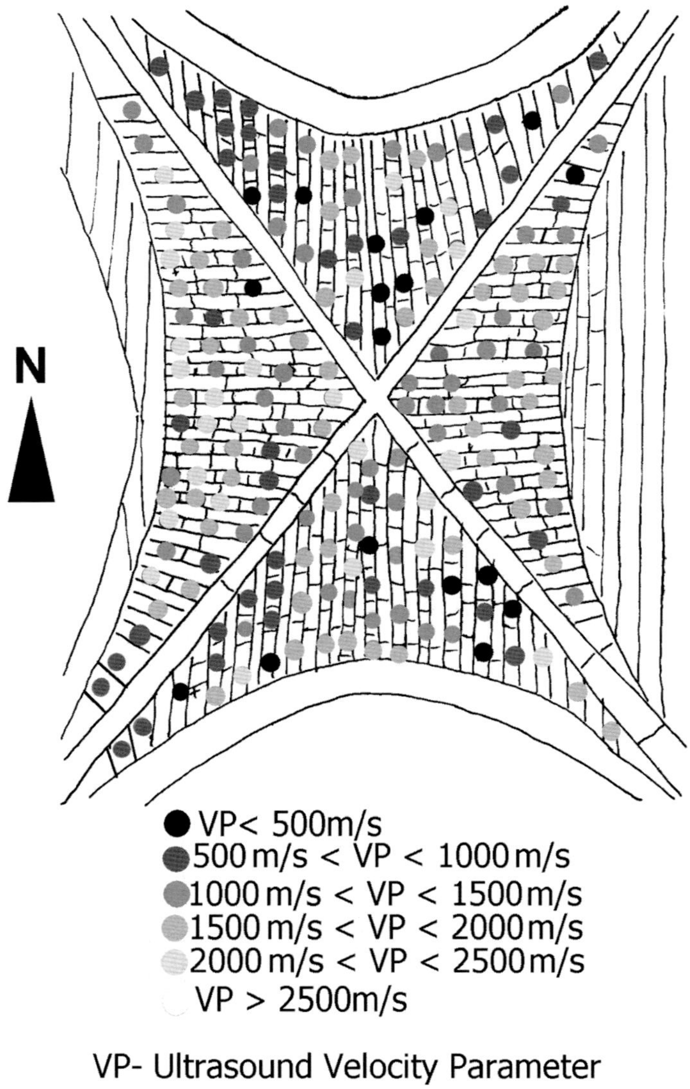


**Fig. 2.** Schematic representation of a vault with indication of the stone blocks studied by means of indirect ultrasound method and respective ultrasound velocity parameters. The vault is located near the Chapel Nossa Senhora da Piedade da Terra Solta (east side of the cloister).

of highly severe decay material exists and has an average thickness of $1.6 \pm 0.3$ cm.

### *Samples subjected to heat-induced laboratory tests*

Regarding the most notable stone decay pathology observed in the cloister, the chromatic alteration, heat-induced tests confirm that, for this type of stone, colour modification occurs (Fig. 3a–c) as result of exposure to a sudden increase in temperature. This colour modification is followed, as in the real situation, by granular disaggregation and the formation of fractures leading to important mass losses as temperature increases (Fig. 3c), mainly for temperatures above 500 °C. Heating also causes darkening and intensification of the red and yellow hue (Fig. 4a–c). This intensification is observed up to 250 °C. At higher temperatures and up to 500 °C these hues are almost stabilized (Fig. 4d). The areas where the reddening is most intense are mainly observed around inter- and intragranular pores and fissures (Fig. 5a–c).

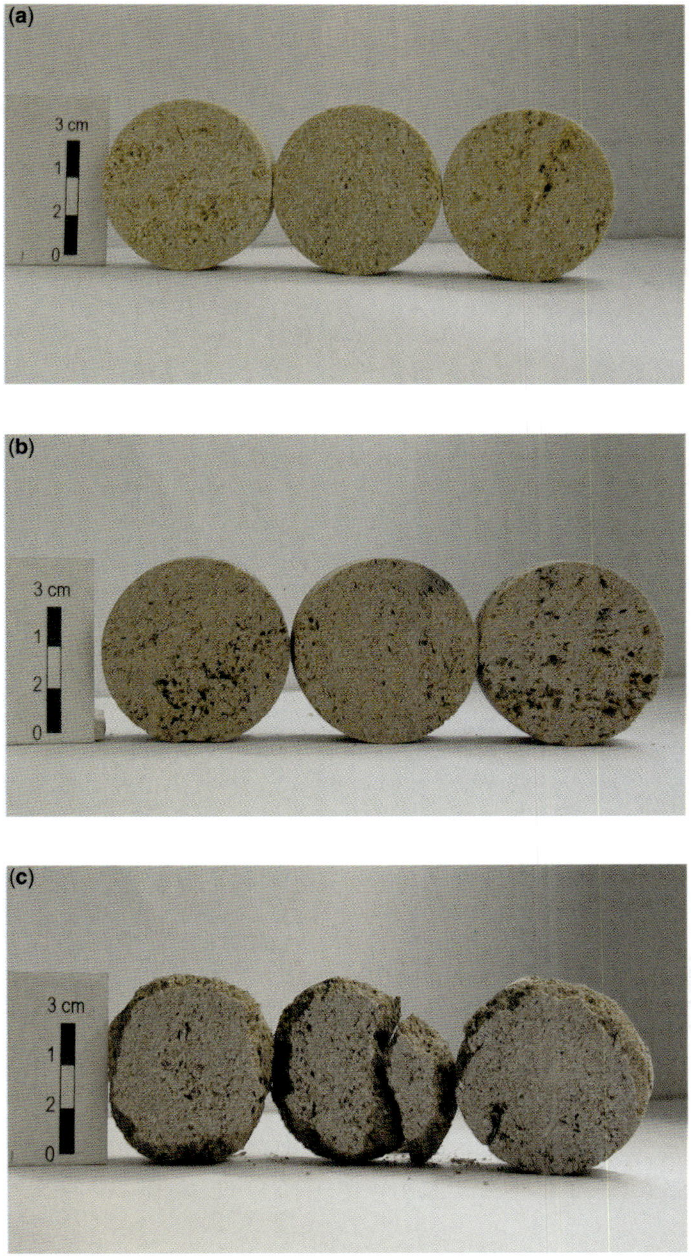


**Fig. 3.** (**a**) Samples of 'calcário gresoso' after heating to 150 °C. (**b**) Samples of 'calcário gresoso' after heating to 250 °C. (**c**) Samples of 'calcário gresoso' after heating to 600 °C.

The bulk mineralogy of 'calcário gresoso' samples is shown in Table 3 and comprises calcite (dominant mineral), quartz and alkali feldspar. Detrital minerals in this stone only occur in trace amounts. Dolomite and plagioclase are observed only in some samples and this is related to the heterogeneity of the stone. The presence of hematite is minimal. According to XRD the average percentage of phyllosilicates is very low and they are not detected in the majority of the samples.

The $<2$ μm fraction mineralogy (Table 3) varies significantly between untreated (quarry) samples

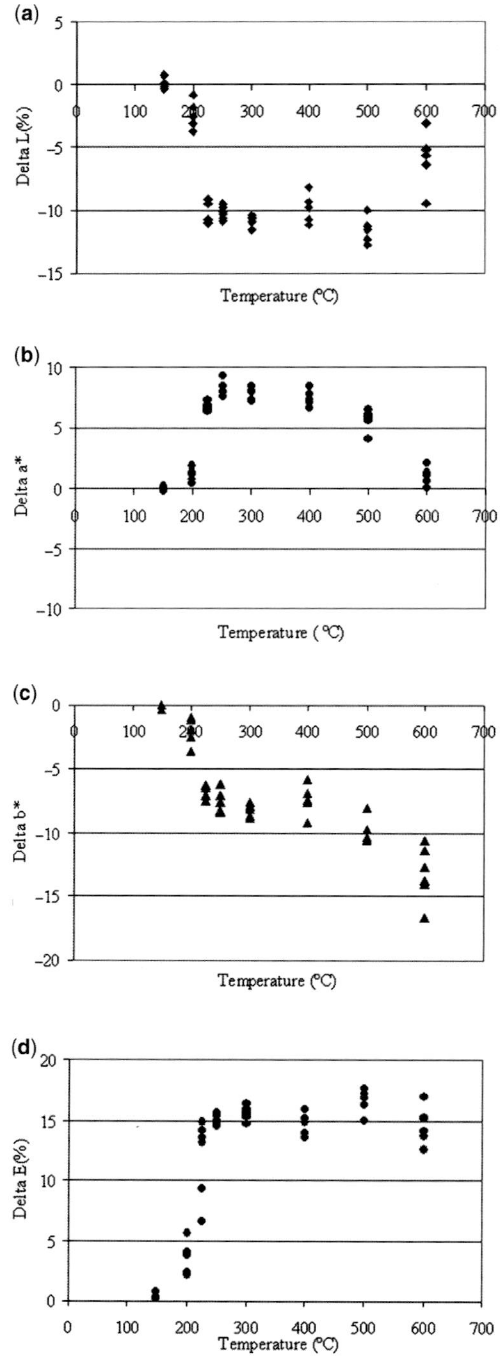


**Fig. 4.** (**a**) Chromatic co-ordinate $L$ variations values ($\Delta L^* = L^*_{\text{final}} - L^*_{\text{initial}}$). (**b**) Chromatic co-ordinate $a^*$ variations values ($\Delta a = a^*_{\text{final}} - a^*_{\text{initial}}$). (**c**) Chromatic co-ordinate $b^*$ variations values ($\Delta b^* = b^*_{\text{final}} - b^*_{\text{initial}}$). (**d**) $E$ variations values ($\Delta E = \sqrt{\Delta L^2 + \Delta a^2 + \Delta b^2}$).

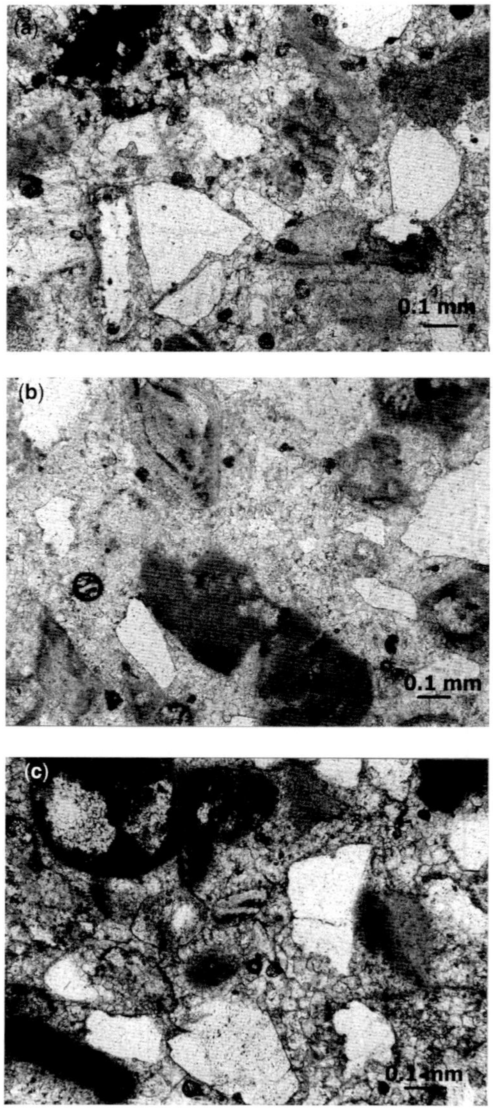


**Fig. 5.** (**a**) Thin section from sample of 'calcário gresoso' heated to 150 °C. (**b**) Thin section from sample of 'calcário gresoso' heated to 300 °C. (**c**) Thin section from sample of 'calcário gresoso' heated to 500 °C.

and heated samples. In samples subjected to heating (Table 3) hematite is dominant, followed by clay minerals. Goethite occurs only in one sample heated to 250 °C. In samples heated to 250 and 300 °C or 600 °C the dominant iron oxyhydroxide is disordered hematite, which contrasts with the control samples where the dominant iron oxyhydroxide is goethite. This indicates the thermal transformation of goethite to hematite by heating.

**Table 3.** *Estimation of the mineralogical composition by DRX of some representative samples (untreated and subjected to heating in the laboratory) of 'calcário gresoso'*

| Temperature (°C) | | Bull rock | | | | | | | | | <2 μm fraction (%) | | | | |
|---|---|---|---|---|---|---|---|---|---|---|---|---|---|---|---|
| | Samples | Cal | Dol | Qz | Fk | Pl | Mica | Clay | Goe | Hem | Ill | Sm | Kao | Goe | Hem |
| 250 | CSG13B | +++ | – | tr | tr | – | – | – | – | tr | + | + | + | + | ++ |
| | CSG14B | +++ | – | + | + | – | – | – | – | tr | + | + | + | – | +++ |
| | CSG15B | +++ | – | + | tr | tr | tr | – | – | tr | + | + | + | – | ++ |
| 300 | CSG16B | +++ | – | + | tr | – | – | – | tr | – | + | + | ++ | – | ++ |
| | CSG17B | +++ | tr | + | tr | tr | – | – | – | tr | + | + | ++ | – | + |
| | CSG18B | +++ | tr | tr | tr | – | – | – | – | – | + | + | ++ | tr | ++ |
| 600 | CSG25B | +++ | tr | + | tr | tr | – | – | – | tr | + | – | – | – | +++ |
| | CSG26B | +++ | tr | + | tr | – | – | – | – | tr | + | – | – | – | +++ |
| | CSG27B | +++ | tr | + | tr | – | – | – | – | tr | ++ | – | – | – | ++ |
| | CG1§ | +++ | – | + | tr | tr | tr | – | – | – | ++ | tr | + | ++ | – |
| | CG2§ | +++ | – | + | + | tr | tr | – | – | – | + | + | + | ++ | – |
| | CG3§ | +++ | – | + | tr | tr | – | – | – | – | + | + | + | ++ | – |

Cal, calcite; Dol, dolomite; Qz, quartz; Fk, K-feldspar; Pl, plagioclase; Goe, goethite; Hem, hematite; Ill, Illite; Sm, Smectite; Kao, Kaolinite; –, not detected; tr, <5%; +, 5–25%; ++, 25–75%; +++, >75%.
§Control samples.

The $^{57}$Fe Mössbauer spectra of the total sample clearly show that $Fe^{2+}$ in the carbonate structure is preserved up to 600 °C. Data on the <2 μm fraction of the samples further show that the oxidation of $Fe^{2+}$ in smectites is almost complete at 250 °C. At 600 °C no $Fe^{2+}$ is observed in silicate structures. At least part of the initial $Fe^{2+}$ in smectites, which is oxidized, is converted to hematite. The degree of crystallinity of the iron oxides clearly increases with temperature. These results agree with those shown in the XRD patterns.

### *Heat-induced modification of monument samples*

Similar mineralogical and chemical analyses were performed on 16 samples collected from the cloister. These samples presented different chromatic alteration levels. The estimation by XRD of the bulk rock and the <2 μm fractions of monument samples can be found in Table 4. Calcite is the dominant mineral followed by quartz. Dolomite, K-feldspar, plagioclase and mica are present in

**Table 4.** *Estimation of the mineralogical composition by DRX of the monument samples of 'calcário gresoso'*

| Samples | Bull rock | | | | | | | | | | <2 μm fraction (%) | | | | |
|---|---|---|---|---|---|---|---|---|---|---|---|---|---|---|---|
| | Cal | Dol | Qz | Fk | Pl | Mica | Clay | Goe | Hem | Gy | Ill | Sm | Kao | Goe | Hem |
| CL1 | ++ | – | ++ | tr | tr | – | – | – | tr | tr | ++ | ++ | + | – | tr |
| CL2 | ++ | – | + | tr | tr | tr | – | – | tr | tr | ++ | ++ | + | – | tr |
| CL3 | ++ | – | + | tr | tr | tr | – | – | – | tr | ++ | ++ | + | – | – |
| CL4 | ++ | tr | + | + | – | – | – | – | – | – | ++ | ++ | + | – | – |
| CL5 | ++ | tr | + | + | tr | tr | tr | – | tr | tr | ++ | tr | tr | – | ++ |
| CL6 | ++ | tr | + | tr | tr | tr | tr | – | tr | tr | ++ | + | + | – | + |
| CL7 | ++ | – | + | + | – | – | – | – | – | tr | ++ | ++ | + | – | – |
| CL8 | ++ | tr | + | + | – | – | – | – | – | tr | ++ | ++ | + | – | – |
| CL9 | ++ | – | + | + | tr | tr | – | tr | – | tr | ++ | + | + | ++ | – |
| CL9A | ++ | tr | ++ | ++ | tr | tr | – | – | – | tr | ++ | + | + | + | – |
| CL10 | ++ | tr | ++ | + | tr | tr | tr | tr | – | – | + | + | ++ | ++ | – |
| CL11 | ++ | – | + | tr | tr | – | – | – | – | – | ++ | + | + | – | – |
| CL12 | ++ | tr | + | + | tr | – | – | – | – | – | ++ | ++ | ++ | tr | – |
| CL13 | +++ | tr | tr | tr | – | – | – | – | – | tr | ++ | + | ++ | ++ | – |
| CL14 | +++ | tr | tr | tr | – | – | – | tr | – | tr | + | tr | ++ | ++ | – |
| CL15 | ++ | tr | + | tr | tr | – | – | tr | – | tr | ++ | ++ | ++ | tr | – |

Cal, calcite; Dol, dolomite; Qz, quartz; Fk, K-feldspar; Pl, plagioclase; Goe, goethite; Hem, hematite; Gy, gypsum; Ill, illite; Sm, smectite; Kao, Kaolinite.–, not detected; tr, <5%; +, 5–25%; ++, 25–75%; +++, >75%.

some samples in low quantities. Clay minerals, goethite and hematite appear in very low quantities in four out of the 16 studied samples. Gypsum occurs also in very low amounts ($<4\%$) in these monuments stones. The other minerals, except hematite, are inherited minerals from the quarries. Illite, smectite and kaolinite, as well as iron oxides (goethite and hematite), were the minerals identified in the $<2$ μm fraction (Table 4). The average percentages show that illite is the dominant clay mineral followed by smectite and kaolinite. Goethite occurs in seven out of the 16 studied samples, while hematite is present only in three samples.

## Discussion and conclusions

The mineralogies of the samples submitted to heat-induced tests show that stone colour modification is mainly due to the thermal transformation of iron minerals: goethite into hematite and with an increasing $Fe^{3+}/(Fe^{2+} + Fe^{3+})$ fraction of the iron not included in the carbonate structure, as well as with the average grain size of the $Fe^{3+}$ oxyhydroxides. Even in a stone with a very low iron content (0.56% w/w of $Fe_2O_3$) colour modifications are easily observed for temperatures above 250 °C.

The presence of smectite in monument samples is very important as a fingerprint of the temperature achieved during the fire. According to Brown & Brindley (1980) the interlayer water is removed from the smectite structure by heating at 300–350 °C and the original 15 Å reflection is replaced by one at 10 Å. Furthermore, as pointed out by Velde (1992), temperatures ranging from 200 to 300 °C are needed to remove the glycol and water from the interlayer site in the smectites. In fact smectite is present both in quarry samples and in samples subjected to 250 and 300 °C in laboratory tests. It also occurs in all monument samples, as reported in Tables 3 and 4, thus confirming that the temperatures reached during the fire were below 350 °C, as at this and higher temperatures smectite could not be identified by XRD. As such, these results clearly explain the absence of new mineralogical phases related to high temperatures, the so-called 'high-temperature crystalline phase' made up of Ca, Al-silicates like gehlenite ($Ca_2Al_2SiO_7$) or mullite ($Al_6Si_2O_{13}$).

Through the laboratory studies it was possible to establish that colour measurements are useful to indicate the degree of reaction caused by a sudden increase in temperature. However, these measurements alone cannot be considered a good fingerprint of the temperature that affected the monument stones as colour tends to stabilize when a specific temperature is achieved.

Granular disintegration associated, in this specific case, with chromatic alteration clearly reveals that the thermal variation exceeded the elastic limit of the minerals and the cohesive strength between the grains. Any satisfactory explanation of this phenomenon should include consideration of the stone polycrystalline microstructure. Indeed, the mineralogy of 'calcário gresoso' comprises calcite (dominant mineral), quartz and alkali feldspar. Heating of this stone must therefore result in the uneven expansion of these minerals. As a consequence damage could result from the highly anisotropic thermal expansion coefficient $\alpha$ of calcite, i.e. extreme expansion parallel and contraction normal to the crystallographic *c*-axis, coupled with the presence of quartz that not only presents different linear expansion along the long axis v. the expansion of the short axis, but also a relatively high total volume increase with temperature (Galán 1991; Winkler 1997; Royer-Carfagni 1999; Siegesmund *et al.* 2000*a*, *b*). As a result, significant stresses are produced and the modulus of elasticity is reduced, leading to a progressive loss of cohesion along grain boundaries.

From ultrasound measurements it is possible to conclude that areas affected by fire, and subsequently by granular disintegration, are severely damaged, without any special pattern to the ultrasound velocities. This situation occurs for fire temperature (estimated) lower than 300–350 °C. The situation is not so uncommon, even for temperatures below 100 °C, as mentioned by several authors for stones such as marble during daily/seasonal cycles of temperature (Galán 1991; Winkler 1997; Zeisig *et al.* 2002). Where spalling is observed in areas that were exposed to fire, this is thought to relate to the low thermal conductivity of the stone, which, in turn, produces strong temperature gradients between the exposed surface and the subsurface. Under these conditions, shearing stresses are developed co-incident with greatest thermal gradient, where the compressive forces in the surface layer of the stone exceed compressive strength so that the surface layer relaxes and expands. As a result, the surface layer may shear and a spall, scale or blister forms (Galán 1991; Hajpál 2002). At stone edges, heating can converge from more than one direction and contour scaling can occur (Hajpál 2002). Spalling and granular disintegration can also be related to factors other than fire action, especially as gypsum was identified in decay products. The occurrence of gypsum is thought to relate to air pollution, where sulphatation processes (reaction between $SO_2$ gas deposited on or in the stone and the stone that occurs in the presence of a catalyst and under high relative humidity) are responsible for gypsum formation. Gypsum is also the main constituent of black and brown crusts. Guanine was also identified in the spalling samples and its origin is probably related to the biochemical action of the numerous pigeons that live inside the cloister.

Microcracks and fractures also occur as a result of dimension and shape change induced by differential linear and volumetric expansion of confined elements (Winkler 1997). Facaoaru & Lugnani (1993) considered that the harmful effect of fire on stone materials results from the development of fractures that are at first restricted to the weathering layer and then extended to the entire stone surface, followed by significant changes in mechanical properties. It is also important to note, however, that, as indicated in this study, fire damage probably accelerates the development of other forms of decay, as well as triggering new ones, especially chromatic alteration.

This study has been partially financed by the Portuguese Project POCTI/CTA/38339/2001, by IPPAR (Programa de estudos integrados da Sé de Lisboa) and by Centro de Petrologia e Geoquímica do Instituto Superior Técnico.

## References

ALLISON, R. J. & BRISTOW, G. E. 1999. The effects of fire on rock weathering: some futher considerations of laboratory experimental simulation. *Earth Surface Processes and Landforms*, **24**, 707–713.

ALLISON, R. J. & GOUDIE, A. S. 1994. The effect of fire on rock weathering: an experimental study. *In*: ROBINSON, D. A. & WILLIAMS, R. B. G. (eds) *Rock Weathering and Landform Evolution*. Wiley, Chichester, 41–56.

BROWN, G. & BRINDLEY, G. W. 1980. X-Ray diffraction procedures for clay mineral identification. *In: Crystal Structures of Clay Minerals and their X-ray Identification*, Mineralogical Society Monograph, 305–359.

CANAS, J. F. 1997. A Igreja de São Domingos de Lisboa. *Monumentos*, **6**, 68–71.

CHAKRABARTI, B., YATES, T. & LEWRY, A. 1996. Effect of fire damage on natural stone work in buildings. *Construction and Building Materials*, **10**, 539–544.

CHILINGAR, G. V., BISSELL, H. J. & FAIRBRIDGE, R. W. 1967. *Carbonate Rocks. Physical and Chemical Aspects*. Developments in Sedimentology 9B. Elsevier, New York.

DAHLIN, E. 1998. The problems of conserving fire-damaged stone from the cathedral of Nidaros. *In*: CIABACH, J. (ed.) 6th *International Congress on Deterioration and Conservation of Stone*. Nicholas Copernicus University Press, Torun, 732–739.

DIONÍSIO, A., RODRIGUES, M. *ET AL.* 2005. Study of heat induced colour modifications in monument stones. *International Journal for Restoration of Buildings and Monuments*, **4**, 199–210.

DIONÍSIO, M. A. A. R. D. 2002. *A degradação da pedra em edifícios históricos. O caso da Sé de Lisboa*. PhD thesis, Lisbon Technical University.

EHLING, A. & KÖHLER, W. 2000. Fire damaged natural bulding stones. *In*: RAMMLMAIR, D., MEDERER, J., OBERTHÜR, T., HEIMANN, R. B. & ENTNGHAUS, H. (eds) *Applied Mineralogy in Research Economy, Technology, Ecology and Culture (ICAM2000)*, Volume 2. A.A. Balkema, Rotterdam, 975–978.

FACAOARU, I. & LUGNANI, C. 1993. Contributions to the diagnosis of stone and concrete historical structures using non-destructive techniques. *In*: Rilem/ Unesco (eds) *Proceedings of the 21st International Congress on Conservation of Stone and Other Materials*, Volume 1, E&FN Spon, London, 238–251.

GALÁN, E. 1991. The influence of temperature changes on stone decay. *In*: ZEZZA, F. (ed.) *Weathering and Air Pollution*. First Course. Community of Mediterranean Universities, University School of Monument Conservation, Mario Adda Editore, Bari, 119–129.

GOUDIE, A. S., ALLISON, R. J. & MCLAREN, S. J. 1992. The relations between modulus of elasticity and temperature in the context of the experimental simulation of rock weathering by fire. *Earth Surface Processes and Landforms*, **17**, 605–615.

HAJPÁL, M. 1999. Behaviour of Sandstones of Historical Monuments Under Thermal Influence. Periodica Polytechnical Budapest, 207–218.

HAJPÁL, M. 2002. Changes in sandstones of historical monuments exposed to fire or high temperature. *Fire Technology*, **38**, 373–382.

HAJPÁL, M. & TÖRÖK, A. 2004. Mineralogical and colour changes of quartz sandstones by heat. *Environmental Geology*, **46**, 311–322.

KIESLINGER, A. 1954. Brandeinwirkungen auf Natursteine. *Schweizer Archiv*, **9**, 305–308.

MALAGA-STARZEC, L., LINDQVIST, J. E. & SCHOUENBORG, B. 2002. Experimental study on the variation in porosity of marble as a function of temperature. *In*: SIEGESMUND, S., WEISS, T. & VOLLBRECHT, A. (eds) *Natural Stone, Weathering Phenomena, Conservation Strategies and Case Studies*. Geological Society, London, Special Publications, **205**, 81–88.

RIVIERA BLANCO, J. 2001. De la Carta de Venecia a la Carta de Cracóvia. *In: De Varia Restauratione. Teoria e Historia de la Restauración Arquitectónica*, R&R, Valladolid, 199–206.

ROYER-CARFAGNI, G. F. 1999. On the thermal degradation of marble. *International Journal of Rock Mechanics and Mining Sciences*, **36**, 119–126.

SIEGESMUND, S., ULLEMEYER, K., WEISS, T. & TSCHEGG, E. K. 2000*a*. Physical weathering of marbles caused by anisotropic thermal expansion. *International Journal of Earth Sciences*, **89**, 170–182.

SIEGESMUND, S., WEISS, T. & TSCHEGG, E. K. 2000*b*. Control of marble weathering by thermal expansion and rock fabricks. *In*: FASSINA, V. (ed.) *9th International Congress on Deterioration and Conservation of Stone*, Volume 1. Elsevier, Venice, 205–213.

TÖRÖK, A. & HAJPÁL, M. 2005. Effect of temperature changes on the mineralogy and physical properties of sandstones. A laboratory study. *Journal for*

*Restoration of Buildings and Monuments*, **4**, 211–218.

VELDE, B. 1992. *Introduction to Clay Minerals – Chemistry, Origins, Uses and Environmental Significance*. Chapman & Hall, London.

WEISS, T., SIEGESMUND, S. & FULLER, E. R. 2002. Thermal stresses and microcracking in calcite and dolomite marble via finite element modelling. *In*: SIEGESMUND, S., WEISS, T. & VOLLBRECHT, A. (eds) *Natural Stone, Weathering Phenomena, Conservation Strategies and Case Studies*. Geological Society, London, Special Publications, **205**, 89–102.

WINKLER, E. M. 1997. *Stone in Architecture. Properties, Durability*. Springer, Berlin.

ZEISIG, A., SIEGESMUND, S. & WEISS, T. 2002. Thermal expansion and its control on the durability of marbles. *In*: SIEGESMUND, S., WEISS, T. & VOLLBRECHT, A. (eds) *Natural Stone, Weathering Phenomena, Conservation Strategies and Case Studies*. Geological Society, London, Special Publications, **205**, 65–80.

# The weathering and weatherability of Basílica da Estrela stones, Lisbon, Portugal

C. A. M. FIGUEIREDO, L. AIRES-BARROS, M. J. BASTO, R. C. GRAÇA & A. MAURÍCIO

*Centre of Petrology and Geochemistry, CEPGIST, IST, Av. Rovisco Pais, 1049–001, Lisbon, Portugal (e-mail: nickfig@popsrv.ist.utl.pt)*

**Abstract:** This paper presents a study of stone decay on the Basílica da Estrela, the most famous 18th century monument in the city of Lisbon, Portugal. It was built with Jurassic and Cretaceous limestones from the surroundings of Lisbon. Different approaches were used to establish the typology, causes and processes of the major weathering forms. Limestone samples from ancient quarries, salt efflorescences and disintegrated stone material from the interior of the church were characterized by chemical, mineralogical and petrographical analyses. Limestone physical properties related to fluid percolation were also determined. Detailed surveys of stone decay phenomena were carried out on the monument. Textures of grey-level images representative of the weathering forms were analysed by image analysis through covariance and granulometry operators. An HIDSPEC computational hydrogeochemical model, phase and graphical diagrams, and multivariate statistical analysis were used for water–rock interaction studies. Physical weathering forms prevail inside the church. The yellow Cretaceous limestone is the most deteriorated stone. This observation compares well with its intrinsic properties. The weathering is determined by the stone structures, such as stilolytes and fossils, and architectural features (geometry and surface finish). Soluble salts such as trona and thenardite were only found in a very small area inside the church.

The Basílica da Estrela is the most important 18th century monument in the city of Lisbon, Portugal. Its construction began in 1779, by order of Queen Mary I, and it was finished 11 years later in the neo-classical style with some baroque elements (Figueiredo 1999).

Nowadays it is located in the centre of the city in a moderately polluted area about 15 km from the sea. Several types of Jurassic and Cretaceous limestone extracted from quarries near Lisbon (western Lisbon–Sintra area) were used in its construction. Some of the ancient quarries are still available for sampling to enable laboratory investigations of stone characteristics. The interior of the Basílica da Estrela is covered with beige, greyish-blue, rose and ochre limestones. These limestone varieties are locally known as 'Lioz' (beige–white limestone), 'Encarnadão' de Negrais (red–pink limestone), 'Amarelo de Negrais' (yellowish limestone) and 'Azul de Sintra' (the greyish-blue limestone).

Over the last 26 years, since 1979, the exterior and interior have been cleaned and repaired on six occasions (1979, 1981, 1982, 1984, 1986–1987 and 2000–2001). The main problem has been the infiltration of rainwater through the terrace of the church.

When used in a building, in addition to natural influences, stone is subject to damage induced by the quarrying and stone preparation process, building processes, inappropriate choice material for a given purpose, and urban and industrial pollution (Amoroso & Fassina 1983). This means that cultural heritage cannot be preserved for future generations without proper care and protective measures. Owing to the fact that cultural heritage objects are unique, one of the main problems to be solved when investigating them scientifically is to reduce as much as possible any damage caused by the scientific study itself. Because of this, laboratory determination of chemical, mineralogical and petrophysical properties of the building materials was restricted to samples from historic quarries. In addition, image analysis was used to create a mathematical model of surface that could be used to assess the textural characteristics of unweathered and flat decayed stone surfaces. These surfaces were built in 'Amarelo de Negrais' Cretaceous limestone and line the walls located inside the church (Fig. 1). Detailed visual/tactile observation shows that these flat vertical panels are more or less deteriorated depending on their position inside the church. In spite of the chemical and physical heterogeneity of natural building stones, it is hoped that the image analysis will establishing clear relationships between the intrinsic properties of the stones and the decay processes affecting them, both in the laboratory and on the monument.

*From*: PŘIKRYL, R. & SMITH, B. J. (eds) *Building Stone Decay: From Diagnosis to Conservation*. Geological Society, London, Special Publications, **271**, 99–107.

**Fig. 1.** General view of the panels located inside the Basílica da Estrela and built with the yellow limestone (structure in the middle of the photograph).

## Methodology

To establish the typology, causes and processes of the major weathering forms observed different approaches were used. The technological characterization of stone samples obtained from ancient quarries was based on chemical, mineralogical, petrographical and physical studies.

The chemical composition of the limestones and weathering products (salt efflorescences and disintegrated stone material) collected from the interior of the Basílica da Estrela was obtained by a variety of chemical analytical techniques (gravimetric analysis, colorimetry and atomic absorption spectrometry (AAS/AES)). X-ray diffraction (XRD) and Fourier transform infrared spectroscopy (FTIR) techniques were used for mineralogical characterization. Thin sections of the limestones were used to determine their textural and mineralogical characteristics.

European (prEN 1936 and prEN 1925) and French (NF B 10–504 1973) standards were used to determine limestone properties related to fluid transport in porous media: open porosity ($N_a$,%), total porosity ($N_t$,%), Hirschwald coefficient ($S_{48}$,%) and capillarity coefficient ($C$, g m$^{-2}$. s$^{0.5}$).

On the monument itself detailed surveys of exterior and interior stone decay phenomena were carried out. Digital image processing and analysis of photographic records were used as a non-destructive and contact-free methods to characterize the weathering of stone surfaces. An image analysis approach based on the Mathematical Morphology operators of covariance and granulometry of grey-tone images (Serra 1982; Coster & Chermant 1985; Benali 1986) was used for the quantitative analysis of textures of static grey-tone CCD (charge-coupled device) video camera images representative of flat stone surfaces. The rationale behind the choice of these morphological operators and corresponding measures is related to their properties and in particular to the nature of the problem at hand. Visual inspection showed that weathering processes break down the order and organization of textures of stone surfaces introducing changes in the topography (roughness) and optical characteristics of the textures of the surfaces. Because stone surfaces exhibit a wide range of textures, different morphological approaches were required for their quantitative description. This reasonable hypothesis enables the application of the granulometric and covariance morphological operators to the analysis of the texture of grey-tone images corresponding to optical images of the macroscopic stone flat panels with different degrees of weathering. Grey-tone image granulometry allows the size distribution of the texture elements to be estimated (Fig. 2). This provides information about the dominant size of the texture elements in an initial characterization of the texture. In seeking to measure an image texture variations in grey levels within the image are quantified using information on the size and spatial arrangements of the texture elements (Benali 1986; Tomita & Tsuji 1990). The size distribution of the dimensions of the stone texture elements can be directly obtained from the grey-level images. The granulometry analysis of the texture of the grey-tone images was performed using the concept of morphological opening (erosion followed by dilation) of a function ($f(x)$) (Serra 1982; González & Woods 1997; Soille 2003; Figueiredo *et al.* 2005). Given a numerical function $f(x)$ (the grey-level image), its granulometric distribution (cumulative distribution function, CDF, curves in Fig. 2) by opening of increasing size $\lambda$, with $\lambda \in R_0^+$, denoted $\psi(f, \lambda)$, is given by the following equation (Serra 1982; Coster & Chermant 1985; Benali 1986):

$$\psi(f, \lambda) = \frac{\int_Z f(x)\mathrm{d}x - \int_Z f_{\lambda B}(x)\mathrm{d}x}{\int_Z f(x)\mathrm{d}x}, \ \psi(f, 0) = 0$$

where $Z$ is the measuring field and $f_{\lambda B}(x)$ is the opening of the function $f(x)$ by the structuring

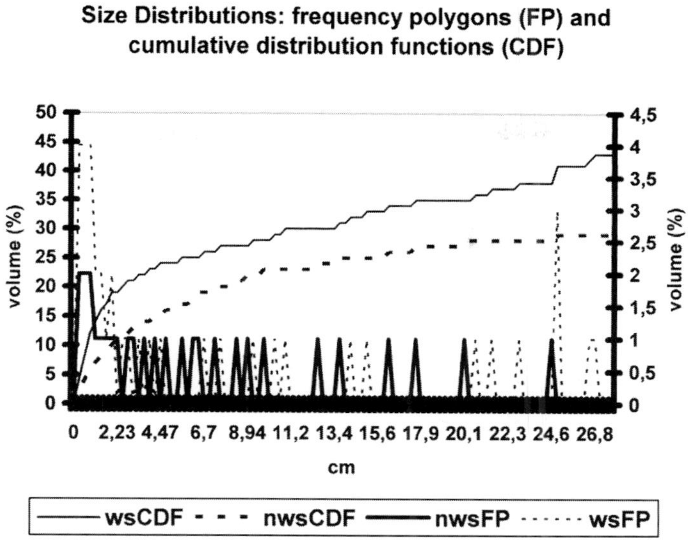


**Fig. 2.** Experimental histograms and distribution function plots obtained from grey-level images of unweathered (nws) and weathered (ws) stones.

element $B$ of size $\lambda$ centred in point $x$, such that:

$$f_{\lambda B}(x) = \sup \inf\{f(z):y \in \check{B}_x, z \in B_y\}$$

where sup and inf are, respectively, the supremum and infimum of a function; $\check{B}_x$ and $B_y$ are the structuring elements and are symmetrical centred at points $x$ and $y$, respectively. The opening has granulometric properties since it is increasing, anti-extensive and idempotent. The opening plays a key role in the construction of morphological and grey-level image filters. This operation cuts the narrow isthmuses and suppresses the small islands and the sharp peaks of the function $f(x)$ (grey-level image). For digital images $\lambda \in N_0$ and the density law, i.e. the granulometric density function (FP curves in Fig. 2), denote $\Delta\psi(f, \lambda)$, which is given by:

$$\Delta\psi(f, \lambda) = \psi(f, \lambda + 1) - \psi(f, \lambda).$$

This gives the volume relative variation by opening of $f(x)$ with respect to a structuring element $B$ of a given size $\lambda$. This granulometric is useful in principle because it can help in discriminating different textures associated with different weathering states of the observed panels.

In probability theory, the covariance $Cov(h)$ of a stationary random function $f(x)$ (the grey-level image) is given by the classical definition (Serra 1982):

$$Cov(h) = E\{[f(x) - p][f(x+h) - p]\} = E[f(x)f(x+h)] - p^2$$

where $p = E(f(x))$ is the mathematical expectation of a random function and $h$ is the distance between the points $x, (x+h)$. In various applications, however, the object investigated does not directly correspond to a particular set, i.e. to a black or white function, as optical densities or reflectance are continuous functions. In such cases, the density of light $f(x)$ can be partitioned into classes by thresholds of successive levels. So the continuous function $f(x)$ can, in practice, be approximated by a direct sum of step functions $a_i k_i(x)$ such that:

$$f(x) = \sum_{i=1}^{n} a_i k_i(x), \quad k_i(x)k_j(x) = \begin{cases} 0 & \text{if } i \neq j \\ 1 & \text{if } i = j \end{cases}$$

where $a_i$ is the mean grey intensity of the grey class $i$.

When applied to grey-tone images the covariance allows estimation of, in a given direction, the average size of texture elements, their spatial distribution and the mean degree of discontinuity of the whole texture. If the covariance is estimated for more than one direction a factor of the texture anisotropy can be calculated. When studying several objects, the experimental covariance function should be standardized (experimental correlation function, $Corr(h)$):

$$Corr(h) = \frac{Cov(h)}{Cov(0)}$$

where $-1 < Corr(h) < 1$.

This enables the results obtained for different directions within an image, and also

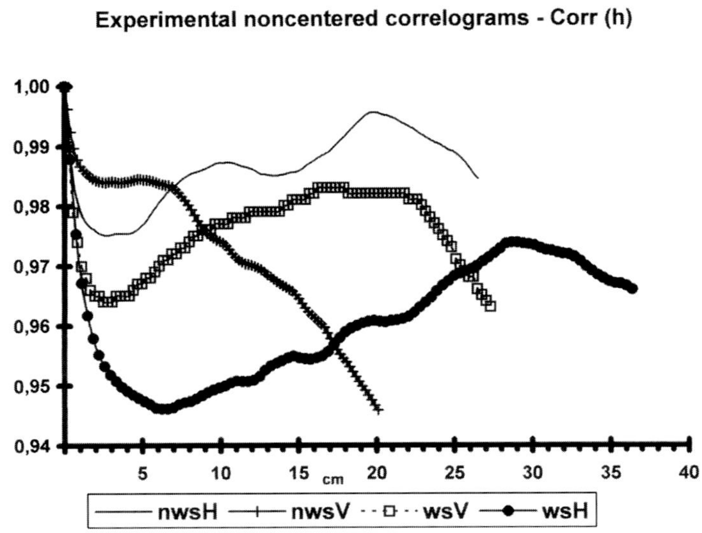


**Fig. 3.** Experimental horizontal (H) and vertical (V) correlogram (*Corr*(*h*)) plots for unweathered (nws) and weathered (ws) stones.

results between different images, to be compared (Fig. 3).

Water–rock interaction studies based on rain and percolating/seepage water collected, respectively, at the terrace and on the interior of the Basílica da Estrela were also carried out. An HIDSPEC computational hydrogeochemical model (Carvalho & Almeida 1989) was used to calculate saturation indexes (S.I.) for some minerals commonly involved in the weathering process of carbonate monument stones. Phase and graphical diagrams and multivariate statistical analysis (principal component analysis – PCA) of some hydrogeochemical significance were also used.

The results that are presented in this paper were then correlated with data obtained in previous studies (Figueiredo 1999; Figueiredo *et al.* 1999). These studies were based on laboratory thermal fatigue and dry and wet ageing tests, microclimate characterization, atmosphere pollution sampling, historical and document surveys of the monument made with particular insights into the cleaning activity, and repair works performed in the past on the interior of the Basílica da Estrela.

## Building materials

### *Petrography*

Macroscopically the limestones used in the Basílica da Estrela are bioclastic and microcrystalline. In thin section, according to the Folk and Dunham classification (Pettijohn 1975), the beige–white limestone ('Lioz', WL) is a biomicrosparite–biomicrosparrudite, wackstone–packstone, while the red–pink limestone ('Encarnadão' de Negrais, RPL) is a biopelsparite, packstone–grainstone, and the yellowish variety ('Amarelo de Negrais', YL) is a biomicrudite, wackstone. The blue and most outstanding limestone ('Azul de Sintra', BL) is a greyish-blue limestone with a granoblastic texture.

### *Chemical composition*

According to the results of chemical analyses summarized in Table 1, the beige–white, rose–pink and

**Table 1.** *Chemical analysis in weight per cent of the four limestones (WL, white; RPL, red–pink; YL, yellow; BL, blue limestone)*

| | Samples | | | |
|---|---|---|---|---|
| | WL | RPL | YL | BL |
| $SiO_2$ | 0.92 | 1.88 | 11.21 | 2.48 |
| $Al_2O_3$ | 0.17 | 0.33 | 3.00 | 0.61 |
| $Fe_2O_3$ | 0.03 | 0.11 | 1.00 | 0.31 |
| MgO | 0.33 | 0.30 | 0.41 | 0.84 |
| CaO | 55.06 | 54.47 | 47.67 | 52.74 |
| $Na_2O$ | 0.02 | 0.01 | 0.03 | 0.02 |
| $K_2O$ | 0.00 | 0.02 | 0.23 | 0.03 |
| MnO | 0.00 | 0.01 | 0.01 | 0.01 |
| $TiO_2$ | 0.00 | 0.00 | 0.08 | 0.02 |
| L.O.I. | 43.02 | 42.68 | 36.64 | 42.28 |
| Total | 99.55 | 99.78 | 100.28 | 99.34 |

L.O.I., loss on ignition.

**Table 2.** *Physical properties concerned with fluids transfers for the four limestone types (WL, white; RPL, red–pink; YL, yellow; BL, blue limestone)*

| Samples | $N_a$ (%) | $N_t$ (%) | $S_{48}$ (%) | $C$ (g m$^{-2}$. s$^{0.5}$) |
|---|---|---|---|---|
| WL | 0.55 | 1.63 | 30.06 | 0.38 |
| RPL | 0.71 | 1.48 | 40.65 | 0.66 |
| YL | 1.36 | 2.52 | 44.88 | 1.53 |
| BL | 0.79 | 1.36 | 61.43 | 1.71 |

the greyish-blue limestones are pure, with more than 95% of calcium carbonate and less than 3% of silica. The yellowish variety is, however, slightly dolomitic and clayey.

### *Petrophysical properties*

Values of properties related to fluid transport in porous media are presented in Table 2. These are open porosity ($N_a$,%), total porosity ($N_t$,%), Hirschwald coefficient ($S_{48}$,%) and capillarity coefficient ($C$, g m$^{-2}$. s$^{0.5}$). According to Bajare & Svinka (2000), these results indicate that these limestones are resistant to decay processes due to water–rock interaction. Concerning the absorption of water by capillary action, these limestones present a behaviour characteristic of porous heterogeneous network rocks (Mertz 1991). The results compare well with those derived from the thermal ageing accelerated tests indicating that the yellow limestone shows a susceptibility to weathering higher than the other limestones studied. Regarding the results derived from thermal ageing accelerated tests, these limestones can be ranked according to their weatherability as follows (Figueiredo *et al.* 1999): YL < BL < RPL < WL, where < means less weather stable. According to these results, the white variety (WL) has the best durability of all limestones studied. In contrast, the yellow limestone (YL) is the most prone to weathering.

## Stone decay

### *Weathering form survey, chemical and mineralogical results*

Detailed survey of stone decay phenomena was carried out on the outside and inside of the Basílica da Estrela in order to establish the typology, causes and processes of the major weathering forms observed. Several weathering forms were identified. Biological colonization of specific areas of the main façade and terrace was recorded on the outside. The presence of a blackish-brown crusting covering sheltered zones of the dome and the fissuring and cracking of some terrace stones were also noticed. According to the survey, the most deteriorated stone of Basílica da Estrela is the yellow Cretaceous limestone ('Amarelo de Negrais', YL) that was used as the lining material on panels (Fig. 1) located in the transept and nave inside the church.

Physical-weathering forms such as granular disintegration, flakes, scales and spalling prevail inside the monument. The weathering products (salt efflorescences and disintegrated stone material) at the bottom or over the panels built with the 'Amarelo de Negrais' (YL) were analysed by stereomicroscopy, chemical analytical techniques, XRD and FTIR. The results obtained are included in Tables 3 and 4.

**Table 3.** *Mineralogical composition of the unweathered yellow limestone and the weathering material collected over the panels located inside the church and built with this stone*

| Mineralogy | Samples | |
|---|---|---|
| | Unweathered yellow limestone | Weathered material |
| Calcite | ν | ν |
| Dolomite | + | + |
| Quartz | + | + |
| Kaolinite/Halloysite | +* | + |
| Silicate | – | • |
| Maghemite | t | + |
| Hematite/Goethite | – | t |
| Trona | – | + |
| Thenardite | – | + |

ν, very abundant; •, abundant; +, present; t, traces; –, not detected; *only halloysite was identified.

**Table 4.** *Chemical analysis in weight per cent of the weathering material collected at the bottom (sample 1) and over (sample 2) the panel located inside the church and built with the yellow limestone*

| | Sample 1 | Sample 2 |
|---|---|---|
| $SiO_2$ | 31.15 | 15.51 |
| $Al_2O_3$ | 7.31 | 3.72 |
| $Fe_2O_3$ | 1.69 | 1.82 |
| MgO | 7.27 | 6.24 |
| CaO | 28.38 | 31.07 |
| $Na_2O$ | 3.55 | 0.87 |
| $K_2O$ | 1.75 | 1.02 |
| MnO | 0.02 | 0.03 |
| $TiO_2$ | 0.30 | 0.32 |
| L.O.I. | 19.30 | 38.60 |
| Total | 100.72 | 99.20 |

L.O.I., loss on ignition.

However, chemical weathering forms, mainly due to calcite reprecipitation forming large white zones on the lining stones, have also been found inside the church. Some calcitic concretions (mainly on the floor of the elevated choir of the church) and stalactites could be seen. Soluble salts, including gypsum, commonly involved in the weathering processes of carbonate stone monuments were practically non-existent. Only trona ($Na_3H(CO_3)_2 \cdot 2H_2O$) and thenardite ($Na_2SO_4$) precipitation (Table 3) were found in a very small area localized at the bottom of only one panel built with the yellow limestone (see Fig. 1).

The Basílica da Estrela is located in a moderately polluted area. Chemical analyses were carried out periodically on both rain and seepage waters, collected over 3 years on a weekly basis, both outside (terrace) and inside (high choir), in order to establish the main source or sources for the soluble salts detected inside the Basílica da Estrela.

Rain and seepage waters are chemically very different. Rainwaters (ϕ) belong to $SO_4$–Cl–Ca and Cl–$SO_4$–Na type (Fig. 4) and show a strong sea-water influence. pH values range from 5 to 7, conductivity values are around 90 μS $cm^{-1}$, and total mineralization lies between 40 and 100 mg $l^{-1}$. All rainwater samples were undersaturated with respect to many minerals, showing, for instance, decreasing saturation index (S.I.) values for gypsum and calcite (gypsum > calcite) (Fig. 5).

Seepage waters (▮) belong to $HCO_3$–Na type (Fig. 4), showing a rather narrow range of variation in composition all over the monitoring time of the monument. With pH between 7 and 12, conductivity about 700 μS $cm^{-1}$, and total mineralization between 200 and 3000 mg $l^{-1}$, seepage waters are much more mineralized than rainwaters. Seepage

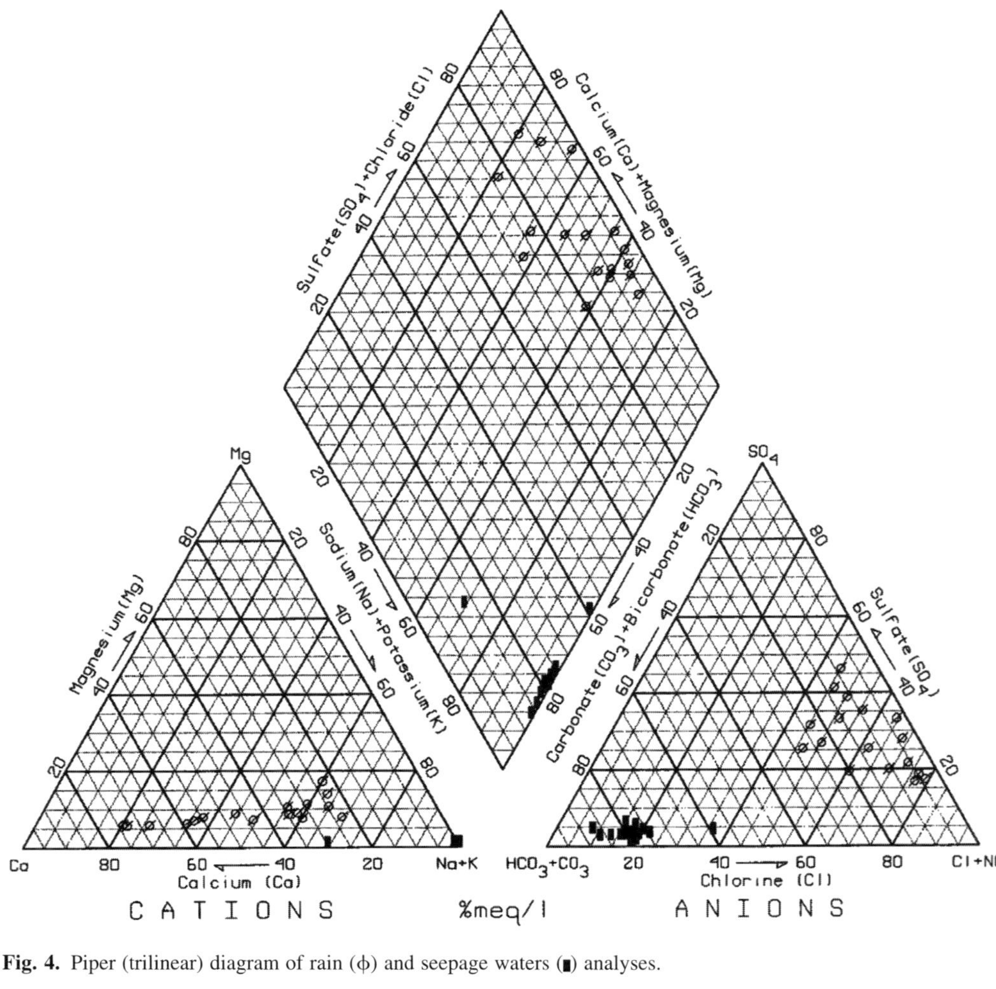


**Fig. 4.** Piper (trilinear) diagram of rain (ϕ) and seepage waters (▮) analyses.

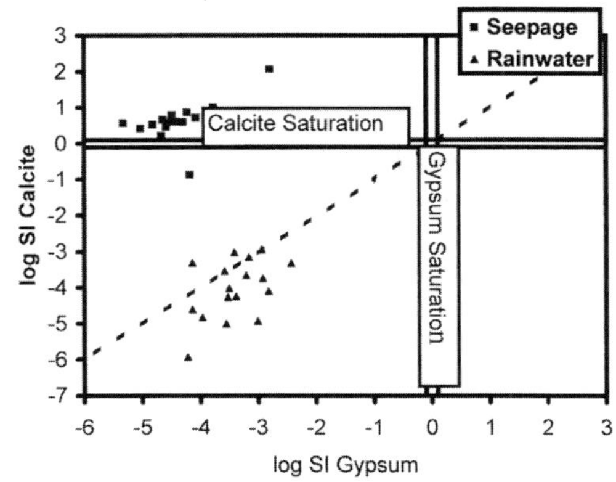


**Fig. 5.** Calcite and gypsum saturation indexes (S.I.) diagram for rain and seepage waters.

waters show higher S.I. values for most of the minerals, but saturation and supersaturation is reached only with respect to calcite (Fig. 5) (Langmuir 1971; Magalhães *et al.* 1997).

The plot of rain and seepage water chemical composition onto a calcite and gypsum S.I. diagram (Fig. 5) is consistent with the large white stain zones of calcite precipitation and small stalactites and stalagmites observed inside the church (Auger 1989; Lewin 1989; Livingstone 1992). Since seepage waters are undersaturated with respect to trona and thenardite, these salts could be precipitated only through seepage evaporation (Arnold & Zehnder 1989; Begonha *et al.* 1995; Goudie & Viles 1997; Winkler 1997). Only a local source and/or enrichment of salt solution in alkaline elements (see Table 4) could explain the very small and confined occurrence of these salts. This fact could be related to cleaning and repair (maintenance/restoration) works performed in the last few years (Arnold & Zehnder 1989), given that the environmental conditions (percolating water composition and thermo-hygrometric values: Figueiredo 1999) for precipitation of these salts are the same all over the interior of the Basílica da Estrela. The efflorescences found inside the Basílica da Estrela could result from allochthonous sources such as environmental and human interventions, and autochthonous ones related to water–rock and mortar interactions (Arnold & Zehnder 1989; Mazor 1998).

### *Image analysis*

The granulometric (Fig. 2) and covariance (Fig. 3) analyses were applied to representative grey-level images of every rock pathology and weathering state. The results obtained indicate that the deterioration processes of the yellow limestone ('Amarelo de Negrais', YL) are controlled by the texture and architectural (geometry and surface finish) characteristics of the stone. These processes generate new stone surfaces with a widespread granulometry (Fig. 2). The frequency polygons (FP curves in Fig. 2) or the cumulative distribution function (CDF curves in Fig. 2) estimated for unweathered (nws) and weathered (ws) stones allow the degree of damage of the panels to be characterized. The size of the texture elements and their frequency in the image were accurately estimated (Fig. 2). The changes of texture due to weathering induce an increase in the frequency of small-size texture elements (an average size of less than 6 cm) associated with mechanical process (sanding and powdering). The appearance of new texture elements of intermediate size and an increase in frequency of large-size (an average size larger than 23 cm) texture elements due to chemical–physical processes related to the dissolution and reprecipitation of calcite were also observed. The calcite reprecipitation produces large white zones on the surface of the lining stone of the panels.

The horizontal (H) and vertical (V) correlograms ($Corr(h)$) estimated for unweathered (nws) and weathered (ws) stones are presented in Figure 3. Comparing them, it can be seen that the remarkable texture anisotropy observed for texture elements with an average size of less than 6 cm is profoundly changed by the weathering of the stone surfaces. The synergetic effect of the remarkable intrinsic texture anisotropy characteristic of the unweathered yellow limestone and the architectural features (geometry and surface finish) of the panels promotes the development of the weathering processes, mainly in the vertical direction, the largest dimension of the vertical panels (see Fig. 3). The vertical direction of the panels becomes the preferred direction for the development of ongoing decay processes. Based also on the granulometry and covariance analysis, it was found that the most deteriorated panels are the ones turned to the north.

## Conclusions

The deterioration processes affecting the yellow limestone of the Basílica da Estrela are determined by stone structures (stilolytes, fossils, intraclasts joints and microfractures) and architectural features (geometry and surface finish of the stones).

Calcite dissolution and reprecipitation, confirmed by the existence of secondary calcite deposition as crusts, stalactites and stalagmites, is considered as one of the most important stone decay processes.

Taking into account previous studies (Figueiredo 1999; Figueiredo *et al.* 1999), wetting and drying cycles could also be a major factor in the decay of the stones of the Basílica da Estrela. The interior microclimate of the Basílica da Estrela shows annual air temperature and relative humidity values ranging from 14.6 to 24.7 °C and from 41.4 to 99.9%. When combined with the presence of rainwater percolating through the stone and the dissolution of stone and joint materials, the microclimate brings about changes in water composition mainly due to evaporation and precipitation of selected water components.

Although there is strong evidence of sea-water contribution to rainwater composition, there is no evidence of the influence of seepage water with dissolved chlorides in stone decay in the areas studied inside the church.

Very similar conditions and mechanisms of water–rock interaction are suggested by the relative hydrochemical uniformity revealed by seepage waters collected inside at the elevated choir after penetrating through the roof and percolating behind the panels. A significant uniformity in the contribution of ion sources and water–rock interaction processes is characterized by the enrichment of seepage waters in $K^+$, $Na^+$, $Cl^-$ and $HCO_3^-$ and loss of $Mg^{2+}$, $Ca^{2+}$ and $SO_4^{2-}$.

The stone decay induced by salt deposition cannot be related to trona and thenardite precipitation, taking into account their small quantity and confined spatial distribution inside the church.

This study was partially financed by Centro de Petrologia e Geoquímica/IST FCT subproject DECASTONE.

## References

AMOROSO, G. G. & FASSINA, V. 1983. *Stone Decay and Conservation: Atmospheric Pollution, Cleaning, Consolidation and Protection*. Materials Science Monographs, **11**. Elsevier, Amsterdam.

ARNOLD, A. & ZEHNDER, K. 1989. Salt weathering on monuments. *In*: ZEZZA, F. (ed.) *The Conservation of Monuments in the Mediterranean Basin. Proceedings of the 1st International Symposium on the Influence of Coastal Environment and Salt Spray on Limestone and Marble*. Grafo Edizioni, Bari, 31–58.

AUGER, F. 1989. World limestone decay under marine spray conditions. *In*: ZEZZA, F. (ed.) *The Conservation of Monuments in the Mediterranean Basin. Proceedings of the 1st International Symposium on the Influence of Coastal Environment and Salt Spray on Limestone and Marble*. Grafo Edizioni, Bari, 65–69.

BAJARE, D. & SVINKA, V. 2000. Restoration of the historical brick masonry. *In*: FASSINA, V. (ed.) *Proceedings of the 9th International Congress on Deterioration and Conservation of Stone*, Venice, 3–11.

BEGONHA, A., SEQUEIRA, M. A. B. & GOMES, F. S. 1995. A acção da Água da chuva na meteorização de monumentos graníticos. Universidade do Porto – Museu e Laboratório Mineralógico e geológico, Memória, **4**, 177–181.

BENALI, M. 1986. *Du choix des mesures dans les procedures de reconaissançe des formes et d'analyse de texture*. PhD thesis, École Nationale Supérieure des Mines de Paris, Fontainebleau, France.

CARVALHO, M. R. & ALMEIDA, C. 1989. HIDSPEC, um programa de especiação e cálculo de equilíbrios água/rocha. *Geociências, Revista da Universidade de Aveiro*, **4**, 1–22.

COSTER, M. & CHERMANT, J. L. 1985. *Précis d'analyse d'images*. Éditions du CNRS, Paris, France.

FIGUEIREDO, C. 1999. *Alteração, Alterabilidade e Património Cultural Construído: o caso da Basílica da Estrela*. PhD thesis, Technical University of Lisbon, Portugal.

FIGUEIREDO, C., FIGUEIREDO, P. & AIRES-BARROS, L. 1999. Geoquímica do envelhecimento laboratorial de calcários. *Actas II Congresso Ibérico de Geoquímica/XI Semana de Geoquímica*, Lisbon, Portugal, 193–196.

FIGUEIREDO, C., FIGUEIREDO, P., AIRES-BARROS, L., PINA, P. & RAMOS, V. 2005. Texture analysis of images taken from artificially aged stones: a statistical and structural approach. *International Journal of Restoration of Buildings and Monuments*, **11**, 235–246.

GONZÁLEZ, R. C. & WOODS, R. E. 1997. *Digital Image Processing*. Addison-Wesley, New York.

GOUDIE, A. & VILES, H. 1997. *Salt Weathering Hazards*. Wiley, New York.

LANGMUIR, D. 1971. The geochemistry of some carbonate ground waters in central Pennsylvania. *Geochimica et Cosmochimica Acta*, **35**, 1023–1045.

LEWIN, S. 1989. The susceptibility of calcareous stones to salt decay. *In*: ZEZZA, F. (ed.) *The Conservation of Monuments in the Mediterranean Basin. Proceedings of the 1st International Symposium on the Influence of Coastal Environment and Salt Spray on Limestone and Marble*. Grafo Edizioni, Bari, 59–63.

LIVINGSTONE, R. A. 1992. Graphical methods for examining the effects of acid rain and sulphur dioxide on carbonate stones. *In*: RODRIGUES, J. D., HENRIQUES, F. & JEREMIAS, F. T. (eds) *Proceedings of the 7th International Congress on Deterioration and Conservation of Stone*, **1**, 375–386.

MAGALHÃES, M. C. F., AIRES-BARROS, L & ALVES, L. M. 1997. Thermodynamics of carbonates and sulphates. Applications to stone decay studies – the case of 'Mosteiro dos Jerónimos', Lisboa. *Geociências*, Revista da Universidade de Aveiro, **11**, 139–147.

MAZOR, E. 1998. Allochthonous ions dissolved in recent and fossil groundwaters: identification and origins. *In*: AREHART, G. B. & HULSTON, J. R. (eds) *Proceedings of the 9th International Symposium on Water–Rock Interaction – WRI-9, Taupo, New Zealand*. A.A. Balkema, Rotterdam, 169–172.

MERTZ, J. D. 1991. *Structures de porosité et propriétés de transport dans les grés*. PhD Thesis, Université Louis Pasteur, Strasbourg.

NFB 10-504. 1973. *Pierres calcaires: mesure du coefficient d'absorption d'eau*. Portuguese Institute for Quality, Caparica, Portugal.

PETTIJOHN, F. J. 1975. *Sedimentary Rocks*. 3rd edn. Harper & Row, New York.

PREN. 1925. *Methods of Test for Natural Stone Units – Determination of Water Absorption Coefficient Due to Capillary Action*. Portuguese Institute for Quality, Caparica, Portugal.

PREN. 1936. *Methods of Test for Natural Stone Units – Determination of Real Density and Apparent Density and of Total and Open Porosity*. Portuguese Institute for Quality, Caparica, Portugal.

SERRA, J. 1982. *Image Analysis and Mathematical Morphology*. Academic Press, London.

SOILLE, P. 2003. *Morphological Image Analysis. Principles and Applications*, 2nd edn. Springer, Berlin.

TOMITA, F. & TSUJI, S. 1990. *Computer Analysis of Visual Textures*. Kluwer, London.

WINKLER, E. M. 1997. *Stone in Architecture. Properties, Durability*, 3rd edn. Springer, Berlin.

# The mineralogical and chemical methods in investigations of decay of the Devonian black 'marble' from Dębnik (Southern Poland)

M. MARSZAŁEK

*AGH – University of Science and Technology, Department of Mineralogy, Petrography and Geochemistry, al. Mickiewicza 30, 30-059 Cracow, Poland (e-mail: mmarszal@uci.agh.edu.pl)*

**Abstract:** Optical microscopy, scanning electron microscopy with energy dispersive spectrometry, X-ray diffraction, infrared (IR) spectroscopy, Rock-Eval pyrolysis and gas chromatography combined with mass spectroscopy were used to examine deterioration of the black limestone from Dębnik near Cracow. Owing to its unique colour and good polishing properties the rock is called the 'Dębnik marble'. The samples were taken from various monuments and natural outcrops exposed to weathering. The material is a compact limestone whose black colour is caused by an admixture of bitumens or pyrite. Its horizontal layers are separated by discontinuities filled with clay minerals. Surface exfoliation is one of the damage signs and results in the formation of irregular or lensoidal fractures. The discontinuities provide an easy access for acid rain that in reaction with calcite produces gypsum.

Crystallization of gypsum leads to alveolar weathering, cracking and chipping of the otherwise compact material. The presence of alveoles or surface exfoliation depends on the orientation of stone blocks. When they are cut along the discontinuities, destruction results in exfoliating and cracking. Perpendicular cutting gives rise to the formation of alveoles. The changes affect the original black colour of the stone surface that alters to grey or even white.

The Devonian Dębnik limestone (Givetian), owing to its unique, deep black colour and good polishing properties, is known as the 'Dębnik marble'. The historical quarries of this limestone are located about 20 km from Cracow in Dębnik village. One of the quarries was owned by Carmelite monks and is called the 'Carmelite' quarry (Narkiewicz & Racki 1984; Baliński 1989).

The Dębnik limestone is compact and occurs in three varieties: an homogeneous, micritic limestone; a micritic limestone with fossils; and a nodular limestone that occurs as horizontal layers separated by discontinuities filled with clay minerals (Bromowicz 2001). The Dębnik limestone from the 'Carmelite' quarry is biomicritic with a nodular texture and undulatory bedding, and reveals few microveins and stylolites. Non-carbonate components include K-feldspar, smectite, illite, subordinate pyrite and organic substances, as well as traces of detrital quartz and hydromuscovite (Marszałek & Muszyński 2001). The chemical composition is presented in Table 1. The black colour of the Dębnik limestone is thought to be caused by an admixture of bitumens (Gradziński 1972; Kozłowski & Magiera 1989; Lewandowska 1998) or pyrite (Bednarczyk & Hoffman 1989). The stone releases spontaneously an odour of petroleum if hit with a hammer.

Physico-mechanical properties (apparent density, water absorption ability, frost and abrasion resistance and compressive strength) of the Dębnik limestone are good and the stone generally withstands well the action of atmospheric factors (Bromowicz 2001), although some alteration of its surface can usually be observed.

These features appealed to the Baroque taste for decoration and starting from those times much architectural detail has been made of the Dębnik 'marble'. They can be found mainly in churches, monasteries, chapels and cemeteries (altars, fonts, portals, balustrades, columns, monuments, tombstones and headstones), not only in Cracow but also throughout Poland and even in other countries (Vienna, Graz and Salzburg in Austria and Frankfurt am Main in Germany: Rajchel 2004).

In Cracow the black Dębnik limestone was used in inner and outer architectural elements of many historical buildings. Many portals and altars in the Cracow churches (e.g. St. Mary's, St Peter and Paul's, St Andrew's, St Anna's, St Adalbert's and St John's, and the churches of the Benedictines, Cammaldolites, Capuchins, Dominicans, and Franciscans) have been made of the Dębnik limestone. Some differences in the stages of decay of the outer layers of architectural elements can, however, be observed and include the formation of irregular

*From*: PŘIKRYL, R. & SMITH, B. J. (eds) *Building Stone Decay: From Diagnosis to Conservation.* Geological Society, London, Special Publications, **271**, 109–115.
0305-8719/07/$15.00 

**Table 1.** *Chemical analyses of the Devonian Dębnik limestone from 'Carmelite' quarry (XRF)*

| Sample | $SiO_2$ | $Al_2O_3$ | $Fe_2O_3$ | $TiO_2$ | MnO | MgO | CaO | $Na_2O$ | $K_2O$ | $P_2O_5$ | $(SO_3)$ | (Cl) | (F) | L.O.I. |
|---|---|---|---|---|---|---|---|---|---|---|---|---|---|---|
| KS04 (%) | 1.2 | 0.53 | 0.29 | 0.04 | 0.007 | 0.85 | 53.92 | <0.01 | 0.45 | 0.006 | <0.01 | 0.09 | <0.01 | 41.54 |
| | As | Ba | Bi | Ce | Co | Cr | Cu | Ga | Hf | La | Mo | Nb | Ni | Pb |
| KS04 (ppm) | 3 | 18 | 2 | 19 | 5 | 2 | 1 | 3 | <1 | 2 | 14 | 4 | <1 | <1 |
| | Rb | Sr | Ta | Th | U | V | W | Y | Zn | Zr | | | | |
| KS04 (ppm) | 9 | 249 | 2 | 5 | 1 | 9 | <1 | 3 | 7 | 22 | | | | |

or lensoidal fractures. The outer elements exposed to ground moisture are more affected. In some of them alveoles or surface exfoliation are developed, while most show changes in the original black colour, which turns grey or even white. This bleaching is observed in all stone types.

## Materials and methods of investigations

The material analysed included three types of samples, of natural, both fresh and weathered stone, collected from the 'Carmelite' Dębnik quarry, as well as samples of architectural detail exposed to anthropogenic influences in the historic buildings of Cracow. The research equipment used to analyse the samples included a scanning electron microscope (SEM) with energy dispersive spectrometer FESEM-EDS (Hitachi D-4700 with a Noran Vantage attachment), a FTIR (Fourier transform infrared spectroscopy) Bio-RAD model 165 spectrometer, a GC8000/MD/800 gas chromatograph in combination with a mass spectrometer, pyrolyser Rock-Eval II and an X-ray diffraction (XRD) (Philips X'Pert APD) diffractometer.

The natural, fresh samples were mainly used to identify the non-carbonate components of the stone. These samples were crushed and dissolved in dilute HCl (1:10), providing the first batch of the acid-insoluble grain fraction. The other batch was obtained from the acid-insoluble fraction calcined at 700 °C to remove organic matter. This technique was used to identify and quantify non-carbonate components, and further investigations presented here centred on pyrite and organic substances.

The samples of weathered Dębnik limestone from the 'Carmelite' quarry and those of architectural detail exposed to anthropogenic conditions were collected for identification of changes in composition of the outer layers of the stone. The latter were taken from the tombs of the Rakowice Cemetery and architectural elements of selected churches in Cracow.

The tombstones and headstones at the Rakowice Cemetery show considerable alteration and in places it was possible to collect large fragments of stone. Other objects studied included the Baroque portals of St Adalbert's and St John's churches (Fig. 1). In both cases it was possible to collect only powder samples, collected either on a self-adhesive carbon tape (used in SEM investigations) placed against the portal surface or mechanically scraped from the portal surface onto the tape. All the details studied show the colour changes from black to grey or even white.

The outer layers of architectural details and samples of weathered limestone from the 'Carmelite' quarry, as well as the grain fractions insoluble in 6 M HCl and separated from the weathered limestone, were analysed using optical microscopy (an Olympus BX 71) and the set of methods mentioned above, the other scanning electron microscope (SEM-EDS Jeol 5600 with an ISIS attachment) was also applied. The samples scraped from the portals of St Adalbert's and St John's churches were only analysed with a Jeol 5600 SEM with an ISIS attachment.

## Results and discussion

### *Characteristics of non-carbonate components of the Dębnik 'marble'*

In addition to the major component of calcite, the Dębnik limestone contains an admixture of K-feldspar, smectite and illite, subordinate pyrite and organic substance, as well as traces of detrital quartz and hydromuscovite (Marszałek & Muszyński 2001). The black colour of the stone is caused by an admixture of bitumens or pyrite (Gradziński 1972; Bednarczyk & Hoffman 1989; Kozłowski & Magiera 1989; Lewandowska 1998). The current investigations centred on these two last phases.

The organic compounds yield a complex thermochemolysis (GC-MS) mixture, in which fatty acids (identifiable as methyl esters: Hermosin *et al.* 2004), saturated and non-saturated *n*-alkanes $C_{16}$–$C_{20}$, aromatic hydrocarbons and polycyclic hydrocarbons predominate (Table 2). The FTIR results

(a) (b)

**Fig. 1.** Portals constructed of black Dębnik 'marble'. (**a**) St Adalbert's church. (**b**) St John's church in Cracow.

**Table 2.** *Major compounds identified in Dębnik 'marble' thermochemolysis*

| Peak | Compounds |
|---|---|
| 1 | Nonanoic acid |
| 2 | Decanoic acid |
| 3 | Undecanoic acid |
| 4 | Benzene dicarboxylic acid |
| 5 | Dodecanoic acid |
| 6 | *n*-Hexadecane |
| 7 | Tridecanoic acid |
| 8 | *n*-Heptadecane |
| 9 | Tetradecanoic acid |
| 10 | Anthracene |
| 11 | anteiso-Pentadecanoic acid |
| 12 | *n*-Octadecane |
| 13 | Pentadecanoic acid |
| 14 | Hexadecenoic acid |
| 15 | Hexadecanoic acid |
| 16 | *n*-Eicosane |
| 17 | Pyrene |
| 18 | Octadecenoic acid |
| 19 | Octadecanoic acid |
| 20 | Heneicosanoic acid |

indicate that the organic matter is mainly aliphatic hydrocarbons (characteristic bands around 2930 and 2960 $cm^{-1}$). The correlation between hydrogen (HI) and oxygen (OI) indices in Rock-Eval pyrolysis points to its kerogen character (the type III of kerogen, i.e. derived from humic organic matter of probable algal and bacterial origin). The content of organic substances in the whole rock was estimated by Marszałek & Muszyński (2001) at 0.30 wt% TOC (TOC, total organic carbon) that corresponds to 4.4 wt% in the HCl–insoluble fraction.

Pyrite is represented by two populations of grains: (1) those of very small sizes (up to 0.04 mm); and (2) cuboidal with larger sizes, reaching 2 mm, with the first population dominant (Fig. 2). The total amount of pyrite in the Dębnik limestone is about 2 wt% (Marszałek & Muszyński 2001).

It is known that grey – deep black coloration of sedimentary rocks can be induced by organic substances as well as an iron pigment. Organic matter is a very common component of carbonate rocks, averaging 0.33 wt% TOC (Lewandowska 1998). Three types of organic matter of different chemical composition can be distinguished in sedimentary

**Fig. 2.** SEM micrograph of euhedral pyrite crystals.

rocks: (a) hydrocarbons (pure solvent-soluble organic matter composed of carbon and hydrogen only; (b) asphalt (solid and semi-solid hydrocarbons largely soluble in carbon disulphide; and (c) kerogen (insoluble in solvents; the bulk organic substance in shales and carbonate rocks). The colour change to be expected during weathering or mild metamorphism results from the pigment stability of each the carbon compounds, the hydrocarbons being the least stable carbon pigment, particularly the aromatic hydrocarbons. Generally, the last ones determine the dark colour of sediments. Dark grey and black limestone bleach quite easily to a light grey within a few years of exposure to weathering agents (oxidation environment). Reaction with aggressive solutions increases the lightness towards a very light tone (Winkler 1997).

In addition, the presence of organic matter in a sedimentary basin leads to a reducing environment and can finally result in the reduction of sulphates to hydrogen sulphide with the formation of pyrite (Lewandowska 1998). Some dark grey carbonate rocks may release hydrogen sulphide gas spontaneously if hit with a hammer or if blasted. According to Winkler (1997), the colour of such rocks may change to pure white as the gas escapes from along mineral grain boundaries. On the other hand, iron is the most common and strongest pigment in sedimentary rocks. It can give the stone colours from deep red to orange, yellow, brown, or tan to green, blue and black. Iron occurs in oxidized ferric ($Fe^{3+}$) or reduced ferrous ($Fe^{2+}$) forms. The variety of the colour is a function of the amount and degree of oxidation, so it is related to the ferric/ferrous ratio in the rock. A deep grey–black colour of a stone is connected with the domination of ferrous iron. In this case $Fe^{2+}$ may occur as a finely distributed pigment that is bluish green–black, and results from the presence of pyrite and marcasite ($FeS_2$); the powder or streak of these minerals are actually black, although on recrystallized to larger grains the minerals become yellowish-brass. Limestones often contain minor quantities of deep black, finely disseminated ferrous sulphide, which was formed in a reducing environment. The presence of organic matter in the sedimentary basin can lead to such a reducing environment and, finally, to the creation of pyrite.

Iron pigments in sedimentary rocks are generally unstable if exposed to weathering and light, but the degree of colour stability is hard to predict.

Current results and prior considerations (e.g. Gradziński 1972; Winkler 1997; Lewandowska 1998) allow it to be stated that the black colour of the Dębnik marble is most likely to be caused by an admixture of organic matter.

### *Characteristics of bleached Dębnik 'marble'*

The change of the original black colour into grey or even white is a common sign of weathering of the Dębnik 'marble', both exposed in the 'Carmelite' quarry and in Cracow.

Samples from the 'Carmelite' quarry show that the original colour of the outer layer of natural samples is bleached. The surface often shows karst-like alteration, with signs of leaching, and is covered by fine calcite grains only. In contrast, the changes in appearance of the outer architectural elements exposed directly to atmospheric conditions are expressed in the decoloration of stone surfaces. It is clearly seen on the portals of St Adalbert's and St John's churches, but polishing of the stone returns it to its original, black colour. The effects of polishing, although unintended, can be observed in the portal of St Adalbert's church in a place frequently touched by the faithful, where they step down to the lower level inside the church.

Surfaces of some elements from historical objects are covered by gypsum layers, with a thickness of to about 250 μm. The gypsum develops as fine, euhedral crystals up to about 30 μm. In cross-section (plane polarized light) two alteration layers can be distinguished. The external one (thickness about 200 μm) is composed of microcrystalline gypsum and deposits of dust (glass particles, iron oxides, unburned carbon particles). The internal one (thickness about 50 μm) is composed of micro- to cryptocrystalline gypsum only (Figs 3 & 4). The distribution of selected elements in the outer layer show enrichment in sulphur, silica and iron, but depletion of calcium (Fig. 5). XRD and FTIR analyses confirm the presence of gypsum in the outer layers of the stone. The bleached outer layer of such stone elements is thicker than the

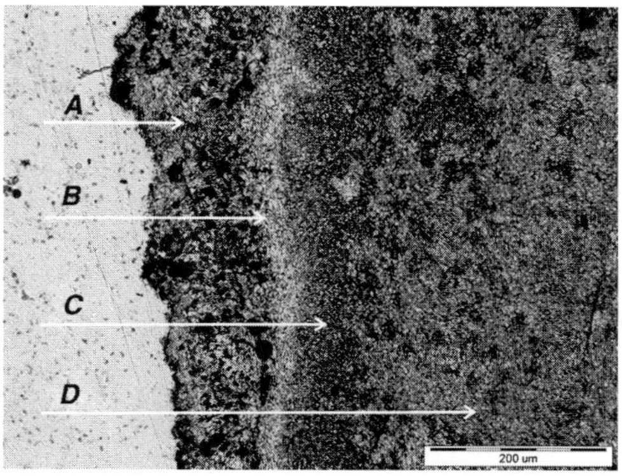


**Fig. 3.** Polished cross-section of a dendritic crust of a headstone, Rakowice Cemetery. **A**, outer layer of alteration composed of microcrystalline gypsum, glass particles, iron oxides and unburned carbon particles; **B**, internal layer of alteration composed of micro- to cryptocrystalline gypsum only; **C**, partly altered 'marble' with fine opaque particles apparently of organic matter; **D**, unaltered 'marble' (optical microscope, 1N).

gypsum layer, so their bleaching cannot be attributed to the presence of gypsum alone. Other architectural elements show signs of leaching and are covered by fine grains of calcium carbonate (sizes from several to 15 μm) or shapeless, cryptocrystalline calcite aggregates. Powder samples, collected from such surfaces (either by scraping or on the adhesive tape as described above), contain only cryptocrystalline calcite aggregates.

Anthropogenic components observed on the surfaces of Dębnik limestone elements (SEM–EDS) also include spherical particles of aluminosilicate glass with a characteristic smooth appearance, as well as spherical particles of iron oxides distinguishable due to their dendritic development. The same particles or their aggregates were observed in the acid-insoluble grain fractions. These types of anthropogenic particles were also seen on the surface of other the limestones exposed to anthropogenic conditions (e.g. Camuffo *et al.* 1983; Manecki *et al.* 1997; Michalik & Wilczyńska-Michalik 1997; Bugini *et al.* 2000; Marszałek 2004). Other widespread components of the samples analysed include colonies of micro-organisms that are particularly abundant in samples from the Rakowice Cemetery.

Some differences in the outer layers of architectural elements were observed. Their presence depends on the rock fabric of the Dębnik limestone variety used, its orientation to bedding and localization in respect to atmospheric factors. Surface exfoliation is one of the signs of the damage and results in the formation of irregular or lensoidal fractures in the nodular limestones with admixtures of clay matter (marly limestones). The surfaces of undulatory or nodular laminations are the result of the penetration of aggressive, atmospheric waters. The fractures become filled with gypsum, a product of the reaction of these waters with the calcium carbonate of the rock. The stone elements cut out of a more massive stone, without sedimentary fabrics, are more resistant to water penetration, and gypsum was recorded only sporadically on the surfaces of both the micritic limestones and the micritic limestones with fauna.

## Conclusions

It is concluded that the discontinuities in one of the type Dębnik limestones – nodular limestone – provide an easy access for rainwater. Exposure to

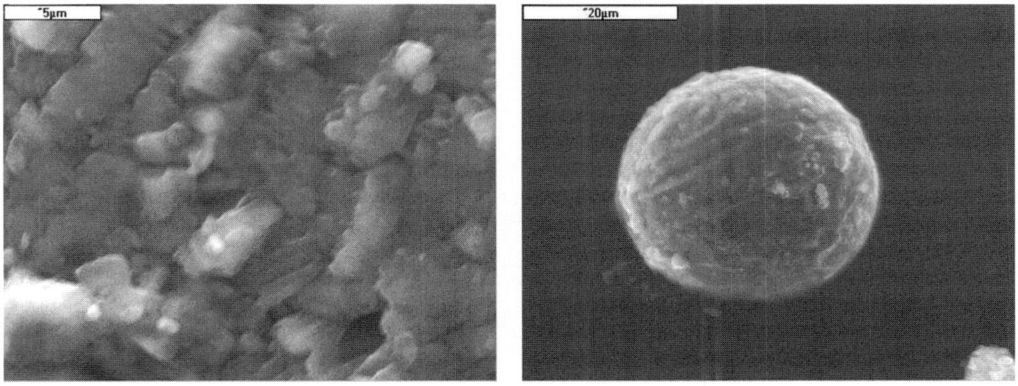


(a) (b)

**Fig. 4.** Anthropogenic components of the outer layer of a headstone, Rakowice Cemetery: (**a**) microcrystalline gypsum and (**b**) spherical particle of iron oxide (SEM).

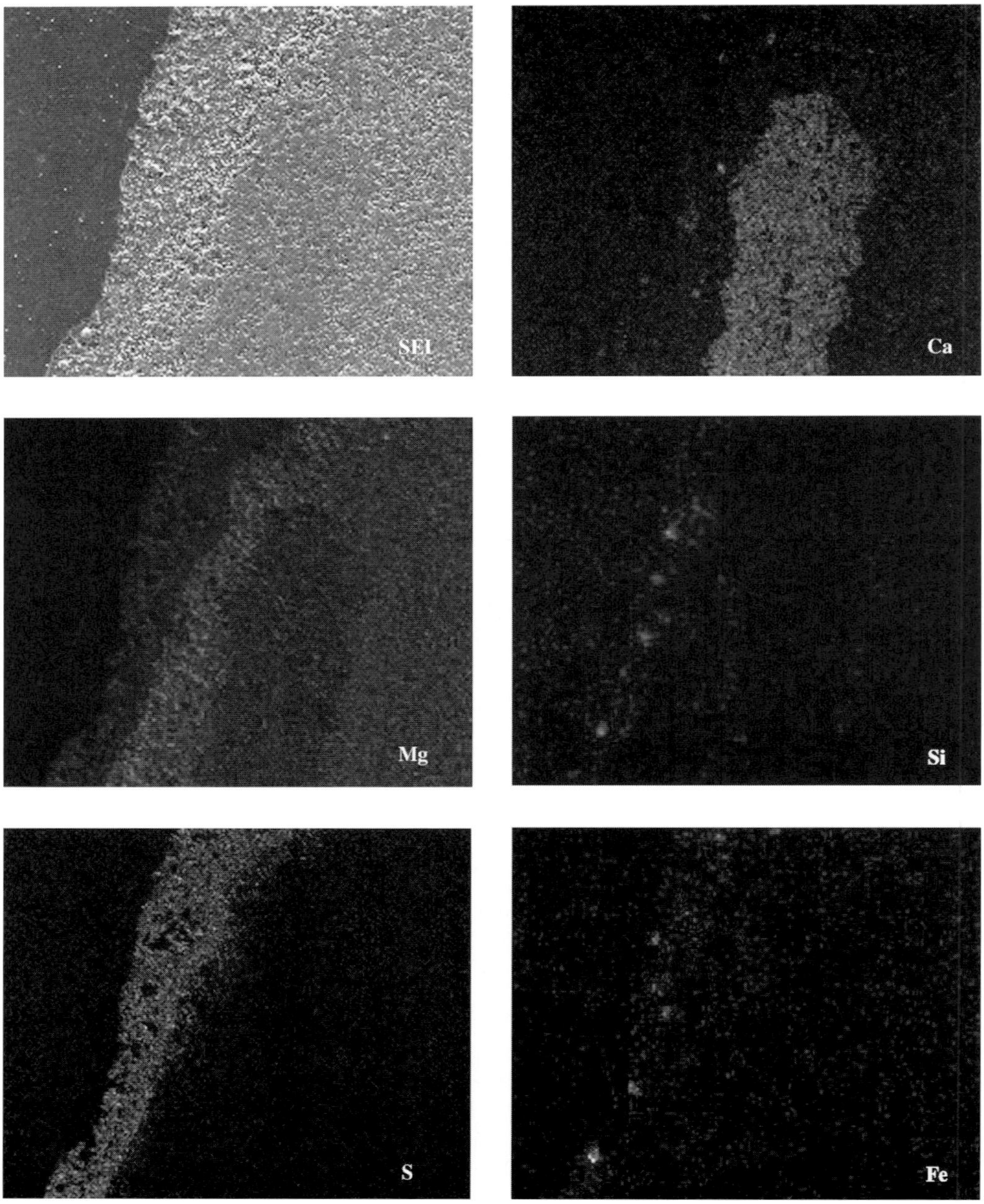


**Fig. 5.** Concentration of selected elements within a cross-section of the crust formed on Dębnik 'marble' (SEM-EDS).

an atmosphere containing $SO_2$ results in the reaction of acid rain with calcite, which infills discontinuities with hydrated calcium sulphate. Crystallization of gypsum leads to alveolar weathering or cracking and exfoliating of the otherwise compact material. The way of cutting the stone blocks controls their further destruction. When the stone is cut along discontinuities, its destruction results in exfoliating and cracking, while perpendicular cutting gives rise to the formation of alveoles. Stone elements cut out of material without sedimentary fabrics are more resistant to water penetration: gypsum was observed only sporadically on the surfaces of both the micritic limestones and the micritic limestones with fauna. In this element mainly recrystallized calcium carbonate was observed.

Stones exposed to external conditions show decoloration of the outer layers, but their bleaching cannot be attributed solely to the presence of gypsum. Sufaces exposed in a non-polluted region show mainly the presence of fine calcite grains. Surfaces exposed to a polluted atmosphere are covered with micro- to cryptocrystalline gypsum and also particles of anthropogenic dust. Among the anthropogenic compounds, aluminosilicate glass spheres and iron oxides were detected.

The methods applied to understanding the reasons for bleaching have not been totally conclusive. In the case of the Dębnik limestone, the alteration is probably also associated with oxidation of organic matter, one of the components of the stone.

This study was supported by the AGH – University of Science and Technology project no. 11.11.140.158. The author's thanks go to C. Saiz-Jimenez and B. Hermosin of the Instituto de Recursos Naturales y Agrobiologia (Sevilla, Spain) for GC-MS analyses. The help of A. Gaweł, M. Kotarba, S. Olkiewicz, A. Skowroński and B. Trybalska of the AGH – University of Science and Technology (XRD, Rock-Eval pyrolysis, FTIR, SEM-EDS analyses) is very much appreciated. I am also grateful to my student, M. Wójtowicz, for collecting some of the samples and XRF analyses.

## References

BALIŃSKI, A. 1989. Biostratygrafia górnego dewonu antykliny Dębnika. (Upper Devonian biostratigraphy of the Dębnik anticline). *In*: RUTKOWSKI, J. (ed.) *Przewodnik LX Zjazdu PTG. [LX Meeting of the Polish Geological Society.]* Polish Geological Society, Kraków, 30–34.

BEDNARCZYK, J. & HOFFMAN, M. 1989. Wapienie dębnickie. (Dębnik limestone). *In*: RUTKOWSKI, J. (ed.) *Przewodnik LX Zjazdu PTG. [LX Meeting of the Polish Geological Society.]* Polish Geological Society, Kraków, 40–46.

BROMOWICZ, J. 2001. Ocena możliwości wykorzystania skał z okolic Krakowa do rekonstrukcji kamiennych elementów architektonicznych. [Application of the building stones of the Cracow area for the masonry reconstructions – evaluation of preferences.] *Mineral Resources Management, Mineral and Energy Research Economy Institute Polish Academy of Science*, **17**(1), 16–73.

BUGINI, R., LAURENZI TABASSO, M. & REALINI, M. 2000. Rate of formation of black crusts on marble. A case study. *Journal of Cultural Heritage*, **1**, 111–116.

CAMUFFO, D., DEL MONTE, M. & SABBIONI, C. 1983. Origin and growth mechanisms of the sulfated crusts on urban limestone. *Water, Air and Soil Pollution*, **19**, 351–359.

GRADZIŃSKI, R. 1972. *Przewodnik geologiczny po okolicach Krakowa. [Geological Guide of the Cracow Region.]* Wydawnictwo Geologiczne (Geological Publishing House), Warszawa.

HERMOSIN, B., GAVINO, M. & SAIZ-JIMENEZ, C. 2004. Organic compounds in black crusts from different European monuments: a comparative study. *In*: SAIZ-JIMENEZ, C. (ed.) *Air Pollution and Cultural Heritage*. Balkema, London, 47–55.

KOZŁOWSKI, R. & MAGIERA, J. 1989. Niszczenie wapieni dębnickich i pińczowskich w zabytkach Krakowa. [Deterioration of the Dębnik and Pinczów limestones in Cracow monuments.] *In*: RUTKOWSKI, J. (ed.) *Przewodnik LX Zjazdu PTG. [LX Meeting of the Polish Geological Society.]* Polish Geological Society, Kraków, 204–208.

LEWANDOWSKA, A. 1998. *Mineralogia skarnów magnezowych grzbietu Dębnika. [Mineralogy of Mg-scarns of the Dębnik hill.]* PhD thesis, Jagiellonian University. Kraków.

MANECKI, A., MARSZAŁEK, M., SCHEJBAL-CHWASTEK, M. & SKOWROŃSKI, A. 1997. Stone decay in some historic buildings of Cracow (Poland) and its reasons. *In*: SULOWSKI, P. & ZEMAN, J. (eds) *ENVIWEATH '96. Folia Facultatis Scientiarum Universitatis Masarykianae Brunensis, Geologia*, **39**, 149–156.

MARSZAŁEK, M. 2004. Deterioration of stone in some monuments exposed to air pollution: a Cracow case study. *In*: SAIZ-JIMENEZ, C. (ed.) *Air Pollution and Cultural Heritage*. Balkema, London, 151–154.

MARSZAŁEK, M. & MUSZYŃSKI, M. 2001. Authigenic K-feldspars in Dębnik limestone (S Poland). *Mineralogia Polonica*, **32**(1), 49–61.

MICHALIK, M. & WILCZYŃSKA-MICHALIK, W. 1997. The influence of air pollution on weathering of building stones in Kraków. *In*: SULOWSKI, P. & ZEMAN, J. (eds) *ENVIWEATH '96. Folia Facultatis Scientiarum Universitatis Masarykianae Brunensis, Geologia*, **39**, 159–167.

NARKIEWICZ, M. & RACKI, G. 1984. Stratygrafia dewonu antykliny Dębnika. [Devonian biostratigraphy of the Dębnik anticline.] *Kwartalnik Geologiczny [Geological Quarterly]*, **28**, 513–546.

RAJCHEL, J. 2004. *Kamienny Kraków. [Stony Cracow – Stone in Cracow Architecture.]* AGH – University of Science and Technology, Kraków.

WINKLER, E. M. 1997. *Stone in Architecture. Properties, Durability*, 3rd edn. Springer, Berlin.

# Effect of long-term changes in air pollution and climate on the decay and blackening of European stone buildings

C. M. GROSSI & P. BRIMBLECOMBE

*School of Environmental Sciences, University of East Anglia, Norwich NR4 7TJ, UK (e-mail: c.grossi-sampedro@uea.ac.uk)*

**Abstract:** This paper reviews the long-term effects of past, present and future air pollution and climate on the decay of stones from historic buildings. It summarizes the historical effects as well as causes and consequences of damage. The most significant change in terms of pollution has been a shift from high levels of sulphate deposition from coal burning to a blackening process dominated by diesel soot and nitrogen deposition from vehicular sources in cities. Blackening of light-coloured fabric eventually reaches a point where it becomes publicly unacceptable. Public opinion can assist the development of aesthetic thresholds and derive limit values for elemental carbon in urban air. Public perception is also affected by the pattern of blackening. This century new climate regimes will cause dramatic changes in blackening patterns by wind-driven rain. Climate changes, most particularly changes in temperature, humidity stress and time of wetness, can also affect the weathering of stone in terms of responses to frost and soluble salts. Future pollution and climate are relevant considerations in the management of historic buildings.

## History and future of air pollution and climate

The history of air pollution and climate is important in understanding the degradation of the built heritage. Buildings may be old, accumulate damage, and be located in cities confronting urban pollution and climate. It is thus necessary to understand not only the factors that damage building stones today, but, in addition, those processes that contribute to building stone decay over the entire lifetime of the structure. This usually involves considering the historical changes in fuels, urban pollution and climate. The future must also be considered, mainly from the perspective of planning and conservation, especially in a world where climate will change and affect both the amount and distribution of damage to historic buildings.

### *Changes in fuel*

Changes in fuel mean different types of smoke and air pollutants. A shift from wood to coal occurred in medieval Britain after the depletion of conveniently usable wood supplies around some major cities. Fears that coal smoke was risky to health, restricted its use in the 13th century mostly to industrial processes such as lime-burning and some metallurgical operations. By the late 16th century household chimneys were common and coal began to be used domestically (Brimblecombe 1987), and in the 19th century it became the dominant fuel of most major European cities. Coal was increasingly used because of new technology (e.g. the steam engine or new metal-refining techniques). The 20th century witnessed changes that are as significant as the transition from wood to coal. The mid century saw a move from solid to liquid fuels. Much of the liquid fuel has been used in the form of petrol by transport. Changes in overall fuel patterns have been determined by large-scale economics rather than by environmental policies (Williams 2004). Liquid fuel volatility has introduced many low molecular weight hydrocarbons into the atmosphere as reactive trace gases. Their reactivity has changed the chemistry of the modern urban atmosphere.

The situation is not identical everywhere. Rather significantly, in Europe diesel fuel has become increasingly important, with a wide use in cars. It has been estimated that during 1965–1999 petrol and diesel use in London increased by around 43 and 175%, respectively (Williams 2004), and in Germany the use of diesel fuel increased by more than 3% in 2004 (Sensfuß *et al.* 2005). Diesel-powered vehicles have changed the composition of the atmosphere yet again. They have introduced large amounts of very fine soot into the air of cities making a significant contribution to the soiling quality of urban air. In the near future greater attention will be focused on the trends and fuel composition used by road traffic.

### *Changes in air pollution*

Fuel changes lead to changes in air pollution. In Roman times documentary comment relates

*From*: PŘIKRYL, R. & SMITH, B. J. (eds) *Building Stone Decay: From Diagnosis to Conservation.* Geological Society, London, Special Publications, **271**, 117–130.
0305-8719/07/$15.00 

combustion processes, both industrial and domestic, to the experience of urban air pollution. It has also been possible to estimate past pollution from observations of visibility, such as fog frequency within diary records (Brimblecombe 1987). In terms of damage to monumental heritage the accumulation of pollutants on ancient building surfaces can reveal changes in their exposure. Typically, these show a transition from deposits of wood smoke to coal and diesel particles (e.g. Ausset *et al.* 1998). The particle morphology changes along with the amount of organic carbon present. Organic carbon concentrations in crusts are increasing as a result of ongoing depositions from the modern urban atmospheres where there is a large amount of traffic (Hermosin *et al.* 2004).

Despite the rapid development of smoke abatement law over the last 100 years early control was generally ineffective, except where peculiar local pressures ensured a change. Unfortunately, 19th century interest in smoke in cities was not accompanied by any widespread enthusiasm for monitoring of air pollutants. Measurements that remain were sporadic and often inaccurate, although there are some reliable measurements, usually of deposited solids and solutes in industrial towns. These hint that the concentrations of soot and $SO_2$ 100 years ago were probably high by modern-day standards. In European cities the concentrations started to decline in the 20th century. This occurred relatively early in London (Brimblecombe 1977) and a little later in cities such as Berlin and Paris. In some other cities it was delayed until the 1970s. The reasons for the decline in $SO_2$ and smoke concentrations also varied over time. Initially, they tended to be the result of dilution as cities expanded with the development of an urban transport system. Cities that have only recently experienced a decline in pollutant concentrations often regard legislation as the driving force for improvement. It is clear with some particular pollutants that legislation is likely to be a key control. However, there are often parallel social and technical forces that alter the concentrations of air pollutants. As an example, the wide use of electricity in households has meant a declining use of coal in the 20th century. Similar changes were underway in cities of North America and Berlin, although the maximum concentrations may not have been experienced until well into the 20th century (Sherwood & Bumbaru 1991).

Increasing emissions of nitrogen oxides and volatile organic compounds (especially from the automobile) have dominated much of the 20th century. The result has been the evolution of photochemically polluted urban air. Here important secondary pollutants are produced by reactions in the atmosphere. Most typical is the production of ozone in urban air. Initially, this seemed a peculiar problem of Los Angeles, but it is now found throughout the world. Although ozone characterizes photochemical smog, oxidation products such as nitrogen dioxide or nitric acid may be more important in terms of the attack on stone buildings and metals (Kirkitsos & Sikiotis 1995).

Harrison (2004) reported a substantial reduction in particulate matter in Europe during the latter half of the 20th century due to cuts in emissions from commercial, domestic and institutional sources and mostly related to reductions in coal combustion. Today road traffic is the main contributor, especially at roadside ground levels, although it is also important to the urban background, with diesel being the main source of elemental and organic carbon. In Western Europe the control of $SO_2$ emissions has been effective, whereas $NO_x$ is more difficult and so is still comparatively high. The European Union (EU) vehicle emission standards will probably help to reduce particulate matter emissions (Harrison 2004).

Ambient air quality standards have been designed to protect human health. Other issues have generally been covered by secondary standards as opposed to primary standards for health (Livingstone 1996). In the EU aspirations to include cultural heritage within air pollution regulations have not been fully realized. Some difficulty arises in creating laws to protect heritage from pollutants because the mechanisms for material damage are so different to that of health. Many health effects are driven by short-term concentration, while damage to materials accumulates over long periods. Here it is the cumulative flux that is often more important than short-term concentration because buildings are exposed for many hundreds of years.

## *Changes in climate*

Air pollution has often been the principal focus of our concern over building damage in the 20th century. Nevertheless, weather has always been regarded as an important factor in degradation. Nicholas Hawksmoor was able to write that weathering was a result of time, weather and smoke (Brimblecombe 2000). The word 'weathering' continues to be used to describe building damage, drawing attention to the importance of climatic factors.

The effect of climate change on buildings has been of less interest. Nevertheless, climates of the past were quite different. The weather in the Early Modern Period was so different that the climatologist Hubert Lamb popularized the term *Little Ice Age*. While modern scholarship has made this period less easy to define, in many areas of Europe the temperatures may have been cooler. It is also possible that it was a good deal

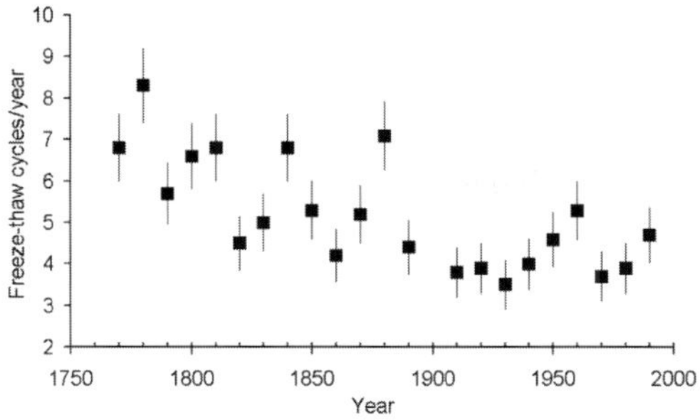


**Fig. 1.** Decadal means of the freeze–thaw cycles in central England (Brimblecombe & Camuffo 2003).

stormier, and Glasser *et al.* (1999) have argued there were periods of notable storminess, in the Adriatic Sea and in Spain.

Much of the early work in establishing past climate was based on descriptive records, but a more quantitative picture is possible for temperature changes over the last 300 years. Often daily records survive for much of this period, so one can estimate the likely number of freeze–thaw cycles (Brimblecombe & Camuffo 2003). Long-term changes in the annual number of freeze–thaw events for central England are shown in Figure 1. Here, colder weather two centuries back meant that there were more freeze–thaw cycles. The largest number of cycles is typically experienced by climates that hover close to 0°. It is likely that northern climates may have become so cold in past centuries that they experienced fewer freeze–thaw cycles. However, in more southerly latitudes the implication is that in past centuries it was likely that freeze–thaw events were more frequent.

## Damage to monuments

### *Weathering*

Gradual damage to stone is often termed 'weathering'. As mentioned before the term 'weathering' reflects the view that climate is a key factor in damage, despite the fact that the earliest uses of the word in the 1500s tended to emphasize the positive benefits of exposure to weather or a sense of drying. However, architects and geologists saw the more negative aspects of weathering in its ability to wear and disintegrate, such that 300 years ago architects were convinced that buildings were destroyed by 'time, smoke and weather' (Brimblecombe 2000).

Often the most evident signs of climate weathering in cooler regions is frost damage. This is caused when wet stone freezes and the associated volume change causes the outer layers to shatter. Wetting and drying cycles, particularly where there are salts in the stone, can also cause similar damage. Salts are most typically found close to the ground, but high up on buildings they can be delivered by occasional salty rains that are likely to occur in near-coastal regimes during periods of very high winds. The high winds can drive rain deep into sheltered parts of buildings that normally remain dry. Gale force wind can cause direct damage, by removing tiles and blowing weak parts of building down. Lightning strikes, although rare, also cause damage particularly to protruding parts of buildings, such as steeples (Brimblecombe 2005). Repeated cycles of heating by sun radiation and cooling may cause temperature gradients leading to the decay of stone surfaces. Marbles and granites may be more susceptible than porous stone to heating. Calcite in marbles expands along one crystallographic axis and contracts in the other two, whereas the main granite minerals have different thermal expansion coefficients (Honeycombe 1990).

### *Pollution*

Since Roman times there have been records of complaints about the soiling of buildings. However, the widespread use of coal in 17th century London led to a much more pronounced problem. Architects such as Wren and Hawksmoor were concerned at the thick black layers of sulphate that covered buildings exposed to coal smoke.

It is difficult to ascertain, in a quantitative way, the changing degradation to building materials on these century-long timescales. Nevertheless, many people argue that the rate of degradation became more rapid in the 20th century. However, before accepting this one should remember that erasure of features is not only irreversible, it is a cumulative effect. This latter point may well obscure our understanding, because so often we observe cumulative damage when infact knowledge of the rate of damage at a given time would be more useful. Despite the difficulties in long-term measurements, damage from urban air pollutants can be shown to be important by comparing the rate of deterioration of monuments in urban and rural areas where clear differences usually emerge (BERG 1989).

The correlation between air pollution and damage may not always be simple. In many cities the amounts of corrosive primary pollutants have been decreasing, for example $SO_2$ or smoke, perhaps from as early as the turn of the 20th century. This is hardly in line with rapid and increasing degradation of buildings and monuments regarded as typical of the century. Improvements in the quality of urban air have not necessarily been matched by improvements in the rate of degradation of the built environment. Some reasons for this can be postulated. The total load of corrosive pollutants in the urban atmosphere may well have decreased, but it is possible that some components may not be decreasing, for example ozone and nitrogen oxides or the components of photochemical smog. These pollutants can enhance the degradation of building materials (Johansson *et al.* 1986) or act as catalysts that increase the effectiveness of the attack of traditional pollutants. Moreover, building materials possess some kind of 'memory' and the damage appearing now represents pollutant deposition that occurred in the past (BERG 1989).

In recent times there has been a shift away from the high levels of sulphate deposition. Now deposits are dominated by diesel soot and nitrogen compounds. This means that crusts found on buildings in the future may show a more pronounced organic chemistry. The oxidation of deposited organic material can change the colour of building surfaces. The decrease of soot concentrations in urban areas can also lead to a self-cleaning of historic buildings by rain and wind (e.g. Davidson *et al.* 2000).

### *Stone damage*

The impact of atmospheric pollution, especially $SO_2$ on stone, has been studied in detail because often treasured old buildings are built from this material. The main impacts of air pollutants on stone are chemical deterioration (material or stone damage) and soiling or blackening (aesthetic damage, e.g. Grossi & Brimblecombe 2002).

Dry deposition of $SO_2$, mainly coming from the combustion of fossil fuels, and the subsequent oxidation into sulphate was until recently the prevalent decay mechanism in urban areas. For the oxidation reaction to take place two key characteristics are necessary. First, moisture must be present on the surface or in near-surface pores. Secondly, there must be an oxidation reaction to convert $SO_2$ to sulphuric acid or to convert an intermediate sulphite salt to a sulphate (calcium sulphate dyhydrate). Other gases present in urban atmospheres, such as $NO_x$, can also deposit on stone surfaces. In the presence of moisture, $NO_x$ can enhance $SO_2$ oxidation (Johansson *et al.* 1988).

The open porosity and specific surface area of the stone influence moisture transfer and determine to a great extent the deposition of pollutants on the stone. Also, the surface roughness can affect the deposition of atmospheric particles on the stone. Where the reaction products from pollutant attack are soluble salts these can migrate to the interior of the stone and can contribute to its deterioration. Salts on building surfaces also maintain high humidity inside the stone, enhancing the deposition of further pollutants.

Carbonate-bearing stones are the most sensitive to the effects of pollution. In a $SO_2$ polluted environment, carbonate stones that are not heavily rain-washed commonly develop a hard surface layer of gypsum, blackened by incorporation of soot particles (Cooke & Gibbs 1994). The growth of this gypsum within the stone can exert pressures that physically destroy the stone fabric in a variety of ways (gypsum has a larger molar volume than calcium carbonate). The surfaces of carbonate stones heavily and frequently washed by rainwater can also suffer some dissolution process. This dissolution is more rapid if the stone has had a prior exposure to air polluted with $SO_2$ (Cooke & Gibbs 1994).

Carbonate stones containing dolomite – $CaMg(CO_3)_2$ – react with $SO_2$ to produce both calcium and magnesium sulphate. Calcium sulphate forms a layer of gypsum and soot. The more soluble magnesium sulphate can penetrate further, leading to additional damage for the crystallization of this type of salt. Calcareous sandstones are attacked by atmospheric sulphur acids and tend to weather severely in highly polluted areas because the dissolving of small amounts of calcite can release many sand grains (Honeycombe 1990).

'Siliceous stones, such as quartz-based sandstones, are very resistant to the sulphur acids in the air, but they can become very dirty (Honeycombe 1990)'.

The attack by acidic gases can be important on some types of roofing slates. If the slate contains calcite, $SO_2$ dissolved in rainwater forms an acid that can be held by capillarity within the slates lap and attack the calcite. The crystallization of gypsum as a result of this reaction can cause further damage. If the slate contains pyrite (iron sulphide) in an unstable form and some calcite, rainwater can react with the pyrite to form a sulphur acid that attacks the calcite and may lead the slate to collapse (Honeycombe 1990).

Granites used for building are supposed to suffer little damage by acidic pollutants. Sabbioni (2003) reported that two different types of damage layers can be found on granite. Some are gypsum crusts, where all constituents derive the deposition of air pollutants. Others correspond to clay–calcitic layers whose constituents originate from the original rock and must be considered as weathering layers as they are a natural evolution of granite. Schiavon (2000) suggested that in humid temperate climates $SO_2$ from air pollution plays a dual role in the weathering of granitic building stones, promoting both sulphate precipitation and kaolinization of feldspars. Some of the $Ca^{2+}$ ions needed for gypsum crystallization may derive from plagioclase weathering.

## *Aesthetic damage: blackening*

The decreasing concentrations of acidic pollutants over the past decades mean less reaction and chemical damage to stone surfaces. This increases the significance of the blackening and has raised the importance of aesthetic considerations.

Soiling or blackening of building materials is a visual nuisance resulting from the darkening of exposed surfaces by the accumulation of particulate matter (Haynie 1986). It is related to the surface area covered by carbonaceous fine particles, which contain dark elemental carbon. Consequently, the deposition of urban particulate matter onto the surfaces of buildings has long been a cause of concern. The poet Horace was especially annoyed by the blackening of religious buildings in Ancient Rome (Brimblecombe 2000). By the mid 1600s, those concerned with the appearance and beauty of architecture, such as the diarist John Evelyn, were particularly worried about the ugliness of London's soot-encrusted buildings. The carbonaceous deposits are retained on surfaces and have high optical absorptivity (Lanting 1986), so are very effective at blackening building surfaces.

During the second half of the 20th century blackening remained an important issue (Newby *et al.* 1991). As the use of coal in many cities has declined dramatically, diesel soot has become the significant source of elemental carbon. Despite continued concern, in some locations decreasing atmospheric soot concentrations have meant that in recent years there has been less darkening, and in some cases rainfall removal has made buildings much cleaner (Davidson *et al.* 2000). Biological activity, perhaps supported by an ongoing increase in organic pollution, may also contribute significantly to stone blackening (Viles & Gorbushina 2003).

## *Rate of blackening*

The blackening process can be measured as a reflectance change and can be described by the expression $R_t = R_o - (R_o - R_\infty)(1 - \exp(-kt))$, where $R_t$ is reflectance at time $t$ and indicates the degree of blackening, $R_o$ is the initial reflectance of the stone and $R_\infty$ represents the final blackening that would be shown by the material after a very long time. The value $k$ – time constant – indicates the rate of blackening and most probably depends on the concentration and deposition velocity of the different types of atmospheric particles and removal processes. This bounded exponential fit seems to explain better than other models the blackening process for long data records (Brimblecombe & Grossi 2004).

Recently, colour variations have often been determined with a colorimeter, rather than a reflectance meter. These results are reported using the CIEL$^*a^*b^*$ and CIEL$^*C^*h$ systems, which give a good representation of the human sensibility to colour. $L^*$ is lightness, which ranges from black to white; $a^*$(red–green) and $b^*$(blue–yellow) are the chromatic co-ordinates. $C^*$ is the chroma, saturation or colour purity, and $h$ is the hue angle in the colour wheel. This approach gives more information than a simple reflectance measurement. Experimental measurements on stone show there are both significant changes in $L^*$ and the colour co-ordinate $b^*$ after a period of outdoors exposure. The colour changes in $b^*$ indicate a yellowing process and can be faster than the blackening. The yellowing may arise from different processes including sulphation (Grossi *et al.* 2000, 2007) or the deposition or oxidation of organic materials or iron (Simon & Snethlage 1996). In the near future it may be that in an atmosphere more dominated by organic pollutants the yellowing process may be more important, depending on the material types being affected.

## *Perception of blackening*

Carey (1959) and Hancock *et al.* (1976) reported than when 0.2% of the area of a white surface is covered by dark particulate matter, the difference between soiled and unsoiled areas can be perceived by the human eye. Lanting (1986) indicated that an area coverage of 2% would trigger off probable complaints, which would be serious at a coverage of 5%. More recent work (e.g. Bellan *et al.* 2000) claimed that observers are only able to detect that a sample is becoming soiled once surface coverage by black carbon particles has reached 2.4%. However, with buildings in context the issues are more complex.

Research on 'aesthetics of soiling', including studies on public perception of blackening of historic building, have considered public feelings about the appearance of the buildings. The perception of blackening depends on the individual and general conditions of the local environment (Newby *et al.* 1991). Blackening of historic buildings can be a visible nuisance or can be aesthetically beneficial. When people look at old buildings there may be an expectation that they should appear darkened and possess a kind of patina (Matteini 2005). Thus, darkened surfaces may be valued and have an aesthetic quality that can enhance the appeal of the building. Light and moderate darkening around architectural details can improve the visual appearance of the building by increasing contrast and enhancing shadowing effects, while at the same time adding a pattern of blackening that was originally absent. However, Andrew (1992) reported that heavy soiling leads to a uniform blackening, reduces the visual information on architectural details, and ultimately obscures the colour, texture and any shadowing effect.

Recently 'public perception' of blackening has been examined using *in situ* questionnaires at different historic European buildings, which presented a range of blackening and occupied different surroundings (Brimblecombe & Grossi 2005). The questionnaires examined impressions of the building, surroundings, sensation of dirtiness, apparent causes of soiling, need of cleaning and feeling of colour. The historic buildings struck respondents as: 'magnificent, beautiful, nice, impressive' and in the case of some buildings 'antique, old'. An impression of age seems a frequent response to civic buildings, while cathedrals tend to be seen as 'magnificent'.

Some opinions that the building was 'dirty' were evident at some historic edifices that exhibited strong contrasts between clean and blackened (and occasionally green) areas or the whole façades covered with dark soot. Despite some level of comment on this blackening, even at these darkened buildings 'discoloration' was not the most striking impression. When prompted, viewers often perceived façades as 'dirty'. However, sometimes they answered that 'it's not dirty, it's old' or 'it's old, ancient, naturally aged ...'. Some even suggested that the building showed the 'patina of time' or commented 'it's dirty, but looks better like that because it seems antique'. Visitors were understandably more sensitive to changes in lightness than in hue or chroma, which remain subtle with regard to stone.

## *Setting aesthetic thresholds*

Management of both air pollution and buildings requires a knowledge of the degree of blackening that is aesthetically tolerable. This issue was explored as part of the EC funded CARAMEL project (ENV4-CT2000-00029). More than 900 respondents from nine historic sites were asked to choose values of lightness from a greyscale (Fig. 2), that best represented the shade of the building. The choices indicated the perceived lightness (reflectance) of the building rather than an estimate of the reflectance of stones. This perceived lightness ($L_p$) is strongly related to the probability of regarding the building as 'dirty', as seen by the sigmoid form in Figure 3.

The results in Figure 3 show that even if a building is almost white a small percentage of respondents (about 6%) would nevertheless still find it to be 'dirty'. This may not simply be because they perceive it to be blackened, but psychologically they relate it to other issues, for example 'that window

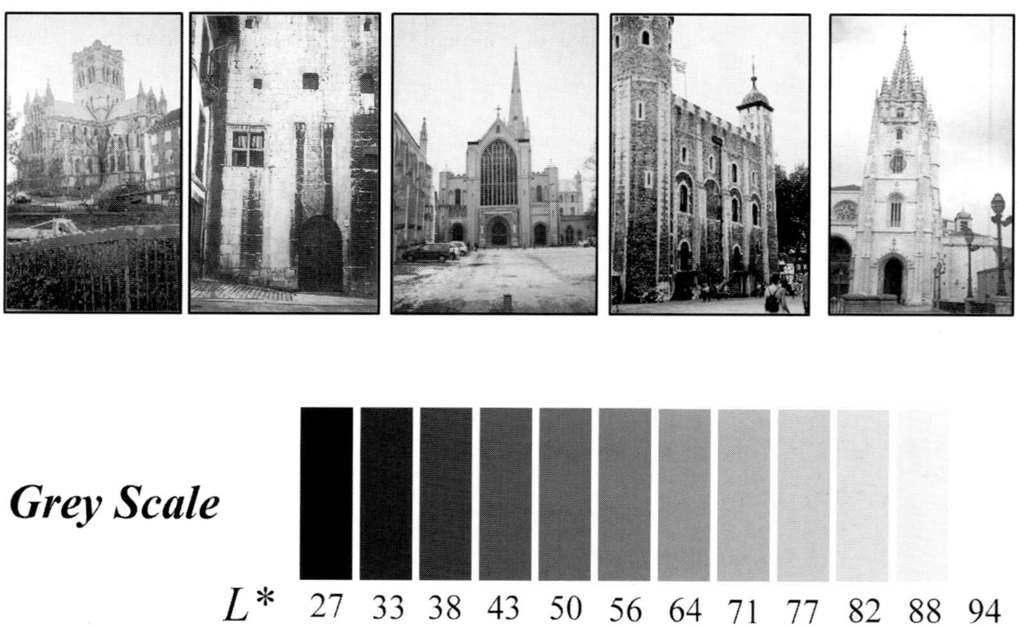


**Fig. 2.** Some historic buildings where the perception of blackening was investigated and the greyscale used to determine visitor's perceived lightness ($L_p$). From left to right: Roman Catholic Cathedral 'St. John the Baptist' (Norwich, UK); Palacio 'Marqués de Sta. Cruz' (Oviedo, Spain); Norwich Gothic Cathedral; White Tower (Tower of London, UK); and Cathedral of Oviedo (Grossi & Brimblecombe 2004*a*).

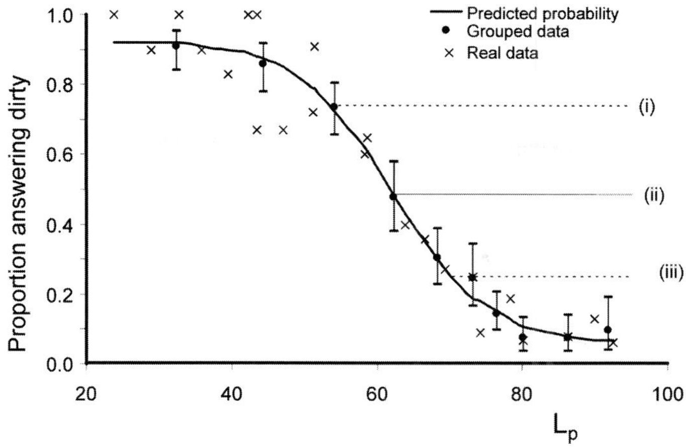


**Fig. 3.** Probability of finding the façade dirty when prompted as a function of the perceived lightness ($L_p$). The line is the best fit in the form of a logistic regression. The crosses represent raw data. The bars are binomial errors. Points i, ii and iii were determined by solving differential equations (Brimblecombe & Grossi 2005).

is broken', 'there is some litter around', the 'colour around the door varies', the 'portico needs restoration' and so on. In the same way, if a historic building is very dark some people (about 7%) seem to find it 'not dirty' and argue that it is simply that 'it's old, it's aged, has the patina of time or has character' rather than being blackened.

When the sigmoid curve is differentiated with respect to perceived lightness it is possible to define various parts of the curve and argue where likely thresholds might occur (see Brimblecombe & Grossi 2005). The lower roughly flat part of the curve in Figure 3 (i.e. $L_p > 70.5\%$) describes a region where there is less than a one in four chance of the façade being judged as dirty. If the surface is perceived to be darker than this one can see a rapid change in opinion. Arguments (fully explored in Brimblecombe & Grossi 2005) such as this have been used to suggest a range of potential thresholds for the acceptance of blackening.

Thresholds for acceptable levels of blackening offer the potential for setting allowable concentrations of elemental carbon (EC) in the atmosphere. The relation between perceived grey-scale values at the sites studied by Brimblecombe & Grossi (2005) and EC concentrations are shown in Figure 4. This figure also indicates the position of five suggested thresholds using several approaches, such as mathematical, statistical or administrative. It is possible to see that when EC concentration reaches 10 $\mu$g m$^{-3}$, the sites fail to attain satisfactory lightness regardless of the threshold criterion. Where EC is 2 $\mu$g m$^{-3}$ things are much better, with most of the criteria being satisfied, and incline one to think of an acceptable level for the exposure of buildings in urban areas in the range 2–3 $\mu$g m$^{-3}$. However, these values were not proposed as adoptable standards, but suggested that a semi-quantitative approach is possible. No doubt any levels ultimately adopted would need to reflect local political and cultural concerns.

The analysis above avoids mentioning of exposure time to the pollutants and has set a limit in terms of air pollution concentration alone. There is some justification in this because while buildings change colour over time this tends to be an exponential process and after some years of exposure to pollution the colour can reach a fairly constant level (Brimblecombe & Grossi 2004). The work also has been limited to light coloured stone, but this is assumed to be the most sensitive

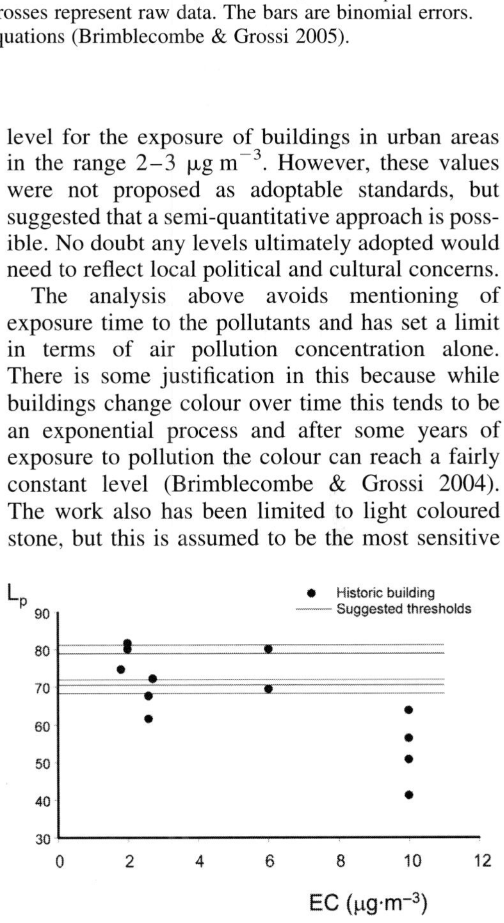


**Fig. 4.** Mean values of perceived lightness ($L_p$) at the historic building sites (as dots) as a function of assigned elemental carbon (EC) concentrations. Lines represent the different suggested aesthetic thresholds (Brimblecombe & Grossi 2005).

**Fig. 5.** Images for desktop exercises on aesthetic of soiling patterns. Left: pedimented window frame to simulate soiling patterns. Right: examples of designed images and corresponding real façades used for guidance (White Tower, Tower of London) (Grossi & Brimblecombe 2004*b*).

to aesthetic change, as darker coloured stone can probably accept higher soot loads.

## *Blackening patterns*

Blackening is not typically present as an homogeneous layer that covers an entire façade. The patterns have long been regarded as offensive. In Ben Jonson's (1606) savage play *Volpone; Or, The Fox* we read of the character Volpone described: 'like an old smoked wall, on which the rain ran down in streaks!'. Despite the negative views of many blackening patterns there is little research on public perception. Grossi & Brimblecombe (2004) studied the acceptability of various blackening patterns using two desktop exercises, with a methodology similar to those used in studies of the psychology of art (Pickford 1972). A range of computer-simulated soiling patterns were placed on a simple architectural element; a pedimented window (Fig. 5). In the first exercise people were asked to arrange the images from the 'most to the least acceptable' pattern. This first study hinted at the importance of certain features, which seemed to be driving the choices. A second exercise tested the importance of these parameters.

'Soiling' acceptance depends on low levels of blackening and uniform distributions (Fig. 6). Some patterns that create shadowing effects have been considered to be more acceptable. Others cause strong negative reaction, more generally those that obscure architectural forms, such as vertical streaking, lumpiness and to some extent the fractal dimension of the feature (Grossi & Brimblecombe 2004b).

Clearly, it is necessary to balance decisions based on the perceived lightness against those derived from views about disfiguring patterns. Although this is hard to assess, offensive patters can easily dominate visitor experience. Managing the

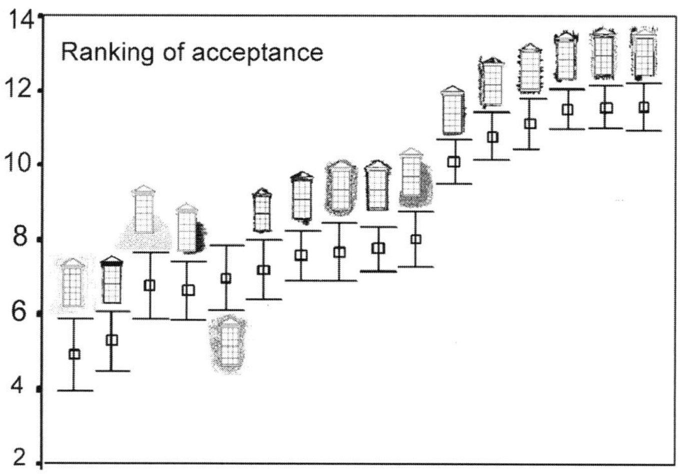


**Fig. 6.** Ranking of acceptance of simulated blackening patterns: from '1', more acceptable, to '16', less acceptable (Grossi & Brimblecombe 2004*b*).

appearance of historic buildings require particular attention to soiling patterns. These may need to be a special focus when selective approaches are adopted to stone cleaning (as at the Tower of London).

## Effects of climate change

In the last century climate has often seemed less important than air pollution as a determinant of damage to building materials. The reduction in acidic air pollutants in urban areas means that frost, rain or wind can be more dominant as weathering processes than in the recent past. Although the predicted changes for future temperature or precipitation seem small they can be amplified in some mechanisms of damage. Frost damage and salt weathering seem likely to be sensitive to climate change over the next century (Brimblecombe *et al.* 2006a).

Concern about climate change and heritage in the UK has been investigated through regional workshops in the east and the NW of England, sampling the main concerns of local managers, advisers and field officers about conservation and management issues (Cassar 2005). When commenting on climate risks to buildings, flooding was rated as the most important issue and extreme weather to be of great importance to the fabric of buildings; this included coastal loss, fluvial flooding, storminess and extreme winds, and rain as the greatest threats to historic buildings and their content. Temperature was considered to be a factor of some importance, thermal shock being judged as more significant than its actual level. Changes in soil moisture content leading to subsidence and heave were considered of some concern. In the case of buried archaeology, coastal loss, flooding and changes in height of water table seemed to be of the highest concern, whereas the effects of heavy rain raised lesser concerns. Few worries were expressed about pest and diseases, and health and safety. The predicted changes in temperature and humidity were considered unlikely to affect the buried archaeology.

Viles (2002) reviewed the implications of climatic changes for the 21st century and mentioned four aspects that are likely to have an impact on stone damage: (1) atmospheric composition (e.g. $CO_2$ and other trace gases concentration) and basic climatic attributes (e.g. temperature); (2) seasonal–decadal variability of climate (e.g. extreme events); (3) changes on terrestrial and oceanic systems (e.g. effects on biotic communities, sea levels, soil chemistry and ground water); and (4) human activities (e.g. building practices or use of land).

More recently the NOAH's ARK proiect (http://noahsark.isac.cnr.it) has identified the following groups of climatic parameters as relevant to stone decay (Brimblecombe *et al.* 2006*b*): (1) temperature derived parameters (i.e. freeze–thaw); (2) water-derived parameters (precipitation, humidity cycles, time of wetness); (3) wind-derived parameters (i.e. wind-driven rain, salt); and (4) pollution-derived parameters (such as $SO_2$, $NO_2$, particulates or pH). The magnitude of changes in these parameters is being estimated through the extensive availability of output from future climate models. NOAH's ARK has particularly used the Europe daily output from the HadCM3a2 scenario.

### *Weathering*

*Temperature-derived parameters.* The influence increasing temperature on the deterioration process might be seen as relatively slight because it is hard to imagine that just few degrees would lead to a significant change in the rate of deterioration of heritage. However, there are factors that serve to enhance the impact of small changes. The number of freeze–thaw cycles is especially sensitive to temperature and the likely reduction in freezing across much of Europe in the future will lower the potential for frost shattering of porous building stone. However, in the far north, increasing temperatures threatens to melt the upper layers of permafrost or to induce freeze–that cycles that can disrupt the structure of soils and damage archaeological and paleoecological remains well preserved in the permafrost (Davis *et al.* 2000; Viles 2002). As an example, calculations from the Hadley model suggest a fourfold increase in the number of days above freezing at Narssarssuaq in Greenland (Brimblecombe *et al.* 2006*b*).

Viles (2002) pointed out that most physical temperature-related weathering processes require not only cycling of temperature to produce decay, but also moisture. Changes in rainfall may be critical, altering the water supply. In the NOAH's ARK project we have defined a climatic parameter named 'wet–frost days' as number of days of freezing weather (i.e. below 0 °C) that follow days of rain. Different parts of Europe will experience different changes. Viles (2002) also hints at the possibility of future wet–frost increase in northern hemisphere high latitudes.

The increase of temperature might also be paralleled by an increase in solar radiation that may accelerate deterioration of organic materials, such as stone conservation treatments or paint coatings. Changes in temperature can also affect wetting–drying cycles and therefore the deposition rate of acidic gases (Brimblecombe *et al.* 2006*a*).

Salt weathering will also respond to change in temperatures in different ways (Viles 2002). Increasing temperature might increase the solubility of some salts, but also encourage evaporation which helps promote crystallization. The precipitation of salts in different states of hydration is also temperature (and humidity) dependent along with thermal expansion of hydrated or dehydrated salts produced from supersaturated solutions. Viles (2002) also mentions the possibility that increasing aridity in some vulnerable areas may encourage evaporation and movement of salts.

Climate change is also predicted to affect individual organisms, populations, species distributions, and ecosystem composition and function (Viles 2002; Brimblecombe 2005).

*Water-derived parameters (humidity and precipitation)*

- Change of relative humidity/moisture. For most materials increases in relative humidity cause an increase in deterioration rate. This often comes about through prolonged times of wetness, higher deposition rates of pollutants and more favourable conditions for microbiological activities. Early analysis hints at much drier mid-summers in Central Europe in the future (Fig. 7), which may reduce damage to buildings (Brimblecombe 2005). However, stone is vulnerable to damage from hygroscopic salts, when the humidity oscillates between high and low values. The predicted decrease of humidity might mean that daily variations in humidity are more likely to cross critical values such as 75.5% RH, where sodium chloride changes from a solution to a crystalline state (Brimblecombe *et al.* 2006*a*). Furthermore, historic buildings have a great intimacy with the ground and porous stones can draw water into the building structure and lose it to the environment by surface evaporation. Changes in soil moisture might result in greater salt mobilisation and consequent damaging crystallization on decorated surfaces.
- Time of wetness. Time of wetness is related to time of coverage by a thin layer of water and is useful to describe water on building surfaces. The most common transformation from meteorological parameters is to assume that is related to high-humidity (i.e. >80%) conditions occurring at temperatures high enough to guarantee that liquid water does not freeze (i.e. >0 °C) (Brimblecombe *et al.* 2006*a*). Time of wetness is predicted to decrease slightly over the next century in the output from the HADCM3a2 model. The seasonal changes are somewhat complex, where drier summers mean less surface wetness (Fig. 8). However, warmer winters result in freezing conditions being less common in future so times of wetness will increase in winter months. This picture may well mean that high pollutant loads in the winter season will be more damaging in the future. However, hopefully acidic pollutants within cities will continue to decline.
- Change of precipitation. The Hadley model suggests rainfall in general is often likely to decrease slightly in Europe over the next century, particularly in the summer months (Brimblecombe *et al.* 2006*b*). However, when looking at the predicted maximum daily rainfall one finds a future with more individual days that are much rainier. The frequency of very rainy days is predicted to increase, at many European sites, over the next century. Predicted maximum daily rainfall amounts also increase. Many

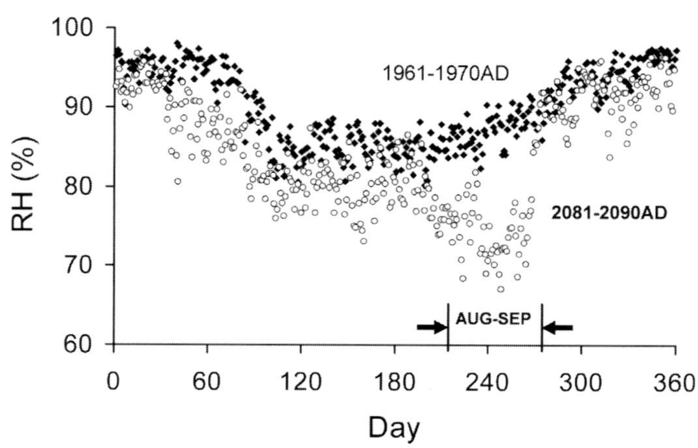


**Fig. 7.** Predicted seasonal humidity cycles for two different decades in Central Europe using the HadCM3a2 output. The figure illustrates the drier summers that may confront heritage at the end of the 21st century (Brimblecombe 2005).

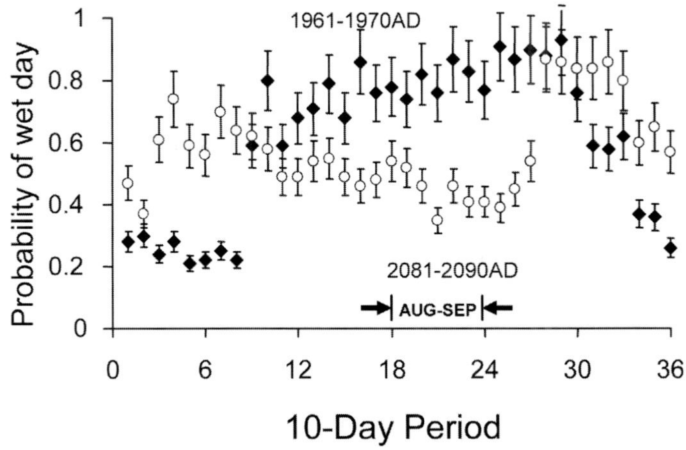


**Fig. 8.** Predicted seasonal *time of wetness* for two different decades in Central Europe using the HadCM3a2 output. The figure illustrates drier summers at the end of the 21st century. However, warmer winters and less freezing conditions might increase the time of wetness in winter months.

processes affecting the conservation of historic sites do not occur at regular intervals, but are concentrated in episodic outbursts (Viles 2002) related to the presence of extreme events, such as severe storms, carrying both water and dust (Brimblecombe 2005). There are fears that the frequency of flooding events will increase in many areas of Europe. However, this has to be addressed on a river-by-river basis because of the considerable spatial variability of rainfall. Increased winter rainfall and storminess, particularly in exposed areas, is resulting in water-saturated building fabric. The problems are serious enough to worry the European Insurance Industry (see Mills 2005).

Precipitation can also affect the damage caused by wet deposition by dissolution of surface layers of materials. Erosion and delivery of acidity are important aspects of the role played by precipitation. Changes in the chemical composition, and especially pH, can affect the deterioration rate of building material also.

*Wind-derived parameters*

- Change in wind velocity – Wind-driven rain. An increase in wind velocity affects the deterioration of materials in several ways. Increased eddies and flows around historic buildings can alter the deposition rates of both gaseous and particulates pollutants and strengthen the effect of driving rain. A very serious effect may be the increased transport of sea salt inland, which can substantially enlarge the areas along sea coasts affected by marine aerosols. As mentioned above, changes in the intensity and direction of wind-driven rain can alter the patterns of disfiguring soot deposits and make buildings less appealing because of rain streaking (these jagged features usually extend down from protrusions and are often deemed as unattractive: see Grossi & Brimblecombe 2004*b*).

In NOAH's ARK initial analyses on a broad European scale indicate only minor changes in wind-driven rain during the 21st century. However, some relevant changes must be hidden by the coarse-scale considered HADCM3 (Brimblecombe *et al.* 2006*b*).

*Pollution*

*Stone damage.* Viles (2002) commented on the impact of future air pollution. It will be dominated by local processes, but as revealed in the NOAH project these are in decline. It may be that the urban atmosphere will be increasingly dominated by organic materials while traditional pollutants such as the sulphur and nitrogen (ultimately) oxides will decrease.

*Blackening.* The 21st century offers the potential for dramatic changes in the blackening patterns due to new climate regimes, most particularly through changes in wind-driven rain. In today's urban environments, where it is likely particle concentrations will decrease, as a result of tightening legislation urban historic buildings could self-clean, but may develop new patterns of darkening. Therefore future blackening patterns will be a balance between accumulation and redistribution: (1) the accumulation will be mostly influenced by atmospheric elemental carbon concentrations, surface roughness and time of wetness; and (2) the redistribution will be dominated by wind-driven rain, precipitation amount and wind direction. A seasonal rainfall increase may also encourage micro-organisms growth, which itself can produce widespread blackening (Viles 2002).

Moreover, today and in the near future cleaner atmospheres, perhaps more dominated by organic pollutants, may result in a yellowing process being of greater concern (Grossi *et al.* 2007). Urban atmospheric deposits richer in oily organics and poorer in elemental carbon are liable to produce brownish–yellowish coatings on urban building stones increasingly noticed in places like the Tower of London. The oxidation of soot on the surfaces of crusts can produce HULIS, which has a brownish colour (e.g. Graber & Rudich 2006).

### *Others*

There are many other aspects of climate change that can affect stone decay. Viles (2002) mentions *sea-level rise*, which can lead to an increase of marine salt damage in near-coastal sites; the alteration in the *depth* and *composition of groundwater* that can also change the effects of soluble salts and even *social change*. The adaptation of humans to global warming will provide some major impacts on stone deterioration, such as the reduction for indoors extensive heating but the increase for internal air conditioning in hotter climates, the reduction in the use of de-icing salts (or urea) in freeze-prone roads, the change of architectural styles or the use of different materials or the use of 'environmental friendly' techniques as a result of worries about human contribution to global warming, etc. Important though these factors are, it is not easy to assess how this complex array of change will affect our architectural heritage in the future.

## Conclusions

The long lifetime of European historic buildings exposes them to very significant changes in pollution and climate. In the past frost damage was important, but in many European locations looks set to decrease in the face of rising temperatures. Air pollution control has substantially decreased the exposure of buildings to traditional acid air pollutants. This means a significant shift from high levels of sulphate deposition, through to a blackening process dominated by diesel soot and nitrogen deposition from combustion sources in cities. In terms of porous stone surfaces this has led to a transition from gypsum-rich crusts through to more organic layers and a concomitant potential for greater biological activity. The coming century offers the potential for even more dramatic changes through new climate regimes, most particularly in changes in humidity stress, time of wetness and wind-driven rain. These will further alter the way in which pollutants attack historic buildings.

Studies on 'aesthetics of soiling' show a complex relationship between blackening and architectural perception. Sometimes soiling can be aesthetically beneficial as many old buildings display a dark layer that enhances the appeal. However, blackening of light coloured fabric eventually reaches a point where it becomes publicly unacceptable and raises pressure for cleaning. Converting these observations into air pollution standards implies a translation from physics and chemistry aspects to the world of values that presents considerable challenge. However, public perception of the lightness of building stones suggests aesthetic thresholds to the darkening of buildings. These aesthetic thresholds can suggest limit values for elemental carbon in the air (perhaps in the range 2–3 $\mu g\ m^{-3}$), such that significant buildings do not become unacceptably discoloured. Developments of this kind contribute to the regulation of non-health aspects of air pollution and aid decision making in the management of significant buildings. Patterns of blackening also affect the perception of buildings, and in future changes in wind-driven rain are likely to redistribute black material.

Key climate factors that are likely to be relevant to building damage are: temperature, which affects the potential for freeze–thaw events and microbiological activity; humidity and time of wetness, which controls the deposition of pollutants, salt damage and microbiological activity; wind velocity, which affects deposition rates, transport of sea salts and blackening patterns; and precipitation, which causes flooding and the transport of pollutants.

The review suggests our attention must focus on a new range of issues and new balances between physical and aesthetic damage. We enter a century where it is climate that will place buildings under new threats.

This paper has benefited from EU funding within the projects CARAMEL (ENV4-CT-2000-0002) and NOAH's ARK (CT-2003-501837)

## References

ANDREW, C. 1992. Towards an aesthetic theory of building soiling. *In*: WEBSTER, R. G. M. (ed.) *Stone Cleaning and the Nature, Soiling and Decay Mechanisms of Stone*. Donhead, London, 63–81.

AUSSET, P., BANNERY, F., MONTE, M. D. & LEFÈVRE, R. A. 1998. Recording of pre-industrial atmospheric environment by ancient crusts on stone monuments. *Atmospheric Environment*, **32**, 2859–2863.

BELLAN, L. M., SALMON, L. G. & CASS, R. 2000. A study on the human ability to detect soot deposition onto works of art. *Environmental Science and Technology*, **34**, 1946–1952.

BERG. 1989. *The Effects of Acid Deposition on Buildings and Building Materials in the United Kingdom*. HMSO, London.

BRIMBLECOMBE, P. 1977. London air pollution, 1500–1900. *Atmospheric Environment*, **11**, 1157–1162.

BRIMBLECOMBE, P. 1987. *The Big Smoke*. Methuen, London.

BRIMBLECOMBE, P. 2000. Air pollution and architecture, past, present and future. *Journal of Architectural Conservation*, **6**, 30–46.

BRIMBLECOMBE, P. 2005. The NOAH's ARK Project. The impact of future climate change on cultural heritage. *THE EGGS. Newsletter & Information service of the E.G.U*, **12**, July, 31–33. http://www.the-eggs.org/articles.php?id=70.

BRIMBLECOMBE, P. & CAMUFFO, D. 2003. Long-term damage to the built environment. *In*: BRIMBLECOMBE, P. (ed.) *The Effects of Air Pollution on the Built Environment*. Air Pollution Reviews, Volume 2. Imperial College Press, London, 1–30.

BRIMBLECOMBE, P. & GROSSI, C. M. 2004. The rate of darkening of material surfaces. *In*: SAIZ-JIMENEZ, C. (ed.) *Air Pollution and Cultural Heritage*. A. A. BALKEMA, Rotterdam, 193–198.

BRIMBLECOMBE, P. & GROSSI, C. M. 2005. Aesthetic thresholds and blackening of stone buildings. *Science of the Total Environment*, **349**, 175–189.

BRIMBLECOMBE, P., GROSSI, C. M. & HARRIS, I. 2006*a*. The effect of long term trends in dampness on historic buildings. *Weather*, **61**, 278–281.

BRIMBLECOMBE, P., GROSSI, C. M. & HARRIS, I. 2006*b*. Climate change critical to cultural heritage. *In*: FORT, R., ÁLVAREZ DE BUERGO, M., GÓMEZ-HERAS, C. & VÁZQUEZ-CALVO, C. (eds) *Heritage, Weathering and Conservation*. A. A. BALKEMA, Rotterdam.

CAREY, W. F. 1959. Atmospheric deposits in Britain – a study of dinginess. *International Journal of Air Pollution*, **2**, 1–26.

CASSAR, M. 2005. *Climate Change and the Historic Environment*. UCL Centre for Sustainable Heritage, London.

COOKE, R. U. & GIBBS, G. B. 1994. *Crumbling Heritage?* National Power plc, Swindon.

DAVIDSON, C., TANG, W., FINGER, S., ETYEMEZIAN, V., STRIEGEL, F. & SHERWOOD, S. 2000. Soiling patterns on a tall limestone building: Changes over 60 years. *Environmental Science and Technology*, **34**, 560–565.

DAVIS, J. L., HEGINBOTTOM, J. A. *ET AL*. 2000. Ground penetrating radar surveys to locate 1918 Spanish flu victims in permafrost. *Journal of Forensic Sciences*, **45**, 68–76.

GLASSER, R., BRÁZIL, R. *ET AL*. 1999. Seasonal temperature and precipitation fluctuations in selected parts of Europe during the Sixteenth Century. *Climatic Change*, **3**, 169–200.

GRABER, E. R. & RUDICH, Y. 2006. Atmospheric HULIS: How humic-like are they? A comprehensive and critical review. *Atmospheric Chemistry and Physics*, **6(3)**, 729–753.

GROSSI, C. M. & BRIMBLECOMBE, P. 2002. The effect of atmospheric pollution on building materials. *Journal of Physique*, **12**, Pr10-197-210.

GROSSI, C. M. & BRIMBLECOMBE, P. 2004*a*. Aesthetics and perception of soiling. *In*: SAIZ-JIMENEZ, C. (ed.) *Air Pollution and Cultural Heritage*. A. A. Balkema, Rotterdam, 199–208.

GROSSI, C. M. & BRIMBLECOMBE, P. 2004*b*. Aesthetics of simulated soiling patterns on architecture. *Environmental Science & Technology*, **38**, 3971–3976.

GROSSI, C. M., ESBERT, R. M., ALONSO, F. J., VALDEÓN, L., ORDAZ, J. & DÍAZ-PACHE, F. 2000. Colour changes and reactivity to $SO_2$ of some cladding stones at the 'Gran Tetare del Liceu' (Barcelona, Spain). *In*: *9th International Congress on Deterioration and Conservation of Stone, Venice, 19–24 June*. Elsevier, Amsterdam, 323–328.

GROSSI, C. M., BRIMBLECOMBE, P., ESBERT, R. M. & ALONSO, F. J. 2007. Color changes in architectural limestones from pollution and cleaning. *Color Research and Application*, in press.

HANCOCK, R. P., ESMEN, N. A. & FURBER, F. P. 1976. Visual response to dustiness. *Journal of Air Pollution Control Association*, **26**, 54–57.

HARRISON, R. M. 2004. Key pollutants – airborne particles. *Science of the Total Environment*, **334–335**, 3–8.

HAYNIE, F. H. 1986. Theoretical model of soiling of surfaces by airborne particles. *In*: LEE, S. D., SCHNEIDER, T., GRANT, L. D. & VERKERK, P. J. (eds) *Aerosols*. Lewis Publishers, Boca Raton, FL, 951–959.

HERMOSIN, B., GAVIÑO, M. & SAIZ-JIMENEZ, C. 2004. Organic compounds in black crusts from different European monuments: a comparative study. *In*: SAIZ-JIMENEZ, C. (ed.) *Air Pollution and Cultural Heritage*. A. A. Balkema, Rotterdam, 47–55.

HONEYCOMBE, D. B. 1990. Weathering and decay of Masonry. *In*: ASHURST, J. & DIMES, F. G. (eds) *Conservation of Building and Decorative Stones*. Butterworth/Heinemann, Oxford (paperback 1998 edition, part 1, 153–178).

JOHANSSON, L.-G., LINDQVIST, O. & MANGIO, R. E. 1988. Corrosion of calcareous stones in humid air containing $SO_2$ and $NO_2$. *Durability of Building Materials*, **5(3 & 4)**, 439–449.

KIRKITSOS, P. & SIKIOTIS, D. 1995. Deterioration of Pentelic marble, Portland limestone and Baumberger sandstone in laboratory exposures to gaseous nitric-acid. *Atmospheric Environment*, **29**, 77–86.

LANTING, R. W. 1986. Black smoke and soiling. *In*: LEE, S. D., SCHNEIDER, T., GRANT, L. D. & VERKERK, P. J. (eds) *Aerosols*. Lewis Publishers, Boca Raton, FL, 923–932.

LIVINGSTONE, R. A. 1996. Air pollution standards for architectural conservation. *In*: BAER, N. S. & SNETHLAGE, R. (eds) *Saving our Architectural Heritage*. Wiley, Chichester, 371–387.

MATTEINI, M. 2005. *Le patine. Genesi, significato, conservazione*. Workshop organized by M. MATTEINI. NARDINI (ed.) Istituto per la Conservazione e Valorizzazione dei Beni Culturali del CNR.

MILLS, E. 2005. Insurance in a climate of change. *Science*, **309**, 1040–1043.

NEWBY, P. T., MANSFIELD, T. A. & HAMILTON, R. S. 1991. Sources and economic implications of building soiling in urban areas. *Science of the Total Environment*, **100**, 347–65.

PICKFORD, R. W. 1972. *Psychology and Visual Aesthetic*. Hutchinson Educational, London.

SABBIONI, C. 2003. Mechanisms of air pollution damage to stone. *In*: BRIMBLECOMBE, P. (ed.) *The Effects of Air Pollution on the Built Environment*. Air Pollution Reviews, Volume 2. Imperial College Press, London, 63–106.

SCHIAVON, N. 2000. Granitic building stone decay in an urban environment: a case of authigenic kaolinite formation by heterogeneous sulphur dioxide attack. *In*: FASSINA, V. (ed.) *9th International Congress on Deterioration and Conservation of Stone, Venice, 19–24 June*. Elsevier, Amsterdam, 411–421.

SENSFUß, F., SEYDEL, P. *ET AL*. 2005. *Annual Survey. Efficient Use of Energy*. BWK – Energie-Fachmagazin, **57**, 25–131.

SHERWOOD, S. I. & BUMBARU, D. 1991. Historical urban $SO_2$ levels. *The Journal of Preservation Technology Bulletin*, **23**, 72.

SIMON, S. & SNETHLAGE, R. 1996. Marble weathering in Europe – Results of the Eurocare–Euromarble exposure programme 1992–1994. *In*: *8th International Congress on Deterioration and Conservation of Stone, Berlin, 30 September–4 October*, Möller Druck und Verlag GmbH, Berlin, Germany, 159–166.

VILES, H. A. 2002. Implications of future climate change for stone consolidation. *In*: SIEGESMUND, S., WEISS, T. & VOLLBRECHT, V. (eds) *Natural Stone, Weathering Phenomena, Conservation Strategies and Case Studies*. Geological Society, London, Special Publications, **205**, 407–418.

VILES, H. A. & GORBUSHINA, A. A. 2003. Soiling and microbial colonisation on urban roadside limestone: a three year study in Oxford, England. *Building and Environment*, **38**, 1217–1224.

WILLIAMS, M. 2004. Air pollution and policy – 1952–2002. *Science of the Total Environment*, **334–335**, 15–20.

# Modelling of the calcareous stone sulphation in polluted atmosphere after exposure in the field

R.-A. LEFÈVRE[1], A. IONESCU[2], P. AUSSET[1], A. CHABAS[1], F. GIRARDET[3] & F. VINCE[1,2]

[1]*Laboratoire Interuniversitaire des Systèmes Atmosphériques (LISA), Université Paris XII, 94010 Créteil, France (e-mail: lefevre@lisa.univ-paris12.fr)*

[2]*Centre d'Etudes et de Recherche en Thermique, Environnement et Systèmes (CERTES), Université Paris XII, 94010 Créteil, France*

[3]*Expert Centre pour la Conservation des Biens Culturels, Ecole Polytechnique Fédérale de Lausanne (EC-EPFL), 1015 Lausanne, Switzerland*

**Abstract:** Parisian Lutetian and Val-de-Loire Turonian Richemont limestone tablets were exposed, sheltered and unsheltered from rain, for up to 3 years in Paris and Tours, respectively. Sulphur concentrations below the stone surfaces were measured from powders obtained by milling the stone in successive steps of 0.1 mm. In tablets exposed to rain, measured sulphur concentration remains equal to the stone background concentration, implying that the sulphur deposited between rain events is leached by the next event. In contrast, in tablets sheltered from rain, the sulphur concentration in the first layer below the stone surface increases non-linearly with time. Sulphation does not, however, penetrate more than 0.2 mm. A sigmoidal Hill curve provides a good fit with changes in measured sulphur concentration over time within the first layer of each sheltered stone. This model reveals a cumulative phenomenon of sulphation, characterized by a saturation level that obstructs deeper penetration of sulphur within the stone. The model shows the same type of time evolution of sulphation for both stones, but with different coefficients; these coefficients are related to the atmospheric environment of exposure and to the different intrinsic properties of each stone.

In polluted atmospheres, calcareous stones undergo many phenomena; among them, sulphation is of paramount importance (Camuffo *et al.* 1982, 1983; Camuffo 1984; Ausset *et al.* 1996, 1999). This sulphation proceeds from the stone surface towards two directions (Lefèvre & Ausset 2002): above the surface, by development of a gypseous black crust, and below the surface, by in-depth sulphation. These two phenomena have been the subject of many descriptions and analyses, but relatively few studies have attempted to model them (Tran Thi Ngoc Lan *et al.* 2005). The present study focuses on modelling in-depth sulphation of limestone, and involves four steps: firstly, the exposure of stone tablets in polluted atmospheres followed by quantification of subsequent sulphur concentration with depth in the stone, analysis of the evolution of sulphation over time and, finally, establishment of a predictive model for in-depth sulphation development.

## Material and exposure protocol

Two series of eight samples of two calcareous stone types were placed on buildings and exposed to atmospheric conditions in Paris (Parisian Lutetian limestone) and Tours (Turonian Richemont limestone) for up to 3 years, in such a way that some were sheltered from and others exposed to rain. The Paris site was located in a pedestrian area subject to the background air pollution of the city, at the top (40 m high) of the northern tower of Saint Eustache Church (Fig. 1a). The Tours site was located at the first floor level (5 m high) of the Psalette Cloister on the northern flank of Saint Gatien Cathedral (Fig. 1b), and subject only to the low air pollution levels of this city located in the Loire Valley.

The Parisian Lutetian limestone, the so-called 'Pierre de Courville', was used for the construction of the most important monuments in Paris (e.g. Notre-Dame Cathedral, Louvre, Saint Eustache Church) and of the Haussmannian buildings. It is grey and fine grained, with a porosity of 19%. Its mineralogical composition is mainly calcite, with limited quantities of silica and clay minerals. Its chemical composition consists of 90% CaO, 7% $SiO_2$, 2% $Al_2O_3$ and 1% MgO. The Turonian Richemont limestone has been used in the restoration of many monuments of the Loire Valley. Its

*From*: PŘIKRYL, R. & SMITH, B. J. (eds) *Building Stone Decay: From Diagnosis to Conservation.* Geological Society, London, Special Publications, **271**, 131–137.
0305-8719/07/$15.00 

(a)

(b)

**Fig. 1.** Exposure sites: (**a**) northern tower of the Saint Eustache Church (Paris); (**b**) Psalette Cloister on the northern flank of the Saint Gatien Cathedral (Tours).

porosity is higher (27%) than the Parisian limestone. It is a siliceous limestone with 94% CaO and 3% $SiO_2$.

The stone samples (tablets) were 10 × 10 × 2 cm in size and were obtained by cutting with a diamond saw the fresh stone (never previously exposed to atmospheric pollution) without any polishing of the exposed surfaces. On each building, one set of tablets was exposed with no shelter from rain on a S-facing rack inclined at 45° (Fig. 2a). The angle was chosen to maximize receipt of precipitation and the direction to maximize incident solar radiation, thus reducing time of wetness of the samples. The second set of tablets was placed vertically in a box naturally ventilated through an open bottom and a 5 cm slit between the cover and the walls (Fig. 2b). In Paris, the experiment started on 4 October 2000 and ended on 18 September 2003. In Tours, the experiment started on 8 September 2001 and ended on 16 April 2004. Samples from each set of stones were removed after 1, 2, 4, 6, 12, 18, 24 and 36 months (Table 1).

The $SO_2$ concentration in the air during the experiment was provided by two air quality monitoring networks: 'Airparif' in Paris and 'Ligair' in Tours (Table 1). During the experiment $SO_2$ concentration at the Paris site was, on average, 6 times higher than at the Tours site (9.8 v. 1.6 $\mu g\ m^{-3}$). Rainwater was not collected.

Gypsum ($CaSO_4 \cdot 2H_2O$) is produced by reactions between sulphur compounds ($SO_2$, $SO_3$, $H_2SO_4$, etc.) and water (liquid or vapour) in the atmosphere and calcite ($CaCO_3$) contained in stone. In this experiment gypsum development was measured by measuring the sulphur content of the exposed stone tablets. The distribution of sulphur concentrations under the surface of tablets was determined by pyrolysis and infrared elemental analysis (Leybold-Heraeous CSA 2003) of powder samples obtained by precise milling of the stone in progressive steps of 0.1 mm down to a depth of 2.5 mm on a surface of $10^{-3}\ m^2$ (see details in Ausset *et al.* 1996). For each tablet, three different holes and a minimum of three analyses per step were performed. The representativeness is considered to be ±5%.

(a)

(b)

**Fig. 2.** Exposure conditions of stone samples: (**a**) unsheltered from rain inclined 45° facing south; (**b**) sheltered from rain, vertically, in a naturally ventilated box (viewed without its cover).

**Table 1.** *$SO_2$ mean air concentrations in Paris and Tours. Duration of exposure, $SO_2$ mean concentrations in the air of exposure sites during experiment and sulphur concentration in the first step (0.1 mm) below the surface of the Parisian and Turonian limestones exposed sheltered from rain in Paris and Tours. Theoretical months and effective days of exposure are mentioned at the top of the table*

| | Months | 0–1 | 0–2 | 0–4 | 0–6 | 0–12 | 0–18 | 0–24 | 0–30 | 0–33 | 0–36 |
|---|---|---|---|---|---|---|---|---|---|---|---|
| Paris | Days of exposure | 35 | 64 | 126 | 182 | 364 | 559 | 777 | – | – | 1079 |
| | $[SO_2]$ ($\mu g\ m^{-3}$) (air) | 7.6 | 8.2 | 11.4 | 11.7 | 9.5 | 10.8 | 9.7 | – | – | 9.8 |
| | [S]% (sheltered stone, first step) | 0.078 | 0.075 | 0.06 | 0.11 | 0.21 | 0.34 | 0.70 | – | – | 1.10 |
| Tours | Days of exposure | 31 | – | 142 | 178 | 366 | 553 | 733 | 899 | 981 | – |
| | $[SO_2]$ ($\mu g\ m^{-3}$) (air) | 1.1 | – | 1.7 | 1.6 | 1.7 | 1.8 | 1.8 | 1.8 | – | – |
| | [S]% (sheltered stone, first step) | 0.039 | – | 0.044 | 0.049 | 0.083 | 0.088 | 0.132 | 0.136 | 0.120 | – |

## Results for stones sheltered from and to rain

Throughout the duration of the experiment, and particularly at its end (1079 days in Paris and 733 days in Tours), the sulphur concentration in the tablets exposed to rain remained equal to the stone background concentration (Fig. 3). This implies that sulphur deposited between rain events was leached by the next event.

In tablets sheltered from rain, sulphur concentration in the outermost layer below the stone surface increased with time (Fig. 4, Table 1). On average, it was 10 times higher in the Parisian limestone than in the Richemont, reflecting the higher concentration in $SO_2$ of the atmosphere of Paris (about 6 times) and the different mineralogical and petrophysical properties of the two stones (e.g. Ca concentration, porosity).

Sulphation was not, however, observed to penetrate more than 0.1 mm into the tablets, despite the increased concentration in the outermost layer. Below 0.1 mm, the sulphur concentration corresponded to the mean natural background concentration in the stone: 0.06% for Parisian and 0.04% for Richemont limestones.

## Modelling the evolution over time of sulphation in time for sheltered limestone

Sulphation is a complex physico-chemical phenomenon that cannot be easily expressed as a mechanistic model. Therefore, empirical models were fitted to the sulphur-enrichment measurements from the two sets of sheltered tablets (Fig. 5). Generally, empirical models are useful for improving the understanding of a phenomenon, for predicting its further evolution and for designing new experiments.

A sigmoidal Hill curve (also known as the variable slope sigmoid) provided a good fit for the evolution in time of the measured sulphur concentrations. This model was previously selected as the best-fitting one for soiling, when modern Si–Ca–Na glass was exposed under sheltered conditions at the same test site in Paris (Lombardo *et al.* 2005). It is interesting to note that the same model, with only model coefficients changed, also describes the evolution of soiling of modern glass exposed at other sites, characterized by other atmospheric conditions (Ionescu *et al.* 2006).

The analytical form of the Hill equation is expressed as:

$$S(t) = B + \frac{K}{1 + (M/t)^H}$$

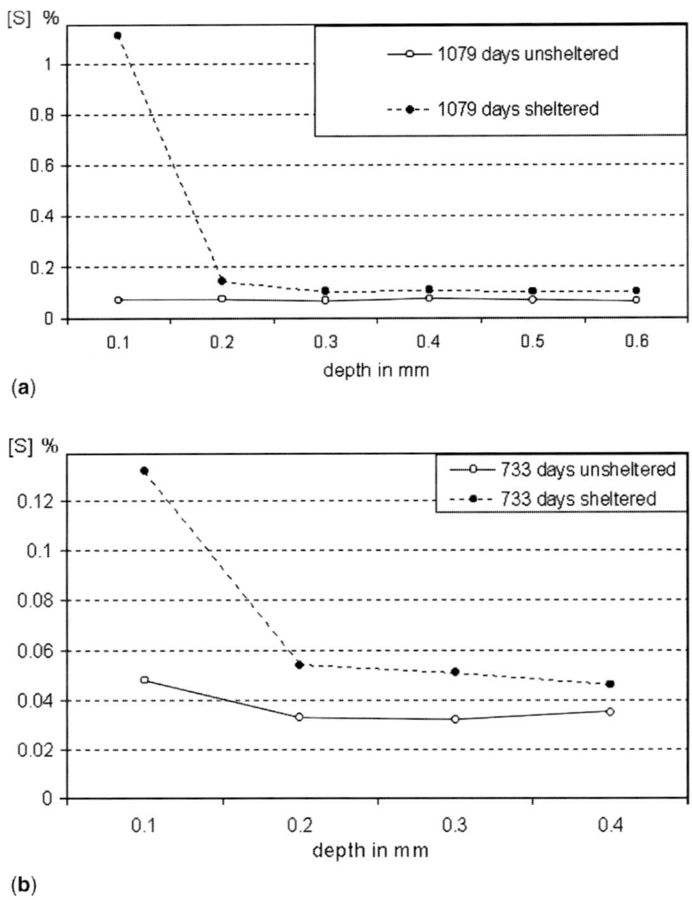


**Fig. 3.** Sulphur concentration with depth in samples exposed to rain, compared to samples sheltered from rain: (**a**) in Paris, over 1079 days; (**b**) in Tours, over 733 days.

where $S(t)$ is sulphation evolution in time $t$; $B$ (bottom) is the initial level of sulphation; $K$ (span) is Top–Bottom, where Top corresponds to the maximum curve asymptote (saturation), or level of response produced after infinite sulphation; $M$ (half-life) is the time when the response (sulphation) is half way between the Top and Bottom– that is it corresponds to the curve inflection; and $H$ (Hill slope) is the maximum slope of the curve at time $M$ – it is used as a measure of the evolution rate.

Using the set of measurements for each location, the four model coefficients ($B$, $K$, $M$ and $H$) were calculated by a classical non-linear regression (see Bates & Watt 1988; Bevington & Robinson 1992) and the 95% confidence intervals were calculated for each estimation (Saporta 1990).

The model coefficients (Table 2) show that the predicted saturation level ($B + K$) is significantly higher for the Lutetian limestone exposed in Paris, $2.1 \pm 0.7$ [S]% (Fig. 5a), than the corresponding one for Richemont limestone exposed in Tours, $0.15 \pm 0.06$ [S]% (Fig. 5b). The existence of a saturation level within the first 0.1 mm layer agrees well with the fact that sulphation does not penetrate more than 0.2 mm depth, and saturation in time seems to be related to a limitation in space (in depth). Half of the saturation level of sulphation $M$ is predicted to be reached sooner in the Tours' experiment than in the Paris one (after $481 \pm 198$ v. $1056 \pm 281$ days). Values for the initial level of sulphation, $B$, and the maximum evolution rate, $H$, are close in both cases.

As well as the soiling of the modern glass exposed in a polluted atmosphere, the Hill's model of sulphation reveals a cumulative phenomenon, characterized by a saturation level. The model's coefficients are related to the atmospheric

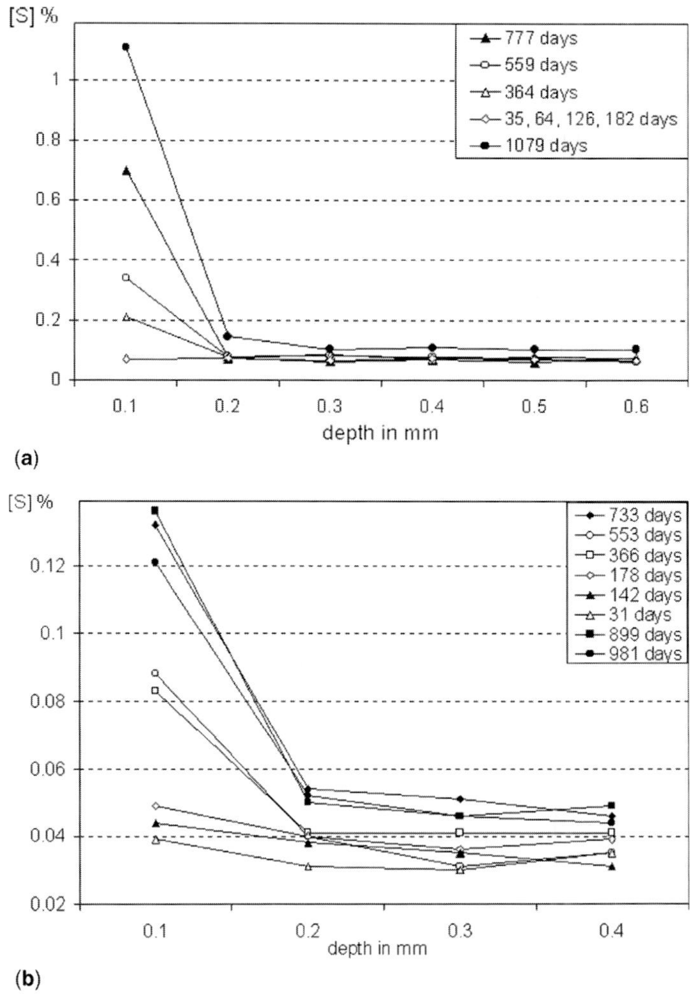


**Fig. 4.** Sulphur concentration with depth below the surface of limestone, for different exposure periods under sheltered conditions in: (**a**) Paris (Lutetian limestone); (**b**) Tours (Turonian limestone).

environment of exposure (e.g. Paris, Tours, Athens, Krakow, Prague, Rome) and to the intrinsic properties of each exposed material (e.g. Lutetian or Richemont limestone, modern glass).

## Conclusions

A quantification, analysis and modelling of the sulphation of two different stones (Parisian Lutetian and Turonian Richemont limestones) were achieved by means of a 3-year field exposure trial of the two limestones in the areas where each is used for construction or restoration.

Limestone tablets exposed to rain are characterized by a constant sulphur concentration (the background level of the stone) that is maintained through repeated leaching by rain.

For exposure tablets sheltered from rain, measurements revealed increasing sulphur enrichment over time. This sulphation does not, however, appear to penetrate more than 0.2 mm into the tablets, despite increasing concentration in the outermost 0.1 mm. According to the measurements and modelling, limestone sulphation has a saturation-limited evolution in time and space (depth). Results from other trials show that sulphation of limestone and soiling of modern glass follow the same pattern of evolution in time, which can be described by the Hill model. This model reveals a cumulative phenomenon of sulphation and of soiling, characterized by a saturation level. The model's coefficients are

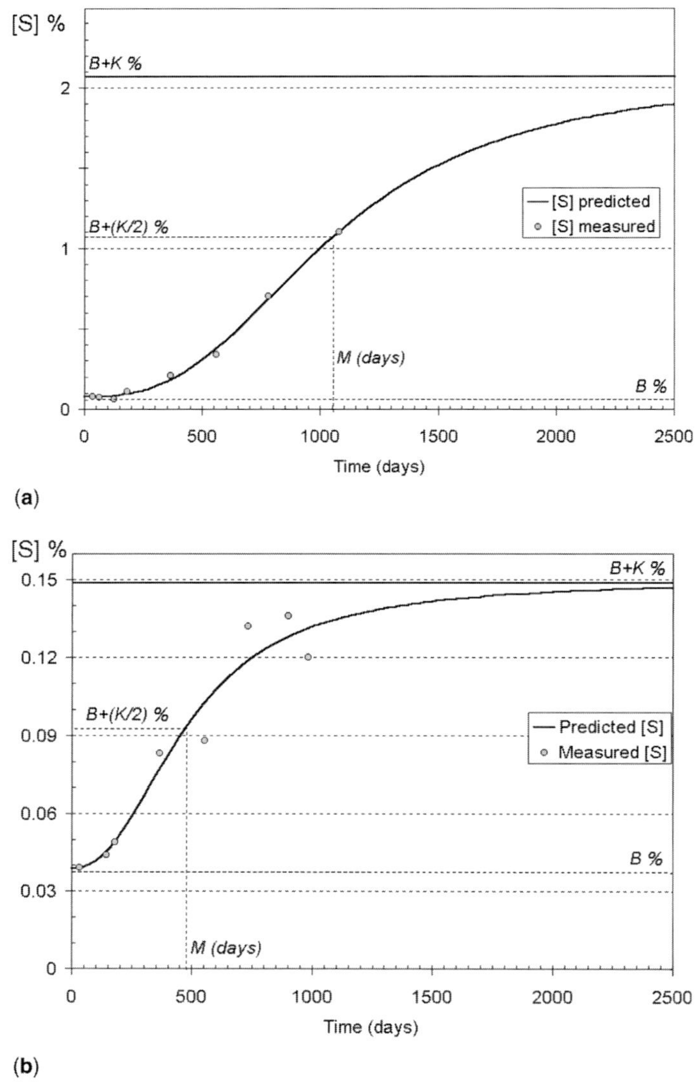


**Fig. 5.** Measured and predicted changes in sulphur concentration with depth v. time for sheltered limestone tablets: (**a**) Paris experiment (Lutetian limestone); (**b**) Tours experiment (Turonian limestone).

related to the atmospheric environment of exposure and to the intrinsic properties of each exposed material.

New exposure trials with other types of limestone exposed in various environments for longer duration might be undertaken to verify the Hill model of sulphation. In terms of the soiling of modern glass, the composition of which is more or less standardized worldwide, new exposure trials in various environments and for longer

**Table 2.** *Model coefficients for Lutetian and Turonian limestone. Hill's model coefficients and 95% confidence interval for Paris (Lutetian limestone) and Tours (Turonian limestone)*

| | *B* ([S]%) | *K* ([S]%) | *M* (days) | *H* |
|---|---|---|---|---|
| Paris | 0.08 ± 0.02 | 2.0 ± 0.74 | 1056 ± 281 | 2.73 ± 0.55 |
| Tours | 0.04 ± 0.01 | 0.11 ± 0.05 | 481 ± 198 | 2.29 ± 1.67 |

duration will not only verify the Hill model but also provide new data for the calculation of dose–response functions.

This study benefited of funding from the French Agency for the Environment and Energy Monitoring (ADEME) within the frame of the PRIMEQUAL Programme.

## References

AUSSET, P., CROVISIER, J. L. & DEL MONTE, M. 1996. Experimental study of limestone and sandstone sulphation in polluted realistic conditions: the Lausanne Atmospheric Simulation Chamber. *Atmospheric Environment*, **30**, 3197–3207.

AUSSET, P., DEL MONTE, M. & LEFÈVRE, R.-A. 1999. Embryonic sulphated black crusts in Atmospheric Simulation Chamber and in the field: role of the carbonaceous fly-ash. *Atmospheric Environment*, **33**, 1525–1534.

BATES, D. M. & WATT, D. G. 1988. *Nonlinear Regression Analysis and its Applications*. Wiley, New York.

BEVINGTON, P. R. & ROBINSON, D. K. 1992. *Data Reduction and Error Analysis for the Physical Sciences*, 2nd edn. McGraw-Hill, New York.

CAMUFFO, D. 1984. The influence of run-off on weathering of monuments. *Atmospheric Environment*, **18**, 2273–2275.

CAMUFFO, D., DEL MONTE, M. & SABBIONI, C. 1982. Wetting deterioration and visual features of stone surfaces in urban area. *Atmospheric Environment*, **16**, 2253–2259.

CAMUFFO, D., DEL MONTE, M. & SABBIONI, C. 1983. Origin and growth mechanisms of the sulphated crusts on urban limestone. *Water, Air and Soil Pollution*, **19**, 351–359.

IONESCU, A., LEFÈVRE, R.-A. & CHABAS, A. *ET AL.* 2006. Modeling of soiling based on silica–soda–lime glass exposure at six european sites. *Science of the Total Environment*, **369**, 246–255.

LEFÈVRE, R.-A. & AUSSET, P. 2002. Atmospheric pollution and building materials: stone and glass. *In*: SIEGESMUND, S., WEISS, T. & VOLLBRECHT, A. (eds) *Natural Stone, Weathering Phenomena, Conservation Strategies and Case Studies*. Geological Society, London, Special Publications, **205**, 329–345.

LOMBARDO, T., IONESCU, A., LEFÈVRE, R.-A., CHABAS, A., AUSSET, P. & CACHIER, H. 2005. Soiling of silica–soda–lime float glass in urban environment: measurements and modeling. *Atmospheric Environment*, **39**, 989–997.

SAPORTA, G. 1990. *Probabilités, analyse des données et statistique*. Technip, Paris.

TRAN THI NGOC LAN, NGUYEN THI PHUONG THOA, NISHIMA, R., TSUJINO, Y., YOKOI, M. & MAREDA, Y. 2005. New model for the sulphation of marble by dry deposition. Sheltered marble – the indicator of air pollution by sulphur dioxide. *Atmospheric Environment*, **39**, 913–920.

# Decay of natural stones caused by fire damage

J. SIPPEL[1,2], S. SIEGESMUND[1], T. WEISS[1], K.-H. NITSCH[1] & M. KORZEN[3]

[1]*Geoscience Centre, University Göttingen, Goldschmidtstrasse 3, 37077 Göttingen, Germany (e-mail: ssieges@gwdg.de)*

[2]*GFZ-Potsdam, Telegrafenberg, 14473 Potsdam, Germany*

[3]*Federal Institute for Materials Research and Testing (BAM), Unter den Eichen 87, 12205 Berlin, Germany*

**Abstract:** Almost every representative ancient building suffered from a fire during its history. Therefore, several limestones, sandstones, a gypsum, granites, tuffs, an orthogneiss and two marbles have been tested to analyse the effect of fire. Thermal expansion measurements up to 1000 °C reveal that every rock shows a specific expansion behaviour. Variations are caused by the single crystal thermal expansion properties of rock-forming minerals and by different damage processes. In silicate rocks, intragranular fracturing is the predominant damage phenomenon. Carbonate rocks show, at low temperatures, a behaviour mainly controlled by the anisotropic expansion of calcite. At higher temperatures, mineral reactions, such as decarbonatization, are directly evidenced by sudden jumps in thermal expansion curves. If water is present, a second stage of deterioration follows fire damage: the huge volume increase due to portlandite formation from decarbonized CaO causes severe scaling at the outermost surface of limestone when exposed to the environment. Small amounts of silicates in carbonate rocks may improve the stability of those rocks due to dicalciumsilicate formation. At high temperatures, an increase in the expansion coefficient may be explained by partial melting for some rock types. Phase changes (e.g. quartz) are monitored by a sudden increase in the expansion coefficient. Investigations on gypsum reveal that dehydration reactions reduce fire temperatures in the vicinity of gypsum rocks significantly. In general, all experiments show that samples are severely damaged after being subjected to fire. Real fire tests show that the penetration depth of heat and the associated damage types vary as a function of lithology. While for granites, cracks in feldspars predominate, the firing of limestone causes a scaling of the outermost layer. The investigations may lead to an improved assessment of natural building stones that have been damaged by fire. Implications can also be drawn for the recent use of façade panels made of natural building stones in case of a future fire.

Catastrophic fires are a frequent damage phenomenon on historical sites and buildings, artworks, sculptures, etc. Much of the observed worldwide destruction of these monuments can be ascribed to war, natural catastrophes, terrorist attacks, technical defects or vandalism. Different applied materials such as mortars for masonry or rendering, ceramic roof tiles or the large variety of natural stones, to name a few, may exhibit completely different deterioration features as a consequence of fire impacts. There is still a lack of any unequivocal scientific or conservation approach for materials damaged by fire that may be used as a methodological guideline for the planning and execution of repair and maintenance. A scientific approach mainly based on the behaviour of the material constituting the artwork guarantees to preserve the cultural heritage.

The first systematic study of rocks suffered from fire was carried out by Kieslinger (1954), who documented the damage of many buildings and objects made of natural stone in Vienna after World War I (e.g. Fig. 1). For historic buildings, investigations are essential that correlate reductions of strength and changes in appearance of natural rocks due to temperature impacts and associated variations in mineralogy and fabric. Cracking, scaling and even fragmentation are the result of expansion and contraction cycles, while changes in colour are controlled by mineralogical phase changes (Kieslinger 1954; Goudie *et al.* 1992; Allison & Goudie 1994; Chakrabati *et al.* 1996; Hajpál 2002; Hajpál & Török 2004).

Once key parameters leading to a certain degradation phenomenon are defined they can be used to predict fire resistance of different types of rocks. In order to understand specific processes and their consequences caused by fire a large number of different sedimentary, magmatic and metamorphic rock types were investigated. The main aim was to characterize related mineral reactions and the causes of strength reduction as a function of thermal impact. Special emphasis was placed on thermally induced mineral reactions such as the transformation from

*From*: PŘIKRYL, R. & SMITH, B. J. (eds) *Building Stone Decay: From Diagnosis to Conservation*. Geological Society, London, Special Publications, **271**, 139–151.
0305-8719/07/$15.00 

**Fig. 1.** Fire damage by extensive spalling (Kieslinger 1954). (**a**) Deep-reaching spalls in red quartz sandstone (Buntsandstein). Column in a burnt-out building next to Mainz Cathedral (1953). (**b**) Surficial spalling in limestone; column at the organ choir, St Stephans Cathedral, Vienna, Austria (1945).

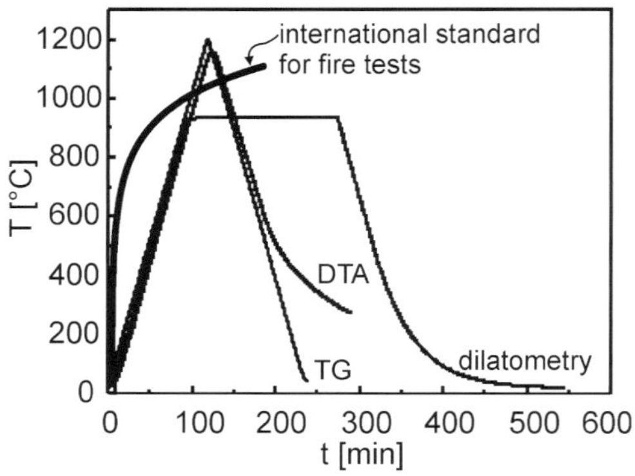


**Fig. 2.** Temperature development for the different thermal analyses (DTA, TG, thermal expansion), as well as for the fire tests carried out according to DIN 4102-8 (2003).

low-quartz to high-quartz at 573 °C, dehydration and dehydroxylation processes (clay minerals, mica, gypsum), decarbonatization reactions and formation of portlandite during and after cooling, melting and sintering processes, oxidation processes (in particular the formation of hematite) and thermal expansion as a result of heat impact. Finally, some selected natural stones were exposed to fire tests following the international standard fire curve adopted by ISO 834-1 (1999) to characterize changes and damage owing to fire.

## Experimentation

Thermally induced mineral reactions were detected by means of differential thermal analysis (DTA) and thermogravimetry (TG), each measurement covering temperatures from 20 to 1200 °C with a heating and subsequent cooling rate of 10 °C $min^{-1}$ (Fig. 2). Thermally induced changes of the modal composition of each rock were characterized by X-ray diffractometry (XRD). Based on the results of the DTA and TG, a number of rock types were selected for thermal expansion measurements up to a maximum temperature of 950 °C (samples of 7 mm in diameter and 20 mm length, heating rate of 10 °C $min^{-1}$, maximum temperature for 3 h at a constant level). The thermal expansion was determined for the direction perpendicular to foliation or bedding, respectively, and in some cases also parallel to this layering.

Structural and mineralogical alterations were determined by thin sections under a polarizing microscope as well as by scanning electron microscopy. Changes of the pore space including the total porosity and the distribution of pore radii were analysed by Hg-porosimetry (see Doveton 1997).

Finally, three selected rock types (samples of 200 × 200 × 200 mm) were subjected to small-scale fire tests in a furnace according to DIN 4102-8 (2003) at the Federal Institute for Materials Research and Testing (BAM). Only one surface of the sample was exposed to the fire. Temperatures inside the sample were measured by means of four thermocouples with distances to the fire impact of 25, 75, 125 and 175 mm, respectively.

Ultrasonic wave velocities were measured before and after the fire tests to determine the fire-induced crack growth (see compilation in Siegesmund 1996 and Siegesmund *et al.* 1999).

## Rock samples

Eighteen different rock types were chosen for the experiments to predict fire resistance (Table 1). These selected stones consist of magmatic, sedimentary as well as metamorphic rocks, that is, silicate, carbonate and sulphate rocks. With respect to the mineralogical composition, the transformation of quartz, the decarbonatization of calcite and/or dolomite, and the dehydration of gypsum and the possible formation of melts at high temperatures were expected to play a decisive role during increasing temperatures. The contents of hydroxide ions in clay minerals or micas are another important parameter. Dehydration reactions are known from clay and mica minerals, and may lead to shrinking effects or an enlargement of the unit cell (e.g. Mazzucato *et al.* 1999; Ehling & Köhler 2000). In cases where no mineral decomposition occurs, a linear thermal expansion may control a length change with increasing temperatures. Different

**Table 1.** *Characterization of natural building stones with respect to their mineral composition and porosity*

| Stone type | Name, abbreviation | Modal composition (XRD) | Description | φ (%) |
|---|---|---|---|---|
| Granite | Kösseine, KÖSS | Pl, Kfs, Qz, Bt, Chl, Ms | bluish grey – no distinct foliation | 0.11 |
| Rhyolite | Löbejün, LQ | Pl, Kfs, Qz, Ms, Chl, Hem | reddish matrix – porphyroclasts (Qz, Kfs, Pl) – unfoliated | 4.52 |
| Syenite | Blue Pearl, BLPE | Kfs, Pl, Bt, Qz, Amph, Ap | dark grey and black – strongly foliated | 0.07 |
| Ignimbrite | Rochlitz, RI | Qz, Kaol, Pl, Hem | red and orange – weakly foliated (prolate shaped pores) | 22.76 |
| Tuff | Weibern, WT | Qz, Kfs, Anl, Di, Bt, Ms, Chl | light beige – weakly layered | 38.49 |
| Orthogneiss | Verde Andeer, VA | Qz, Ms, Kfs, Pl | green (Fe-rich Ms) and white – mylonitic foliation | 0.89 |
| 'Fruchtschiefer' | Theuma, THEU | Qz, Ms, Pl, Bt, Chl | grey matrix – black Cor-pseudomorphoses (Ms) – strongly foliated | 0.95 |
| Sandstone | Obernkirchen, OBKI | Qz, Kaol, Pl, Ms | yellow – fine grained – siliceous bound – weakly layered | 16.68 |
| Sandstone | Wesersandstein, grey GRAWE | Qz, Pl, Kfs, Ms, Chl | grey – fine grained – siliceous bound – strongly layered | 4.41 |
| Sandstone | Wesersandstein, red ROWE | Qz, Pl, Kfs, Ms, Hem, Chl | red – fine grained – siliceous bound – strongly layered | 5.33 |
| Calcitic sandstone | Anröchte, ANSF | Cc, Qz, Glau, Chl | green – fine grained – calcitic bound – bedding marked by fossil biofragments | 6.59 |
| Limestone | Eibelstadt, EI | Cc, Qz, Pl | grey – micritic calcite dominates – lenses of brownish clay – strongly layered | 4.64 |
| Limestone | Thüste, THKA | Cc, Qz | pale yellow – porous – bedding marked by micritic peloids | 20.94 |
| Travertine | Bad Langensalza, TRAV | Cc | light brownish – bedding marked by brown-lined pores | 9.12 |
| Calcite marble | Cava Ortensia, C1 | Cc, Ms (<2%) | white with grey veins – irregular grain boundaries – weakly foliated | 0.21 |
| Calcite marble | Cima di Gioia, C2 | Cc | white – grey dolomitic veins – straight grain boundaries – weakly foliated | 0.26 |
| Dolomite marble | Thassos, GTH | Dol, Pl | white – interlobate grain boundaries – weakly foliated | 0.54 |
| Gypsum | Ührde, GIPS | Gy, Anh | light grey (Gy) and dark grey (Anh) layered | 1.10 |

Anh, anhydrite; Anl, analcime; Amph, amphibolite; Ap, apatite; Bt, biotite; Cc, calcite; Chl, chlorite; Cor, cordierite; Di, diopside; Dol, dolomite; Gy, gypsum; Glau, glauconite; Hem, hematite; Kaol, kaolinite; Kfs, K-feldspar; Ms, muscovite; Pl, plagioclase; Qz, quartz. φ, porosity.

temperature-induced volume changes of adjacent minerals may produce stress concentration along grain boundaries and thermally induced microcracks. These processes are much more pronounced for minerals with strongly anisotropic thermal properties (e.g. calcite). Besides the texture (lattice preferred orientation), other structural features such as foliation, grain shape, grain-boundary geometries and porosity have to be considered as well when evaluating the thermal resistance and its directional dependence of a rock. Pre-existing microcracks and pores, for instance, may compensate for the expansion of minerals to a certain degree (Siegesmund *et al.* 2000; Zeisig *et al.* 2002).

## Mineralogical observations

The DTA curve (Fig. 3a) of the sandstone ROWE (see Table 1) is more or less representative for almost all quartz-bearing silicate rocks and exhibits the transition of low- to high-quartz as an endothermic but reversible reaction between about 565 and

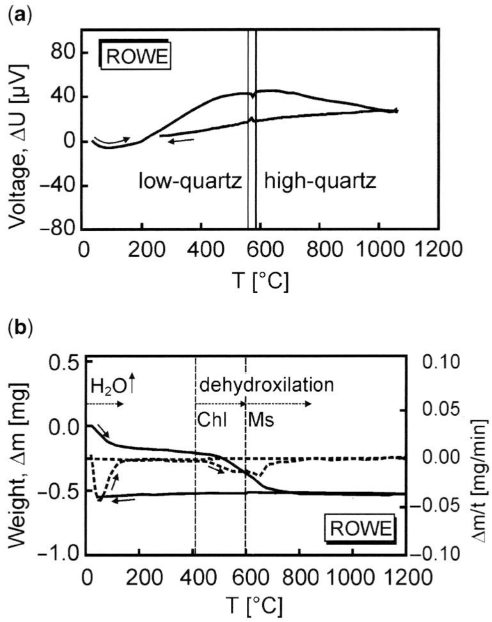


**Fig. 3.** Thermal analyses for Wesersandstein, red: (**a**) DTA and (**b**) TG. Initial weight 27.284 mg; atmosphere air (solid curve); DTG (dotted line; not referred to in the text).

580 °C. For rocks with only small amounts (<5%) of quartz, the phase transformation is not observed (sample BLPE and WT see Table 1, Fig. 4) or cannot be detected due to the simultaneous dehydroxylation of kaolinite (sample RI). Also the TG curve of ROWE (Fig. 3b) clearly documents that the adsorptively bound water evaporates at temperatures below 150 °C, while the dehydroxylation of phyllosilicates requires higher temperatures (altered biotite about 200 °C, kaolinite about 400 °C, chlorite about 400 °C or muscovite >500 °C).

The dehydroxylation of phyllosilicates like chlorite, biotite, muscovite or certain clay minerals, especially illite, is mostly associated with the release of $Fe^{2+}$. This iron will be mostly oxidized causing a red discoloration of the rock. This formation of hematite was proved by means of XRD and corresponds to observations made by other authors (e.g. Hajpál & Török 2004).

Endothermic reactions at temperatures above 850 °C observed for the granite KÖSS and the orthogneiss VA may indicate melting processes. However, neither corresponding thin sections nor XRD measurements could prove the existence of glassy material inside the heated samples. However, the mineralogical composition of both rocks is suitable for grain-boundary melting (see also Blatt & Tracy 1996; Hall 1996).

For carbonate rocks, heating leads to the decomposition of calcite and dolomite between 650 and 900 °C depending on the $CO_2$ partial pressure. The decarbonatization, for example, in the case of the calcite marble C1, corresponds to a single-stage endothermic reaction accompanied by weight loss of around 44% (Fig. 5a). The decarbonatisation of the dolomite marble GTH is a two-stage endothermic process starting with the decomposition of $MgCO_3$ and subsequently proceeding to that of $CaCO_3$. Finally, the remaining CaO reacts at the retrograde track at temperatures of below 600 °C with atmospheric water to form portlandite $Ca(OH)_2$ (Fig. 5b). The formation of portlandite is an exothermic reaction (DTA) with a slight increase of weight (TG). If only pure helium is present, that is a water-free atmosphere, portlandite formation is absent (Fig. 5b). The two rock samples ANSF and THKA, which contain quartz and calcite, exhibit dicalciumsilicate ($\beta$-$C_2S$) after heat treatment, a reaction product of CaO and $SiO_2$ at higher temperatures.

The crucial process for the gypsum Ührde is the two-stage dehydration to halfhydrate (HH) and then to anhydrite III (AIII) between 100 and 200 °C. Finally, a weak exothermic reaction roughly beginning at 330 °C has to be interpreted as the transformation from A III to A II.

## Thermal expansion

Two parameters are used to characterize the linear thermal expansion behaviour of the rocks: (i) the thermal expansion coefficient, $\alpha$, as a measure of the length change with temperature [$mm\,m^{-1} \cdot 10^{-3}\,K^{-1}$] and (ii) the residual strain, $\varepsilon_r$, after heating and subsequent cooling [$mm\,m^{-1}$]. The latter is a result of microcrack formation as well as the growth of pre-existing microcracks and can be used as a quantitative expression of thermally induced damages.

### *Silicate rocks*

Owing to the transition of low- to high-quartz all quartz-bearing rocks experience a sudden and remarkable expansion at temperatures just below 600 °C (Fig. 6). Although this phase transformation is reversible, a striking positive residual strain can be recognized for the rhyolite and the granite after cooling from maximum temperatures of around 700 °C (Fig. 6b, c), whereas temperatures of 500 °C result in much lower values of residual strain. Consequently, all quartz-bearing rocks show varying degrees of intragranular microfractures within quartz grains. In the granite KÖSS a high proportion of quartz grains is intensively fractured, while in the

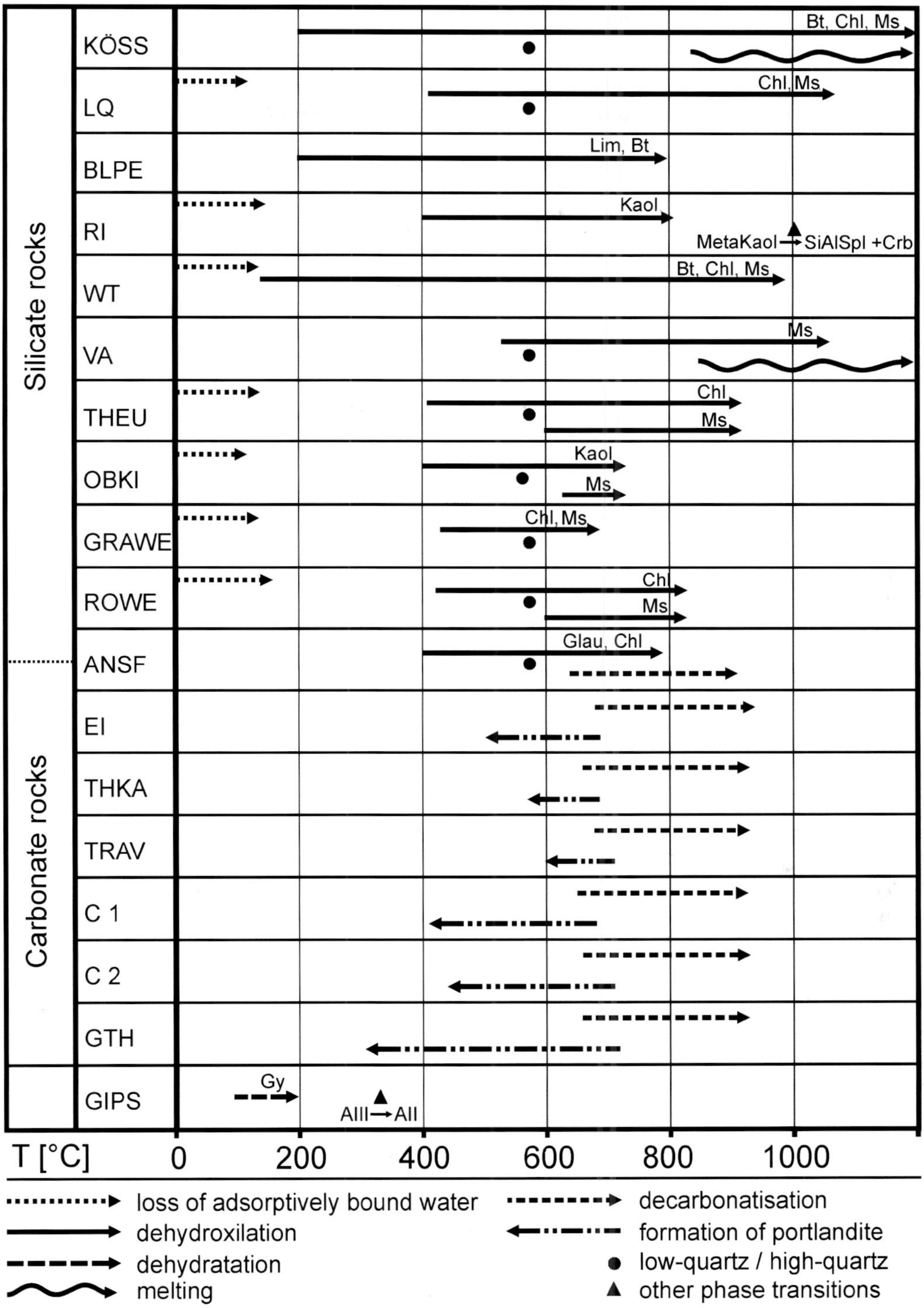


Abbreviations: AIII - Anhydrite III, AII - Anhydrite II, Bt - Biotite, Chl - Chlorite, Crb - Cristobalite, Gy - Gypsum, Glau - Glauconite, Kaol - Kaolinite, Lim - Limonite, MetaKaol - Meta-Kaolinite, Ms - Muscovite, SiAlSpl - Silicon Aluminium Spinel.

**Fig. 4.** Results of differential thermal analyses (DTA) and thermogravimetry (TG) measurements for all samples.

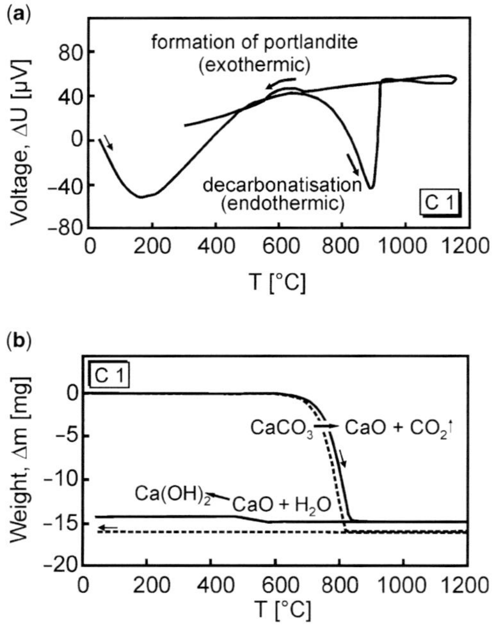


**Fig. 5.** Thermal analyses for the calcite marble Cava Ortensia: (**a**) DTA and (**b**) TG. Solid line, atmosphere air (initial weight 33.982 mg); dotted line, helium (36.638 mg).

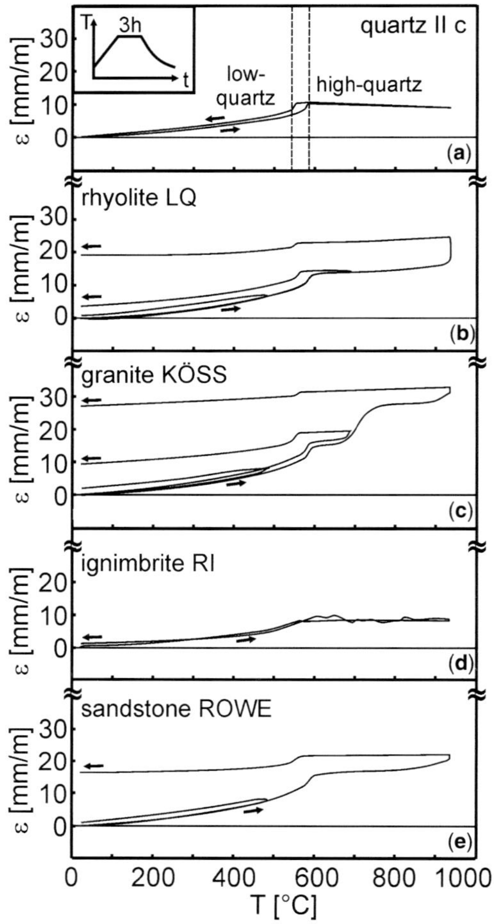


**Fig. 6.** Thermal expansion of quartz parallel to the *c*-axis in comparison with the expansion behaviour of a selection of silicate rocks.

quartzitic sandstone OBKI only grains around pores exhibit microcracks. Moreover, the granite KÖSS and the sandstones GRAWE and ANSF show a remarkable loss of cohesion even when heated up to 500 °C (indicated by values of residual strain of 2.0, 5.5 and 3.0 mm $m^{-1}$, respectively). Fredrich & Wong (1986) referred to intragranular microfractures in feldspars of heated granites, whereas Hajpál & Török (2004) described intergranular microfractures in quartzitic sandstones for temperatures below 500 °C. Residual strains in silicate rocks are significantly higher at temperatures above 700 °C, except for porous rocks (higher porosities than 15%) like RI (Fig. 6d) and OBKI. The strong expansion occurring above 700 °C is explained by a progressive crack growth and finally a higher degree in transgranular microcracks. This progressive microfracturing could either be a one-stage (rhyolite LQ and sandstone ROWE) or a two-stage (granite KÖSS and orthogneiss VA) expansion process (Fig. 6). For the sandstone ROWE, the impact of temperatures above 500 °C causes an increasing total porosity and a remarkable shift to larger pore radii, whereas at temperatures below 500 °C the pore space does not change significantly (not shown here).

One result of the thermal expansion measurements up to 950 °C is that for silicate rocks in general a higher expansion coefficient, $\alpha$, is associated with a higher residual strain, $\varepsilon_r$, and vice versa (Fig. 7). Furthermore, for the investigated rocks a lower initial porosity corresponds to a higher residual strain. The thermal expansion and the residual strain of the mylonitic orthogneiss VA is strongly anisotropic, with the highest values perpendicular to the foliation and significantly lower values parallel to the foliation. Among the sedimentary rocks the calcite-bearing sandstone (ANSF) shows the strongest expansion and the highest residual strain, although its porosity is higher than that of the sandstones ROWE and GRAWE. In the case of the ignimbrite RI, temperatures of 950 °C result in a very low residual strain. Nevertheless, this rock shows disintegration traces in the form of open microcracks, particularly in the cryptocrystalline matrix.

## *Carbonate rocks*

The expansion curves as a function of temperature for the calcite marble C1 and the dolomite marble

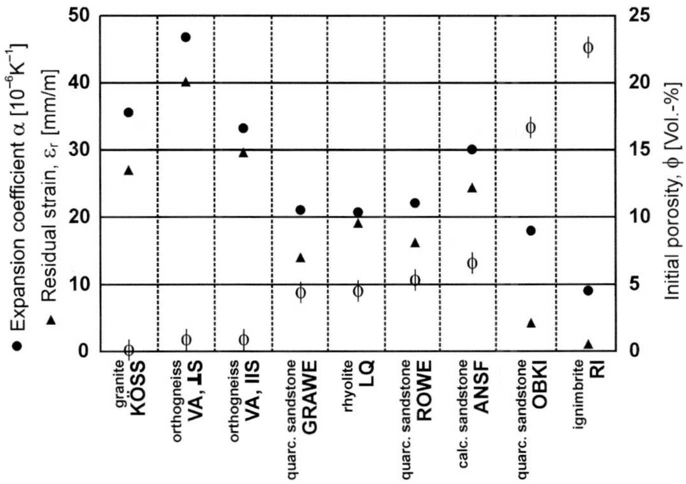

**Fig. 7.** Thermal expansion behaviour of silicate rocks heated up to 950 °C compared to their initial porosity. The mylonitic orthogneiss was the only rock tested in both the direction parallel to foliation and perpendicular to foliation.

GTH clearly document the effect of the decarbonatization that produces a remarkable contraction above 800 °C (Fig. 8a, c). In an atmosphere of low $CO_2$ pressures (e.g. pure helium) or at lower heating rates, the above reported decarbonatization occurs at lower temperatures. Calcite marbles exhibit comparably high expansion coefficients and high residual strains when heated up to 100 °C (Siegesmund *et al.* 2000; Zeisig *et al.* 2002). This thermal sensitivity can also be observed up to 600 °C. For example, the residual strain for the Carrara marble C1 is around 5.4 mm $m^{-1}$ at 500 °C, while the averaged residual strain for silicate rocks is less than 2 mm $m^{-1}$. The most intensive damages are shown by the calcite marble C2 with $\varepsilon_r = 9.5$ mm $m^{-1}$ (Fig. 8b). In this case thermally induced stresses that originated at adjacent grains lead to a total decay, but before decarbonatization starts. The different thermal-induced dilation of C1 and C2 is also expressed by their strength loss, that is the loss in cohesion between the grains indicated by a different tendency to granular disintegration. In contrast to normal calcite crystals, which are translucent, decarbonatized rocks exhibit a rather dull white appearance. Fracture surfaces exhibit: (i) that the loss of cohesion after decarbonatization occurs along irregular surfaces in the case of C1 and along straight grain boundaries for C2; and (ii) that the shape of calcite crystals survives the decomposition to CaO. The dolomite marble GTH is more resistant to heat impacts even at temperatures up to 600 °C (maximum residual strains of 2.1 mm $m^{-1}$, see Fig. 8c). According to the DTA and TG data a two-stage shrinkage is discussed.

In addition, the limestone from Eibelstadt (EI) was chosen for thermal expansion tests. This rock shows a directionally dependent thermal expansion. After decarbonatization and cooling down, an extension is observed perpendicular to bedding whereas a contraction occurs parallel to bedding

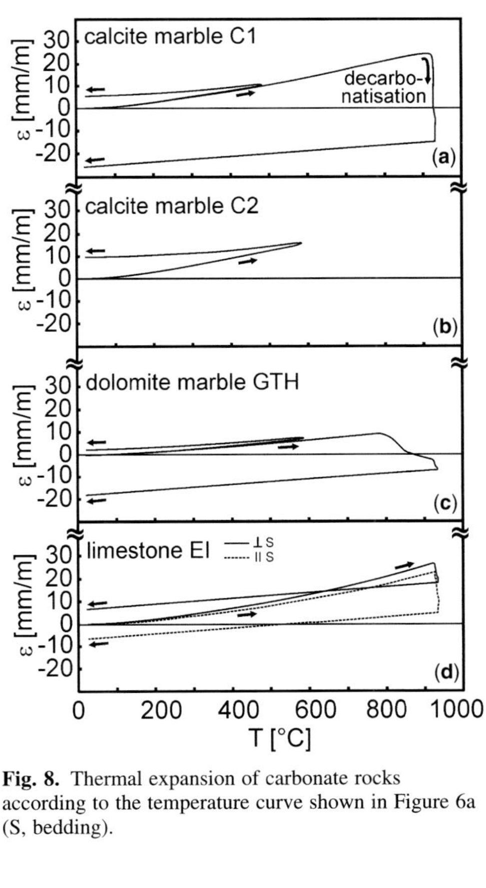

**Fig. 8.** Thermal expansion of carbonate rocks according to the temperature curve shown in Figure 6a (S, bedding).

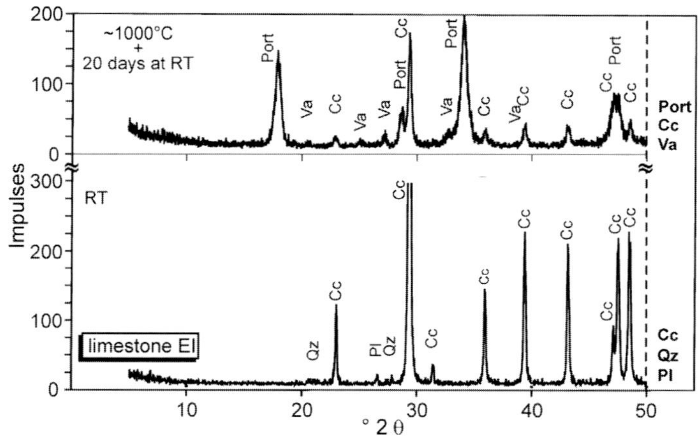


**Fig. 9.** Mineralogical composition of the limestone EI at room temperature (RT), as well as after exposition to temperatures of about 1000 °C and subsequent storage for 20 days at room temperature in normal atmospheric humidity.

(Fig. 8d). Subsequent to the heat impact and days of exposure to room temperature, a phase transformation and accompanied volume increase can be observed for all carbonate rocks (including the calcitic sandstone ANSF). XRD data give evidence of the reaction of CaO with atmospheric water to form portlandite $Ca(OH)_2$. After 20 days of this exposure, the new metastable phase vaterite ($\gamma$-$CaCO_3$) has already formed at the expense of portlandite due to the exchange of OH-groups by atmospheric $CO_2$ (Fig. 9).

## *Sulphate rock Ührde*

Two effects are crucial for the thermal behaviour of the gypsum-bearing rock: (i) the dehydration of gypsum results in a limited expansion followed by a contraction within the temperature range of 180–300 °C; and (ii) the transformation of $CaSO_4$ (A III) to anhydrite (A II) reveals an intense contraction above 800 °C expressed by a negative residual strain of about 20 mm $m^{-1}$ (perpendicular to the bedding) and 35 mm $m^{-1}$. (parallel to the bedding) (Fig. 10). SEM images show that the anhydrite crystals change their habit from prismatic to more isometric shapes when heated.

## Fire damage

Three rocks, the granite Kösseine (KÖSS), the rhyolite Löbejün (LQ) and the limestone Eibelstadt (EI), were selected for small-scale fire tests at the BAM (Bundesanstalt für Materialforschung und -prüfung).

Temperatures inside each sample were recorded with thermocouples (st 1–4 in Fig. 11) at varying distances from the fire-exposed surface. Temperatures inside the furnace, which were recorded by the thermocouple 'ft', were adjusted according to the internationally so-called standard temperature–time curve (CJSO 834-1 1999). When a temperature of around 100 °C was reached at thermocouples 'st 3' and 'st 4' (i.e. at greater distances from the fire) it remained constant for some time as a result of water evaporation. Regardless of rock type, the maximum temperature at the samples' fire-exposed surface differs remarkably from that recorded at the most internal parts of the sample even after

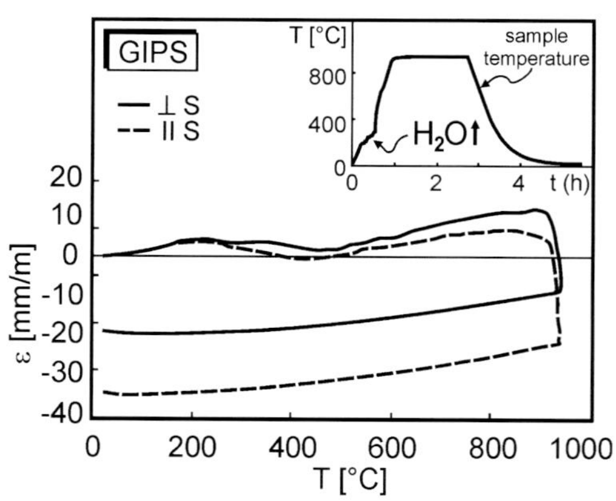


**Fig. 10.** Thermal expansion of the gypsum-bearing rock Ührde (GIPS) for the direction perpendicular to bedding (S) and parallel to S; note that the temperature development during dehydration of gypsum is non-linear despite the constant heating rate given by the oven.

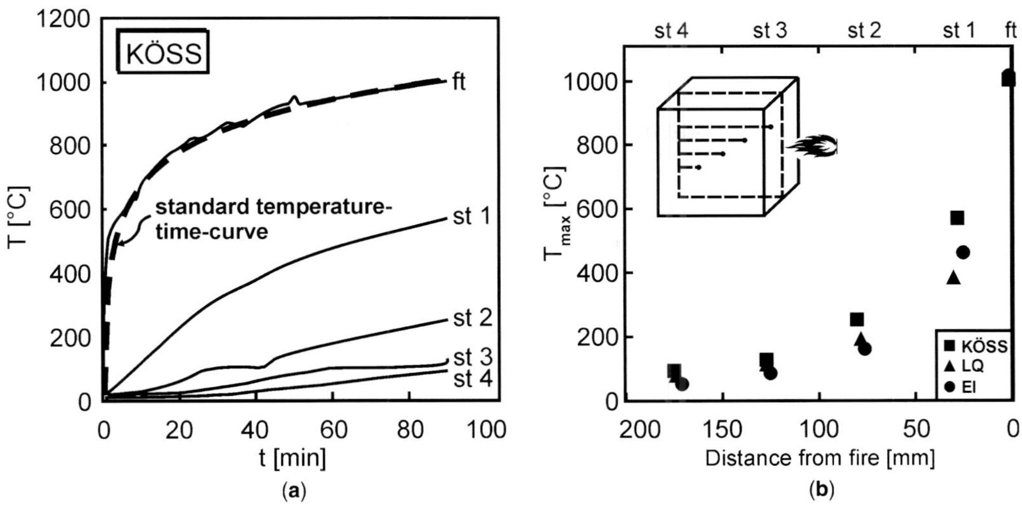


**Fig. 11.** (**a**) Temperature development for the fire experiment with the granite *Kösseine*. (**b**) Maximum temperatures at the end of the fire tests (i.e. after 90 min) as a function of the distance to the fire-exposed surface of the granite KÖSS, the rhyolite LQ and the limestone EI. Thermocouples: ft, furnace temperature; st, sample temperature.

90 min of fire testing (Fig. 11a). However, the degree of this temperature gradient is completely different for the selected samples that document their different thermal conduction properties (Fig. 11b).

The common decay phenomena of the different rocks are colour changes, crack initiation and crack growth. For the granite and the rhyolite a lightening of the fire-exposed surface is obvious, while the clay lenses inside the limestone may change the colour depending on the acting temperatures.

In the case of the granite, an intense micro-fracturing can be observed (Fig. 12a). In addition, numerous smaller arc-like fractures are evident. In the case of the rhyolite, as well as the limestone, thermally induced stresses are released by the formation of fractures mainly parallel to the fire-exposed surface. Finally, a total loss of cohesion along a plane perpendicular to the highest temperature gradient is observed for both rocks (Fig. 12b).

For the limestone, the main fracture plane is oriented parallel to a clay layer documenting the importance of the bedding plane as a pre-existing discontinuity and its control on fracture propagation. Furthermore, the atmospheric humidity supports the formation of portlandite at room temperature, which results in a total collapse of the rock structure in the outermost 4 mm (Fig. 13). This indicates that temperatures exceeded the critical value for the decarbonatization at this fire-exposed surface.

A quantitative measure of pores and cracks within a rock volume is the velocity of ultrasonic waves: a decreasing velocity is related to a higher concentration of cracks or pores. Thus, this method is useful to characterize fire-induced dilation as a result of microfracturing.

Although totally different in terms of mineralogical composition and porosity (see Table 1), the three rocks selected for the fire tests exhibit similar wave velocities in their initial stage ranging from 4.9 to 5.9 km $s^{-1}$ (Fig. 14). As expected, the impact of fire results in reduced velocities. For the granite the reduction of wave velocities is more pronounced directly at the fire-exposed surface than at greater distances, whereas for the rhyolite and the limestone it is more or less comparable. Moreover, comparing the rocks with regard to the difference between velocities before and after the fire tests it can be concluded that the granite is most sensitive to microcracking – a remarkable observation considering the fact that the rhyolite and the limestone show the more clear decay phenomena.

## Discussion

As expected, the parameters controlling the fire resistance of a rock are related to both the mineral composition and the fabric. During heating the transition from low- to high-quartz at about 573 °C determines the first critical temperature level for silicate rocks. In contrast, the dehydroxylation of water- and iron-bearing phyllosilicates explains the colour changes, particularly due to the formation of hematite at lower temperatures. The volume increase by water released during dehydroxylation processes is most probably responsible for crack formation

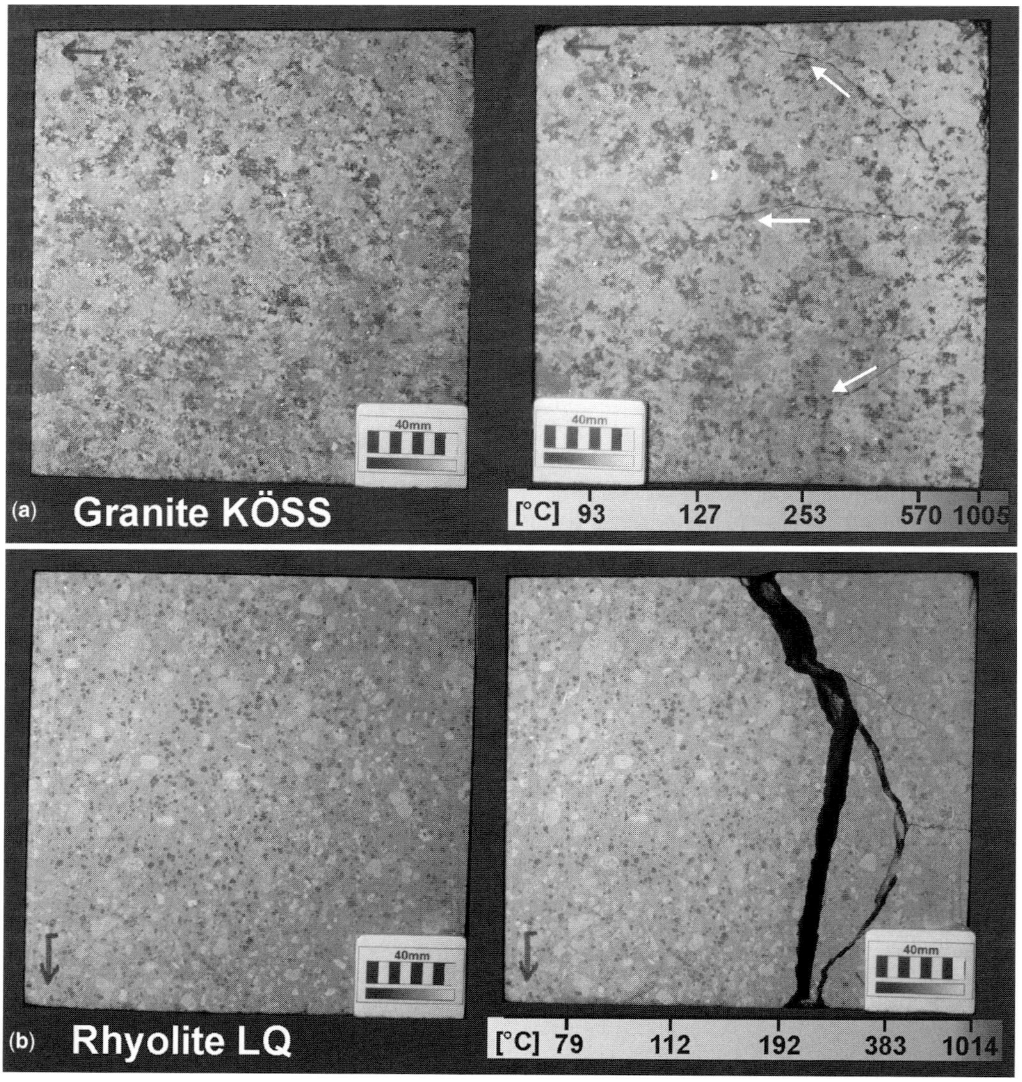


**Fig. 12.** Changes of rock fabric owing to exposure to fire: samples in their initial stage (left-hand side) and after the fire test (right-hand side). Length of each scale bar: 40 mm. Temperatures are maximum values reached at the end of each test. Arrows trace macrofractures in the sample KÖSS.

and crack growth at temperatures of more than 600 °C. This was clearly documented by the higher porosities after such a temperature impact. Another possible explanation for dilation in silicate rocks could be a certain amount of glass resulting from melting processes at grain boundaries.

The response of carbonate rocks to varying temperatures is decisively related to the behaviour of calcite and dolomite and their physical properties. At lower temperatures the strongly anisotropic thermal expansion of these minerals and their grain–grain fabrics may control the stresses along grain boundaries (shear, compression and tensile stresses) that give rise to the observed microcracking. Especially in the case of marbles, the associated decay is expressed by high residual strain values, several times higher than those for silicate rocks (see also Zeisig *et al.* 2002; Weiss *et al.* 2004). At temperatures above 600 °C, however, the disintegration of carbonate minerals is accompanied by the release of $CO_2$ and should, therefore, be responsible for the intense shrinking of most carbonate rocks up to the final formation of CaO and MgO, respectively. Even for the calcitic sandstone, the release of $CO_2$ results in a very large positive residual strain indicating that volume

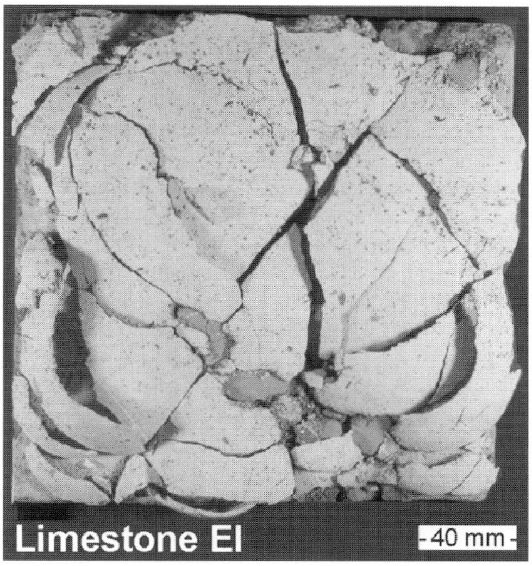


**Fig. 13.** Fire-exposed surface of limestone EI after the fire test and subsequent storage for 3 days at room temperature in normal atmospheric humidity.

effects similar to that of released water can be ascribed to $CO_2$. The formation of portlandite due to the reaction of CaO with water below 600 °C gives rise to a strong volume increase and further decay processes. The formation of portlandite is strongly reduced when quartz may react with CaO to dicalciumsilicate ($\beta$-$C_2S$), which in turn reacts only slowly with water. It is still an open question, how vaterite, the secondary formed $\gamma$-$CaCO_3$ modification, influences the physical properties of a rock.

The dehydration of gypsum led to a temperature decrease in the oven. Zier & Weise (2002) made similar observations from the masonry of a church that was fire damaged: rocks roughcast with gypsum showed less intense damage than rocks without such a plaster. Another process controlling the constitution of this rock seems to be the transformation of anhydrite III to anhydrite II causing an enormous shrinkage.

The foliation or bedding of a rock is one important structural parameter controlling the response of rocks to heat impacts. This is clearly expressed by the directional-dependent expansion of the

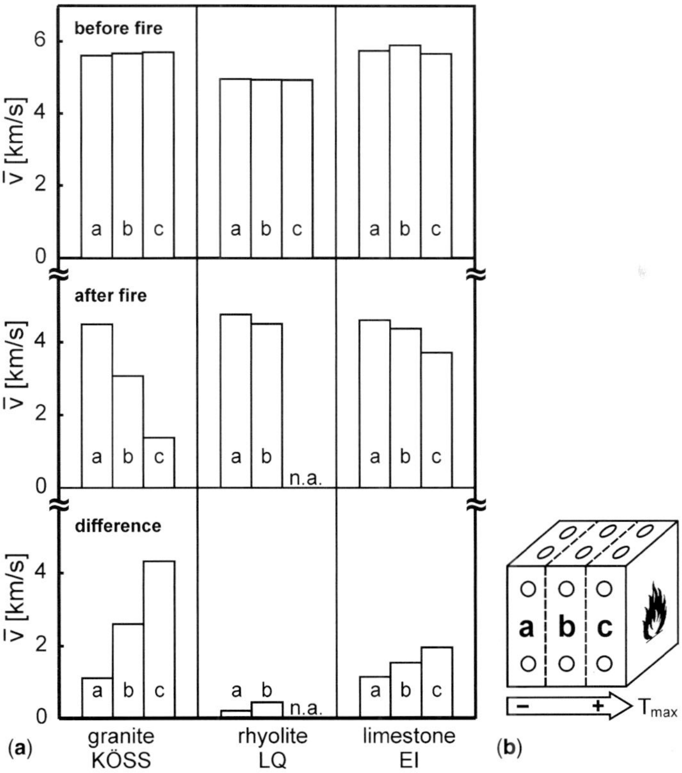


**Fig. 14.** (**a**) Average ultrasonic wave velocities before and after the fire tests, as well as their difference. (**b**) An average velocity was calculated according to measurements at four different positions (small circles) at each section (a, b, c). n.a., values not available; no measurements of parts of LQ that have been completely lost.

mylonitic orthogneiss, which can be explained (i) by microcracks opening preferably along the foliation or (ii) by a pronounced expansion of muscovite normal to the (001)-basal planes triggered while dehydroxylating (Mazzucato *et al.* 1999), as almost all muscovite crystals exhibit a strong preferred orientation of the *c*-axis perpendicular to the foliation.

The influence of the lattice preferred orientation of the dolomite marble can also be derived from the α-value of $10.5 \times 10^{-6}\,K^{-1}$, since the single crystal properties of dolomite are $6.2 \times 10^{-6}\,K^{-1}$ parallel to the *a*-axes and $22.9 \times 10^{-6}\,K^{-1}$ parallel to the *c*-axes. This corresponds to the texture analysis carried out for this rock by Zeisig *et al.* (2002). The difference for the two calcitic marbles (C1 and C2) from the Carrara region are related to the different grain-boundary geometry and its most probable control on their decay behaviour even below the critical temperature of decarbonatization.

In comparison, all different silicate rocks showed that the thermal expansion of the rock-forming minerals can be compensated to a certain degree by the porosity of a rock. Furthermore, preferably oriented pores can lead to an anisotropic thermal expansion as could be observed in the case of the limestone from Eibelstadt.

Another important factor controlled by the porosity is the thermal conductivity because heat conduction in air is different from that in the solid phase of rocks. As a result, from the small-scale fire tests a somewhat larger thermal gradient was observed in the more porous rhyolite than in the granite. Therefore, it was concluded that the scaling along planes perpendicular to the fire-exposed surface of the samples – more intensely developed in the rhyolite and the limestone than in the granite – can be assigned to the temperature gradient. The orientation of the main fracture plane inside the limestone also reflects the influence of the existing bedding planes. In contrast, the detachment of the corners of the cube-shaped granite sample mainly seems to be controlled by the test configuration, which allowed the fire to act on these corners from several sides.

## Conclusions

Evaluation of the response of natural building stones to fire shows that decay phenomena are controlled by both their mineral composition and their fabric. However, the conditions of the fire impact are also of critical importance: temperatures, duration and moisture content. The main conclusions can be summarized as follows:

- The transition of low- to high-quartz at around 573 °C is associated with a sudden change of the single crystal thermal expansion properties originating stresses that are released by the formation of mostly intragranular cracks.
- The degradation of carbonate rocks at temperatures up to 600 °C is related to the strongly anisotropic thermal expansion of calcite and dolomite that causes intergranular microcracking.
- The decomposition of phyllosilicates (at mineral specific temperatures mostly above 400 °C) and carbonates (above 600 °C) is associated with the release of water and $CO_2$, respectively – gas phases that may support microcracking due to their volume increase during heating.
- Colour changes are mostly a result of oxidation processes: a red discoloration owing to the oxidation of $Fe^{2+}$ (previously released by the dehydroxilation of phyllosilicates) is most prevalent.
- Carbonate rocks that suffered decarbonatization during heating disintegrate further if temperatures decrease to below 600 °C and water is present so that portlandite may form.
- The propagation and orientation of cracks is controlled to a large extent by grain-boundary geometries and the foliation pattern.
- A high porosity may compensate the thermal expansion of minerals to a certain degree, thus increasing the thermal resistance of a rock. On the other hand, a high porosity is related to a larger thermal gradient within a volume of rock that is partially heated, thus it favours scaling.

We gratefully acknowledge support in thermal expansion measurements by S. Webb and H. Büttner, help with the fire simulation from J. König and S. Reimer, and the discussion on fire damage of natural building stones with H.-W. Zier.

## References

ALLISON, R. J. & GOUDIE, A. S. 1994. The effect of fire on rock weathering: An experimental study. *In*: ROBINSON, D. A. & WILLIAMS, R. B. G. (eds) *Rock Weathering and Landform*. Wiley, Chichester, 41–56.

BLATT, H. & TRACY, R. J. 1996. *Petrology. Igneous, Sedimentary and Metamorphic*. Freeman & Co., New York.

CHAKRABATI, B., YATES, T. & LEWRY, A. 1996. Effect of fire damage on natural stonework in buildings. *Construction and Building Materials*, **10**, 539–544.

DIN 4102-8. 2003. *Brandverhalten von Baustoffen und Bauteilen; Kleinprüfstand*. Deutsches Institut für Normung e.V., Beuth Verlag GmbH.

DOVETON, J. H. 1997. *Log Analysis of Petrofacies and Lithofacies*. GFZ Logging Course. Geoforschungszentrum Potsdam.

EHLING, A. & KÖHLER, W. 2000. Fire damaged natural building stones. *In*: RAMMLMAIR, D., MEDERER, J.,

OBERTHÜR, T., HEIMANN, R. B. & ENTINGHAUS, H. (eds) *Applied Mineralogy in Research, Economy, Technology, Ecology and Culture*. ICAM 2000. A. A. Balkema, Rotterdam, Vol. **2**, 975–978.

FREDRICH, J. T. & WONG, T. 1986. Micromechanics of thermally induiced cracking in three crustal rocks. *Journal of Geophysical Research*, **91**, 743–764.

GOUDIE, A. S., ALLISON, R. J. & MCLAREN, S. J. 1992. The relations between modulus of elasticity and temperature in the context of the experimental simulation of rock weathering by fire. *Earth Surface Processes and Landforms*, **17**, 605–615.

ISO 834-1. 1999. *Fire-Resistance Tests – Elements of Building Construction – Part 1: General Requirements*. International Organization for Standardization, Geneva.

HAJPÁL, M. 2002. Changes in sandstones of historical monuments exposed to fire or high temperature. *Fire Technology*, **38**, 373–382.

HAJPÁL, M. & TÖRÖK, A. 2004. Mineralogical and colour changes of quartz sandstones by heat. *Environmental Geology*, **46**, 311–322.

HALL, A. 1996. *Igneous Petrology*. Longman, London.

KIESLINGER, A. 1954. Brandeinwirkungen auf Natursteine. *Schweizer Archiv*, **20**, 305–308.

MAZZUCATO, E., ARTIOLI, G. & GUALTIERI, A. 1999. High temperature dehydroxylation of muscovite-2M1: a kinetic study by in situ XRPD. *Physics and Chemistry of Minerals*, **26**, 375–381.

SIEGESMUND, S. 1996. The significance of rock fabrics for the geological interpretation of geophysical anisotropies. *Geotektonische Forschungen*, **85**, 1–123.

SIEGESMUND, S., ULLEMEYER, K., WEIß, T. & TSCHEGG, E. 2000. Physical wheathering of marbles caused by anisotropic thermal expansion. *International Journal of Earth Sciences*, **89**, 170–182.

SIEGESMUND, S., WEIß, T., VOLLBRECHT, A. & ULLEMEYER, K. 1999. Marble as a natural building stone: rock fabrics, physical and mechanical properties. *Zeitschrift der Deutschen Geologischen Gesellschaft*, **150**, 237–258.

WEISS, T., SIEGESMUND, S., KIRCHNER, D. & SIPPEL, J. 2004. Insolation weathering and hygric dilatation: Two competitive factors in stone degradation. *Environmental Geology*, **46**, 402–413.

ZEISIG, A., SIEGESMUND, S. & WEISS, T. 2002. Thermal expansion and its control on the durability of marbles. *In*: SIEGESMUND, S., WEISS, T. & VOLLBRECHT, A. (eds) *Natural Stone, Weathering Phenomena, Conservation Strategies and Case Studies*. Geological Society, London, Special Publications, **205**, 64–79.

ZIER, H.-W. & WEISE, G. 2005. Brandschäden an Natursteinen – dargestellt am Beispiel des Kirchenbrandes in Riethnordhausen. *WTA-Journal – International Journal for Technology and Applications in Building Maintenance and Monument Preservation*, **1**, 35–63.

# Post-depositional modification of atmospheric dust on a granite building in central Rio de Janeiro: implications for surface induration and subsequent stone decay

B. J. SMITH[1], J. J. McALISTER[1], J. A. BAPTISTA NETO[2] & M. A. M. SILVA[3]

[1]*School of Geography, Archaeology and Palaeoecology, Queen's University Belfast, Belfast BT7 1NN, Northern Ireland, UK (e-mail: b.smith@qub.ac.uk)*

[2]*Departamento de Geografia/FFP, UERJ, Sao Gonçalo, Brazil*

[3]*Departamento de Geologia, UFF, Niterói, Brazil*

**Abstract:** Extensive contour scaling of a 200 year old granite church is associated with the breaching of an apparently iron-rich crust and the widespread deposition of atmospheric dust within the canyon-like streetscape of Rio de Janeiro. Contemporary dust, accumulated dust from within a depression on the building surface, the surface crust and the underlying granite are examined by a combination of total element analysis and sequential extraction, X-ray diffraction and energy dispersive X-ray fluorescence. Results indicate an increase in total organic carbon and marked decrease in pH within the accumulated dust, and a rapid mobilization of anions and cations from the water-soluble and carbonate phases. It is considered that the latter is linked to salt accumulation within and eventual salt weathering of the granite. Post-depositional alteration of the dust is also linked with the de-silicification of clay minerals (illite to kaolinite) and the loss of silica from the amorphous Fe/Mn phase of the accumulated dust under the initially saline and progressively more acidic conditions experienced at the stone–atmosphere interface. This mobilization of silica is associated with the formation of what is, in effect, a thin silica-rich surface crust or glaze. Within the glaze, accessory amounts of extractable iron are concentrated within the amorphous and crystalline Fe/Mn phases at levels that are significantly elevated with respect to the underlying granite, but much lower than in the equivalent phases of the accumulated dust from which it is principally assumed to derive. The protection afforded to the stonework by the crust is not, however, permanent and within the last 15 years it has been possible to observe a rapid increase in the surface delamination of the church close to street level.

Building stone decay in polluted urban environments and subsequent conservation intervention is strongly affected by a range of surface modifications (Smith & Curran 2000). At the most obvious level, the aesthetic damage caused by, for example, the growth of black gypsum crusts has been the major driving force behind widespread campaigns of stone cleaning that typified many European and North American cities in the late 20th century (e.g. Maxwell 1992). However, surface modification can produce a wide variety of chemical and physical responses that strongly control the rate and pattern of stone decay. Curran & Smith (2000) have shown, for example, how reductions in surface porosity/permeability consequent to exposure influence moisture ingress and egress, together with potentially damaging salts held in solution. The effects of surface modification need not, however, be immediately detrimental, and in some cases modification can result in surface induration. It is for this reason that, historically, stone masons encouraged the formation of surface accumulations through, for example, the formation of calcium oxalate by the application of substances such as albumen, casein and other organic materials (Jenkins & Middleton 1988; Lazzarini & Salvadori 1989; Sabbioni & Zappia 1991). The protection afforded by surface induration may not, however, be permanent. This is especially the case with induration produced by the outward migration of iron that precipitates at or near the exposed surface (McAlister *et al.* 2003). This can occur at the expense of weakening the underlying stone by the removal or weakening of iron cement. If the outer crust is breached or delaminates, the stone decays rapidly through the creation of a cavernous hollow. Iron crusts and surface stains are not, however, restricted to iron-rich (or even iron-containing) stonework. In the latter case the source of the iron must be exogenic. Iron-rich, exogenetic crusts have been studied in great detail on natural rock outcrops where they are generally referred to as rock varnishes. In these studies there is concurrence on the importance of wind-blown dusts and their deposition as a major source of iron and other components such as manganese. Controversy continues, however, as to whether iron and manganese are mobilized and subsequently

*From*: PŘIKRYL, R. & SMITH, B. J. (eds) *Building Stone Decay: From Diagnosis to Conservation.* Geological Society, London, Special Publications, **271**, 153–166.
0305-8719/07/$15.00 

precipitated by physico-chemical processes or whether mobilization is via the range of organisms, especially bacteria, that typically colonize these surfaces (Dorn 1998). Debate also persists over the extent to which iron and manganese are mobilized in association with, or as accessories to, other elements. This applies specifically to silica, which ranges from a significant accessory to iron/manganese-dominated varnishes through to distinctive, silica-rich varnishes also referred to as 'silica glazes' (Fisk 1971; Dorn 1998). In contrast to these studies of rock varnishes there has been little detailed research into iron-rich exogenetic varnishes within the built environment. This is despite a general acknowledgement that iron can be mobilized in highly acidic ambient aerosol solutions (Zhu *et al.* 1992; Spokes *et al.* 1994). In turn, this acidification could result from the incorporation of sulphur and nitrogen oxides into dust particles within urban/industrial environments.

It is for the above reasons that the current study set out to investigate what appears to be iron staining of a 200 year old granite church in central Rio de Janeiro that was first reported by Smith & Magee (1990). Within this framework, specific attention is paid to the possible role of surface dust deposition as a source for the stain and any associated surface induration. To accomplish this, samples of the underlying granite, contour scales (including the surface crust), and accumulated and contemporary dusts from the church were analysed. The contemporary and accumulated dust samples were collected specifically to identify post-depositional modification with respect to iron, manganese and silica, plus total and readily oxidizable organic carbon and pH. In addition to total element analysis, iron, manganese and silica were also studied using a selective dissolution technique. This analyses samples after extraction from the water-soluble, exchangeable/carbonate, amorphous Fe/Mn, crystalline Fe/Mn, organic and residual phases, and provides an indication of the conditions required to mobilize the different components within the dusts. Water-soluble cations and anions were also analysed, as salt weathering was previously identified as a key factor in the mechanical weathering of the church (Smith & Magee 1990). Sample mineralogy was examined by X-ray diffraction (XRD) and the crust surface examined by scanning electron microscope fitted with an EDXRF (energy dispersive X-ray fluorescence) spectrometer.

## Background

### *Atmospheric particulate matter*

Particulate matter is the general term used to describe a mixture of solid and liquid droplets dispersed in the atmosphere that originate from both natural and anthropogenic sources. Primary particles are released from sources of generation and secondary particles are formed in the atmosphere as a result of gaseous reactions. These particles become airborne in a gaseous medium, and are subject to diffusion, coagulation, chemical interactions, scavenging and, ultimately, deposition. Larger particles originate from various sources including the breakdown of construction materials, eroded soil, foundry and pulverized coal dusts. Finer particles may include carbon combustion products from incinerators, vehicle emissions, domestic and forest fires, and from sea-salt nuclei. Some particles may be of biological origin, and these include bacterial and fungal spores and pollen. Dust in the upper size fraction ($>1$ μm) is removed from the atmosphere by wet and dry deposition. In wet deposition, particle are incorporated in cloud droplets (rainout) and removed by falling precipitation (washout). Dry deposition is slow and continuous, whereas wet deposition delivers sudden and infrequent concentrations of pollutants in dilute solution (Bloch *et al.* 1980; Georgii & Perseke 1980; Hicks 1981; Colin 1998; Morselli *et al.* 2003). Smaller particles are deposited by coagulation (Junge 1963; Corn 1976). Other phoretic effects that cause dry deposition include thermal collision of air molecules (Brownian movement), temperature differences between stone surfaces and the surrounding atmosphere (thermophoresis), gravitational settling and electrostatic forces (electrophoresis) (Camuffo 1998*a*, *b*). Dust particles formed from the disintegration of larger particles, also referred to as dispersion aerosols, are important in this study since they have a high specific surface area and are therefore capable of adsorbing a wide range of gaseous and particulate pollutants before being deposited on buildings. The final result is a mixture of insoluble and soluble materials that have a very diverse composition (Del Monte & Lefèvre 1998; Garrett 2000; Espinosa *et al.* 2001; Smith *et al.* 2003). Anthropogenic particles from combustion processes are very important pollutants within city environments as they are principally composed of amorphous carbon, alumino-silicates and metals. Abrasive products can also originate from vehicle brakes, clutches and rubber tyres. Gases occur in the atmosphere as primary (e.g. $SO_2$, CO, $CO_2$, volatile organic compounds (VOCs) and NO) and secondary (e.g. $NO_2$ and $O_3$) pollutants, and originate from fossil fuel combustion and biomass burning. Gases such as $SO_2$ and $NO_2$ can also undergo photochemical oxidation to form $H_2SO_4$ and $HNO_3$ that contribute towards environmental acidification (Perros 1998). Nitric acid is strong and highly soluble, and research has shown that $HNO_3(g)$ becomes incorporated in the particulate phase, especially when non-acidic aerosols are present (Goodman *et al.* 2000; Hanisch & Crowley 2001; Metzger *et al.* 2002). Other primary reactions that occur within accumulated particulates could include the hydrolytic and oxidative decomposition of $Fe^{2+}$ silicates. As a consequence of these reactions, it is probable that dust accumulations are a complex chemical and mineralogical mix, in which the same element may be held within a number of phases. Consequently, the importance of an element cannot be assessed solely by its total concentration. Only the study of elemental partitioning within the different phases will provide insights into their participation in, for example, surface induration.

### Selective extraction

Selective extraction is a technique used to quantify elements that are partitioned between different phases within geological materials. Reactions responsible for element partitioning are strongly controlled by redox potential (Eh) and hydrogen ion activity (pH). The former determines element mobility and the latter controls mineral dissolution, precipitation and complexation (Dzombak & Morel 1990). These phases include water-soluble, exchangeable/carbonate, amorphous Fe/Mn, crystalline Fe/Mn, organic and residual. Elements exist in these phases in different physico-chemical forms, which includes exchangeable, occluded, co-precipitated or those bound by secondary oxides, especially amorphous ones, such as iron oxyhydroxides $[Fe(OH)_3 \cdot nH_2O]$. Elements also exist in carbonates, organo-metallic complexes and as ions in the crystal lattices of primary minerals (Chao & Zhou 1983; Ellis & Fogg 1985). Selective extraction exposes these phases to a sequence of solutions of increasing concentration/aggressivity using a stepwise procedure under strict conditions (Chao 1972; Agemian & Chau 1977; Skei & Paus 1979; Tessier *et al.* 1979; Ure *et al.* 1993; Quevauviller *et al.* 1994; Hall *et al.* 1996; McAlister & Smith 1999; McAlister *et al.* 2003, 2005). Consecutive extraction procedures can in turn provide information on potential transport mechanisms, mobilization and transformation of elements under, for example, acidic, alkaline, oxidizing or reducing environmental conditions. Like other analytical techniques, there are some operational problems involved in selective extraction and no general agreement has been reached as to which extractant should be used for a particular phase since matrix effects are involved in heterogeneous processes (Pickering 1981; Van Valin & Morse 1982). The study aim, type of sample and elements of interest determine the extraction protocol to be used. A number of analytical problems can be avoided by using the appropriate extracting solutions and sample : solution ratios (Rauret *et al.* 1989; McAlister *et al.* 2003).

## Field area

### Location and description of the church

The Igreja Nossa Senhora do Carmo is located on the Praça XV de Novembro close to the ferry terminal that links Rio de Janeiro to Niteroí (Fig. 1). The church was built approximately 200 years ago, primarily from light-coloured, medium-grained garnet-rich granite. It fronts on to an extremely busy road and is exposed to high levels of vehicle emissions near ground level that are accentuated by surrounding high buildings that create a corridor/canyon effect. The front of the church is heavily discoloured, although there is streaking in some areas subject to concentrated rainwash. Below 1 m most sheltered areas exhibit a thin black crust. Elsewhere – even on rainwashed areas – there is a widespread brownish discoloration that typically has a covering of carbonaceous dust. When Smith & Magee (1990) first studied this building they observed that this staining was associated with limited patches of surface scaling (2–3 mm in thickness) that exposed the lighter granite substrate. On revisiting the church for the current study, scaling was observed to be much more widespread, especially near street level where discoloration is also most pronounced (Fig. 2).

### Environmental conditions in Rio de Janeiro

Rio de Janeiro, although it lies just within the tropics, experiences a humid, subtropical climate due to its coastal location. Most rainfall (annual average 1100–1800 mm) occurs between December and April. However, relative humidity is consistently high and it can rain at any time during the year. Rainfall acidity is exemplified by precipitation studies carried out in Tijuca National Park within the city, where pH values of between 4.7–6.1 and 3.8–5.4 were recorded by Silva Filho (1985) and de Mello (2001), respectively. These figures reflect reduced air quality principally as a result of vehicle emissions and photo-chemical smogs that contain high concentrations of carbonaceous and sulphate-rich particles (Daisey *et al.* 1987). These aerosols originate from marine sources, industry, construction sites, soil disturbance and weathered stone masonry (Azevedo *et al.* 1999). Brazil also experiences specific pollution derived from the use of anhydrous alcohol (from sugar cane) for fuel and as an additive to gasoline. High concentrations of acetaldehyde

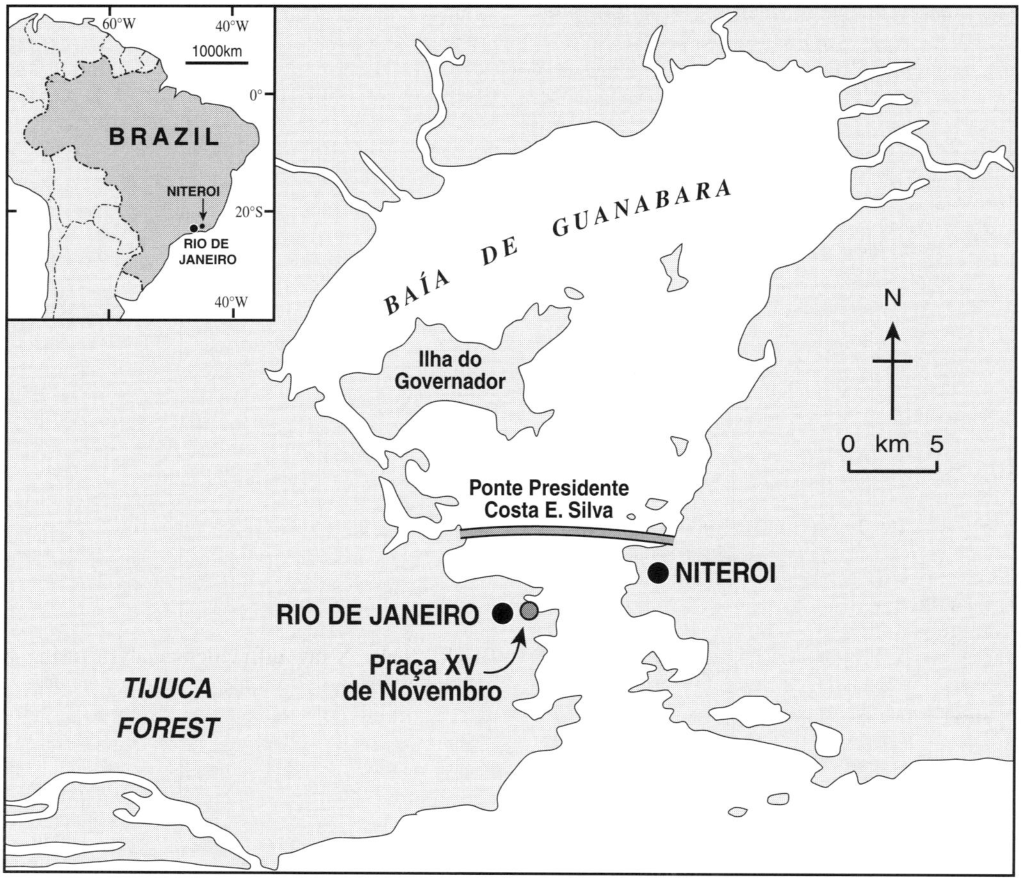


**Fig. 1.** Location map.

compared to formaldehyde in the atmosphere during the summer months have been attributed to this combustion of ethanol (Corrêa *et al.* 2003). These authors also state that high levels of acetaldehyde can be supplemented by subsequent photochemical oxidation of other volatile hydrocarbons. Acetaldehyde and formaldehyde have a significant influence on the formation of other smog components, where a photochemical reaction between nitrogen oxide and ozone results in the formation of nitrate. This in turn reacts with acetaldehyde to form nitric acid and a peroxyl free radical (Grosjean *et al.* 1990, 2002; de Andrade 1998; Nguyen *et al.* 2001; Martins *et al.* 2003).

## Sampling and analysis

Sampling and analysis were designed specifically to address the role of particulate deposition in contributing to surface modification of stonework. Samples of contemporary dust were gently brushed from the surface of the brown-stained façade and from a thick accumulation of dust within a surface depression on the same area of the façade, approximately 1.5 m above street level. The latter is considered to represent a longer-term record of dust deposition together with the effects of any post-depositional, *in situ* modification. Samples of detached contour scales were gently lifted from the church front and the underlying granite was sampled from an area of clean stone exposed by the detachment of a large surface scale. On returning to the laboratory the surface patina was carefully removed from the contour scaling using a diamond-tipped blade attached to a Dremel Multi engraver and prepared for analysis (McAlister *et al.* 2003). However, owing to the thinness of the discoloured surface layer, it was impossible to ensure that the final bulk sample did not contain some of the substrate. Samples were air dried between 30 and 35 °C in a

**Fig. 2.** Photograph showing delamination of contour scales on the street facing façade of Igreja Nossa Senhora do Carmo near to street level.

fan-assisted oven, gently ground using an agate pestle and mortar, and the <63 μm fractions collected by passing them through a nylon mesh sieve. Subsamples (0.500 g) were weighed into acid-washed polypropylene tubes and analysed using a selective extraction technique that was a modification of the protocol used to study iron migration in sandstone (McAlister *et al.* 2003). This method extracted Ca, Mg, Na, K, $Cl^-$, $NO_3^-$, $SO_4^-$, Fe, Mn and Si from the water-soluble phase, and Fe, Mn and Si from the exchangeable/carbonate, amorphous Fe/Mn, crystalline Fe/Mn, organic and residual/siliceous phases of the dust samples. The organic phase of the crust and substrate samples was not extracted because of its very low organic carbon concentrations. The organic phase was extracted using a 4:1 $HNO_3$–HCl solution. This acid mixture was used in a sediment leach PAAR005H microwave digestion technique (Perkin Elmer), but in this study a water bath was used so that consistency of extraction was maintained. Analysis of the residual phase was carried out by total digestion in an $HNO_3$–HF–HCl solution using a Perkin Elmer Microwave Digestion system. Chloride, nitrate and sulphate analysis was carried out using a Dionex Model DX 500 ion chromatograph. Elemental analysis employed a Perkin Elmer AAnalyst 200 atomic absorption spectrometer, and silica was analysed using a Unicam Model 8675 visible spectrometer after development and reduction of molybdosilicic acid (Fanning & Pilson 1973). This method was modified to ensure that all the solutions were around pH 7.0 prior to colour development. Total organic carbon was analysed by loss-on-ignition (Hesse 1971) and oxidizable organic carbon used a chromic acid digestion procedure (Walkely & Black 1934). X-ray diffraction analysis using a Siemens D5000 diffractometer was carried out on powder samples (<63 μm) by preparing random-orientated mounts, and clay (<2 μm) samples were separated from the latter and preferentially mounted using a membrane technique (McAlister & Smith 1995*a*, *b*). Extractions were carried out using a Lab Line orbital shaker and a Grant JB4 water bath, and suspensions were separated using a Heraeus Megafuge 1.0 centrifuge. Morphology and elemental analysis of the crust surface was carried out using a JEOL scanning electron microscope fitted with an energy dispersive X-ray fluorescent (EDXRF) spectrometer. Samples were mounted onto aluminium stubs using an epoxy resin and gold coated prior to analysis.

## Results and Discussion

Results from total element analysis (Table 1) show that the substrate sample contains significantly lower levels for all the analysed elements when

**Table 1.** *Results from total element analysis of surficial dust, surface crust and granite substrate showing ionic concentrations*

| Sample | Ca ($mg\ kg^{-1}$) | Mg ($mg\ kg^{-1}$) | Na ($mg\ kg^{-1}$) | K ($mg\ kg^{-1}$) | Fe ($mg\ kg^{-1}$) | Mn ($mg\ kg^{-1}$) | Al (%) | Si (%) |
|---|---|---|---|---|---|---|---|---|
| Crust | 6750 | 1875 | 20 250 | 32 500 | 8146 | 191 | 7.5 | 25.1 |
| Sub. | 9750 | 1600 | 27 000 | 20 000 | 3308 | 80 | 3.31 | 12.2 |
| Acc. | 10 300 | 4000 | 9750 | 20 000 | 47 996 | 425 | 6.67 | 22.4 |
| Con. | 15 500 | 2250 | 19 750 | 27 500 | 23 313 | 574 | 6.25 | 20.1 |

Con, contemporary dust; Acc., accumulated dust; Sub, substrate (underlying granite).

**Table 2.** *Results from water-soluble extraction of surficial dust, surface crust and granite substrate showing ionic concentrations, together with pH, loss on ignition (LOI) and oxidizable organic carbon (OOC)*

| Sample | Ca ($mg\,kg^{-1}$) | Mg ($mg\,kg^{-1}$) | Na ($mg\,kg^{-1}$) | K ($mg\,kg^{-1}$) | Cl ($mg\,kg^{-1}$) | $NO_3$ ($mg\,kg^{-1}$) | $SO_4$ ($mg\,kg^{-1}$) | Si ($mg\,kg^{-1}$) | pH | LOI (%) | OOC (%) |
|---|---|---|---|---|---|---|---|---|---|---|---|
| Crust | 66 | 10 | 48 | 80 | 2.2 | 2.3 | 6.1 | 20 | | | |
| Sub. | 24 | 4 | 128 | 64 | 3.0 | 0 | 3.1 | 24 | | | |
| Acc. | 540 | 72 | 116 | 64 | 274 | 1484 | 590 | 32 | 4.9 | 18.2 | 6.8 |
| Con. | 3400 | 260 | 955 | 260 | 1705 | 1303 | 3434 | 58 | 7.3 | 7.5 | 3.5 |
| A/C | 0.2 | 0.3 | 0.1 | 0.2 | 0.2 | 1.1 | 0.2 | 0.6 | | | |

compared to the crust and especially to the dust samples. This could suggest low, if any, input of these elements to the dusts and crust from the underlying granite. Elements are released from complex materials according to their mobility. However, to understand where and why element migration has taken place requires the partitioning of elemental components into different phases that can be linked to the conditions controlling their mobility. The most mobile (labile) elements are to be found in the water-soluble and exchangeable/carbonate phases. Non-labile elements are more likely to be found in the Fe/Mn oxide and organic phases or held within the residual, silicate phase of any sample. Results for water-soluble anion and cation concentrations, plus pH, total and oxidizable organic carbon and depletion/enrichment ratios for the dusts, are presented in Table 2.

Post-depositional modification of the accumulated dust is highlighted in both chemical and mineralogical analyses. Total and oxidizable organic carbon concentration show enrichment factors of 2.4 and 1.9, respectively, for the accumulated dust. Comparison of these results shows how non-oxidizable organic carbon is concentrated in the accumulated dust. This is possibly through a combination of the continued accumulation and integration into the dust of carbon from sources such as soot, as well as the loss and relative depletion of other dust components. A consequence of this increase is a pH for the accumulated dust that is 20 times more acidic than that of the contemporary dust.

Examination of water-soluble ions is important when studying weathering mechanisms – especially those associated with salt weathering. However, sulphate and chloride anions may also form relatively insoluble complexes with metals and therefore cause them to desorb and/or precipitate (Basta *et al.* 2005). Results in Table 2 indicate relatively high levels of chloride, nitrate and sulphate ions in the contemporary dust, with chloride and sulphate concentrations 6 times those in the accumulated dust. Contradictory to other anions, the nitrate level is slightly higher in the accumulated. This might suggest that whilst soluble anions are progressively lost following deposition, nitrate is continuously replenished from atmospheric sources. This might in turn suggest that chloride and sulphate ions derive primarily from particulate deposition rather than, for example, the sulphation of existing dust deposits. Water-soluble calcium, magnesium, sodium, potassium and silica concentrations all show significant reductions in the accumulated dust. Values are reduced by a factor of approximately 4 for magnesium and potassium, by 6 and 8 for calcium and sodium respectively and silica is reduced by a factor of 2 (Table 2). In the absence of sources such as road salting, sodium and chloride are presumed to come primarily from marine aerosols. Indeed, the ratio between sodium and chloride in the contemporary dust is 1.79 and this figure correlates well with the ratio of 1.8 that has been used as an index for marine aerosol (Hidy 1984; Sabbioni 1995). Sulphate originates mainly from fossil fuel combustion, whilst nitrate is especially associated with vehicle emissions. Soluble cations and anions in both the accumulated and contemporary dusts are orders of magnitude higher than in the crust and underlying granite. The water-soluble chloride and sulphate in the granite must result from their migration into the stone, most probably in response to intense wetting and quick drying (Kunzel 1995). Interestingly, there is no evidence of nitrate accumulation in the underlying granite. A possible explanation of this may lie in the nature of the sample. This was highly fractured, and may be that highly soluble nitrate salts are periodically flushed from an open network of microfractures, whereas the removal of accumulated chloride and sulphate salts is less efficient. Salt combinations such as $CaSO_4$ and NaCl are very effective in producing contour scaling and granular disaggregation, especially because the presence of the latter is known to enhance the solubility and penetration into stonework of the former (Warke & Smith 2000). Despite this, both calcium and sulphate concentrations are slightly higher in the crust than in the underlying granite, whereas sodium and chloride are higher in the granite. These differences are probably related to the solubility differences between halite and gypsum. However, differences in anion and cation concentration between the crust and underlying granite are inconsistent and, for example, crust and underlying granite samples show similar concentrations of water-soluble silica. It must also be acknowledged that, in absolute terms, differences between crust and underlying granite are very minor when compared to the values recorded for contemporary dust. What is more important is the overall conclusion that following deposition, dusts appear to weather and progressively lose their more soluble components. There is evidence that a limited quantity of salts derived from this dissolution may accumulate in the underlying stonework, and that dust deposits continue to acquire nitrate, possibly from gaseous deposition.

In terms of the less soluble components of deposited dust, total element analysis (Table 1) shows that both iron and silica are higher in the accumulated material. However, the differences may be apparent rather than real, in that any increase could be a product of the more effective loss of more soluble components such as calcium, sodium and potassium from the contemporary dust. The possibility that iron, manganese and silica are also lost from the accumulated dust receives support from the total element analysis of the crust and underlying granite that show concentrations in the crust more than twice those in the underlying granite (Table 1). It should be noted, however, that in the case of silica this represented an increase of 12.9% to a total of 25.1% compared to an increase in iron of 0.48%. The concentration of iron and silica in the crust are also indicated by SEM/EDXRF analysis of the crust surface (Fig. 3). The same technique did not, however, detect significant levels of manganese in the surface crust, and this accords with the low levels recorded by total element analysis and selective extraction. This would seem to suggest that, although manganese is an important accessory element that might migrate and concentrate in conjunction with, in particular, iron, it does not form a key component of the surface crust on this church. By the same token, the crust should not be

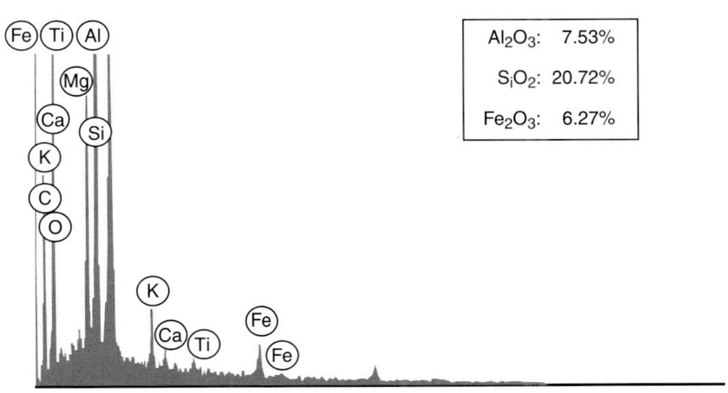


**Fig. 3.** Energy dispersive X-ray fluorescence trace of the surface of a contour scale from Igreja Nossa Senhora do Carmo, emphasizing its high iron, aluminium and silica composition. Unlabelled peaks are for gold coating on the sample.

equated directly with naturally precipitated, manganese-rich rock varnishes described by authors such as Dorn (1998).

The proposed changes in elemental concentrations as dusts weather, especially a possible loss of silica from contemporary dusts, receive support from the mineralogical analyses of the samples (Figs 4 and 5). Analysis by XRD of the <63 μm fractions for both dusts and the underlying granite show major and minor peaks for the common rock-forming minerals (quartz, feldspar and mica) found within granitic rocks. This may indicate that a prime source of the granular components of the dust is a combination of eroded soil and the disintegration products of the granite and gneiss that underlie much of the region and

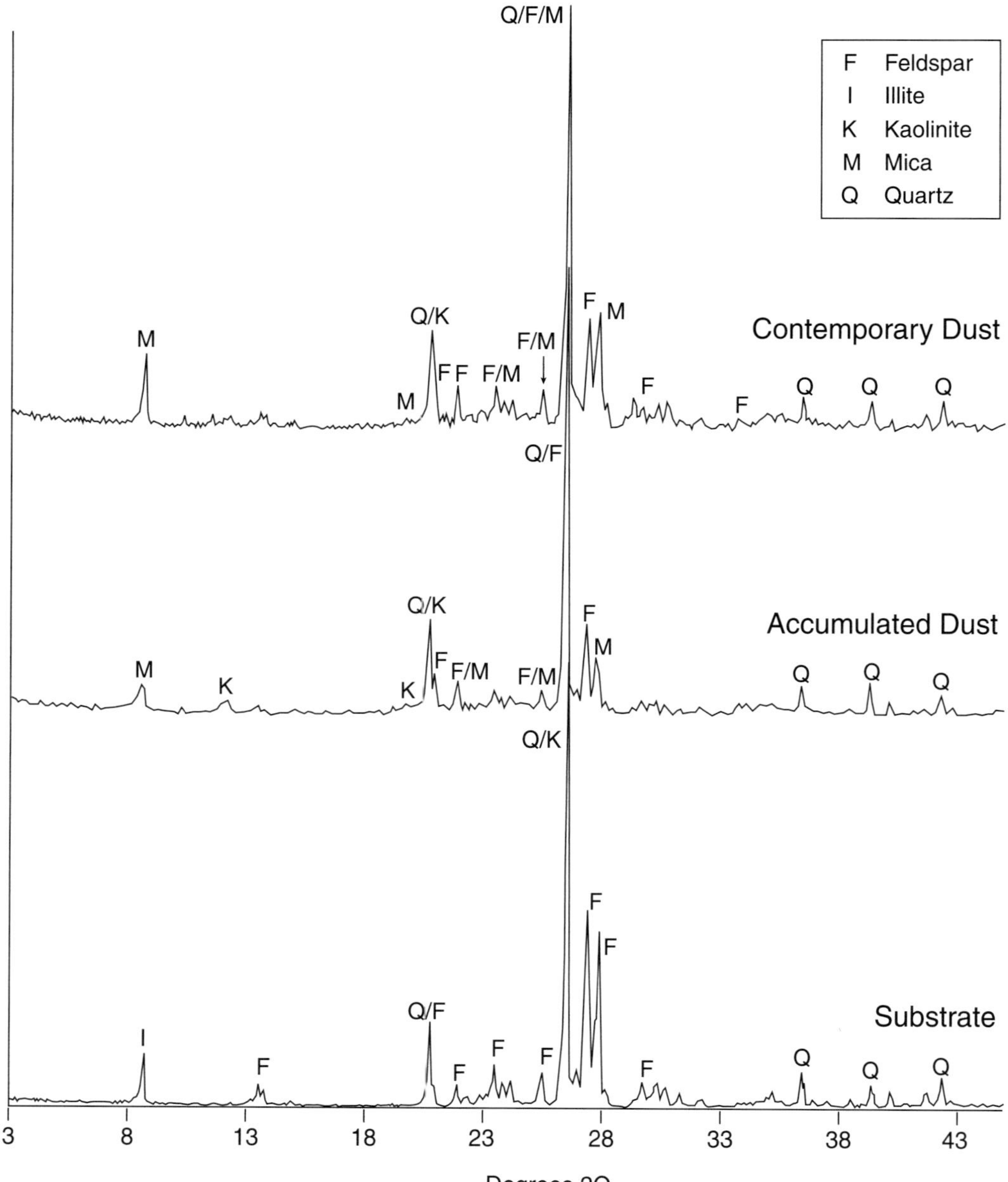


**Fig. 4.** X-ray diffraction traces of the <63 μm fractions from samples of contemporary dust, accumulated dust and a ground sample of the surface crust.

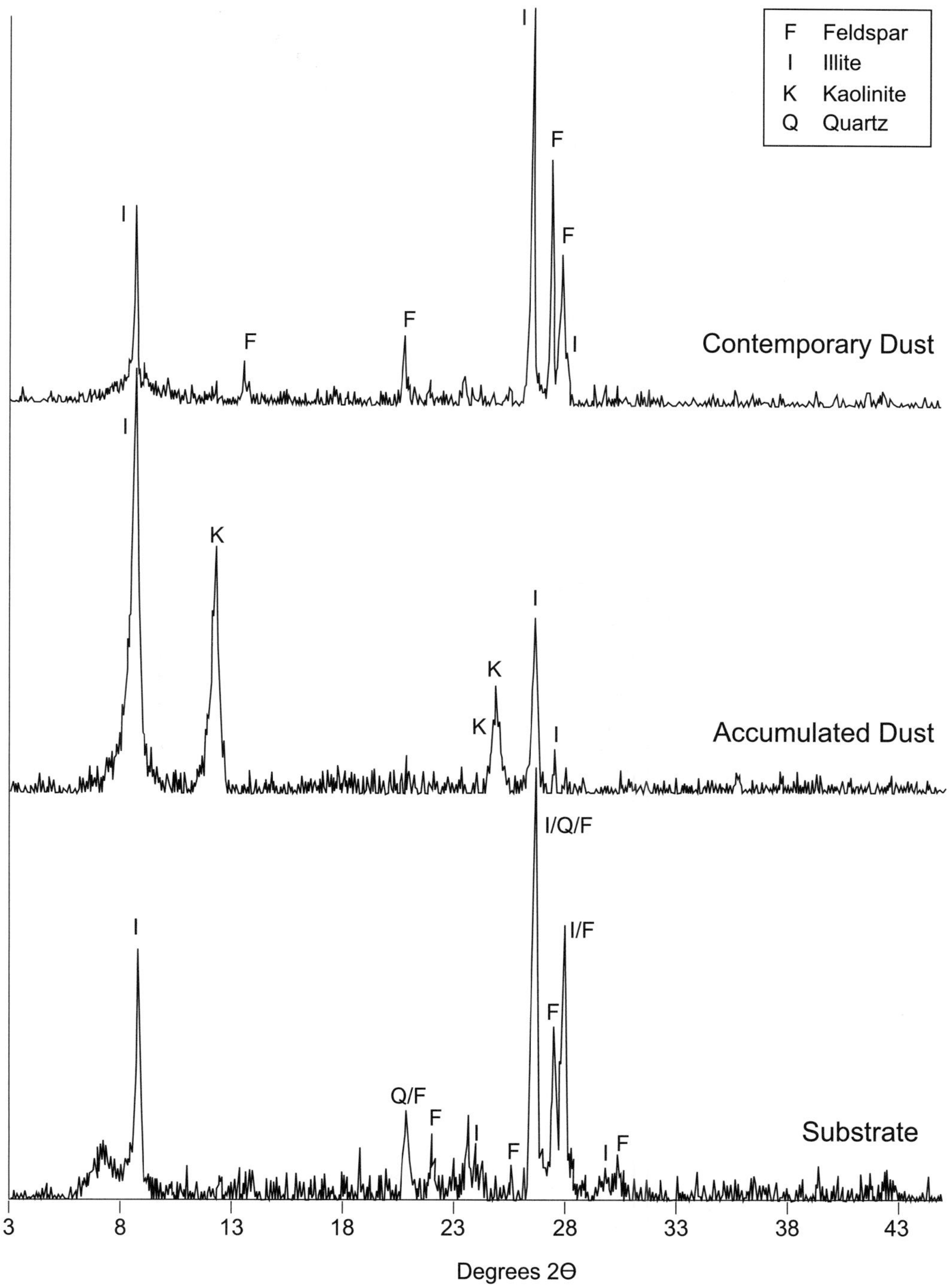


**Fig. 5.** X-ray diffraction traces of the <2 μm fractions from samples of contemporary dust, accumulated dust and a ground sample of the surface crust.

that are used in many of the city's buildings. Results for the <2 μm fraction show an overall similarity between the underlying granite and the contemporary dust, but distinct differences between these and the accumulated dust. The substrate and contemporary dust have major peaks for the clay mineral illite. This has been previously identified as a by-product of the initial hydrolysis of silicate minerals within naturally weathered profiles on the granitic rocks in SE Brazil (Power & Smith 1994). As such it is a common component of the soils of the area. The diffractogram for the accumulated dust has similar peaks for illite, but also major peaks for the simpler 1:1 lattice clay kaolinite. Within subarial weathered materials, kaolinite is widely viewed as the product of the intense chemical weathering of primary silicate minerals and/or a consequence of the de-silicification of more complex 2:1 lattice clay minerals such as illite under continued hydrolysis (Thomas 1994). The hydrolysis of illite in the accumulated dust could also explain the observed decrease in potassium identified by total element analysis and possibly the elevated levels of potassium in the crust (Table 1).

Results of the selective extraction of iron, manganese and silica from the contemporary and accumulated dusts are given in Table 3. These will be used specifically to compare the phase concentrations between the contemporary and accumulated dusts, and to identify the phase composition of the surface crust. In addition to confirming the depletion of the water-soluble phase in the accumulated crust, the data also show that the dramatic reduction in pH of the accumulated dust is associated with the removal of carbonaceous material. This is shown by the significant decreases in iron, manganese and silica concentrations in the exchangeable/carbonate phase of the accumulated dust. In contrast to the carbonate phase, the amorphous and crystalline Fe/Mn phases and the organic phase all exhibit a marked enrichment of iron within the accumulated dust (enrichment ratio of 12.7 in the crystalline phase). There is some enrichment of manganese in the crystalline and organic phases of the accumulated dust, but a slight decrease in the amorphous phase. In all three of these phases manganese concentration is comparatively low and consequently any recorded change is less reliable than for the other elements. However, the low concentrations are, apart from the residual phase, consistent with the very low levels of manganese extracted from the crust. Organic complexation is shown to be just as effective as the crystalline Fe/Mn phase for the retention of iron and manganese in the accumulated dust and for manganese in the contemporary dust. The organic phase is shown to be a very effective sink for iron in the contemporary dust where the concentration is higher by a factor of 9 compared to the crystalline Fe/Mn phase. The data show an increase for silica in the crystalline phase, but a marked reduction (enrichment ratio of 0.6) in the amorphous phase of the accumulated dust.

Data from the crust show high silica and iron concentrations in the amorphous and crystalline

**Table 3.** *Results from sequential extraction of surficial dust, surface crust and granite substrate showing elemental concentrations in different phases*

| | | Con. | Acc. | Crust | Sub. | A/C |
|---|---|---|---|---|---|---|
| Water-sol. | Fe | 1.2 | 0 | 4.0 | 3.6 | |
| | Mn | 11.8 | 3.6 | 0 | 0 | 0.3 |
| | Si | 58 | 32 | 20 | 24 | 0.6 |
| Exc/Carb | Fe | 462 | 210 | 132 | 72 | 0.5 |
| | Mn | 54 | 9.6 | 4.2 | 3.6 | 0.2 |
| | Si | 270 | 66 | 78 | 18 | 0.2 |
| Am Fe/Mn | Fe | 8166 | 17 676 | 1680 | 683 | 2.2 |
| | Mn | 192 | 180 | 9.6 | 5.4 | 0.9 |
| | Si | 3660 | 2110 | 1464 | 606 | 0.6 |
| Cry Fe/Mn | Fe | 1014 | 12 900 | 1140 | 600 | 12.7 |
| | Mn | 60 | 72 | 9.6 | 6.0 | 1.2 |
| | Si | 1506 | 1758 | 996 | 330 | 1.2 |
| Organic | Fe | 9420 | 12 960 | | | 1.4 |
| | Mn | 72 | 90 | | | 1.3 |
| | Si | 85 | 65 | | | 0.8 |
| Residual | Fe | 4250 | 4250 | 5250 | 1950 | 1.0 |
| | Mn | 132.5 | 70 | 168 | 65 | 0.5 |
| | Si | 195 000 | 220 000 | 248 000 | 121 000 | 1.1 |

Con., contemporary dust; Acc., accumulated dust; Sub., substrate (underlying granite); A/C, ratio of results for accumulated and contemporary dusts.
Phases: Water-sol., water-soluble; Exc/Carb, exchangeable/carbonate; Am Fe/Mn, amorphous Fe/Mn; Cry Fe/Mn, crystalline Fe/Mn.

Fe/Mn phases when compared to those in the underlying granite – especially the amorphous phase. However, although these findings are in accord with the high concentrations of these elements found in the same phases of both dusts, it is worth noting that the elevated level of silica in the amorphous phase of the crust contrasts with a silica decrease in the same phase of the accumulated dust when compared to contemporary dust. In contrast to their relationships with the surface dusts, iron and silica concentrations in the crust are significantly higher than in equivalent Fe/Mn phases for the underlying granite (Table 3). The low level of manganese bound in the amorphous Fe/Mn phase of the crust suggests that this element plays little role in crystallization and mineral formation compared to iron and silica.

In terms of total silica, by far the largest components in both dusts and the crust are to be found in the residual/siliceous phase. Within the dusts this confirms the importance of siliceous particles in their general make-up and within the crust it could reflect the inability to remove all underlying siliceous material during sample preparation. In all three cases the concentrations are markedly higher than that observed for the underlying granite, but this may simply reflect the latter's mixed mineralogy and the presence of a high proportion of non-siliceous minerals. Similar concentrations of iron and silica bound by the residual/siliceous phases of the contemporary and accumulated dusts, and similar XRD patterns for primary minerals in powdered samples (Fig. 3), do, however, appear to confirm that they have originated from substantially the same combination of sources. As iron is very immobile in the residual/siliceous phase, it is unlikely that there would have been movement of this element either from the substrate or the dusts towards crust formation.

## Interpretation and conclusions

Examination of contemporary dust deposited on the façade of Igreja Nossa Senhora do Carmo shows it to be a complex mixture of mineral and organic components in which, in particular, the products of atmospheric pollution are held within a number of phases that strongly influence patterns of element availability. Following deposition, the dust is observed to undergo a transformation – that can be likened to chemical weathering – under the warm, moist, acidic conditions experienced at the building stone–atmosphere interface. Most prominent amongst the effects of this weathering is the loss of common anions and cations from the water-soluble and carbonate phases of the dust. Although there is an indication that nitrate continues to be adsorbed by deposited dust in the nitrogen-rich atmosphere generated by intense traffic pollution within the canyon-like streetscape of central Rio de Janeiro. It might be expected that any salt liberated from dust in this humid tropical environment would be rapidly washed from the building surface. However, previous studies (Smith & Magee 1990) have observed that less soluble salts, especially gypsum, can be absorbed and retained by the underlying stonework, where they appear to be instrumental in its eventual surface scaling. Possible explanations for the retention of salts are that buildings may be sheltered from driven rain at the bottom of 'urban canyons', and that the strong localized winds frequently associated with urban streets can encourage rapid drying after wetting. Furthermore, the high specific surface of cities (the ratio of building surface to land area) such as Rio means that average rainfall per unit surface of a building is much less than the average rainfall recorded by meteorologists. In other words, microclimate at the stone surface may differ substantially from the macroclimate conditions used by geographers to classify major climatic zones.

Despite this observation, there seems little doubt that the combination of frequent wetting, high relative humidity and high air temperatures are conducive to chemical weathering – even if the weathering products may not be transferred any great distance before being re-precipitated. Testimony to this is given by the observed weathering of clay minerals such as illite, and possibly primary silicate minerals, to form kaolinite within accumulated dust. The formation of kaolinite in this way is associated with the loss of silica from the original mineral. This alteration is redolent of natural chemical weathering normally ascribed to humid tropical environments. The fact that it is observed in a surface dust deposit, as compared to the more usual (and much older) soil and regolith profiles, is also indicative of the aggressivity of the stone surface environment.

In addition to mineralogical observations, selective extraction of contemporary and accumulated dust, and what appears to be an underlying iron-rich crust, shows that most extractable iron and silica is located within the amorphous and crystalline Fe/Mn phases. Because of the loss of the water-soluble and carbonate components, iron in these two phases is seen to increase during the weathering of the contemporary dust. However, this does not mean that in absolute terms some iron is not lost from these phases of the dust, and the data indicate a real decrease in silica within the amorphous Fe/Mn phase of the accumulated dust compared to contemporary dust. Total silica and the extractable concentrations bound in the

Fe/Mn phases indicate that this element is a very important component of the crust – a conclusion that is verified by EDXRF analysis. Thus, although there is iron accumulation in the crust, especially with respect to the underlying granite, the dominant feature of crust formation appears to be the mobilization and re-precipitation of silica from surface dust deposits. Salinity, especially the presence of sodium chloride (van Lier *et al.* 1960, in Goudie & Viles 1997), is known to enhance silica dissolution and its reaction velocity, and this may be a factor in its initial mobility. This should, however become less significant as the soluble salts are progressively lost from dust that is retained on the surface for any length of time. To compensate for this, as water-soluble components are lost, silica mobility, especially amorphous silica, could be maintained by the increased acidity of the surface dusts in response to the relative accumulation of organic matter. This is because, although enhanced silica mobility often equated solely with extreme alkalinity, the detailed dissolution curve for amorphous silica also indicates greater mobility as acidity increases from neutral (General Electric, 2006). The co-precipitation of iron may give the visual appearance of an iron crust to this 'silica glaze', but it appears that it is an extremely thin silica-rich skin (total silica 25.1%, compared to 12.2% in the underlying granite) that lends surface stability to the stonework. Ultimately, as with other surface crusts (e.g. McAlister *et al.* 2003), the stability produced by this surface induration is not permanent. Salt accumulated prior to crust formation, and augmented by continued absorption through an ever-growing near-surface microfracture network, eventually stimulates the loss of a surface layer that may now only be held together by the original surface crust. In terms of the study building it appears that since the original survey by Smith & Magee (1990), which identified only patchy contour scaling, much of the lower façade has breached this threshold and is at risk of entering a new phase of more rapid decay in which salt weathering exploits a weakened substrate (Fig. 2). Unfortunately, it is not possible to determine whether this recent onset of surface loss results from a gradual stress accumulation over the 200 year life of the building or whether it has been triggered by increased pollution loading of the stonework – especially from vehicle emissions. The fact that scaling is pervasive and concentrated near street level, and does not appear to be influenced by any geological differences between stones, might lead one to favour the latter explanation. In reality, however, any recent stress increase will be superimposed on the effects of decades of exposure in a naturally aggressive weathering environment, and it is this combination of conditions that may ultimately explain the rapid change in the stability of the stonework.

The writers are indebted to J. Simpson and Y. Megarry for their assistance with the analyses, and to G. Alexander for preparing the figures. Financial support was provided by the British Council and the Brazilian Federal Research Council (CAPES).

## References

AGEMIAN, H. A. & CHAU, A. S. Y. 1977. A study of different analytical extraction methods for non-detrital heavy metals in aquatic sediments. *Archives of Environmental Contamination and Toxicology*, **6**, 69–82.

AZEVEDO, D. A., MOREIRA, L. S. & DE SIQUEIRA, D. S. 1999. Composition of extractable organic matter in aerosols from urban areas of Rio de Janeiro city, Brazil. *Atmospheric Environment*, **33**, 4987–5001.

BASTA, N. T., RYAN, J. A. & CHANEY, R. L. 2005. Trace element chemistry in residual-treated soil: key concepts and metal bioavailability. *Journal of Environmental Quality*, **34**, 49–63.

BLOCH, P., ADAMS, F., VAN LANDUYT, J. & VAN GEOTHEM, L. 1980. Morphological and chemical characteristics of individual aerosol particles in the atmosphere. *In*: VERSINO, B. & OTT, H. (eds) *Proceedings of the 1st European Symposium on Physio-chemical Behaviour of Atmospheric Pollutants, Ispra, 16–18 October*. Commission of European Communities, Brussels, 307–321.

CAMUFFO, D. 1998*a*. Dry deposition of pollutants. *In*: LEFÈVRE, R. (ed.) *Sciences and Technologies of the Materials and of the Environment for the Protection of Stained Glass and Stone Monuments*. European Commission, Paris, Report, **14**, 117–131.

CAMUFFO, D. 1998*b*. *Microclimate for Cultural Heritage*. Elsevier, Amsterdam.

CHAO, T. T. 1972. Selective dissolution of manganese oxides from soils and sediments with acidified hydroxylamine hydrochloride. *Journal of the Soil Science Society of America*, **36**, 764–768.

CHAO, T. T. & ZHOU, L. 1983. Extraction techniques for selective dissolution of amorphous iron oxides from soils and sediments. *Journal of the Soil Science Society of America*, **47**, 225–232.

COLIN, J. L. 1998. The wet components of the atmosphere rain and fog. *In*: LEFÈVRE, R. (ed.) *Sciences and Technologies of the Materials of the Environment for the Protection of Stained Glass and Stone Monuments*. European Commission, Paris, Report, **14**, 61–68.

CORN, M. 1976. Aerosols and the primary air pollutants-Nonviable particles-their occurrence, properties and effects. *In*: STERN, A. C. (ed.) *Air Pollution*, Volume 1, 3rd edn. Academic Press, New York, 77–168.

CORRÊA, S. M., MARTINS, E. M. & ARBILLA, G. 2003. Formaldehyde and acetaldehyde in a high traffic street of Rio de Janeiro, Brazil. *Atmospheric Environment*, **37**, 23–29.

CURRAN, J. M. & SMITH, B. J. 2000. Non-destructive testing: the use of permeametry in building stone studies. *In*: RAMMLMAIR, D., MEDERER, J., OBERTHÜR, T., HEIMANN, R. B. & PENTINGHAUS, H. (eds) *Applied Mineralology in Research, Economy, Technology, Ecology and Culture. Volume 2, Proceedings of the 6th International Congress on Applied Mineralogy, ICAM 2000, Göttingen*. Balkema, Rotterdam, 967–970.

DAISEY, J. M., MIGUEL, A. H., DE ANDRADE, J. B., PEREIRA, P. A. P. & TANNER, R. L. 1987. An overview of the Rio de Janeiro aerosol characterisation study. *Journal of the Air Pollution Control Association*, **37**, 15–23.

DE ANDRADE, J. B., ANDRADE, M. V. & PINHEIRO, H. L. C. 1998. Atmospheric levels of formaldehyde and acetaldehyde and their relationship with vehicular fleet composition in Salvador, Bahia, Brazil. *Journal of the Brazilian Chemical Society*, **9**, 219–223.

DE MELLO, W. Z. 2001. Precipitation chemistry in the coast of the Metropolitan Region of Rio de Janeiro, Brazil. *Environmental Pollution*, **114**, 235–242.

DEL MONTE, M. & LEFÈVRE, R. A. 1998. Particulate matter in the urban atmosphere. *In*: LEFÈVRE, R. A. (ed.) *Sciences and Technologies of the Materials and of the Environment for the Protection of Stained Glass and Stone Monuments*. European Commission, Paris, Report, **14**, 99–107.

DORN, R. 1998. *Rock Coatings*. Developments in Earth Surface Processes, **6**. Elsevier, Amsterdam.

DZOMBAK, D. A. & MOREL, F. M. M. 1990. *Surface Complexation Modelling: Hydrous Ferric Iron Oxide*. Wiley, New York.

ELLIS, W. D. & FOGG, T. 1985. *Interim Report: Treatment of Soils Contaminated by Heavy Metals and Hazardous Waste*. Engineering Laboratory, Office of Research and Development, US Environment Protection Agency, Cincinnati, OH.

ESPINOSA, A. J. F., RODRIGUEZ, M. T., BARRAGÁN, F. J. & SANCHEZ, J. C. J. 2001. Size distribution of metals in urban aerosols in Seville (Spain). *Atmospheric Pollution*, **35**, 2595–2601.

FANNING, K. A. & PILSON, M. E. Q. 1973. On the spectrophotometric determination of dissolved silica in natural waters. *Analytical Chemistry*, **45**, 136–140.

FISK, E. P. 1971. Desert glaze. *Journal of Sedimentary Petrology*, **41**, 1136–1137.

GARRETT, R. G. 2000. Natural sources of metals to the environment. *Human and Ecological Risk Assessment*, **6**, 945–963.

GENERAL ELECTRIC. 2006. The study of silica and reverse osmosis. World Wide Web Address: http://www.gewater.com/library/tp/696_A_Study.jsp (accessed January 2006).

GEORGII, H. W. & PERSEKE, C. 1980. Some results in wet and dry deposition of sulphur compounds. *In*: VERSINO, B. & OTT, H. (eds) *Proceedings of the 1st European Symposium on Physio-chemical Behaviour of Atmospheric Pollutants, Ispra, 16–18 October*. Commission of European Communities, Brussels, 410–418.

GOODMAN, A. L., UNDERWOOD, G. M. & GRASSIAN, V. H. 2000. A laboratory study of the heterogeneous reaction of nitric acid on calcium carbonate particles. *Journal of Geophysical Research*, **105**, 29 053–29 064.

GOUDIE, A. S. & VILES, H. A. 1997. *Salt Weathering Hazards*. Wiley, Chichester.

GROSJEAN, D., MIGUEL, A. H. & TRAVARES, T. 1990. Urban air pollution in Brazil: acetaldehyde and other carbonyls. *Atmospheric Environment*, **22**, 1637–1648.

GROSJEAN, D., GROSJEAN, E. & MOREIRA, L. F. R. 2002. Speciated ambient carbonyls in Rio de Janeiro, Brazil. *Environmental Science and Technology*, **36**, 1389–1395.

HALL, G. E. M., VAIVE, J. E., BEER, R. Y. & HOASHI, M. 1996. Selected leaches revisited, with emphasis on the amorphous iron oxyhydroxide phase extraction. *Journal of Geochemical Exploration*, **56**, 59–78.

HANISCH, F. & CROWLEY, J. M. 2001. Heterogeneous reactivity of gaseous nitric acid on $Al_2O_3$, $CaCO_3$ and atmospheric dust samples: a Knudsen cell study. *Journal of Physics and Chemistry*, **105**, 3096–3160.

HESSE, P. R. 1971. *A Textbook of Soil Chemical Analysis*. John Murray, London.

HICKS, B. B. 1981. Wet and dry surface deposition of air pollutants and their modelling. *In*: BARKIN, S. M. (ed.) *Conservation of Historic Stone Buildings and Monuments*. Natural Academy of Sciences, Washington, DC.

HIDY, G. M. 1984. *Aerosol An Industrial and Environmental Science*. Academic Press, London.

JENKINS, I. D. & MIDDLETON, A. P. 1988. Paint on the Parthenon sculptures. *Annual of the British School of Archaeology at Athens*, **83**, 183–207.

JUNGE, C. E. 1963. *Air Pollution and Radioactivity*. Academic Pres, New York.

KUNZEL, H. M. 1995. *Simultaneous Heat and Moisture Transport in Building Components: One and Two Dimensional Calculation Using Simple Parameters*. Fraunhofer Information Centre for Regional Planning of Buildings (IRB), **36**, 69–83.

LAZZARINI, L. & SALVADORI, O. 1989. A reassessment of the formation of the patina called Scialbatura. *Studies in Conservation*, **34**, 2227–2245.

MARTINS, E. M., CORREA, S. M. & ARBILLA, G. 2003. Formaldehyde and acetaldehyde in high street traffic in Rio de Janeiro, Brazil. *Atmospheric Environment*, **37**, 23–29.

MAXWELL, I. 1992. Stone cleaning – for the better or worse? An overview. *In*: WEBSTER, R. G. M. (ed.) *Stone Cleaning and the Nature, Soiling and Decay Mechanisms of Stone*. Donhead, London, 3–49.

MCALISTER, J. J. & SMITH, B. J. 1995*a*. A rapid preparation technique for X-ray diffraction analysis of saprolitic materials. *Microchemical Journal*, **52**, 53–61.

MCALISTER, J. J. & SMITH, B. J. 1995*b*. A pressure membrane technique to prepare clay samples for X-ray diffraction analysis. *Journal of Sedimentary Research*, **A65**, 569–571.

MCALISTER, J. J. & SMITH, B. J. 1999. Selectivity of ammonium acetate, hydroxylamine hydrochloride and oxalate/ascorbic acid solutions for the speciation of Fe, Mn, Zn, Cu, Ni and Al in early Tertiary paleosols. *Microchemical Journal*, **63**, 415–426.

McAlister, J. J., Smith, B. J. & Curran, J. A. 2003. The use of sequential extraction to examine iron and trace metal mobilisation and the case hardening of building sandstone: a preliminary investigation. *Microchemical Journal*, **74**, 5–18.

McAlister, J. J., Smith, B. J., Neto, J. B. & Simpson, J. K. 2005. Geochemical distribution and bioavailability of heavy metals and oxalate in street sediments from Rio de Janeiro, Brazil: a preliminary investigation. *Environmental Geochemistry and Health*, **27**, 429–441.

Metzger, S., Dentener, M., Krol, M., Jeuken, A. & Lelieveld, J. 2002. Gas/aerosol partitioning 2. Global modelling results. *Journal of Geophysical Research*, **107**, 4313.

Morselli, L., Oliveieri, P., Brusori, B. & Passarini, F. 2003. Soluble and insoluble fractions of heavy metals in wet and dry atmospheric depositions in Bologna, Italy. *Environmental Pollution*, **124**, 457–469.

Nguyen, H. T., Takenaka, N., Bandow, H., Maeda, Y., De Oliva, S. T., Botelho, M. M. F. & Tavares, T. M. 2001. Atmospheric alcohols and aldehydes concentrations measured in Osaka, Japan and in Sao Paulo, Brazil. *Atmospheric Environment*, **35**, 3075–3083.

Perros, P. E. 1998. The gaseous components of the atmosphere. *In: Protection and Conservation of the European Cultural Heritage*. European Community, Paris, Research Report, **14**, 55–59.

Pickering, W. F. 1981. Selective chemical extraction of soil components and bound metal species. *Critical Reviews in Analytical Chemistry*, **2**, 233–266.

Power, E. T. & Smith, B. J. 1994. A comparative study of deep weathering and weathering products: case studies from Ireland, Corsica and Southeast Brazil. *In*: Robinson, D. A. & Williams, R. B. G. (eds) *Rock Weathering and Landform Evolution*. Wiley, Chichester, 21–40.

Quevauviller, P., Rauret, G. *et al.* 1994. Evaluation of a sequential extraction procedure for the determination of extractable trace elements in sediments. *Fresenius Journal of Analytical Chemistry*, **349**, 808–814.

Rauret, G., Rubio, R. & Lopez-Sanchez, J. F. 1989. Optimization of the Tessier procedure for metal solid speciation in river sediments. *International Journal of Environmental and Analytical Chemistry*, **36**, 69–83.

Sabbioni, C. 1995. Contribution of atmospheric deposition to the formation of damage layers. *The Science of the Total Environment*, **167**, 49–55.

Sabbioni, C. & Zappia, G. 1991. Oxalate patinas on ancient monuments: the biological hypothesis. *Aerobiologia*, **7**, 31–37.

Silva Filho, E. V. 1985. *Estudo de chuva acida e entradas atmosfêricas de Na, K, Ca, Mg, e Cl na hacia do alto Rio Cachoeira Parque, National da Tijuca-RJ*. Msc thesis, Universidade Federal Fluminense, Niteroi.

Skei, J. & Paus, P. E. 1979. Surface metal enrichment and partitioning of metals in a dated sediment core from a Norwegian fjord. *Geochemica et Cosmochimica Acta*, **43**, 239–246.

Smith, B. J. & Curran, J. M. 2000. Surface modification of building stone in a polluted urban environment: Belfast. *In*: Rammlmair, D., Mederer, J., Oberthür, T., Heimann, R. B. & Pentinghaus, H. (eds) *Applied Mineralology in Research, Economy, Technology, Ecology and Culture. Volume 2, Proceedings of the 6th International Congress on Applied Mineralogy, ICAM 2000, Göttingen*. Balkema, Rotterdam, 67–70.

Smith, B. J. & Magee, R. W. 1990. Granite weathering in an urban environment: an example from Rio de Janeiro. *Singapore Journal of Tropical Geography*, **2**, 144–153.

Smith, B. J., Török, A., McAlister, J. J. & Megarry, Y. 2003. Observations on the factors influencing stability of building stones following contour scaling: a case study of oolitic limestones from Budapest, Hungary. *Building and Environment*, **38**, 1173–1183.

Spokes, L. J., Jickells, T. D. & Lim, B. 1994. Solubilization of aerosol trace metals by cloud processing: a laboratory study. *Geochimica et Cosmochimica Acta*, **58**, 3281–3287.

Tessier, A., Campbell, P. G. C. & Bisson, M. 1979. Sequential procedure for the speciation of particulate trace metals. *Analytical Chemistry*, **51**, 844–851.

Thomas, M. F. 1994. *Geomorphology in the Tropics*. Wiley, Chichester.

Ure, A. M., Quevauviller, Ph., Muntau, H. & Griepink, B. 1993. Speciation of heavy metals in solids and sediments. An account of the improvement and harmonization of extraction techniques undertaken under the auspices of the BCR of the Commission of the European Communities. *International Journal of Environmental Analytical Chemistry*, **51**, 135–151.

Van Lier, J. A., de Bruyn, P. L. & Overbeck, J. Th. G. 1960. The solubility of quartz. *Journal of Physical Chemistry*, **64**, 1675–1682.

Van Valin, R. & Morse, J. W. 1982. An investigation of methods normally used for the selective removal and characterisation of trace metals in sediments. *Marine Chemistry*, **11**, 535–564.

Walkley, A. & Black, I. A. 1934. An examination of the Degtjareff method for determining soil organic matterand proposed modification of the chromic acid titration method. *Soil Science*, **37**, 29–38.

Warke, P. A. & Smith, B. J. 2000. Salt distribution in clay-rich weathered sandstone. *Earth Surface Processes and Landforms*, **25**, 1333–1342.

Zhu, X., Prospero, J. M., Millero, F. J., Savoie, D. L. & Brass, G. W. 1992. The solubility of ferric iron in marine mineral aerosol solutions at ambient relative humidities. *Marine Chemistry*, **38**, 91–107.

# Dilation of building materials submitted to frost action

C. THOMACHOT[1] & N. MATSUOKA[2]

[1]*Groupe d'Etude sur les Géomatériaux et les Environnements Anthropiques et Naturels (GEGENA), University of Reims, 2 esplanade Roland Garros, 51100 Reims, France (e-mail: celine.thomachot@univ-reims.fr)*

[2]*Doctoral Program of Geoenvironmental Sciences, University of Tsukuba, Ibaraki 305–8572, Japan*

**Abstract:** This paper presents the results of laboratory frost weathering of five geomaterials used in stone monuments showing more or less frost damage: two sandstones (France), molasses (Switzerland), and a volcanic tuff and brick (Japan). Samples were submitted to unidirectional freezing simulations during which temperature and dilation were measured. The aim of these experiments was to understand which internal or external factors prevailing on dilation would lead to cracking. Results showed that water supply and repetition of freeze–thaw cycles were most important in the dilation of the materials. They also showed that the materials with the weakest transfer properties by capillary absorption were the most sensitive to frost action.

Frost damage mechanisms have sometimes been related to the freezing of water at 0 °C at atmospheric pressure and a volume expansion of 9%. However, water within a porous medium can remain in a liquid state at temperatures lower than 0 °C through supercooling (Chahal & Miller 1965; Fagerlund 1971; Prick 1995). In a porous network supercooling depends on the pore radius, and the smaller the pore the lower the freezing point (Everett 1961). Thus, in a porous environment with many different pore radii, freezing is not uniform (Thomas 1938) and begins in macropores. In capillary saturation unfrozen water is attracted towards the ice front by cryosuction and it crystallizes in macropores without generating real strain (Powers 1945; Everett 1961; Litvan 1978). In total saturation the volume expansion of the water that freezes exerts a pressure on unfrozen water and generates hydraulic pressures (Powers 1945). Only when all the macropores are frozen can ice be extruded into the finer pores. This extrusion generates so-called capillary pressures that are in proportion to the difference of radius between the macropore and the contraction (Everett 1961). These various pressures create expansion or contraction of the porous network and if it exceeds the resistance of the material cracking occurs. Thus, studying dilation of materials can help in the understanding of rock behaviour when subjected to frost action.

## Material

Several building materials from France, Japan and Switzerland were studied. These were Japanese volcanic tuffs and bricks made at the end of the 19th century from Japan, two kinds of Buntsandstein siliceous sandstones from the east of France and Swiss molasses (Fig. 1). All of these materials are used in monuments and their weathering behaviour is relatively well known (see Table 1 for their petrophysical properties).

### *Usui brick*

*Location.* Samples of bricks were collected from the historical site of the Usui Pass Railway Facility (Gunma prefecture, 100 km NW of Tokyo, Japan). This was an 11.2 km-long railway with tunnels and bridges built of brick in 1893 and linked Tokyo and Nagano (Smith 1997). Many historical buildings and bridges were built of similar bricks at that time, including Tokyo station (1914), Yokohama's red brick warehouses (1908), Biwa aqueduct in Kyoto (1890) and the Hokkaido government office building (1911) (Ito & Chiba 2001). Bricks of the Usui Pass Facility were made in the Nihon Brick Corporation, one of the oldest brick companies in Japan.

*Petrophysical properties.* The brick contained tridymite (high-temperature $SiO_2$), quartz ($SiO_2$), hematite ($Fe_2O_3$), albite ($NaAlSi_3O_8$) and anorthite ($(Ca,Na)Al_2Si_2O_8$), and had no clay minerals. The average value of total porosity was 35%. The porous network was made of three main scales of pores: macropores around 0.1 mm, mesopores around 10 μm and micropores around 1 μm. Average pore threshold determined by mercury porosimetry was 0.37 μm, but the scatter coefficient

*From*: PŘIKRYL, R. & SMITH, B. J. (eds) *Building Stone Decay: From Diagnosis to Conservation.* Geological Society, London, Special Publications, **271**, 167–177.
0305-8719/07/$15.00 

**Fig. 1.** Stone monuments composed of tested materials. (**a**) Bricks of Usui Pass, central Japan. (**b**) Sandstones of Strasbourg Cathedral, eastern France. (**c**) Tuffs of Heiwa-kannon temple, central east Japan. (**d**) Molasses sandstone of Fribourg Cathedral, Switzerland.

$C_d$ was 2.1, indicating a spread pore distribution (Wardlaw *et al.* 1988). Measurements showed fast absorption kinetics. The *B* coefficient relative to the migration of the wet fringe was 15.6 cm $h^{-0.5}$ and the *A* coefficient relative to the weight increase per surface unit was 4.0 g $cm^{-2}$ $h^{-0.5}$ (Thomachot *et al.* 2005). This showed that even though the brick had a large pore-size distribution, the distribution was homogeneous and the porous network was well connected.

*Weathering.* In the tunnels of the historical site of Usui, bricks were entirely covered with soot accumulated during the use of the railway. Bricks of the S-facing walls showed cavernous weathering features possibly resulting from salt weathering, while the N-facing walls showed weathering that could be attributed to frost action. Indeed, a preliminary study showed that angular flake detachment 1–6 cm thick occurred during winter, but most of the hollows were shallower than 2 cm (Thomachot *et al.* 2005). Furthermore, there was no relation of the height of intensive erosion with the capillary rise from the ground, but intensive erosion corresponded to the water seepage from the walls.

## *Ohya tuff*

*Location.* Tuff samples, called Ohya tuffs, came from quarries in Utsunomiya (Tochigi prefecture,

**Table 1.** *Petrophysical properties of materials*

| | | Usui bricks | Buntsandstein Sandstone | | Villarlod molasse | Ohya tuff |
|---|---|---|---|---|---|---|
| | | | Fine | Coarse | | |
| Water porosimetry | Total porosity $N_t$ (%) | 35.0 | 23.5 | 18.0 | 15.5 | 25.0 |
| Mercury porosimetry | Pore threshold $R_a$ (μm) | 0.37 | 4.7 | 6.0 and 0.004 | 1.2 | 0.05 |
| | Dispersion coefficient $C_a$ | 2.1 | 1.1 | 2.2 | 2.6 | 33.3 |
| Capillary absorption | Weight increase *A* (g $cm^{-2}$ $h^{-0.5}$) | 4.04 | 1.26 | 0.18 | 0.25 | 0.13 |
| | Wet fringe migration *B* (cm $h^{-0.5}$) | 15.56 | 8.43 | 1.59 | 2.28 | 0.21 |
| | Kinetics | Very fast | Fast | Slow | Slow | Very slow |

130 km north of Tokyo). They are Miocene green tuffs formed by volcanic activity (−20 Ma). Most of the old buildings of Tochigi prefecture include Ohya tuff. Monuments such as the Heiwa-kannon temple, the Usuki Buddha and the Imperial Hotel in Tokyo were also built of tuffs.

*Petrophysical properties.* The Ohya tuff has a rhyolitic composition of submarine origin characterized by a cemented structure with air-bubble and flow-line features (Akagawa & Fukuda 1991). Its structure includes phenocrysts of plagioclase and quartz in a vitreous groundmass. Clayey clusters of clinoptilolite, chlorite and montmorillonite that can reach 10 cm in diameter replaced volcanic glass and other minerals modified by diagenesis. Average values of total porosity are 25%. The porous network has a fluidal and heterogeneous structure due to the mode of formation. Average pore threshold determined by mercury porosimetry is 0.05 μm, but as the scatter coefficient $C_d$ was 33.3, indicating a very large spread pore distribution, the value of pore threshold has no significance. Measurements showed very slow absorption kinetics. The $B$ coefficient relative to the migration of the wet fringe was 0.21 cm $h^{-0.5}$ and the $A$ coefficient relative to the weight increase per surface unit was 0.13 g $cm^{-2}$ $h^{-0.5}$. Owing to their heterogeneity, tuffs have a large pore-size distribution that explained their slow capillary transfer.

*Weathering.* The high clay content and the slow capillary absorption makes the Ohya tuffs sensitive to water transfer (Matsuoka 2001). Owing to the heterogeneity of its porous network and the presence of clayey clusters that could be easily washed by rains, tuffs usually show differential weathering through the development of Tafoni. It is sensitive to salt weathering (Yamada *et al.* 2005). Cracking that could be attributed to frost action also occurs on the tuff. Akagawa & Fukuda (1991) showed that long-term and very slow freezing lead to the formation of thick ice lenses.

## *Buntsandstein sandstones*

*Location.* Monuments in Alsace (east of France), and especially Strasbourg's cathedral, are typically built of two types of pink Buntsandstein sandstones (Lower Triassic – 300 Ma: Mader 1985) extracted from Alsatian quarries.

*Petrophysical properties of fine-grained sandstone.* The Meules sandstone is a thinly bedded variety with an average total porosity of 23.5%. Bedding within the Meules sandstone is visible macroscopically. It comprises 6–7% clayey concentrations. Quartz and feldspar grains of this stone are on average 60 μm long. They are angular and lie parallel to the bedding. A clay matrix forms aggregates that provide cohesion between grains. Average pore threshold determined by mercury porosimetry is 4.7 μm and the scatter coefficient $C_d$ is 1.1, indicating a narrow pore distribution. In this case the pore threshold is significant. It indicates an homogeneous porous network. Capillary absorption shows fast absorption kinetics. The $B$ coefficient relative to the migration of the wet fringe is 8.43 cm $h^{-0.5}$ and the $A$ coefficient relative to the weight increase per surface unit is 1.26 g $cm^{-2}$ $h^{-0.5}$ (Thomachot & Jeannette 2002). In spite of macroscopic heterogeneity caused by bedding, microporosity associated with the clay matrix controls connectivity of the pore network so that, on the whole, the pore structure is homogeneous.

*Weathering of fine-grained sandstone.* On monuments the fine-grained sandstone usually shows 'shivering', flaking or cracking that could be attributed to frost action or a combination of frost and salt weathering (Thomachot & Jeannette 2002). On surfaces exposed to rain and running water they are usually covered by a black, fine varnish of glossy aspect (Thomachot & Jeannette 2004).

*Petrophysical properties of coarse-grained sandstone.* The Vosgien sandstone is a coarser variety with an average value of total porosity of 18%. It is composed of 80% quartz grains. These are usually massive, ovoid and average 200 μm in length. The grains are cemented by light overgrowths. Although elongated, the grains do not lie in beds. Intergranular spacings are subdivided into irregular large pores generally linked by narrow throats filled with clayey concentrations. They represent a macroporosity that is likely to trap air during water transfer (David *et al.* 1993). Two pore thresholds determined by mercury porosimetry are 6.0 and 0.004 μm and the scatter coefficient $C_d$ is 2.2, indicating a dispersed pore distribution. Capillary absorption shows slow absorption kinetics. The $B$ coefficient relative to the migration of the wet fringe is 1.59 cm $h^{-0.5}$ and the $A$ coefficient relative to the weight increase per surface unit is 0.18 g $cm^{-2}$ $h^{-0.5}$. The porous network of this sandstone is heterogeneous because of the contrasts between the large intergranular pores and the number of microporous zones.

*Weathering of coarse-grained sandstone.* Compared to the fine-grained sandstone, the coarse sandstone does not seem particularly sensitive to weathering. Only in the higher parts of monuments, such as spires, does it seem to damage faster than in other parts, with loose grains on the surface (Thomachot & Jeannette 2002). Compared to the fine-grained sandstone, black varnishes do not appear to develop on the coarse sandstone.

### *Villarlod molasses*

*Location.* The Villarlod molasse is a grey sandstone of the lower Miocene (Burdigalian – 20–16 Ma), extracted near Fribourg in western Switzerland. It was specifically used to build the cathedral in Fribourg.

*Petrophysical properties.* The Villarlod molasses is a calcareous and clayey sandstone with an average value of total porosity of 15.5%. Grains, elongated and angular, were on average 300 μm long (Félix 1993). The average pore threshold determined by mercury porosimetry is 1.2 μm and the scatter coefficient $C_d$ is 2.6, indicating a dispersed pore distribution. Capillary absorption shows slow absorption kinetics. The $B$ coefficient relative to the migration of the wet fringe is 2.28 cm $h^{-0.5}$ and the $A$ coefficient relative to the weight increase per surface unit is 0.25 g $cm^{-2}$ $h^{-0.5}$.

*Weathering.* The Villarlod molasses are sensitive to weathering, especially to swelling (Gonzalez & Scherer 2004). The weak cementation between the grains leads to the loss of grains. They show weathering features due to salt weathering such as flaking, as well as cracking and detachment due to frost action.

## Apparatus

To study dilation of materials subject to frost action it was decided to perform unidirectional freezing, closer to the natural conditions where only one face is exposed to temperature changes.

Samples of 5 × 5 × 15 cm dimension were isolated in a thermal insulator, except for the top surfaces. On the top surface a Peltier cooling plate was placed to control the freezing temperature while the bottom surface was stood in water. Thus, the experiments reproduced the layout of a wall, except that freezing progressed vertically rather than horizontally.

To follow dilation along the sample, five strain gauges were attached in staggered rows at 2, 4, 6, 8 and 10 cm depths from the top, on one side of the surface (Fig. 2). To follow temperature changes seven thermocouples were placed at 1, 2, 4, 6, 8, 10 and 12 cm depths from the top, after drilling 3 mm holes 2 cm from the side surface. The whole apparatus was placed in a cold room. Before freezing, the cold room and the cooling plate were maintained at 5 °C until thermal equilibrium was reached. During the experiment the temperature of water at the base of the sample was kept constant at +3 °C.

To estimate the heave amount of the segment represented by each strain gauge (2 cm), the original strain values were multiplied by 2 cm, except for the 2 cm-deep gauge that represented the uppermost 3 cm segment. Total heave was computed by adding the heave amounts of each segment. It represented the dilation of the sample from the top down to 11 cm. As the freezing front did not reach deeper than 11 cm during experiments, it was assumed that calculated heave was representative of the whole sample.

Two samples of each material were concurrently submitted to continuous freezing then to diurnal freeze–thaw cycles: one sample was at capillary saturation and the other was saturated under vacuum (total saturation: 100% of the open porosity was filled with water).

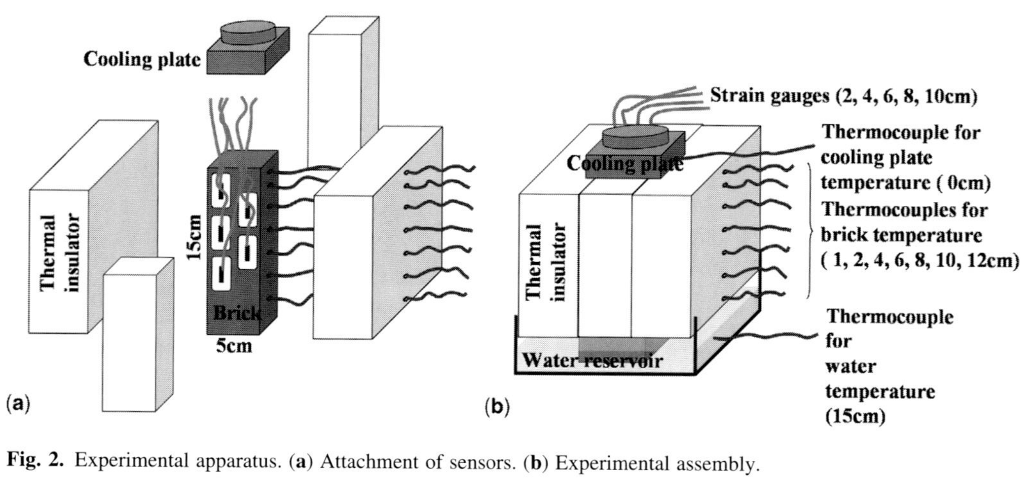


**Fig. 2.** Experimental apparatus. (**a**) Attachment of sensors. (**b**) Experimental assembly.

## Experimental conditions

Freezing conditions were chosen according to climatic data recorded at the Japanese site of Usui (Thomachot *et al.* 2005). Monitoring showed a combination of short seasonal freezing of around $-3$ °C, which were finished within 2 weeks, and repetitive diurnal freezing of a minimum $-2$ to $-5$ °C. These data were representative of a typical winter with seasonal freezing. They did not fit exactly with all the sites but they could be used as a reference.

### *Continuous freezing*

For the continuous freezing simulation, the temperature of the cooling plate was maintained at $-3$ °C for 3 days according to field data. Then, to assess the effect of freezing intensity, the cooling plate was maintained at $-5$ °C for 3 days and then at $-8$ °C for 3 days. Total duration of the experiment was 9 days.

### *Diurnal freezing*

After this simulation, samples were submitted to freeze–thaw cycles with a period of 24 h: four cycles from $+2$ to $-3$ °C and then four cycles from $+3$ to $-5$ °C, as recorded in the tunnel. A further four cycles from $+5$ to $-8$ °C were simulated to assess the effect of a more intensive freezing and to compare results with those of continuous freezing at $-8$ °C. Each freezing and thawing phase was maintained for 12 h.

Experimental cooling and thawing rates were as fast as the cooling plates could work. These climatic conditions simplified the field data where freezing and thawing were progressive. Before each test, dry samples were submitted to temperature variations to assess thermal dilation of each material. This thermal dilation was used to comment on the results, and thus, the final results showed dilation as a result of water movements, liquid, vapour, ice or change of state. Two tests were performed for each type of freezing and to determine the reproducibility of data; the second test of each pair was presented in the following phase.

## Results

The dilation recorded at each depth of samples was plotted over time, as well as temperature of the cooling plate (Figs 3 & 4). Dilation was expressed in strain value ($10^{-4}$) without dimension. Values did not exceed $15 \times 10^{-4}$ for all materials, except the coarse sandstone that reached $140 \times 10^{-4}$ when subjected to diurnal freezing at total saturation. A common behaviour of these materials was that at the first stage of water freezing below the cooling plate, contraction occurred below the ice front and was probably induced by upward migration of water by cryosuction. Contraction was largest close to the cooling plate at the shallowest strain gauge (2 cm depth) and decreased with depth.

According to their dilation behaviour to frost action, materials could be divided into three groups. The common point for the materials of each group was their capillary kinetics (Table 2).

The first group, including the Japanese brick and the fine Buntsandstein sandstone, had fast capillary kinetics. Whatever the type of freezing, they did not show visible cracking and dilation at capillary saturation. Furthermore, expansion at continuous freezing was lower than at diurnal freezing.

The second group, which included only the coarse Buntsandstein sandstone, had slow capillary kinetics. Visible cracking occurred only during diurnal freezing at total saturation. On samples tested at capillary saturation, dilation was significant and expansion at continuous freezing was still lower than at diurnal freezing.

The third group, including the Ohya tuff and the Villarlod molasse, had very slow capillary kinetics. Visible cracking occurred during freezing of both rocks at total saturation. Dilation was also significant at capillary saturation. Only for this group was expansion during continuous freezing higher than during diurnal freezing.

## Discussion

### *Moisture condition*

Two moisture conditions were tested, with capillary saturation and total saturation. At the end of the experiments, although drying of materials was minimized, the vacuum saturated materials had slightly dried. This indicated desaturation of near-surface macropores. In contrast, materials tested at capillary saturation slightly increased their moisture content during the experiments. Cryosuction and water supply from the tank contributed to the increase of moisture content.

*Group 1: fast capillary kinetics materials.* Any freezing dilation was insignificant on bricks (less than $1 \times 10^{-4}$) or fine sandstone subjected to simple capillary absorption for a few days. Only fully saturated bricks showed significant expansion ($\leq 10 \times 10^{-4}$ during diurnal freezing, $\leq 4 \times 10^{-4}$ during continuous freezing). This means that in natural conditions frost is effective only on bricks or fine sandstone regularly supplied with water for a long time as they have time to saturate by diffusion of air through water.

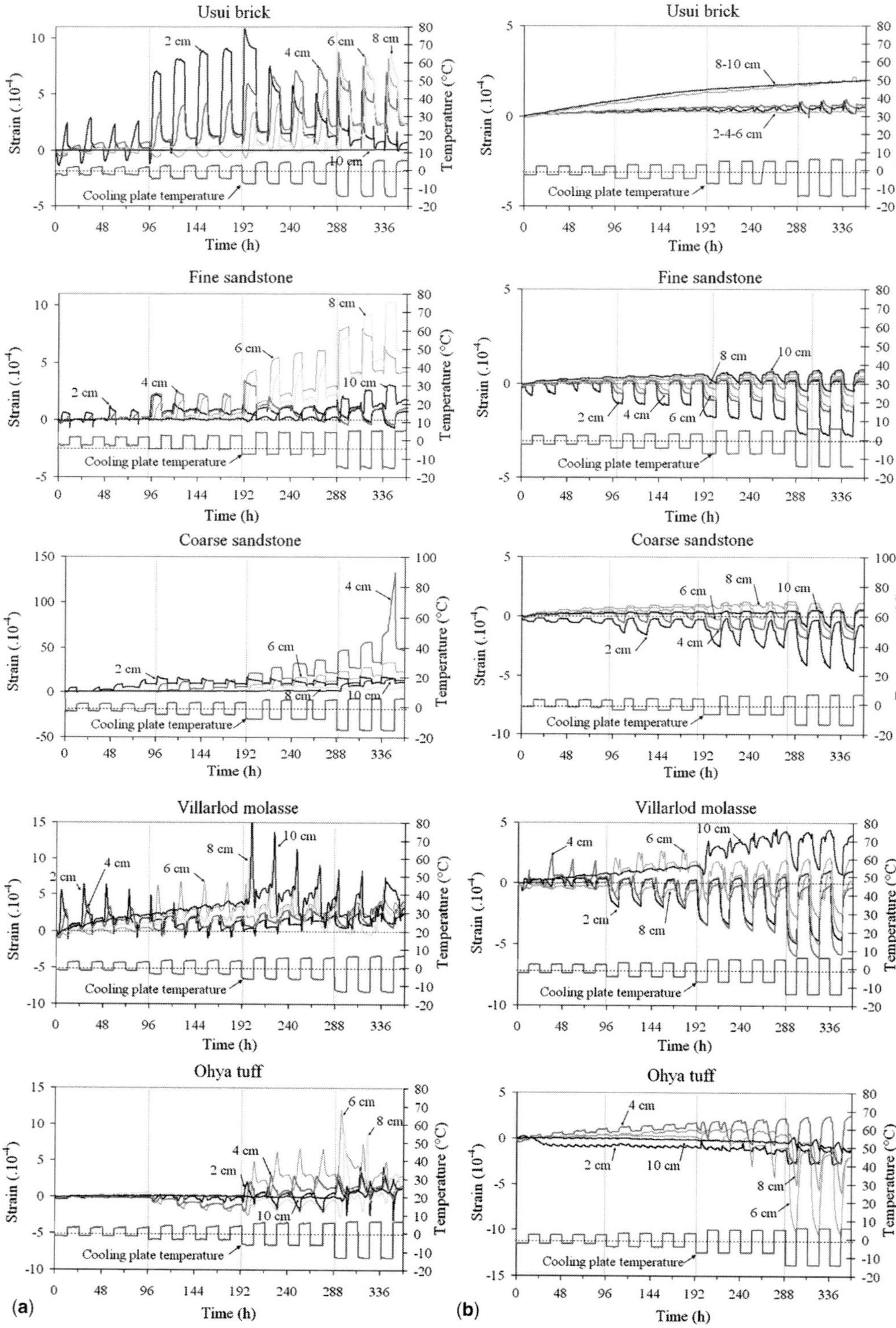


**Fig. 3.** Strain changes of materials submitted to diurnal freezing at (**a**) total saturation and (**b**) capillary saturation. (Note that for better legibility, the vertical scale of saturation differs from one stone to the next.)

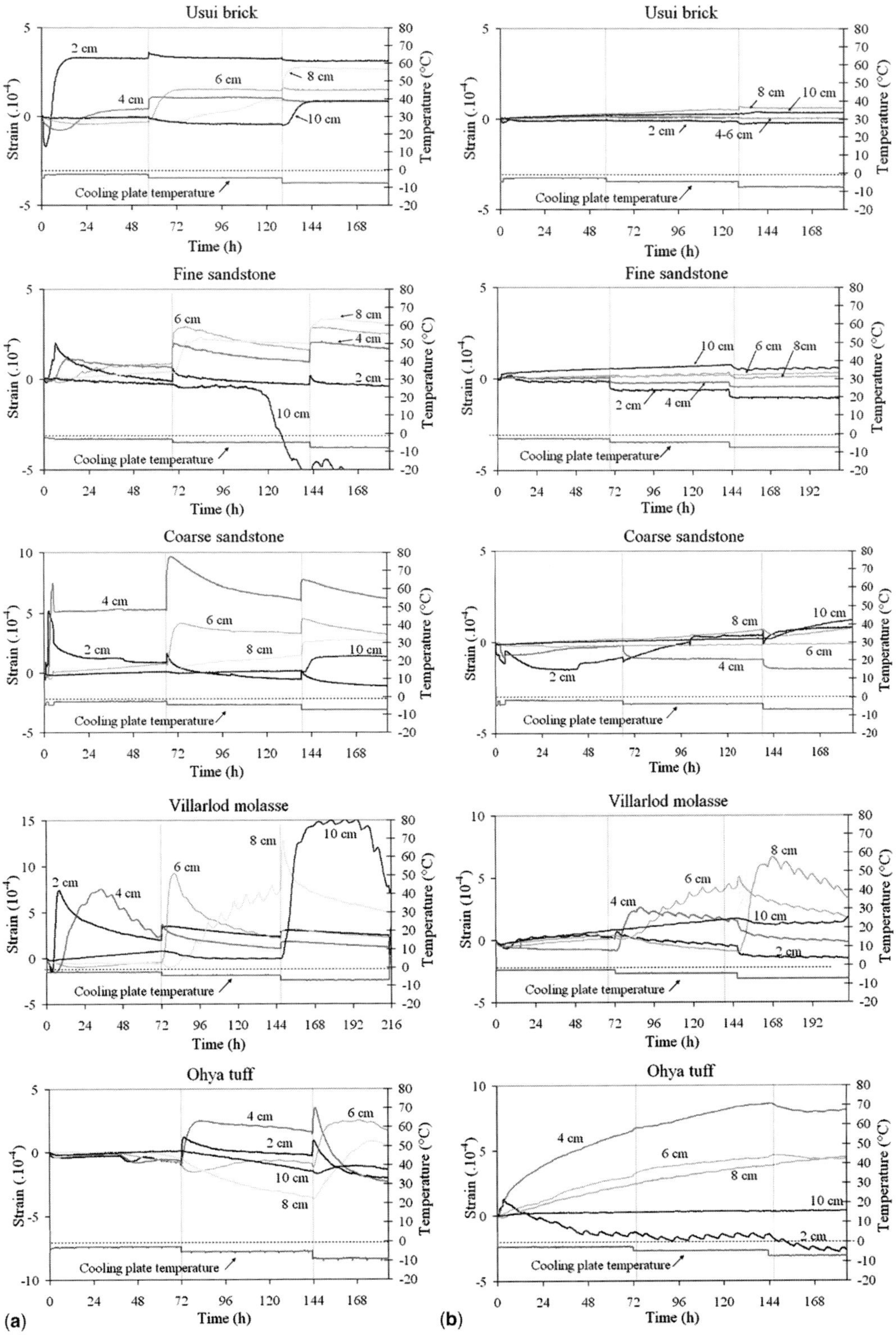


**Fig. 4.** Strain changes of materials submitted to continuous freezing at (**a**) total saturation and (**b**) capillary saturation. (Note that for a better legibility, the vertical scale of saturation differs from one stone to the next.)

**Table 2.** *Groups of materials and their response to frost action*

| | Group material | Capillary kinetics | Response to frost action |
|---|---|---|---|
| 1 | Usui brick<br>Fine sandstone | Fast | No visible cracking<br>Insignificant dilation at capillary saturation<br>Dilation at continuous freezing less than at diurnal freezing |
| 2 | Coarse sandstone | Slow | Cracking (diurnal freezing at total saturation)<br>Dilation at capillary saturation<br>Dilation at continuous freezing less than at diurnal freezing |
| 3 | Ohya tuff<br>Villarlod molasse | Very slow | Cracking (both freezing at total saturation)<br>Dilation at capillary saturation<br>Dilation at continuous freezing greater than at diurnal freezing |

*Group 2: slow capillary kinetics materials.* The dilation of coarse sandstone tested at capillary saturation was more significant than that of the first group, but most dilation occurred in fully saturated samples. Expansion reaching $140 \times 10^{-4}$ during diurnal freeze–thaw cycles and $10 \times 10^{-4}$ during continuous freezing occurred for both at 4 cm from the top of the sample (Figs 3 & 4). Cracking also occurred at 4 cm on samples tested at total saturation and diurnal freezing.

*Group 3: very slow capillary kinetics materials.* Freezing dilation was significant on the Ohya tuff and the Villarlod molasses at any moisture condition. Vacuum-saturated samples were cracked during both types of freezing. In these rocks with slow capillary kinetics and a large amount of clay minerals, water movements induced by the freezing process should be able to generate pressures higher than the resistance of the materials. However, diagrams of total heave showed that the total amount of dilation could be higher at capillary saturation than at total saturation, especially during continuous freezing (Ohya tuff, Fig. 5). Maximum heave has to be located and repeated at the same depth to lead to cracking. Large heave does not always mean cracking, but could reflect the distribution of ice lenses at different depths. The slow transfer properties of tuff and molasses favour the growth of a thick ice lens at one depth. It means that in natural conditions, high initial water content is not necessary to induce large dilation of the material.

## *Type of freezing*

*Group 1: fast capillary kinetics materials.* When brick or fine sandstone was fully saturated the magnitude of expansion depended on the location of the freezing front in the sample. During continuous freezing, expansion followed frost penetration (down to 11 cm depth), whereas during repetition of diurnal freeze–thaw cycles expansion occurred and increased in all the frozen part but mainly near the surface (maximum of ice front penetration up to 7 cm). Because of a longer freezing time, continuous freezing allowed deeper migration of the freezing front, but led to lower total heave than diurnal freeze–thaw cycles (Fig. 5).

*Group 2: slow capillary kinetics materials.* Diurnal freeze–thaw cycles in the coarse sandstone at total saturation caused a large amount of heave, close to 450 μm, while continuous freezing caused a total heave of 30 μm (Fig. 5). In rocks with slow capillary kinetics and a small amount of clay minerals, the geometry of the porous network is important. In the coarse sandstone where intergranular spacings are subdivided into macropores linked by narrow throats filled with clayey concentrations, extrusion of ice and water movements are restricted. Thus, repetition of diurnal freeze–thaw cycles would create high expansion and lead to cracking.

*Group 3: very slow capillary kinetics materials.* Both types of freezing led to cracking of vacuum-saturated materials of this group. Slow water transfer and a large amount of clays prevent water movement and create pressure that leads to expansion during repetition of freeze–thaw cycles as well as during continuous freezing.

## *Intensity of freezing*

For all materials at continuous freezing significant total heave started at −3 °C and increased with intensity of freezing due to the progressive penetration of the ice front (Fig. 5b). On the other hand, the depth of maximum expansion was different depending on the stones (Fig. 4b).

With diurnal freezing, significant heave started during the −5 °C cycles for the first and second groups. In the third group, significant heave of the Ohya tuff started during the −8 °C cycles, while significant heave of the Villarlod molasses started

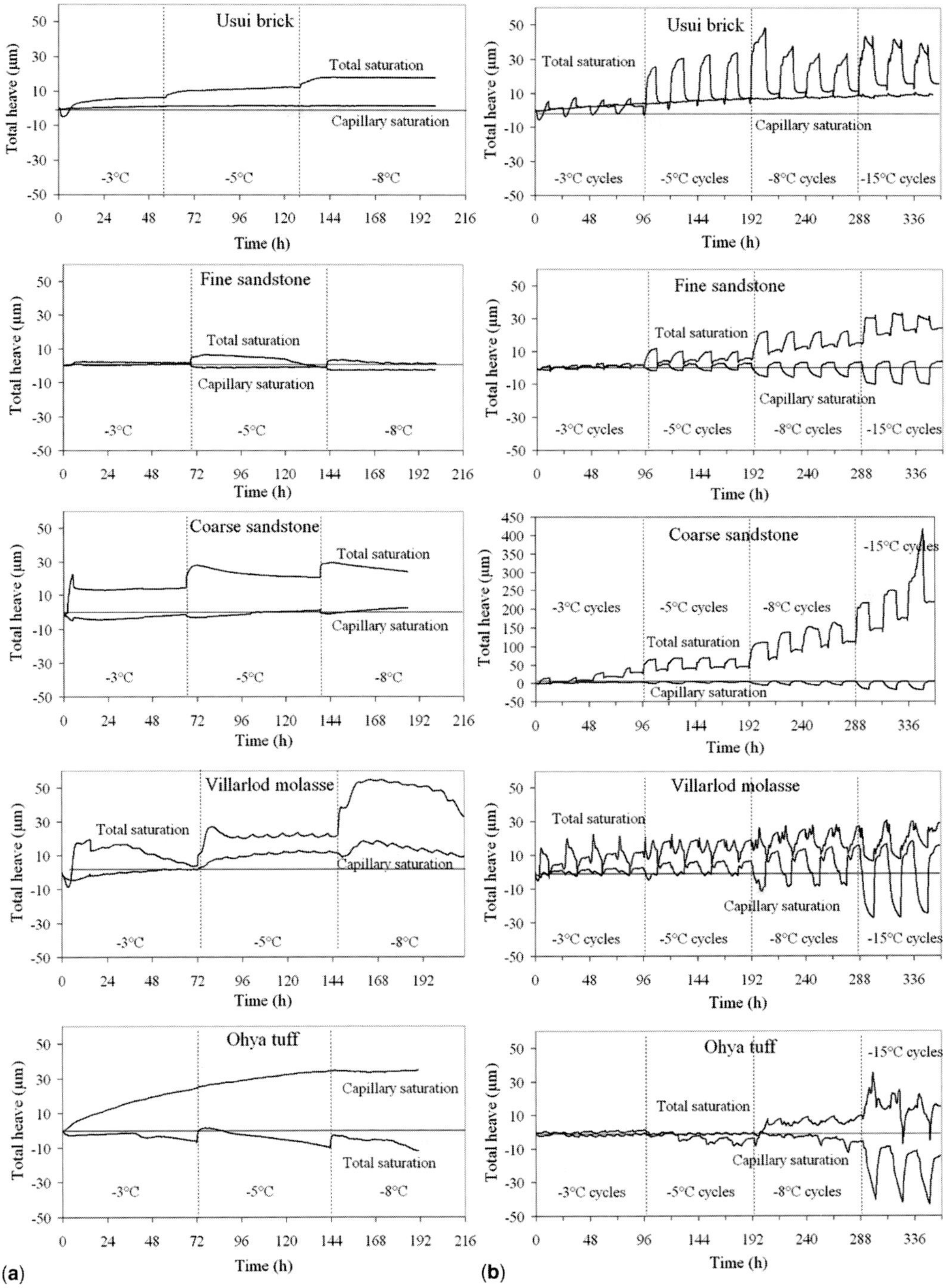


**Fig. 5.** Total heave changes of materials submitted to (**a**) diurnal freezing and (**b**) continuous freezing. (Note that the vertical scale for the coarse sandstone at total saturation is exaggerated.)

during the −3 °C cycles. On the other hand, maximum expansion did not occur at the same intensity as freeze–thaw cycles and at the same depth in the samples. In the fine sandstone, the coarse sandstone and the Ohya tuff, maximum dilation occurred during the −15 °C cycles, respectively at 6, 4 and 6 cm depth. In the Usui brick, maximum expansion occurred at 2 cm during the −5 °C cycles, while in the Villarlod molasses maximum expansion occurred at 4 cm during the −8 °C cycles (Fig. 3a).

Whatever the tested materials, more intensive freezing leads to higher heave amount (Fig. 5). But, according to the type of porous network, progression of the ice front and associated dilation is different for each material. It is difficult to know which petrophysical characteriztic causes these differences.

## Conclusions

Submitted to the same experimental conditions, materials with different or close petrophysical properties had mixed dilation responses to frost action. Testing with different moisture conditions, types and intensity of freezing allowed them to be grouped according to their petrophysical properties and their dilation behaviour.

Three groups were determined. One group included materials with relatively fast capillary kinetics. Continuous freezing of these materials does not induce significant dilation. These rocks are most sensitive to the repetition of diurnal freeze–thaw cycles with low a freezing temperature (≤−15 °C). Stronger or more numerous than 15 diurnal freeze–thaw cycles with minimum freezing temperatures ranging progressively from −3 to −15 °C are necessary to create visible cracks.

The third group regroups rocks with very slow capillary kinetics. They are also sensitive to the repetition of freeze–thaw cycles. Less than the previous 15 diurnal freeze–thaw cycles leads to cracking. But continuous freezing is also significant and leads to cracking more easily.

The second group is halfway between the other two. Capillary kinetics are intermediate between the two groups. This group is most sensitive to the repetition of diurnal freeze–thaw cycles that leads to their cracking. The number of cycles necessary to create cracks is also intermediate between the two groups. On the other hand, continuous freezing at −3, −5 and −8 °C leads to significant expansion, but is not enough to create visible cracks.

This study showed that materials with the weakest transfer properties by capillary absorption are most sensitive to continuous freezing, while materials with fast capillary kinetics are most sensitive to the repetition of freeze–thaw cycles. The faster the capillary kinetics, the more intensive has to be the freezing to lead to significant expansion. These results confirm the major role of water migration in frost weathering.

The authors would like to thank B. Rousset from the Expert Centre of Lausanne for providing the Swiss rocks and the Japan Society for the Promotion of Sciences (JSPS) for financing this study.

## References

AKAGAWA, S. & FUKUDA, M. 1991. Frost heave mechanisms in welded tuff. *Permafrost and Periglacial Processes*, **2**, 301–309.

CHAHAL, R. S. & MILLER, R. D. 1965. Supercooling of water in glass capillaries. *British Journal of Applied Physics*, **16**, 231–239.

DAVID, C., DAROT, M. & JEANNETTE, D. 1993. Pore structures and transport properties of sandstone. *Transport in Porous Media*, **11**, 161–177.

EVERETT, D. H. 1961. The thermodynamics of frost damage to porous solids. *Transactions of the Faraday Society*, **57**, 1541–1551.

FAGERLUND, G. 1971. Degré critique de saturation: un outil pour l'estimation de la résistance au gel des matériaux de construction. *Matériaux et Constructions RILEM*, **4**(23), 271–283.

FÉLIX, C. 1993. *Propriétés physiques et mécaniques du grès molassique bleu de la carrière de Villarlod.* Non published report EC/FX/98–06.1. Laboratoire de l'Expert-Center pour la conservation du patrimoine bâti, Lausanne.

GONZALEZ, I. J. & SCHERER, G. W. 2004. Effect of swelling inhibitors on the swelling and stress relaxation of clay bearing stones. *Environmental Geology*, **46**, 364–377.

ITO, K. & CHIBA, M. 2001. Railway stations and local communities in Japan. *Japan Railway & Transport Review*, **28**, 4–17.

LITVAN, G. G. 1978. *Freeze–Thaw Durability of Porous Building Materials, in Durability of Building Materials and Components*. Special Technical Publication, **STP 691**, American Society for Testing and Materials, 455–463.

MADER, D. 1985. Aspects of fluvial sedimentation in the lower triassic Buntsandstein of Europe. *Lecture Notes in Earth Sciences*, **4**.

MATSUOKA, N. 2001. Microgelivation versus macrogelivation: towards bridging the gap between laboratory and field frost weathering. *Permafrost and Periglacial Processes*, **12**, 299–313.

POWERS, T. C. 1945. TA working hypothesis for further studies of frost resistance of concrete. *Journal of the American Concrete Institute*, **16**, 245–272.

PRICK, A. 1995. Dilatometrical behaviour of porous calcareous rock samples subjected to freeze–thaw cycles. *Catena*, **25**, 7–20.

SMITH, R. A. 1997. The Usui Toge Railway of the Shin-etsu Line 1893–1997. Impact of railways on

Japanese Society & Culture. *Japan Railway & Transport Review*, **13**, 28–33.

THOMACHOT, C. & JEANNETTE, D. 2002. Evolution of the petrophysical properties of two types of Buntsandstein sandstone subjected to simulated freeze–thaw conditions. *In*: SIEGESMUND, S., WEISS, T. & VOLLBRECHT, A. (eds) *Natural Stone, Weathering Phenomena. Conservation Strategies and Case Studies*. Geological Society, London, Special Publication, **205**, 19–32.

THOMACHOT, C. & JEANNETTE, D. 2004. Effects of iron black varnish on petrophysical properties of building sandstone. *Environmental Geology*, **47**, 119–131.

THOMACHOT, C., MATSUOKA, N., KUCHITSU, N. & MORII, M. 2005. Frost damage of bricks composing a railway tunnel monument in Central Japan: field monitoring and laboratory simulation. *In: Natural Hazards and Earth System Sciences*. EGU Special Publication, **5**, 465–476.

THOMAS, W. N. 1938. *Experiments on the Freezing of Certain Buildings Materials*. Department of Scientific and Industrial Research, (Great Britain), Building Research Station, Technical Paper, **17**.

WARDLAW, N. C., MCKELLAR, M. & LI, Y. 1988. Pore and throat size distributions determined by mercury porosimetry and by direct observation. *Carbonates and Evaporites*, **3**(1), 1–15.

YAMADA, Y., AOKI, H., TAKAHASHI, M. & MATSUKURA, Y. 2005. Effect of rock properties on rates of salt weathering: a laboratory experiment. *Journal of the Japan Society of Engineering and Geology*, **46**(2), 72–78.

# The effects of wetting and drying, and marine salt crystallization on calcarenite rocks used as building material in historic monuments

G. F. ANDRIANI & N. WALSH

*Dipartimento di Geologia e Geofisica, Università degli Studi di Bari, Via Orabona 4, 70125 Bari, Italy (e-mail: nwalsh@geo.uniba.it)*

**Abstract:** The results of a study of the effect of marine salt crystallization on the physical and mechanical properties of Plio-Pleistocene calcarenites cropping out in southern Italy are presented here. Owing to their workability, aesthetic appeal and availability, the calcarenites have been widely used as building stones in many historic monuments. Samples of medium-grained packstones and fine-grained packstones–wackestones were prepared for the salt crystallization test defined by EN 12370, using sea water instead of a 14% solution of $Na_2SO_4 \cdot 10H_2O$. To determine the effect of imbibition alone on the performance of the calcarenites, the same procedure was followed with distilled water without soluble salts. Microfabric analysis, evaluation of index parameters and grain-size distribution were carried out as well. Particular attention was given to pore-size distribution by mercury intrusion porosimetry (MIP), loss of weight and uniaxial compressive strength determined before and after the tests, and after every five cycles of complete immersion in sea water and distilled water. The results suggest that detailed information on fabric and pore network are indispensable to predicting the weatherability of rocks. Crystallization tests that involve the complete immersion of the samples in a saline solution are not effective for an understanding of the real importance of salt damage on soft and porous calcarenites owing to a significant incidence of imbibition in accelerating deterioration rates and in influencing patterns and intensity of weathering.

It is widely accepted that the presence of soluble salts in porous systems, such as building stones and mortars, produce deterioration even though the process and mechanism of salt weathering are not fully understood (Rijniers *et al.* 2003). In recent years, many researchers have focused on the relationship between salt deterioration mechanisms, microclimate and decay patterns, and several methods have been developed for a qualitative and quantitative assessment of the behaviour of building material contaminated with salts (Weyl 1959; Evans 1970; Arnold & Zehnder 1989; Richardson 1991; Bell 1993; Camuffo 1995; Goudie & Viles 1997; Scherer 1999; Přikryl *et al.* 2003; Lubelli *et al.* 2004; Tuğrul 2004). An interesting overview on salt-induced deterioration of inorganic porous material is reported by Charola (2000). Further details concerning several salt damage mechanisms and corresponding theoretical models can be found in Rodriguez-Navarro & Doehne (1999). Although many mechanisms have been proposed in the salt deterioration of porous materials, crystallization from supersaturated salt solutions and hydration of salts that can exist in more than one hydration state appear to be the most important decay processes (Price 1996; Chatterji 2000).

Within the pores and other cavities of a stone, the growth of salt crystals and the volume increase in the formation of hydrated crystals from their anhydrous precursors produce tensile stresses, which can create new cracks and/or the extension and widening of existing microcracks and pores, causing crumbling and powdering of the stone. These disruptive stresses and decay patterns depend on the type of salt or salt mixture and on the rock fabric, and they are strongly influenced by the relative humidity (RH) and temperature. Although it is now clear that rocks with a large number of micropores connected to macropores are the most susceptible to salt damage and that crystallization pressure is inversely related with pore size, several theories have been developed to estimate the range of pore sizes in which crystal growth takes place. By analogy with the thermodynamics of frost damage in porous media as developed by Everet (1961), Fitzner & Snethlage (1982) state that crystallization occurs, primarily, in larger pores and, after these have been filled, continues in the connected small pores. Arnold & Zehnder (1989) affirm that crystal growth begins in pores with a size of about 1–10 μm, whilst Rodriguez-Navarro & Doehne (1999) assert that NaCl crystallization takes place within the smaller pores between 0.01 and 0.1 μm.

Many historic buildings and monuments located in southern Italy were built from locally quarried calcarenites. These rocks belong to a

*From*: PŘIKRYL, R. & SMITH, B. J. (eds) *Building Stone Decay: From Diagnosis to Conservation*. Geological Society, London, Special Publications, **271**, 179–188.
0305-8719/07/$15.00 

Plio-Pleistocene formation (Calcarenite di Gravina) overlapping the Mesozoic–Cenozoic limestone successions (exceeding 3000 m in thickness) of the Apulian foreland (Ricchetti *et al.* 1988). The Calcarenite di Gravina Formation constitutes continuous exposure of intrabasin biocalcarenites and biocalcirudites and terrigenous calcarenites, with a carbonate content between of 90 and 99%. These transgressive deposits of limited thickness (up to 80 m thick) are composed of several lithofacies, namely conglomerates at the proximal end (rocky coast) that are progressively replaced by sands offshore (Pomar & Tropeano 2001). The calcarenites are soft and porous, and suffer much from deterioration as a result of weathering (Andriani & Walsh 2002, 2003). The type and rate of weathering of this building stone depend on the petrographic and physical properties (fabric, porosity, etc.), position within a monument or building, geographical location and air pollution level (Ordòñez *et al.* 1997). Particularly in coastal areas of Apulia, salt weathering can be considered the most important cause of stone decay, especially where historic monuments and buildings are close to the coastline and partially or completely submerged by sea water. Among the latter, the archaeological site of Egnazia (from Greek name 'Gnathia'), an ancient settlement dating back to the Bronze Age (13th–12th century BC), is notable for its remains of a Messapian necropolis and Roman port.

In this context, the salt crystallization test outlined in EN 12370 (1999) was performed on samples of medium-grained packstones and fine-grained packstones–wackestones, collected from two quarry districts ('Gravina' and 'Massafra') located along the SW margin of the Murge plateau (Festa 2003) (Fig. 1). The test was carried out using sea water (salinity 35.07 ppt) instead of a 14% solution of $Na_2SO_4 \cdot 10H_2O$, and, separately, using distilled water. Petrographic and physical characteristics (fabric, total and effective porosity, water absorption, pore-size distribution), uniaxial compressive strength and dry weight loss (DWL) were evaluated on fresh and tested specimens.

This study provides a quantitative and qualitative assessment of the separate and combined actions of imbibition and salt crystallization on soft and porous, fine- and medium-grained calcarenites, with the aim of highlighting the fundamental role of fabric in affecting the physical behaviour and decay patterns of two different varieties belonging to the same rock unit.

## Description of fabrics

Large representative square prism blocks of two varieties of calcarenite were selected in different quarry districts of Apulia: whitish, bioclastic medium-grained packstone and light yellow, biolitoclastic fine-grained wackestone–packstone from the areas of Gravina and Massafra, respectively. Both varieties are composed of low-magnesium calcite (>98%) with an insoluble residue of mainly clay minerals with negligible quartz, feldspars, gibbsite and goethite. At the mesoscopic scale, the calcarenites are light, very porous and without visible structural features.

For each of the two varieties of calcarenite, thin sections were prepared, parallel and at right angles to the bedding. Fabric observations and photographs of the thin sections were undertaken with a Leitz Diaplan polarizing microscope and 35 mm-camera (Fig. 2). The general fabric is

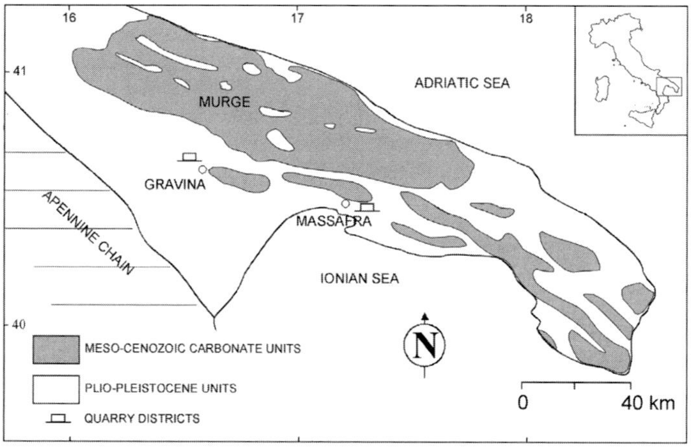


**Fig. 1.** Geographical location of the quarry districts.

**Fig. 2.** Fine- and medium-grained calcarenites used in the experiment. On the right, images of petrographic thin sections in plane-polarized light; on the left, images of laboratory prepared transversal sections of the specimens.

typical of loosely packed calcarenites, well (fine calcarenite) to moderately graded (medium calcarenite), with a self-supporting framework of skeletal remains of marine organisms and subangular and rarely subrounded lithoclasts of limestone, derived from the weathering of the carbonate basement. The micrite matrix is present in small quantities in the medium calcarenite, while it represents the main textural element of the fine calcarenite in the mud-matrix-supported texture (wackestone).

The carbonate skeletal grains consist of fragments and, rarely, whole shells, and include benthic and planctonic foraminifera, echinoderms, bryozoa, lamellibranchs, gastropods, calcareous algae and serpulid worm tubes. Some of these were dissolved after deposition leaving empty casts with thin recrystallized micrite envelopes representing only the outer layers of the shells. The skeletal remains constitute between 20 and 40% of the fine calcarenite volume, and 50 and 70% of the medium calcarenite volume. Lithoclasts are at some points not distinguished from recrystallized bioclasts; they lack any recognizable inner structure and were probably formed by intense micritization. The degree of cementation is very low. The carbonate cement is not evenly distributed, and mostly occurs at the contact between grains or on the grain walls in open pore spaces as a fine encrustation of calcite microcrystals. Early stages of cementation are also found where internal cavities exist. The cement disseminated in the matrix is hardly distinguishable from the recrystallized micrite, especially in the medium calcarenite. Sparry calcite, in both interparticle pore spaces and skeletal moulds, is uncommon.

Considering Choquette & Pray's (1970) classification scheme, the major contribution to the total porosity is made by intergranular porosity. Intragranular porosity and moldic porosity are especially typical of the medium calcarenite.

## Experimental methods

The experimental procedure utilized in this research was subdivided into different phases. In the first phase of the study, following the standard test procedures outlined in ISRM (1979) and EN 1926 (1999), dry density ($\delta_d$), total porosity ($n$) and uniaxial compressive strength in the dry state ($\sigma_n$) were determined on fresh cylindrical calcarenite specimens with a diameter of 50 mm and height of 100–150 mm, which were prepared from the quarry rock blocks. The volume of the specimens was calculated from micrometer caliper measurements. An average of several readings for each dimension, with an accuracy of 0.05 mm, were used in the calculation of bulk volume. As regards the specific gravity ($G$), reference was made to a value of 2.70 on the basis of the chemical composition of the rocks.

Degree of saturation ($Sr$) and water absorption ($w_a$) were measured by the weight of the calcarenite specimens completely immersed and suspended in water applying Archimede's principle ('an immersed body is buoyed up by a force equal to the weight of the fluid it displaces'). Three samples for each variety were used in this procedure. The cylindrical specimens were dried for 24 h in a ventilated oven at a temperature of 105 °C. After drying, the specimens were cooled for 30 min and weighed. Then the specimens were immersed and suspended in distilled water at 20 °C and weighed at fixed intervals of time. The suspended weights were determined to the nearest 0.01 g.

It was possible to determine degree of saturation ($Sr$) at every instant of the test as follows:

$$Sr = \frac{m_{st} - m_{s0}}{\delta_w n V}$$

where $m_{s0}$ is the mass of the specimens suspended in water at the initial time of the test; $m_{st}$ is the mass of the specimens suspended in water at any time of the test; $\delta_w$ is the density of the distilled water at 20 °C; $n$ is the total porosity, obtained indirectly from $\delta_d$ and $G$ ($n = 1 - \frac{\delta_d}{G\delta_w}$); and $V$ is the total volume of the specimen.

At atmospheric pressure, the test was stopped when the degree of saturation reached 80%. On fresh cylindrical specimens characterized by a volume of $2.76 \times 10^{-4}$ m$^3$ (diameter 0.053 m; height 0.125 m), this value was obtained in 14 days for the fine calcarenite and 5 days for the medium calcarenite. The specimens were then saturated completely under vacuum (5.0 Pa) without removing them from the water basket. With this procedure a value of $Sr$ (%) = 100 was obtained for all the varieties. Full saturation demonstrates that pores in the rock particle systems are interconnected and continuous; it follows that porosity is an effective porosity. At the same time, the water absorption was determined at every instant of the test as follows:

$$w_a = \frac{Sr\delta_w n}{\delta_d}.$$

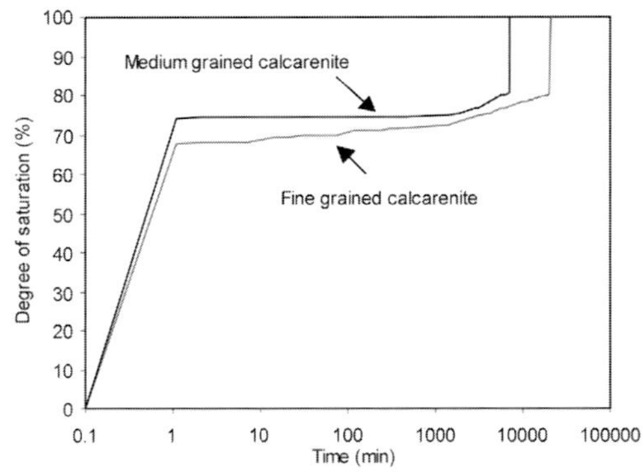


**Fig. 3.** Degree of saturation v. time. The specimens were suspended in distilled water at 20 °C and weighed at fixed intervals until the degree of saturation reached about 80%. The specimens were then saturated completely under vacuum.

Degree of saturation v. time for the calcarenite specimens suspended in distilled water is illustrated in Figure 3. It is interesting to note that most water is absorbed quickly (between 1 and 2 min), following which water absorption and degree of saturation increase very slowly. Similar trends were observed for the studied varieties, with an increase of the degree of saturation occurring in slightly longer times in the fine calcarenites. The experimental data are summarized in Table 1.

To obtain loose material for the grain-size analyses, three saturated cylindrical samples of each calcarenite variety were subjected to 30 freeze–thaw cycles and then disaggregated by hand to avoid breaking of bioclasts and lithoclasts. The loose material was dried in an oven at 105 °C for 24 h and afterwards sieved using sieve sizes ranging from 2.00 to 0.063 mm. The remaining fine fractions (passing 230, ASTM series) were examined through sedimentation analysis. A comparison of the cumulative curves obtained for the two varieties is shown in Figure 4. On the basis of the uniformity coefficient ($C_u$), the fine calcarenite is poorly graded ($C_u = 1.7$) while the medium calcarenite is moderately graded ($C_u = 3.8$).

**Table 1.** *Physical and mechanical properties of the calcarenite varieties belong to the Calcarenite di Gravina Formation*

| Physical–mechanical properties | Fine-grained calcarenite | | | Medium-grained calcarenite | | |
|---|---|---|---|---|---|---|
| | min | max | mean | min | max | mean |
| Specific gravity, $G_s$ | | | 2.70 | | | 2.70 |
| Dry density, $\delta_d$ (Mg m$^{-3}$) | 1.40 | 1.66 | 1.52 | 1.32 | 1.50 | 1.39 |
| Saturation density, $\delta_{sat}$ (Mg m$^{-3}$) | 1.88 | 2.05 | 1.96 | 1.85 | 1.97 | 1.90 |
| Porosity, $n$(%) | 38.5 | 48.2 | 43.7 | 45.5 | 52.0 | 49.5 |
| Water absorption, $w_a$ (%) | 23.2 | 34.5 | 28.9 | 30.9 | 40.2 | 36.4 |
| Degree of saturation, $Sr$ (%) | | | 100 | | | 100 |
| Uniaxial compressive strength (dry), $\sigma_n$ (MPa) | 1.44 | 2.65 | 2.20 | 1.81 | 2.74 | 2.34 |

In the second phase of the study, the salt crystallization test outlined in EN 12370 (1999) was performed on the same specimens of the calcarenites previously utilized to calculate the index parameters. In order to better comprehend the performance of the calcarenites under conditions simulating the marine environment and to estimate the contribution of sea-salt weathering on rock deterioration, sea water (salinity: 35.07 ppt), sampled along the Adriatic coast of southern Italy, instead of a 14% solution of $Na_2SO_4 \cdot 10H_2O$ was used in the test. The salt content of the sea water used for testing is mainly made up of about 75% sodium chloride, 13% magnesium sulphate and 11% magnesium chloride. To evaluate the effect of imbibition alone on the behaviour of the calcarenites, the same procedure was also followed with distilled water without soluble salts. In the two tests, the dry weight loss (DWL) was evaluated at the end of each soaking and drying cycle as the percentage loss in dry weight of the calcarenite samples. The uniaxial compressive strength was determined every five full cycles. At the end of the 15th cycle before determining DWL and uniaxial compressive strength, the calcarenite specimens were dried at 105 °C for 24 h and cooled at ambient temperature for 2 h. A total of 24 specimens (15 of which were utilized in the salt crytallization test and nine subjected to the same procedure with distilled water) from each calcarenite variety were used in this phase of the research. The results obtained are shown in Tables 2–5.

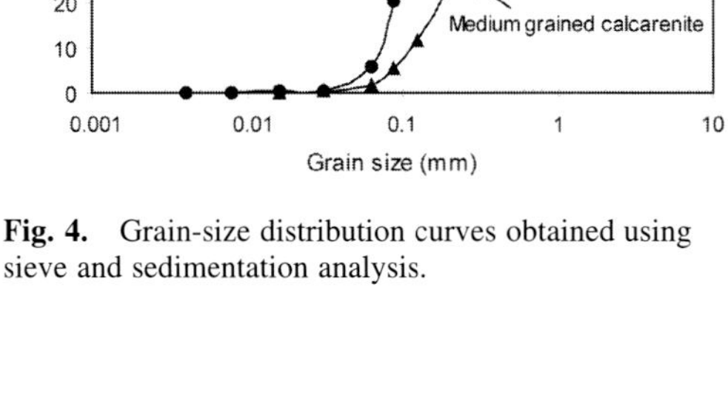


**Fig. 4.** Grain-size distribution curves obtained using sieve and sedimentation analysis.

In order to characterize the pore network of the calcarenites and the changes induced in rock porosity because of sea-salt contamination and imbibition, in the third phase of the study mercury intrusion porosimetry (MIP) was used as outlined by ASTM D4404 (1984). In particular, total intrusion volume, total pore area, pore-size distribution, average pore diameter and complementary measurements of porosity ('effective' porosity) were determined on oven-dried samples of about 2.5 g, using a Micromeritics porosimeter (AutoPore IV series, model 9500). The analyses were performed at low (3.44–345 kPa) and high pressure (0.1–228 MPa) on fragments of irregular shape detached from fresh and tested specimens utilized in the first and second phases of the research. Within the limitations of the operating conditions and the applied method, as well as the simplification of textures and real geometry of pores (Dullien 1992), the MIP was capable of measuring pore diameters ranging from 420 to 0.005 μm.

In order to assess quantitatively the changes induced by salt crystallization and imbibition in the pore network, a comparison of the pore-size distribution of the fresh and tested samples was carried out. The experimental values obtained using MIP are provided in Tables 6 and 7. At the same time, by means of optical petrographic microscopy analysis on thin sections of the fresh and tested samples, a qualitative evaluation of the fabric change induced by salt crystallization and imbibition was performed. The research was completed

**Table 2.** *Dry weight loss (%) and uniaxial compressive strength (MPa) on different samples of the fine-grained calcarenite subjected to the soaking and drying test with sea water (salinity 35.07 ppt)*

| Sample number | Dry weight loss DWL (%) | | | Uniaxial compressive strength (MPa) | | |
|---|---|---|---|---|---|---|
| | cycle 5 | cycle 10 | cycle 15 | cycle 5 | cycle 10 | cycle 15 |
| 1 | 2.71 | 2.48 | 1.34 | – | – | 1.64 |
| 3 | 2.58 | 2.29 | 0.32 | – | – | 1.23 |
| 5 | 1.73 | 1.79 | 0.12 | – | – | 0.64 |
| 6 | 1.61 | 1.70 | 0.06 | – | – | 1.07 |
| 7 | 2.36 | 2.36 | 0.36 | – | – | 0.96 |
| 8 | 2.45 | 2.10 | – | – | 1.52 | – |
| 9 | 2.33 | 2.24 | – | – | 1.39 | – |
| 10 | 2.33 | 2.18 | – | – | 1.35 | – |
| 11 | 1.94 | 2.00 | – | – | 2.39 | – |
| 12 | 2.85 | 2.45 | – | – | 2.72 | – |
| 13 | 2.41 | – | – | 2.37 | – | – |
| 14 | 2.95 | – | – | 2.03 | – | – |
| 15 | 2.33 | – | – | 1.58 | – | – |
| 16 | 2.40 | – | – | 2.25 | – | – |
| 17 | 2.12 | – | – | 2.98 | – | – |
| Mean value | 2.34 | 2.16 | 0.44 | 2.24 | 1.87 | 1.11 |

**Table 3.** *Dry weight loss (%) and uniaxial compressive strength (MPa) on different samples of the fine-grained calcarenite subjected to the soaking and drying test with distilled water*

| Sample number | Dry weight loss DWL (%) | | | Uniaxial compressive strength (MPa) | | |
|---|---|---|---|---|---|---|
| | cycle 5 | cycle 10 | cycle 15 | cycle 5 | cycle 10 | cycle 15 |
| 74 | −0.06 | −0.12 | −0.18 | – | – | 1.49 |
| 75 | −0.11 | −0.17 | −0.43 | – | – | 0.99 |
| 76 | −0.11 | −0.12 | −0.19 | – | – | 1.15 |
| 79 | −0.12 | −0.18 | – | – | 1.30 | – |
| 80 | −0.20 | −0.26 | – | – | 1.99 | – |
| 82 | −0.12 | −0.18 | – | – | 1.67 | – |
| 83 | −0.09 | – | – | 2.56 | – | – |
| 85 | −0.12 | – | – | 1.64 | – | – |
| 87 | −0.14 | – | – | 2.17 | – | – |
| Mean value | −0.12 | −0.17 | −0.27 | 2.12 | 1.65 | 1.21 |

by observations of the visual appearance of the specimens utilized in the study.

## Discussion and conclusions

Before discussing the data obtained from the experiments, it is important to underline two points. It is well known that the durability of a stone is mainly dependent on its strength and on the geometry and topology of its pore network. In order to assess the influence of fabric and pore system on rock weatherability caused by the separate and combined actions of water imbibition and salt crystallization, laboratory tests were conducted on two calcarenite types, belonging to the same rock unit, which show, as can be seen in Table 1, a certain similarity in terms of such properties as porosity, dry density and uniaxial compressive strength in the fresh state, but have different textural features and pore-size distributions. Even though in many laboratory tests where the effects of different salts have been analysed NaCl crystallization produced little damage compared with other salts such as sulphates, for the salt crystallization test sea water was utilized instead of a 14% solution of $Na_2SO_4 \cdot 10H_2O$. This choice is explained by the fact that at many Apulian sites, especially where historic monuments and buildings are close to the coastline, NaCl crystallization has been found to be the most important cause of salt damage.

In the test with sea water, DWL data, summarized in Tables 2 and 4, show, at the end of the first five cycles, an average increase of 2.34 and

**Table 4.** *Dry weight loss (%) and uniaxial compressive strength (MPa) on different samples of the medium-grained calcarenite subjected to the soaking and drying test with sea water (salinity 35.07 ppt)*

| Sample number | Dry weight loss DWL (%) | | | Uniaxial compressive strength (MPa) | | |
|---|---|---|---|---|---|---|
| | cycle 5 | cycle 10 | cycle 15 | cycle 5 | cycle 10 | cycle 15 |
| 45 | 1.20 | 0.46 | −0.46 | – | – | 2.37 |
| 46 | 1.32 | 0.56 | −0.45 | – | – | 1.90 |
| 47 | 1.03 | 1.13 | −0.38 | – | – | 1.68 |
| 50 | 1.92 | 0.93 | −0.53 | – | – | 1.72 |
| 51 | 1.39 | 0.80 | −0.24 | – | – | 2.33 |
| 52 | 1.80 | 0.62 | – | – | 2.39 | – |
| 53 | 1.35 | 0.80 | – | – | 2.18 | – |
| 55 | 2.10 | 0.71 | – | – | 2.20 | – |
| 58 | 1.20 | 1.06 | – | – | 2.57 | – |
| 59 | 0.92 | 0.72 | – | – | 1.98 | – |
| 60 | 0.59 | – | – | 2.20 | – | – |
| 61 | 2.30 | – | – | 2.32 | – | – |
| 64 | 1.98 | – | – | 2.82 | – | – |
| 65 | 1.37 | – | – | 2.47 | – | – |
| 68 | 0.98 | – | – | 1.99 | – | – |
| Mean value | 1.43 | 0.78 | −0.41 | 2.36 | 2.26 | 2.20 |

**Table 5.** *Dry weight loss (%) and uniaxial compressive strength (MPa) on different samples of the medium-grained calcarenite subjected to the soaking and drying test with distilled water*

| Sample number | Dry weight loss DWL (%) | | | Uniaxial compressive strength (MPa) | | |
|---|---|---|---|---|---|---|
| | cycle 5 | cycle 10 | cycle 15 | cycle 5 | cycle 10 | cycle 15 |
| 99 | −0.31 | −0.37 | −1.22 | – | – | 2.19 |
| 102 | −0.24 | −0.28 | −0.96 | – | – | 2.26 |
| 103 | −0.27 | −0.31 | −1.10 | – | – | 2.33 |
| 104 | −0.23 | −0.30 | – | – | 2.45 | – |
| 105 | −0.24 | −0.31 | – | – | 2.68 | – |
| 108 | −0.24 | −0.35 | – | – | 2.77 | – |
| 109 | −0.27 | – | – | 2.89 | – | – |
| 153 | −0.21 | – | – | 2.78 | – | – |
| 166 | −0.27 | – | – | 2.18 | – | – |
| Mean value | −0.25 | −0.32 | −1.09 | 2.61 | 2.63 | 2.26 |

**Table 6.** *Results of the mercury intrusion porosimetry (MIP) performed on fresh and tested samples of the fine-grained calcarenite variety*

| Sample | Fresh | Tested with sea water | | | Tested with distilled water | | |
|---|---|---|---|---|---|---|---|
| | | cycle 5 | cycle 10 | cycle 15 | cycle 5 | cycle 10 | cycle 15 |
| Total intrusion volume (mL $g^{-1}$) | 0.2518 | 0.2222 | 0.2465 | 0.2427 | 0.2155 | 0.2817 | 0.2430 |
| Total pore area ($m^2$ $g^{-1}$) | 1.874 | 1.681 | 2.041 | 1.889 | 1.787 | 1.734 | 1.299 |
| Average pore diameter (μm) | 0.5374 | 0.529 | 0.483 | 0.754 | 0.566 | 0.650 | 0.748 |
| Effective porosity by MIP (%) | 37.44 | 36.60 | 39.04 | 38.27 | 31.90 | 42.61 | 37.60 |
| Pore-size diameter, $d > 30$ μm (%) | 6.3 | 7.1 | 8.2 | 9.7 | 6.5 | 6.7 | 7.2 |
| 30 μm $> d >$ 10 μm (%) | 41.4 | 36.2 | 33.9 | 31.6 | 40.3 | 40.6 | 36.8 |
| 10 μm $> d >$ 1 μm (%) | 31.2 | 35.8 | 38.0 | 39.8 | 31.6 | 33.3 | 35.7 |
| 1 μm $> d >$ 0.1 μm (%) | 17.4 | 16.8 | 16.4 | 16.2 | 17.2 | 16.4 | 17.2 |
| Pore-size diameter, $d <$ 0.1 μm (%) | 3.7 | 4.1 | 3.5 | 2.7 | 4.4 | 3.0 | 3.1 |

**Table 7.** *Results of the mercury intrusion porosimetry (MIP) performed on fresh and tested samples of the medium-grained calcarenite variety*

| Sample | Fresh | Tested with sea water | | | Tested with distilled water | | |
|---|---|---|---|---|---|---|---|
| | | cycle 5 | cycle 10 | cycle 15 | cycle 5 | cycle 10 | cycle 15 |
| Total intrusion volume (mL $g^{-1}$) | 0.2621 | 0.2751 | 0.2802 | 0.2989 | 0.2941 | 0.2723 | 0.3231 |
| Total pore area ($m^2$ $g^{-1}$) | 1.454 | 1.423 | 1.469 | 1.433 | 1.477 | 1.453 | 1.456 |
| Average pore diameter (μm) | 0.7434 | 0.7824 | 0.7963 | 0.8345 | 0.7967 | 0.8146 | 0.8878 |
| Effective porosity by MIP (%) | 39.2 | 41.3 | 41.5 | 41.9 | 41.5 | 40.7 | 44.3 |
| Pore-size diameter, $d > 30$ μm (%) | 48.1 | 49.8 | 50.8 | 50.6 | 50.2 | 51.0 | 52.8 |
| 30 μm $> d >$ 10 μm (%) | 19.2 | 19.0 | 18.0 | 18.4 | 17.4 | 16.6 | 16.7 |
| 10 μm $> d >$ 1 μm (%) | 16.8 | 16.3 | 16.4 | 16.2 | 17.6 | 18.6 | 17.5 |
| 1 μm $> d >$ 0.1 μm (%) | 12.6 | 12.2 | 11.5 | 11.7 | 11.4 | 10.6 | 9.9 |
| Pore-size diameter, $d < 0.1$ μm (%) | 3.3 | 2.7 | 3.1 | 3.1 | 3.4 | 3.2 | 3.1 |

1.43% for the fine- and medium-grained calcarenites, respectively. This gain in weight is related to crystal growth in the voids. At the end of the first five cycles, crystallization pressure had, in fact, produced no destructive tensile stresses over the void surfaces because the salts did not completely fill the voids. At the same time, it can be seen that the compressive strength did not seem to change for each calcarenite variety. Between the fifth and 10th cycles, both varieties began to lose weight, although the DWL at the end of the 10th cycle is still positive, 2.16 and 0.78% for the fine- and medium-grained calcarenites, respectively. The weight loss is certainly due to the detachment of grains that are less strongly bound to the samples and is not due to dissolution phenomena, which can occur only with difficulty in such a short time. The compressive strength in this interval decreased by 15.0 and 3.4% for the fine- and medium-grained calcarenites, respectively. The weight loss, like the loss of strength, was more marked between the 10th and 15th cycles so that at the end of the test the DWL was 0.44% (fine calcarenite) and −0.41% (medium-grained calcarenite), and the compressive strength decreased by 49.5% in the fine-grained calcarenite and by only 6.0% in the medium-grained calcarenite. The negative DWL value obtained for the medium-grained variety indicates that at the end of the test the samples analysed were characterized by a dry weight that is slightly lower than that found in the fresh state, before testing.

In the test with distilled water (Tables 3 and 5) the DWL values were negative from the beginning, indicating a weight loss from the first cycles of soaking and drying. At the end of the test, DWL was −0.27 and −1.09%, with a greater weight loss in the medium-grained variety. Also in this test the detachment of grains in the samples tested was more evident between the 10th and 15th cycles for both varieties. While the loss of material is more marked in the medium-grained variety, owing to the presence of course grains loosely bound by scarce cementation on the external surface of the samples, the compressive strength decreased more rapidly with the increasing number of cycles in the fine-grained variety. The latter shows a decrease in compressive strength of 3.6 and 25.0% at the end of the fifth cycle and 10th cycle, respectively, and 45.0% at the end of the test. The medium calcarenite, however, shows no reduction of compressive strength before the end of the 10th cycle. A 3.4% decrease of compressive strength was recorded only at the end of the test, after the 15th cycle.

Comparison of obtained data suggest that while it is clear that water imbibition has a negative influence on the overall resistance of the calcarenites, the real role of the growth of the NaCl crystals inside the pores in the weakening of the rocks is not equally clear.

The effects of deterioration mechanisms, whether induced by water imbibition or by salt crystallization, are certainly more marked in the fine-grained variety in terms of the reduction of compressive strength, although the medium-grained variety shows a greater loss of dry weight.

The low resistance of the fine-grained calcarenite is a result of the low cement/matrix ratio and a lesser degree of cementation with respect to the medium-grained calcarenite. In addition, in the medium-grained calcarenite, the calcite cement is concentrated at the grain contacts, while in the fine-grained calcarenite most of the cement, which is represented by microcrystalline calcite, occurs within the micrite matrix. The compressive strength does not seem to be influenced by the packing and number of the grain contacts as the medium variety is more loosely packed. To this we can add that grains of medium size contribute to increasing rock strength because they tend to have a more massive carbonate structure and fewer internal cavities than fragments from fine and more delicate shells.

With regard to the MIP analysis results and the changes in the pore distribution observed during the tests, the data obtained from the two calcarenite types show a similar trend. Total intrusion volume, total pore area, average pore diameter and effective porosity show an upward trend as test cycles in sea water increase. The data obtained do not permit a reliable percentage estimate of the increases observed for each property. However, the increases do seem more marked in the medium-grained variety. In both varieties, the main change observed in the porous system with the increased number of cycles in sea water concerns the constant increase of pores with a diameter greater than 30 μm. The same phenomenon is found for pores between 1 and 10 μm, but only in the fine-grained variety. For the other dimensional intervals a reduction of the percentages is generally observed with the increased number of sea-water cycles. Overall, therefore, the analysis of the data shows there is a genesis of larger pores from smaller pores. A similar observation can be made for the samples submitted to the test with distilled water, although some of the trends for certain pore-size classes ($<10$ μm) are not so clear. In any case, the study carried out by MIP is limited because it cannot provide information about macropores greater than 420 μm, which in the medium-grained variety form some 15% in volume of the total porosity of the rock.

At the end of the laboratory tests, the calcarenite specimens of the two varieties showed no variation either of colour or of aesthetic appearance at the mesoscopic scale. This observation was confirmed by the analysis of thin sections of the tested calcarenites by optical microscopy. No new type of porosity or changes in pore geometry were observed, perhaps these phenomena effect the pores of smaller dimensions than those that can be examined by the resolving power of the optical microscopy used in the study.

To conclude, the study shows that soft and porous calcarenites are so susceptible to the action of water that any salt crystallization test involving the complete immersion of the samples in a saline solution can be considered largely inappropriate as an experimental salt durability test.

## References


Andriani, G. F. & Walsh, N. 2002. Physical properties and textural parameters of calcarenitic rocks: qualitative and quantitative evaluations. *Engineering Geology*, **67**, 5–15.

Andriani, G. F. & Walsh, N. 2003. Fabric, porosity and water permeability of calcarenites from Apulia (SE Italy) used as building and ornamental stone. *Bulletin of Engineering Geology and Environment*, **62**, 77–84.

Arnold, A. & Zehnder, K. 1989. Salt weathering on monuments. *In*: Zezza, F. (ed.) *The Conservation of Monuments in the Mediterranean Basin*. Grafo, Bari, 31–58.

ASTM D4404. 1984. *Standard Test Method for Determination of Pore Volume and Pore Volume Distribution of Soil and Rock by Mercury Intrusion Porosimetry*. ASTM Standards, **12.01**, 744–748.

Bell, F. G. 1993. Durability of carbonate rock as building stone with comments on its preservation. *Environmental Geology*, **21**, 187–200.

Camuffo, D. 1995. Physical weathering of stones. *The Science of the Total Environment*, **167**, 1–14.

Charola, A. E. 2000. Salts in the deterioration of porous material: an overview. *Journal of the American Institute of Conservation*, **39(3)**, art2. World Wide Web Address: http://aic.stanford.edu/jaic/articles/jaic39-03-002_appx.html.

Chatterji, S. 2000. A discussion of the paper 'Crystallisation in pores' by G. W. Scherer. *Cement and Concrete Research*, **30**, 669–671.

Choquette, P. W. & Pray, L. C. 1970. Geologic nomenclature and classification of porosity in sedimentary carbonates. *AAPG Bulletin*, **54**, 207–250.

Dullien, F. A. L. 1992. *Porous Media – Fluid Transport and Pore Structure*, 2nd edn. Academic Press, San Diego, CA.

EN 12370. 1999. *Determination of Resistance to Salt Crystallization*. European Standards for Natural Stones prepared. Comité Européen de Normalisation Brussels Technical Commision, 246.

EN 1926, 1999. *Determination of Compressive Strength*. European Standards for Natural Stones prepared. Comité Européen de Normalisation Brussels Technical Commision, 246.

Evans, I. S. 1970. Salt crystallisation and rock weathering: a review. *Revue de Géomorphologie Dynamique*, **19**, 155–177.

Everet, D. H. 1961. The thermodynamics of frost damage to porous solid. *Transactions of the Faraday Society*, **57**, 1541–1551.

Festa, V. 2003. Cretaceous structural features of the Murge area (Apulian Foreland, Southern Italy). *Eclogae Geologicae Helvetiae*, **96**, 11–22.

Fitzner, B. & Snethlage, R. 1982. Über Zusammenhange zwischen salzkristallisationsdruck und Porenradienverteilung. *Group Petrography Newsletter*, **3**, 13–24.

Goudie, A. S. & Viles, H. 1997. *Salt Weathering Hazards*. Wiley, Chichester.

ISRM. 1979. Suggested methods for determining water content, porosity, density, absortion and related properties and swelling and slake-durability index properties. *International Journal of Rock Mechanics and Mining Science & Geomechanical Abstracts*, **16**, 141–156.

Lubelli, B., van Hees, R. P. J. & Brocken, H. J. P. 2004. Experimental research on hygroscopic behaviour of porous specimens contaminated with salts. *Construction and Building Materials*, **18**, 339–348.

ORDÒÑEZ, S., FORT, R. & GARCÌA DEL CURA, M. A. 1997. Pore size distribution and the durability of a porous limestone. *Quarterly Journal of Engineering Geology*, **30**, 221–230.

POMAR, L. & TROPEANO, M. 2001. The Calcarenite di Gravina Formation in Matera (Southern Italy): New insights for coarse-grained, large-scale, cross-bedded bodies encased in offshore deposits. *AAPG Bulletin*, **85**, 661–689.

PRICE, C. A. 1996. *Stone Conservation: An Overview of Current Research*. Getty Conservation Institute, Santa Monica, CA. World Wide Web Address: http://www.getty.edu/conservation/publications/pdf_publications/stoneconservation.pdf.

PŘIKRYL, R., LOKAJÍČEK, T., SVOBODOVÁ, J. & WEISHAUPTOVÁ, Z. 2003. Experimental weathering of marlstone from Přední Kopanina (Czech Republic) historical building stone of Prague. *Building and Environment*, **38**, 1163–1171.

RICCHETTI, G., CIARANFI, N., LUPERTO SINNI, E., MONGELLI, F. & PIERI, P. 1988. Geodinamica ed evoluzione sedimentaria e tettonica dell'Avampaese apulo. *Memorie della Società Geologica Italiana*, **41**, 57–82.

RICHARDSON, B. A. 1991. The durability of porous stones. *Stone Industries*, **12**, 22–25.

RIJNIERS, L. A., HUININK, H. P., PEL, L. & KOPINGA, K. 2003. Salt crystallization in porous materials and its implications for stone decay. *In*: *EUROMAT 2003, Symposium P2 – Materials and Conservation of Cultural Heritage*. World Wide Web Address: http://expert-center.epfl.ch/publications/euromat_2003/P2_07.pdf.

RODRIGUEZ-NAVARRO, C. & DOEHNE, E. 1999. Salt weathering: influence of evaporation rate, supersaturation and crystallisation pattern. *Earth Surface Processes and Landforms*, **24**, 191–209.

SCHERER, G. W. 1999. Crystallisation in pores. *Cement and Concrete Research*, **29**, 1347–1358.

TUĞRUL, A. 2004. The effect of weathering on pore geometry and compressive strength of selected rock types from Turkey. *Engineering Geology*, **75**, 215–227.

WEYL, P. K. 1959. Pressure solution and the force of crystallization, a phenomenological theory. *Journal of Geophysical Research*, **64**(11), 2001–2025.

# Stone properties and weathering induced by salt crystallization of Maltese Globigerina Limestone

E. ROTHERT[1], T. EGGERS[2], J. CASSAR[3], J. RUEDRICH[1], B. FITZNER[4] & S. SIEGESMUND[1]

[1]*Geoscience Centre, University Göttingen, Goldschmidtstrasse 3, 37077 Göttingen, Germany (e-mail: e.rothert@wes.de)*

[2]*Geographical Institute of the University Göttingen, Göttingen, Germany*

[3]*Institute for Masonry and Construction Research, University of Malta, Malta*

[4]*RWTH Aachen, Germany*

**Abstract:** Most monuments and buildings in the Maltese Islands are constructed of the local Globigerina Limestone. Today, this Globigerina Limestone shows considerable damage in many buildings, particularly through alveolar weathering, which is frequently very intense. Owing to Malta's marine environment, salt crystallization in the stone's pore spaces has been recognized as the main weathering process responsible for the deterioration of the country's monuments. In order to obtain more information on the fabric-dependent weathering processes of Globigerina Limestone, detailed analyses were carried out. Globigerina Limestone samples obtained from stone types with two different known qualities were characterized according to petrographical, geochemical and physical properties. These included porosity, pore radii distribution and tensile strength, as well as water and humidity transport properties. Investigations by means of salt crystallization tests on quarry samples of both stone types reinforced the idea that the extent of salt weathering depends on salt type and concentration and pore-space properties. Visible weathering damage was recorded and evaluated for a representative monument (the Church of Santa Marija Ta' Cwerra in Siggiewi) by means of a monument mapping method, which was carried out twice over a period of 9 years (1995 and 2004). The identified weathering forms were also correlated with a previously developed weathering model for Globigerina Limestone. According to the results of the mapping, salt analyses carried out on samples from the church and salt-loading tests on quarry samples, there exists a significant correlation between visible damage and salt load. The zoning of weathering damage is obviously related to different salt concentrations. The zone with severe weathering damage is characterized by a high concentration of halite. Consequently, salt weathering represents the main damage process for the Globigerina Limestone of Malta.

The Maltese archipelago of Malta, Comino and Gozo lies in the central Mediterranean Sea approximately 90 km south of Sicily. During the geological epochs of the Oligocene and Miocene 30–5 Ma ago, extensive sedimentation took place in this area. This sedimentation led to the laying down of beds of lime- and mudstones (Pedley *et al.* 2002).

The Maltese Globigerina Limestone Formation is one of five main formations, and varies in thickness from 20 to over 200 m. The material used for building is located stratigraphically in the lower part of the Globigerina Limestone Formation, called the Lower Globigerina Limestone. During the deposition of the sediments that eventually formed this stone, far away from the continent and below the action of waves, only minor but variable amounts of clay in suspension were carried from a distant land source. The Globigerina Limestone also contains numerous shells, algae and planktonic fossils, mainly the planktonic foraminifera Globigerina (Pedley *et al.* 2002).

As can be widely seen in the Maltese Islands, the local limestone has always been used as the predominant building material. The Maltese prehistoric temples, which were constructed approximately 6000 years ago, bear testimony to this (Cassar & Vannucci 2001). Between 1530 and 1798 the Order of the Knights of St John built kilometres of fortifications to protect the island from the expanding Ottoman Empire. Fortifications, impressive churches (Fig. 1) and palaces were built of the local building stone during this period. The capital city of Valletta is included in the UNESCO World Heritage List, as are the prehistoric temples. Even today, the local building stone is still much in demand. Many modern buildings are constructed using the Globigerina Limestone (Fig. 2a) and many local quarries are still active (Fig. 2b).

*From*: PŘIKRYL, R. & SMITH, B. J. (eds) *Building Stone Decay: From Diagnosis to Conservation.* Geological Society, London, Special Publications, **271**, 189–198.
0305-8719/07/$15.00 

**Fig. 1.** The Baroque parish church of Siggiewi is one of numerous impressive Maltese buildings constructed from Globigerina Limestone.

**Fig. 2.** (**a**) Modern buildings in Globigerina Limestone show that to this day this limestone is still one of the most important building materials in Malta. (**b**) This quarry, which is still active today, shows the extraction of Globigerina Limestone. The limestone blocks, here exposed to rain and sun for a short period, were previously left for much longer and led to a hardening of the stone surface.

## Aim of this paper

This paper discusses salt-induced weathering of the Maltese Globigerina Limestone. On the basis of work carried out in the past (Fassina *et al.* 1996; Fitzner *et al.* 1996; Torfs *et al.* 1996) it has been found to be necessary to obtain more information about the damage processes and to emulate site salt loading and its relation to the stone fabric and different stone types. Thus, this work has had a multifaceted approach. One part consisted of the detailed investigation of the two main qualities of Globigerina Limestone, derived from different layers of a quarry near the village of Siggiewi, which lies in the main quarry area of the Maltese Islands. Here, detailed petrographical and fabric determinations, and numerous petrophysical analyses, have been carried out on these fresh samples to supplement previous work (Vannucci *et al.* 1994; Galan *et al.* 1996; Cassar 1999, 2002) and also to subsequently characterize the weathering resistance of these two limestone types with respect to salt content. In addition, the local occurrence of weathering forms and their distributions were meticulously recorded on two separate occasions for the Church of Santa Marija Ta' Cwerra, also in Siggiewi, where the weathering problem owing to salt load is clearly evident. In the past, detailed scientific analyses of salt load within this building had also been carried out (Fassina *et al.* 1996). The results from laboratory weathering resistance tests with halite, thenardite and epsomite have now been compared with prevailing damage phenomena at the church that have been attributed to salt load. An important part of this work has also been the comparison of the mapping of this church carried out in 1995 with that carried out in 2004, to determine the rate of stone damage on this building. This work has also been useful to help determine materials and methods used for recent conservation work on the four external façades of the church.

## Deterioration of Globigerina Limestone

Today it is often, although not always, the case that the stone in older buildings in Malta is badly deteriorated. Frequently, the main deterioration phenomenon is alveolar weathering. A model developed for Globigerina Limestone deterioration some years ago explains that the weathering process initiates by the dissolution and re-precipitation of the mineral calcite, which at first leads to the formation of a thick and compact superficial crust (Vannucci *et al.* 1994; Fitzner *et al.* 1996). Areas of stone that demonstrate a loss of calcite are observed to become weaker. In addition to this, the main weathering process responsible for the deterioration of the building stone has been recognized as salt crystallization in the pore spaces of this very porous limestone (Cassar 2002). For this reason, parts of the hardened and compact surface fall off. The main source of the salt is the surrounding marine environment (Torfs *et al.* 1996). The $SO_x$ loads that derive from urban pollution and lead to the development of sulphates can here be mostly disregarded, except in areas downwind from power stations (Torfs *et al.* 1996). The weathering intensity varies because of local differences in the salt types and content in buildings, and because of differing quality and weathering resistance of the natural building stone itself (Cassar 2002).

Local stoneworkers distinguish macroscopically between two building stone qualities: 'Franka' and 'Soll'. Whereas 'Soll' represents *bad* quality building material, 'Franka' tends to resist the local environmental conditions well. In the fresh state, the two types cannot be distinguished visually, although a geochemical test and the pore-radii distribution may help to identify the two qualities (Farrugia 1993; Fitzner *et al.* 1996; Cassar 1999; Cassar & Vella 2003). In addition, on abandoned quarry faces 'Franka' and 'Soll' can be seen to differ in their weathering intensities. It can also be seen that in buildings, and occasionally in quarries, both types can sometimes coexist in the same horizontal layer, forming local areas with different geometries and shapes. As the weathering of the good-quality Globigerina Limestone leads to hardening of the stone surface, fresh quarry stones were in the past exposed to rain and sun for a long period before being utilized (Fig. 2b). This practice has been abandoned, primarily for economic reasons.

## Sampling

At the start of the present testing programme, quarry owners were asked to supply two different stone types – what they considered to be 'Franka' (i.e. *good* stone) and 'Soll' (i.e. *bad* stone). These will be identified in this paper as Type I (*good* stone) and Type II (*bad* stone). Four standard stone blocks, measuring 229 × 260 × 610 mm, of each type were supplied.

## Analytical methods

Petrographic analyses (in polarized light) on standard thin sections of both Globigerina Limestone types were performed to obtain a qualitative description of different fabric parameters of the investigated rock samples (e.g. mineralogical composition, properties of detrital and authigenic components). Previous work had concentrated on only one type of Globigerina Limestone (Vannucci *et al.* 1985; Galan *et al.* 1996). To analyse the bulk rock composition, X-ray fluorescence (XRF) was carried out (cf. Cassar 1999). For a quantitative determination of pore-size distribution, mercury porosimetry was used (cf. Brakel *et al.* 1981).

The total accessible porosity of the two stone types was also characterized – samples were thus measured using buoyancy weighing. The dry weight, the water-saturated weight and the weight immersed in water of cubic samples (65 × 65 × 65 mm) were. To determine the directional dependence of capillary water absorption, the same device as for the buoyancy weighing and cubes with the same dimensions were used, but were only dipped 1 cm into water. To analyse the degree of saturation (S-value) of free moisture absorption, the sample weight after 24 h of water immersion was again measured. The S-value represents the ratio between the free capillary water uptake and the maximum uptake under vacuum.

The water vapour diffusion resistance value, $\mu$, of the limestones was studied using the wet-cup method. This value characterizes the diffusion resistance of a porous material compared to an equal inactive air film. Slices of the stones (ϕ 40 × 10 mm) were used as covers on the cups. The relative humidity difference caused moisture to flow through the porous material from the side with higher moisture (inside at 100%) to the side with lower relative humidity (outside at 50%). The moisture flow was obtained by weighing the cups at various times until a steady state was reached.

The tensile strength ($\sigma_z$) was determined by means of the 'Brazilian test', which involves disc-shaped specimens. The diameter of the samples was 40 mm and the length was 20 mm. To obtain the average value, a minimum of four samples were used. A constant strain rate of $0.3 \times 10^{-6}$ mm $s^{-1}$ ($\approx 10^{-5}$ $s^{-1}$) was applied.

Ultrasonic velocity measurements were carried out on cubic rock samples (65 × 65 × 65 mm). Transient times of ultrasonic pulses (piezoceramic transducers, resonant frequency 1 MHz) were measured in three orthogonal directions using the pulse transmission technique (Birch 1960, 1961).

The thermal expansion behaviour of the investigated samples was measured on cylindrical specimens (ϕ 15 × 50 mm). This was determined as a function of temperature. For this dilatation experiment, a heating cycle from 20 to 90 °C was employed. The experimental set-up allows simultaneously investigation of six samples. The thermal expansion coefficient, $\alpha$, expresses the volume change of a material as a function of temperature.

Hygric expansion was determined on cylindrical samples (ϕ 15 × 50 mm), which were preconditioned at 30% relative humidity and room temperature. Afterwards, the samples were completely immersed in distilled water. The accuracy of the incremental displacement transducer is 1.0 μm.

## Results

### *Rock fabric*

The investigated samples of both types of Globigerina Limestone can be described as soft and almost pure limestone with a pale cream–yellow colour. They are fine grained and homogenous. In thin sections, it was confirmed that in both types large concentrations of randomly distributed and non-orientated microfossils, mainly round planktonic globigerinae and some elongated forms, occur. They often make up 80 vol.%, whereas the micritic matrix is only about 20 vol.%. Finely dispersed iron oxides and iron hydroxides, mainly limonite, can also be observed. No clear differences were distinguished between the two stone types, as had already been recognized in previous studies (Cassar 1999).

The limestone fabric is grain-supported with a micritic matrix, and can be described as foraminiferal packstone, although wackestones also occur (Cassar 2004). The pore space is formed by intergranular pores, secondary solution pores and often by empty fossil chambers. As can be seen from the geochemical investigations (X-ray fluorescence) the Globigerina Limestone contains minor amounts of $SiO_2$ and $Al_2O_3$ phases (Table 1). These are a result of sparse quartz and clay minerals (cf. Cassar 1999, 2002; Galan *et al.* 1996).

**Table 1.** *Properties of two fresh samples of Globigerina Limestone*

| Stone properties | Type I | Type II |
|---|---|---|
| Bulk composion [wt%] | | |
| CaO | 52.2 | 51.9 |
| $SiO_2$ | 2.70 | 3.90 |
| $Al_2O_3$ | 1.20 | 1.10 |
| MgO | 0.85 | 0.86 |
| $Fe_2O_3$ | 0.73 | 0.59 |
| Porosity [vol%] | 36.46 | 34.59 |
| Pore-radii distribution [%] | | |
| 0.001–0.01 μm | 0.00 | 0.71 |
| 0.01–0.1 μm | 5.56 | 10.04 |
| 0.1–1 μm | 27.85 | 47.02 |
| 1–10 μm | 66.59 | 41.91 |
| >10 μm | 0.00 | 0.33 |
| Average pore radii [μm] | 1.06 | 0.56 |
| Capillary water uptake (*w*-value) [kg $m^2 \cdot h^{-0.5}$] | 6.74 | 8.73 |
| Water vapour diffusion resistance, μ | 7.78 | 7.83 |
| Saturation degree | 0.69 | 0.76 |
| Tensile strength [MPa] | 2.96 | 2.83 |
| Ultrasonic velocity, $V_p$ [km $s^{-1}$] | 2.95 | 2.84 |
| Thermal dilatation coefficient [$K^{-1}$] | 2.32 | 4.51 |
| Hygric expansion [mm $m^{-1}$] | 0.20 | 0.25 |

### *Physical properties*

Both types of Globigerina Limestone are very porous with a relatively low tensile strength and high water absorption (Table 1). The effective porosity for both investigated types is comparable at approximately 35 vol.%, whereas distinct differences in the pore-radii distribution occur. Type I has, at 67%, a high proportion of pores in the range above 1 μm. In contrast, Type II has only 42% of its pores in this range. Correspondingly, the pore volume of pores smaller than 1 μm is 33% for Type I, whereas Type II has a very high proportion of these pores, at 58%. Concerning the water absorption coefficient of Globigerina Limestone, both investigated types show a very high W-value ranging between 6.74 and 8.73 kg $m^2$ $h^{-0.5}$. Thus, they can be described as rapidly absorbing stones. Site observations indicate that large amounts of water are absorbed during heavy rain. There is practically no run-off from the walls, even though guttering is absent in most Maltese buildings. The saturation value, i.e. the ratio between capillary water uptake (*w*) and effective porosity of the Globigerina Limestones, is, at 0.69 and 0.76, relatively low.

The physical properties are dependent on the limestone fabric and prove the presence of only slight diagenetic hardening. Otherwise, the two stone types show few differences. For example, the tensile strengths of the selected stone types differ slightly between 2.96 and 2.83 MPa. Even the ultrasonic velocities (in this case the compressional wave velocities, $V_p$) are almost comparable: 2.95 and 2.84 km $s^{-1}$. With regards to thermal dilation, however, a notable difference can be detected. The dilation coefficient of Type I is, at 2.32 $K^{-1}$, almost half that of Type II at 4.51 $K^{-1}$. A special phenomenon seen in both samples is a pronounced hygric expansion. This notable property is attributed to the presence of small amounts of swelling clay minerals, in particular the minerals smectite and illite-smectite (Vannucci *et al.* 1994).

## Damage mapping

The monument mapping method is a phenomenological, but meaningful, tool for the non-destructive registration of decay features. The stone surface is examined visually and observable changes are compared to the original condition of the building stone. The applied mapping method was based on a classification scheme proposed by Fitzner & Kownatzki (1990, 1991), Fitzner *et al.* (1992, 1995) and Kownatzki (1997).

The current mapping was carried out on the Church of Santa Marija Ta'Cwerra and focused primarily on a comparison with preceding investigations, which had also included a detailed mapping of all four external façades of the church in 1995 (Fassina *et al.* 1996; Fitzner *et al.* 1996; Torfs *et al.* 1996). This comparison was aimed at determining and classifying the progress of deterioration with time.

### *Church of Santa Marija Ta' Cwerra*

The church is located in the centre of Siggiewi, a village in the SW of Malta about 3 km from the coast. A building in this same location dates back to the 16th century, while the present monument was rebuilt in the 18th century. It is a small free-standing church of 10 × 10 m square and is built entirely of the local Globigerina Limestone. Only the lower courses have been covered with plaster, presumably to stop progressive deterioration. The weathering response of Globigerina Limestone to salt loading is a significant phenomenon at this church.

Over the last decade this monument has been extensively investigated. These studies included, besides the mapping of damage forms, several types of analyses aimed at understanding and quantifying the salt load (Fassina *et al.* 1996; Fitzner *et al.* 1996; Torfs *et al.* 1996). The damage recorded in a detailed monument mapping exercise 9 years previously (Fitzner *et al.* 1995) has now been compared with recent mapping (2004) to illustrate the changes in the damage forms and intensities after these years.

The work by Fitzner *et al.* (1995) determined the back-weathering rates. As the back-weathering provides insufficient information about the local damage distribution on heterogeneous back-weathered façades, a detailed mapping of damage forms was additionally performed. The main advantage of this approach is that the damage observed can be attributed to a damage phase of the weathering model by Vannucci *et al.* (1994) and Fitzner *et al.* (1996). Thus, an index of each individual stone conforming to the state of weathering is possible and a better correlation for future works can be derived.

### *Mapping results and correlations to weathering model*

The predominant damage phenomena occurring at the church were seen to be 'relief' and 'back-weathering'. In the model for the damage processes suggested by Vannucci *et al.* (1994) and Fitzner *et al.* (1996), these authors distinguished five phases of the damage development (Fig. 3). The classification of the current mapping was harmonized with the damage phases of this weathering model.

The original, non-weathered stone surface was one of the mapping forms identified. Following the damage model, at this stage of preservation, the formation of a superficial crust by re-precipitation of dissolved calcite can be seen to be taking place (phase 1, Fig. 3).

Slight–medium 'back-weathering' in the form of alveolar weathering was also observable. This is the mapped form called 'initial relief'. In this case, the stone surface is back-weathered through the formation of neighbouring cavities. Following the damage model by Vannucci *et al.* (1994) and Fitzner *et al.* (1996), this state represents damage phases 2 and 3 (both phases could not be distinguished on site). Local back-weathering with the formation of cavities can be traced back to cracking and/or partial loss of the crust owing to mechanical stress provoked by salt crystallization. Further accumulation of salt behind the crust leads to a detachment of the stone material in the form of granular disintegration and flaking within the cavities. The cause of the preferred back-weathering of the cavities is probably the increase of evaporation in areas where the crust has been lost.

The mapped form 'advanced relief' describes a weathering state where material loss and the formation of alveoli is very pronounced (phase 4 of the damage model). Also, connection of the

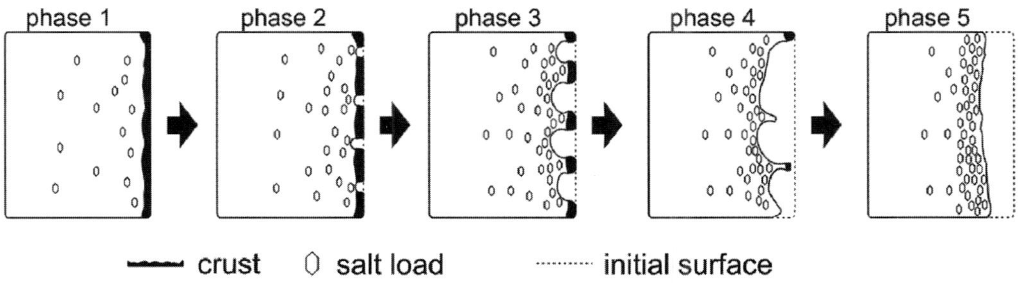


**Fig. 3.** Schematic representation of damage development, divided into five distinct phases (modified from Fitzner *et al.* 1996; explanation in text).

alveoli occurs. The septa of the honeycomb structure here are severely back-weathered, but are still recognizable.

The mapped form of 'back-weathering' represents the final deterioration state (phase 5 of the damage model) and follows after the 'advanced relief' weathering. The septa of the honeycombs are here totally back-weathered so that a more or less plain surface has developed and the edges of the blocks become rounded. Although back-weathering represents the final state of decay in the damage model, progressive material loss in form of flaking and granular disintegration is still subsequently observable on site.

In Figure 4, the distribution of damage forms is shown for the south façade of the church. The uppermost part of this façade is characterized by original and undamaged stone surfaces. However, in the lower parts of the wall the predominant damage phenomena are 'relief' and 'back-weathering' (Fig. 5). The middle part of the mapped façade is dominated by 'initial' and 'advanced relief' forms. Furthermore, a zoning of the relief forms can be observed. 'Advanced relief' is characterized by irregular back-weathered stone surfaces, mainly located in the lower middle part, while 'initial relief' development is observed in the upper middle part. Material loss in the areas where both relief forms occur is characterized by granular disintegration and crumbling in the cavities.

Very severe damage can be noticed in the lower parts of the walls, directly above the plastered lower courses (Fig. 4). In this part of the masonry

**Fig. 4.** Mapping of damage forms on the Church of Santa Marija Ta' Cwerra (south façade; 2004) reveals a clear distribution of weathering phenomena. Three zones in vertical order can be distinguished. Severe damage in the form of back-weathering occurs in the area immediately above the plastered lower courses. Further up, the middle zone is seen to have severe to moderate damage. Here alveolar weathering additionally affects numerous building stones. With the exception of a few building stones, the uppermost zone has remained in relatively good condition. These stones have developed a red–brown patina typical of good local building stone. There are, however, areas with organic deposition.

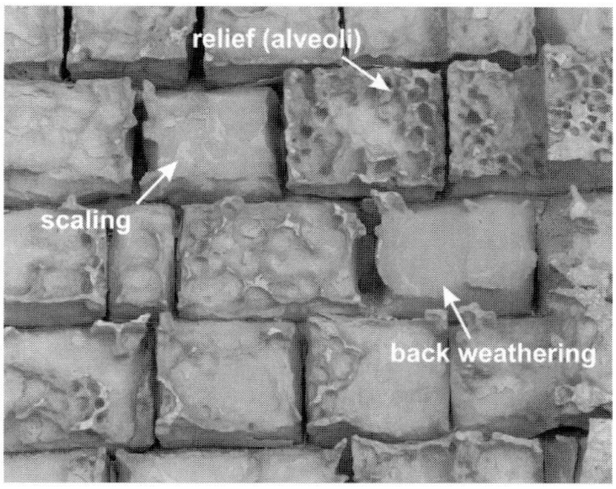


**Fig. 5.** Detachment and loss of stone material characteristic of the south façade. Relief in the form of alveoli and back-weathering by scaling are the prevalent weathering forms.

'back-weathering' dominates, and stone decay occurs by flaking and contour scaling. Flakes and larger detached parts of the stone surface are characterized by a bright brownish colour.

The intensity of stone loss evaluated during the mapping has been subdivided according to the estimated depth of the back-weathered surface, namely, slight (<3 cm), moderate (3–5 cm) and severe (>5 cm). Back-weathering of the whole stone surface can be observed on the lower part of the church, above the plastered lower courses. However, the comparison between the mappings from 1995 (Fitzner *et al.* 1995) and 2004 (Rothert 2004) showed that the weathering intensity had changed only slightly over the years.

### *Damage mechanisms*

The distribution of decay features on the south façade indicates a significant influence of moisture by capillary water uptake. The most severe deterioration is in fact observed on this façade, whereas the north and east façades are less affected. This is most probably explained by the sun's radiation being a critical instigator of damage. The sun influences the water evaporation rate and, consequently, the capillary suction. However, characteristic damage phenomena were observed on all four façades of the church.

The state of decay of the church can be correlated with salt-loading data from Fassina *et al.* (1996). The reported anion content of the wall masonry shows a typical distribution (Table 2) with sulphates mainly concentrated in the lower parts of the walls, while chlorides and nitrates occur in the upper parts. This salt distribution also indicates vertical capillary rising damp. Thus, the content and distribution of the salts in the wall confirm the results of our mapping. X-ray diffraction analyses by Fassina *et al.* (1996) demonstrated the predominance of halite. In additional, both crystalline phases (thenardite and mirabilite) for sodium sulphate were detected. The origin of the sulphates is probably from the mortar.

**Table 2.** *Water-soluble anion content of drilled core samples at different heights and depths (Fassina et al. 1996)*

| Height [m] | Depth [cm] | $Cl^-$ [%] | $NO_3^-$ [%] | $SO_4^{2-}$ [%] |
|---|---|---|---|---|
| 0.5 | 0–5 | 0.12 | 0.22 | 0.70 |
| 0.5 | 5–15 | 0.10 | – | 0.22 |
| 1.5 | 0–5 | 0.67 | 0.44 | 0.30 |
| 1.5 | 5–15 | 0.76 | 0.27 | 0.18 |
| 2.5 | 0–5 | 2.27 | 0.72 | 0.20 |
| 2.5 | 5–15 | 1.12 | 0.62 | 0.18 |
| 3.5 | 0–5 | 0.49 | 0.56 | 0.19 |
| 3.5 | 5–15 | 0.47 | 0.55 | 0.18 |

## Salt resistance tests

Salt-loading tests in the laboratory were carried out to verify the weathering susceptibility of the two types of Globigerina Limestone. For this purpose crystallization tests by means of halite, thenardite and epsomite were carried out on a number of stone cubes (65 × 65 × 65 mm). The samples were submitted to wetting and drying cycles as follows: loading with 10% salt solution for 4 h, followed by a drying cycle with a duration of 16 h at 60 °C. After cooling to room temperature, the weight change was determined.

The weathering effect of sodium sulphate is considered to be a result of the transformation of the water-free thenardite ($Na_2SO_4$) to the hydrated phase mirabilite ($Na_2SO_4{\cdot}10H_2O$). The salt hydration is coupled with a volume increase of about 300% (Price & Brimblecombe 1994). The transformation of the water-free kieserite ($MgSO_4{\cdot}H_2O$) to the hydrated phase epsomite ($MgSO_4{\cdot}7H_2O$) is also associated with a volume increase of about 173%. These tests were carried out to investigate how hydration pressure may affect the durability of Globigerina Limestone. For sodium chloride (NaCl) crystallization pressure is held to be the main damage factor. Both the expected hydration pressure and the crystallization pressure are greater than the tensile strength of porous natural stone. In addition, the rock fabric may also affect the durability of natural stones, whereas the pore-radii distribution should be of additional importance (Snethlage 1984).

The results of the salt action show that the limestone samples for both stone types submitted to the sodium sulphate test exhibited a slight granular disintegration at the surfaces already after the second cycle. For the samples seen in Figure 6a, back-weathering is most pronounced parallel to pre-existing sedimentary structures, evident after the sixth cycle. The tests were discontinued after a

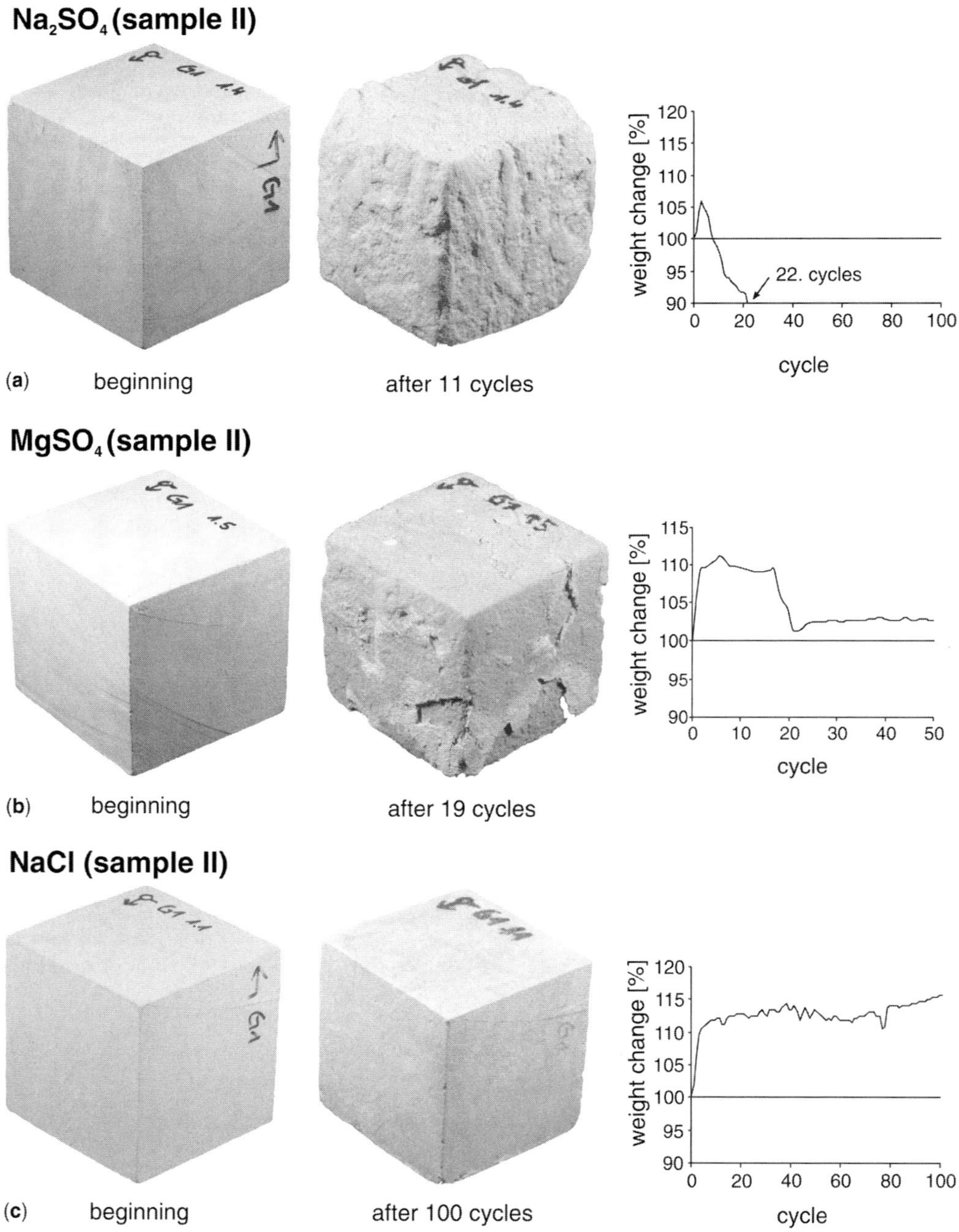


**Fig. 6.** Results of the salt-loading tests. (**a**) Samples that have undergone the sodium sulphate test, at the beginning, and showing definite deterioration after 11 cycles of artificial weathering. (**c**) Largely unchanged sample that has undergone the sodium chloride test, at the beginning, and after 100 cycles. On the right-hand side, the respective change in weight is plotted against the number of loading cycles.

loss in weight of 10%. The overall result was that the stone cubes of Type II resisted 22 or 24 such cycles, whereas the Type I cubes behaved differently. Here, for one sample, the critical weight loss was achieved after 17 cycles, while for another sample this happened after 51 cycles. This indicates that the latter samples were probably more heterogeneous.

The salt-loading tests with magnesium sulphate demonstrated that the Globigerina Limestone also showed a slight granular disintegration at the edges and the surfaces after the second weathering cycle. After the ninth cycle and again after the 19th cycle, an obvious scaling effect occurred (Fig. 6b). A total of 50 salt-weathering cycles were performed, although after the 25th cycle no further macroscopic changes occurred. The changes in weight were at a maximum after the sixth cycle (where an 11% increase in weight was registered), which means that the samples retained a large amount of the $MgSO_4$. After the breakaway of the scales, a decisive weight loss was ascertained.

In contrast, the samples loaded with sodium chloride were only slightly affected, even after 100 test cycles (Fig. 6c). For both investigated stone types an increase in weight was observable, which is again the consequence of salt enrichment. Only slight granular disintegration occurred at the edges of these specimens. Thus, the damage caused by sodium chloride was only slight when compared to the effects of sodium sulphate.

## Conclusion

The damage and the salt content distribution in the walls of the Church of Santa Marija Ta' Cwerra show a distinct correlation. This suggests that salt weathering is the main damage process for Globigerina Limestone on the island of Malta. The salt-loading tests in the laboratory demonstrate damage primarily for sodium sulphate and, to a lesser extent, for magnesium sulphate. On the other hand, the lack of damage by loading with sodium chloride indicates that the high halite content in the building is not the only cause of the observed damage. To obtain more information on the damage processes, further research is necessary to emulate the site loading and its relation to the stone fabric. This is of particular importance because the conservation approach in Malta has changed in recent years. In the past, the Maltese carried out stone replacement because the required natural stone was to be found in great quantities and the stone-working skills still existed. Today, preservation by means of modern conservation methods is accepted and largely carried out. However, to attain an acceptable degree of preservation by this approach, knowledge of the weathering processes and the contamination paths is of fundamental importance.

Thanks go to the reviewer E. Galan and the other reviewer for their comments. The authors would also like to thank C. Huss and D. Zahra for their help during the mapping in Malta. Our work was supported by the Deutsche Forschungsgemeinschaft (Si 438/17-1).

## References

ARNOLD, A. & ZEHNDER, K. 1990. Salt weathering on monuments. *In*: *Advanced Workshop Analytical Methodologies for the Investigation of Damaged Stones*, 14–21 September 1990, La Goliardica Pavese srl, Italy.

BIRCH, F. 1960. The velocity of compressional waves in rocks up to 10 kilobars. Part I. *Journal of Geophysical Research*, **65**, 1083–1102.

BIRCH, F. 1961. The velocity of compressional waves in rocks up to 10 kilobars. Part 2. *Journal of Geophysical Research*, **66**, 2199–2224.

BRAKEL, J. VAN, MODRY, S. & SVATA, M. 1981. Mercury porosimetry: State of the art. *Powder Technology*, **29**, 1–12.

CASSAR, J. 1999. *Geochemical and mineralogical characterisation of the Lower Globigerina Limestone of the Maltese Islands with special reference to the 'soll' facies*. Ph.D. thesis, University of Malta.

CASSAR, J. 2002. Deterioration of the Globigerina limestone of the Maltese Islands. *In*: SIEGESMUND, S., WEISS, T. & VOLLBRECHT, A. (eds) *Natural Stone, Weathering Phenomena, Conservation Strategies and Case Studies*. Geological Society, London, Special Publications, **205**, 25–41.

CASSAR, J. 2004. Composition and property data of Malta's building stone for the construction of a database. *In*: PŘIKRYL, R. & SIEGL, P. (eds) *Architectural and Sculptural Stone in Cultural Landscape*. Karolinum Press, Prague, 11–28.

CASSAR, J. & VANNUCCI, S. 2001. Petrographical and chemical research on the stone of the megalithic temples. *Malta Archaeological Review*, **5**, 40–45.

CASSAR, J. & VELLA, A. J. 2003. Methodology to identify badly weathering limestone using geochemistry: case study on the Lower Globigerina Limestone of the Maltese Islands. *Quarterly Journal of Engineering Geology and Hydrogeology*, **36**, 85–96.

FARRUGIA, P. 1993. *Porosity and related properties of local building stone*. B.E. & A. dissertation, University of Malta.

FASSINA, V., MIGNUCCI, A., NACCARI, A., STEVAN, A., CASSAR, J. & TORPIANO, A. 1996. Investigation on the moisture and salt migration in the wall masonry and on the presence of salt efflorescences on stone surface in the church of Sta. Marija Ta'Cwerra at Siggiewi, Malta. *In*: ZEZZA, F. (ed.) *Origin, Mechanisms and Effects of Salt on Degradation of Monuments in Marine and Continental Environments*. Proceedings, European Commission Research Workshop on Protection and Conservation of the European Cultural Heritage, Bari, Italy. Research Report, **4**, 89–100.

FITZNER, B. & KOWNATZKI, R. 1990. Bauwerkskartierung–Schadensaufnahme an Naturwerksteinen. *Der Freiberufliche Restaurator*, **4**, 25–40.

FITZNER, B. & KOWNATZKI, R. 1991. *Klassifizierung der Verwitterungsformen und Kartierung von Natursteinbauwerken*. Jahresberichte aus dem BMFT-Forschungsprogramm Steinzerfall – Steinkonservierung, **1**, 1–13.

FITZNER, B., HEINRICHS, K. & KOWNATZKI, R. 1992. Classification and mapping of weathering forms. *In*: DELGADO RODRIGUES, J., HENRIQUES, F. & TELMO JEREMIAS, F. (eds) *Proceedings of the 7th International Congress on Deterioration and Conservation of Stone*, 15–18 June, Lisbon, Laboratorio Nacional de Engenharia Civil, Lisbon, Portugal, 957–968.

FITZNER, B., HEINRICHS, K. & KOWNATZKI, R. 1995. Weathering forms-classification and mapping. *In*: SNETHLAGE, R. (ed.) *Denkmalpflege und Naturwissenschaft, Natursteinkonservierung I*. Ernst and Sohn, Berlin, 41–88.

FITZNER, B., HEINRICHS, K. & VOLKER, M. 1996. Model for salt weathering at Maltese Globigerina Limestones. *In*: ZEZZA, F. (ed.) *Origin, Mechanisms and Effects of Salt on Degradation of Monuments in Marine and Continental Environments*. Proceedings, European Commission Research Workshop on Protection and Conservation of the European Cultural Heritage, Bari, Italy. Research Report, **4**, 333–344.

GALAN, E., AIRES-BARROS, L., CHRISTARAS, B., KASSOLI-FOURNARAKI, A., FITZNER, B. & ZEZZA, F. 1996. Representative stones from the Sanctuary of Demeter in Eleusis (Greece), Sta. Marija Ta' Cwerra of Siggiewi (Malta) and Bari (Italy) and Cadiz (Spain) Cathedrals: petrographic characteristics, physical properties and alteration products. *In*: ZEZZA, F. (ed.) *Origin, Mechanisms and Effects of Salt on Degradation of Monuments in Marine and Continental Environments*. Proceedings, European Commission Research Workshop on Protection and Conservation of the European Cultural Heritage, Bari, Italy. Research Report, **4**, 77–85.

KOWNATZKI, R. 1997. *Verwitterungszustandserfassung von Natursteinbauten unter besonderer Berücksichtigung phänomenologischer Verfahren*. Aachener Geowissenschaftliche Beiträge, **22**.

PEDLEY, M., CLARKE, M. H. & GALEA, P., 2002. *Limestone Isles in a Crystal Sea. The Geology of the Maltese Islands*. P.E.G., Malta.

PRICE, C. & BRIMBLECOMBE, P. 1994. Preventing salt damage in porous materials. *In*: ROY, A. & SMITH, P. (eds) *Preprints of the Contributions to the Ottawa Congress, Preventive Conservation – Practice, Theory and Research*. IIC, London, 90–93.

ROTHERT, E., 2004. *Gefügeabhängige Verwitterung von Kalksteinen durch Salzkristallisation im Porenraum*. MSc. thesis, Universität Göttingen.

SNETHLAGE, R. 1984. *Steinkonservierung*. Bayerisches Landesamt für Denkmalpflege, Arbeitshefte, **22**.

TORFS, K., VAN GRIEKEN, R. & CASSAR, J. 1996. Environmental effects on deterioration of monuments: case study of the Church of Sta. Marija Ta' Cwerra, Malta. *In*: ZEZZA, F. (ed.) *Origin, Mechanisms and Effects of Salt on Degradation of Monuments in Marine and Continental Environments*. Proceedings, European Commission Research Workshop on Protection and Conservation of the European Cultural Heritage, Bari, Italy. Research Report, **4**, 441–451.

VANNUCCI, S., Alessandrini, G., CASSAR, J., TAMPONE, G. & VANNUCCI, M. L. 1994. The prehistoric megalithic temples of the Maltese Islands: causes and processes of deterioration of Globigerina Limestone. *In*: FASSINA, V., OTT, H. & ZEZZA, F. (eds) *Conservation of Monuments in the Mediterranean Basin. Proceedings of the 3rd International Symposium, Venice, Italy, Sopritendenza di Beni Artistici e Storici di Venezia, Italy*, 555–565.

VANNUCCI, S., CASSAR, J. & TAMPONE, G. 1985. The treatment of a typical soft limestone with different consolidants: a comparative study. *In*: BOLLETTINO INGEGNIERI, *Florence, Italy, Collegio degli Ingegnieri della Toscana, Florence, Italy*, 3–11.

# Length changes of sandstones caused by salt crystallization

J. RUEDRICH, M. SEIDEL, E. ROTHERT & S. SIEGESMUND

*Geoscience Centre, University Göttingen, Goldschmidtstrasse 3, 37077 Göttingen, Germany (e-mail: jruedri@gwdg.de)*

**Abstract:** Salt crystallization in the pore spaces of building stones can produce significant deterioration. The properties of the salt solution, the salt phases and the climatic conditions, as well as the rock fabric, significantly influence the state of rock weathering. To examine the influences of rock fabric and salt type on salt weathering, detailed investigations were performed on three sandstones. The fabric (mineralogical composition, grain size, etc.) and the petrophysical properties (porosity, pore-size distribution and hygric dilatation) of the sandstones were analysed and correlated with length changes during cyclic salt loading. The salt tests were carried out with two different salt types: (i) sodium sulphate and (ii) sodium chloride. The observed length changes differ for the investigated sandstones. Contractions of the samples, as well as a pronounced residual strain after the applied salt cycles, were observed. Specific deterioration features can be determined for the sandstones independent from the salt types used. However, the decay mechanisms, which lead to a significant deterioration, are different for sodium sulphate and sodium chloride. For sodium sulphate, a strong expansion occurs during the solution uptake cycles. This expansion can be attributed to hydration pressure during the transition from the water-free thenardite to the hydrate phase mirabilite. In contrast, the samples in the sodium chloride test show the main expansion in the drying stage. This can be related to the crystallization pressure caused by the growth of halite.

It has been known for a long time that salt attack leads to the deterioration of porous building stones (Darwin 1839). However, the processes and driving forces of salt-induced deterioration are still under discussion. In the literature many hypotheses have been made to explain how stresses may be created owing to salt crystallization in porous materials. For an overview of this topic see Duttlinger & Knöfel (1993), Charola (2000) and Doehne (2002).

Linear crystal growth pressure, introduced as an idea by Correns & Steinborn (1939) and Correns (1949), can develop if a crystal grows against a surrounding pressure and a thin, supersaturated solution film exists between the pore-surface and the salt crystal (Taber 1916). The pressure that arises is mainly controlled by the supersaturation of the surrounding solution. Crystallization pressures calculated for different salts using the equation from Correns & Steinborn (1939) are frequently cited in the literature (Winkler 1975), although they require extremely high supersaturation ratios of between 2 and 50. Snethlage (1984) suggested that these supersaturation ratios are unrealistically high owing to the multiple possibilities of heterogeneous nucleation within natural rocks. Wellmann & Wilson (1965, 1968) provided another approach based on a thermodynamic model. They assumed that crystal growth first takes place in larger pores, as confirmed by Putnis & Mauthe (2001). More recently, Steiger (2005) also presented an updated discussion on the origin and nature of crystal growth pressure. For stress development within the pore space, the properties of the growing crystals could be of crucial importance (cf. Scherer 1999). Investigations by Zehnder & Arnold (1989) on crystal growth dynamics within a pore (considering the crystallographic properties of the salt type) show that crystal shape and growth rate have an important influence on the pattern and intensity of rock damage. Sunagawa (1981) found that the shape and morphology of a growing crystal depends on the supersaturation rate. According to Rodriguez-Navarro & Doehne (1999) the factors discussed above should control salt crystallization, i.e. the location and distribution depends on the respective salt type. They demonstrated that, under the same experimental conditions, halite may show efflorescence, whereas thenardite crystallizes within the pore space.

The hydration pressure that will develop following an increase in volume upon hydration of a salt was first suggested by Mortensen (1933). Based on the proposed model by Mortensen, Winkler & Wilhelm (1970) calculated the resulting hydration pressures for different salt types. These high pressures may only hold true if they are of an osmotic nature (see the discussion in Duttlinger & Knöfel 1993). They suggest a modified model, based on the literature mentioned by Mortensen, which results in lower hydration pressures. The damage potential could also be controlled by the hydration–dehydration reaction, especially in complex pore spaces of

*From*: PŘIKRYL, R. & SMITH, B. J. (eds) *Building Stone Decay: From Diagnosis to Conservation.* Geological Society, London, Special Publications, **271**, 199–209.
0305-8719/07/$15.00 

natural rocks (Duttlinger & Knöfel 1993). For example, some investigation results give evidence that the dehydration of mirabilite to thenardite takes place over multiple dissolution steps (Charola & Weber 1992; Doehne 1994). On the other hand, the hydration of thenardite will slow down if a thin hydrated film is present on the salt surface. However, for salts with a hydration phase, it is still questionable which process (crystallization or hydration) may have the higher damage potential (Chatterji *et al.* 1979; Sperling & Cooke 1980).

There are also other hypothesis on how salt can account for the damage of natural rocks, for example differences in the thermal dilatation between the salt and the rock minerals (Winkler 1994). The thermal expansion coefficient for halite is four times larger than for quartz (cf. Skinner 1966). If there is a large amount of halite in the pores, temperature changes can also produce stresses within the rocks. In contrast, Pühringer (1983) suggested that the crystallization and recrystallization of thin salt layers may result in shear stresses that lead to an 'erosion' of the substrate.

It is evident that the rock fabric plays a significant role in the degree of weathering (e.g. Fitzner 1969). The question arises as to what extent each fabric element (fabric and/or mineralogical composition) is responsible for the differing stone durability. Fitzner & Snethlage (1982) found that pore-size distribution is of crucial importance. Their investigations on different German sandstones show that samples with a large amount of smaller capillary and micropores (a bimodal pore-size distribution) are extremely susceptible to salt attack (see also Zehnder & Arnold 1989; Rossi-Manaresi & Tucci 1991). Salt resistance tests by Ruedrich *et al.* (2005) also show that rocks with a relatively low tensile strength are less resistant.

Stresses resulting from the growth of salt crystals within the pore space could lead to a volume change of the stone. The fabric decay by salt loading is accompanied by microcracking, which results in a permanent increase in volume and yields important information about the time-dependent deterioration process (Kirchner & Worch 1993; Ruedrich *et al.* 2005). Substantial investigation of these phenomena, including the anisotropy and heterogeneity of rocks, is still lacking.

This paper mainly focuses on the fabric-dependent deterioration induced by salt crystallization in the pore space of building stones. Three sandstone types with varying mineralogical composition, different fabric properties, as well as different pore-radii distribution, were selected: Bad Bentheim, Cotta and Schoetmar (all from quarries located in Germany). Our studies focus on two topics: (i) fabric-controlled effects on salt weathering and (ii) dependence on the salt type. To do this, detailed fabric and petrophysical analysis are combined with length change measurements during salt crystallization. Two different salt types are used: (i) sodium sulphate and (ii) sodium chloride.

## Analytical methods

For the investigations, a reference co-ordinate system with respect to the macroscopically visible bedding was chosen ($X$, $Y$, $Z$). The $XY$-plane marks the sedimentary bedding, while the $Z$-direction is perpendicular to the bedding plane. Conventionally, the $X$-direction is parallel to the lineation. As the investigated sandstones do not show a lineation, the $X$- and $Y$-directions were determined according to given planes (e.g. joints). An arbitrary co-ordinate system was defined if the specimens did not show any macroscopically visible fabric elements. To survey the directional dependence of fabric parameters, investigations were performed on specimens from two mutually perpendicular directions parallel to the $X$- and $Z$-directions.

Petrographic analyses (in polarized light) on standard thin sections were performed for a qualitative description of different grain parameters (e.g. mineralogical composition, properties of detrital and authigenic components).

To characterize the total accessible porosity, samples were measured using buoyancy weighting. The samples were mounted in a basket attached under a balance. The dry mass, the water-saturated mass and the mass immersed in water were determined.

For a quantitative determination of pore-size distribution, mercury porosimetry was applied (cf. Brakel *et al.* 1981).

The tensile strength ($\sigma_z$) was determined by means of the 'Brazilian test', which involves disc-shaped specimens. The samples were 40 mm in diameter and 20 mm in length. In order to calculate the average value, a minimum of four samples were used. A constant strain rate of $0.3 \times 10^{-6}$ mm s$^{-1}$ ($\approx 10^{-5}$ s$^{-1}$) was applied. The tensile strength was measured perpendicular to the $XY$- and $XZ$-planes.

The hygric expansion of the sandstones was determined on cylindrical samples ($\phi$ 15 $\times$ 50 mm), which were preconditioned at 30% relative humidity and room temperature. Afterwards, the samples were completely immersed in distilled water. The accuracy of the displacement transducer is 1.0 μm. The investigations were carried out on samples parallel to the $X$- and $Z$-directions.

## Salt-loading experiments

In order to obtain detailed information about weathering mechanisms, simultaneous length change

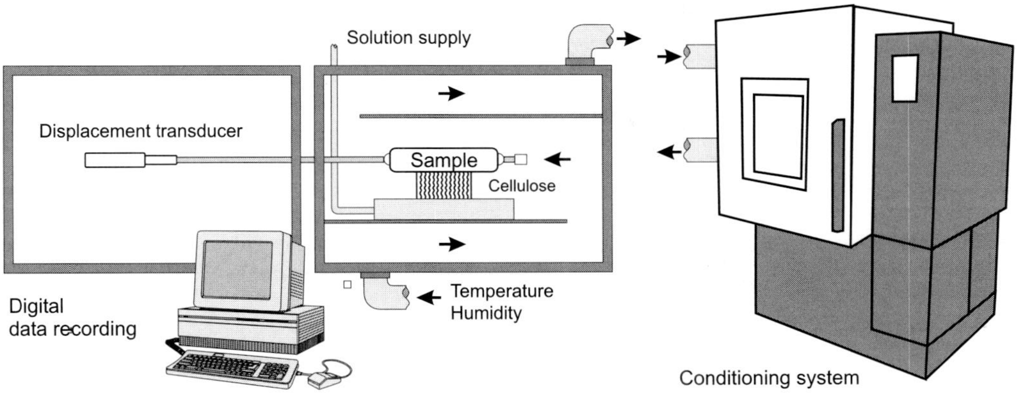


**Fig. 1.** Schematic diagram of the dilatometer set-up with climatic control for the measurement of length change behaviour of porous solids on salt loading.

measurements were performed during salt-loading cycles. For the investigations two salt types: (i) sodium sulphate and (ii) sodium chloride were used.

The measurements were carried out with a specially modified dilatometer (Fig. 1), which allows independent control of temperature and humidity as well as of the solution supply. Moreover, six samples can be analysed simultaneously in one experiment. The length change behaviour of the sandstones was determined on cylindrical samples ($\phi$ 15 $\times$ 50 mm) in two fabric directions: parallel (*X*-direction) and perpendicular (*Z*-direction) to the bedding. The accuracy of the incremental displacement transducers used is 1.0 $\mu$m. For the experiments, a 10% sodium sulphate and a 10% sodium chloride solution was used, respectively. The solution absorption of the samples was obtained over a 4 h period using a capillary uptake through several layers of cellulose. Following this, the samples were dried for over 16 h. The evaporation of the water and, therefore, the supersaturation and crystallization were achieved by a constant low relative humidity of 30% at 20 °C ambient temperature. This approach was chosen to avoid the influence of temperature on the length change and, thus, another possible deterioration mechanism.

From the sodium sulphate solution, two crystal phases can develop. The water-free phase, thenardite ($Na_2SO_4$), and the hydrate phase, mirabilite ($Na_2SO_4 \cdot 10H_2O$). The phase transition from thenardite to mirabilite at water absorption is associated with a volume increase of about 300% (Price & Brimblecombe 1994). This hydration process takes place in a climate range, which is often found in nature (Steiger *et al.* 1998). The conversion at 20 °C takes place at 75% relative humidity, whereas thenardite is stable in the lower, and mirabilite in the higher, humidity range (Fig. 2a). The hydration process takes place over a short period. In contrast, sodium chloride only crystallizes out from a solution above 0 °C (Fig. 2b) in the non-hydrating phase, halite (NaCl).

## Rock fabrics

The rock fabric of the investigated sandstones, which includes size, sorting, roundness of detrital grains as well as grain contacts and cement properties, play a significant role in the changes in petrophysical properties and material behaviour during weathering. The fabric is the result of a complex formation process during sedimentation, compaction, diagenesis and alteration. The investigated sandstones represent diverse lithotypes and are characterized by a variation of fabric elements. A compilation of important parameters as well as petrographic features is given in Table 1. All sandstones show detrital quartz grains as the main mineralogical content with a varying portion of feldspar and clay.

### *Bad Bentheim Sandstone*

The Cretaceous sandstone from Bad Bentheim displays macroscopically an ochre colour, which is caused by a low, but dispersed distribution of iron oxide and hydroxide, for example limonite. The sandstone represents pure quartz sandstone of a coastal deposit and is well sorted. In addition to quartz, clastic feldspar is occasionally observable. The sandstone is well sorted, with the dominant size of clastic grains ranging between 100 and 200 $\mu$m. The grain shapes are more or less rounded. Authigenic kaolinite plates are found at the surfaces

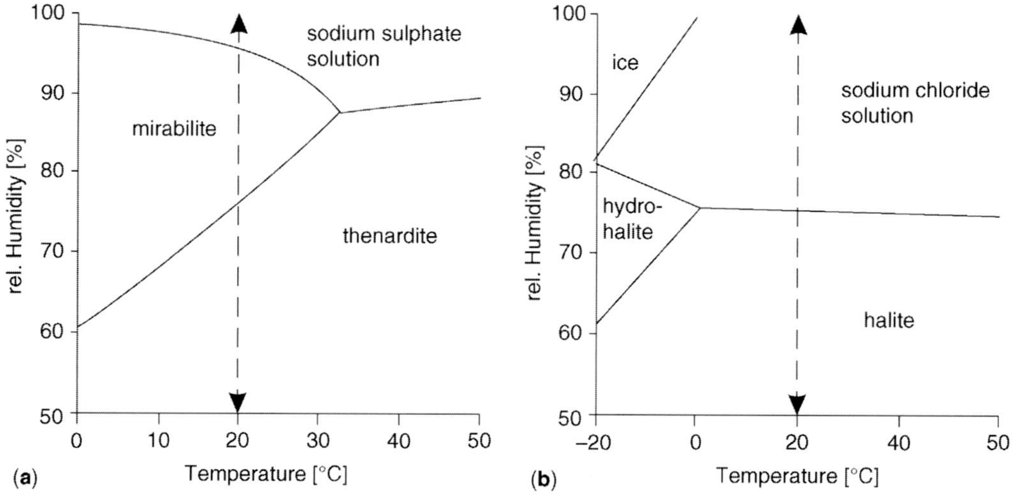


**Fig. 2.** Temperature and relative humidity stability ranges of the salt types used: (**a**) sodium sulphate and (**b**) sodium chloride (the dashed arrows mark the temperature and humidity range realized in the salt-loading experiments).

of clastic grains. The cohesion of the sandstone is mainly caused by quartz cements that appear as syntaxial overgrowth on detrital quartz grains. The contacts of clastic grains indicate only low-pressure solution, and are characterized by predominantly flat and subordinate concave–convex contacts.

### *Cotta Sandstone*

This sandstone is characterized macroscopically by a yellow–ochre colour and a clouded fabric caused by rich, dark organic layers. The sandstone is the result of a coastal deposit during the Cretaceous period. The detrital component is dominated by quartz while feldspar is uncommon. The dominant grain size of the detrital components is about 80–250 μm. Therefore, the sandstone can be regarded as moderately sorted. Dominantly angular grain shapes can be observed. In contrast to the other investigated samples, the clastic quartz grains often show oval shapes with a preferred orientation of long axis parallel to the bedding. Further, the Cotta Sandstone exhibits a large amount of authigenic kaolinite as plates covering clastic quartz grains. In addition, quartz cements as syntaxial overgrowth on clastic grains are common.

### *Schoetmar Sandstone*

The Triassic Sandstone from Schoetmar displays a light green–brownish colour. It represents a deposit from a braided river system. Macroscopically, no bedding is detectable in the investigated sample, thus the sandstone seems to be more or less homogeneous. The clastic composition is dominated by quartz grains, but also by a high content of feldspar and clay-rich lithoclasts. The major grain size of the detrital components ranges between

**Table 1.** *Qualitative fabric properties of the investigated sandstone types (Qz, quartz; Fsp, feldspar; Mc, mica)*

| Sandstone type | Detritus | | | | | Authigen | |
|---|---|---|---|---|---|---|---|
| | Dominant grain size (μm) | Sorting | Roundness | Main phases | Preferred grain-shape orientation | Clay content | Cements |
| Bad Bentheim | 100–200 | Well | Well rounded | Qz, Fsp | None | Very low | Qz |
| Cotta | 80–150 | Moderate | Rounded | Qz, Fsp, Mc, Clay | Pronounced parallel to bedding | High | Qz, Clay |
| Schoetmar | 100–200 | Moderate | Angular | Qz, Fsp, Mc, Clay | Slight parallel to bedding | High | Clay, Qz |

**Table 2.** *Petrophysical properties of the investigated sandstone types*

| Sandstone type | Porosity (vol.%) | Average pore radius (μm) | Frequency of pores of specific pore radii ranges (%) | | | | |
|---|---|---|---|---|---|---|---|
| | | | 0.001–0.01 (μm) | 0.01–0.1 (μm) | 0.1–1 (μm) | 1–10 (μm) | >10 (μm) |
| Bad Bentheim | 24.8 | 9.15 | 2.8 | 3.2 | 2.7 | 10.3 | 81.1 |
| Cotta | 25.7 | 1.73 | 2.8 | 6.9 | 23.9 | 54.7 | 11.7 |
| Schoetmar | 10.3 | 0.07 | 20.2 | 38.1 | 32.3 | 5.4 | 4.0 |

100 and 200 μm. They show angular grain shapes and are moderately sorted. Prolate grain shapes of the clastic quartz and feldspar components are uncommon, whereas a slight preferred shape orientation is observable parallel to the *XY*-plane. The cementation is dominated by clay minerals and subordinate quartz cements as syntaxial overgrowth of clastic quartz grains.

## Selected petrophysical properties

The petrophysical properties of rocks were controlled by their mineralogical composition and the fabric. The petrophysical properties show conspicuous differences owing to the fabric variability of the investigated rocks. For resistance against salt weathering, the pore-space properties, i.e. porosity and pore-size distribution, and the tensile strength of sandstones are of particular importance (Fitzner & Snethlage 1982; Ruedrich *et al.* 2005). The significance of the porosity and the pore-size distribution is based on the fact that they control the solution transport properties of the rock (Snethlage & Wendler 1997). Another important constraint on salt weathering is the resistance against tensile stresses. The stresses induced by salt growth have to exceed the tensile strength before damage can occur. For length change measurements, knowledge of the hygric expansion of the rock material is also required, as it affects the length change induced by salt growth (Snethlage & Wendler 1997). A compilation of the measured petrophysical parameters is shown in Tables 2 and 3.

The investigated samples show differences in the total accessible porosity. The Bad Bentheim and the Cotta Sandstone represent high-porosity materials with values of around 24.8 and 25.7 vol.%, respectively, whereas the sandstone of Schoetmar yields only 10.3 vol.%.

The pore-size distribution of sandstones depends on the rock fabric and is mainly determined by the size of detrital grains and the clay content, as well as diagenetic compaction. The investigated samples exhibit different patterns of pore-size distribution (Fig. 3). The Bad Bentheim Sandstone shows a narrow spaced pore-radii maximum, and is therefore more or less equally porous. The pore-radii maximum is between 10.000 and 25.118 μm. The average pore radius is 9.150 μm. In contrast, the sandstone from Cotta shows a bimodal pore-radii distribution with a maximum from 3.981 to 10.000 μm, and a submaximum between 0.158 and 0.398 μm. The average pore radius is 1.73 μm, and thus between both maxima. The pore-space distribution of the Schoetmar Sandstone strongly differs from the pattern of the other samples, and covers a wide range between micropores and small capillary pores (0.004–0.398 μm). The average pore radius is 0.070 μm, which is significantly smaller than for the other samples.

The tensile strength varies for the different sandstones types between 2.65 and 5.35 N mm$^{-2}$. For both high porous sandstones, the tensile strength is very low at 2.65 and 2.75 N mm$^{-2}$ for the Bad Bentheim, and 2.68 and 3.16 N mm$^{-2}$ for the Cotta sample. In contrast, the sandstone from

**Table 3.** *Data of the tensile strength and hygric dilatation (anisotropy calculated by* $A = (\varepsilon_{max} - \varepsilon_{min})/\varepsilon_{max} \times 100$*)*

| Sandstone type | Tensile strength ($\sigma_z$) | | | Hygric dilatation | |
|---|---|---|---|---|---|
| | Perpendicular to *XY*-plane (N mm$^{-2}$) | Perpendicular to *XZ*-plane (N mm$^{-2}$) | A (%) | Parallel to *X* (mm m$^{-1}$) | Parallel to *Z* (mm m$^{-1}$) |
| Bad Bentheim | 2.65 ± 0.32 | 2.75 ± 0.17 | 3.6 | 0.00 | 0.02 |
| Cotta | 2.68 ± 0.56 | 3.16 ± 0.46 | 15.2 | 0.07 | 0.13 |
| Schoetmar | 5.31 ± 0.20 | 5.35 ± 0.77 | 0.7 | 0.79 | 1.13 |

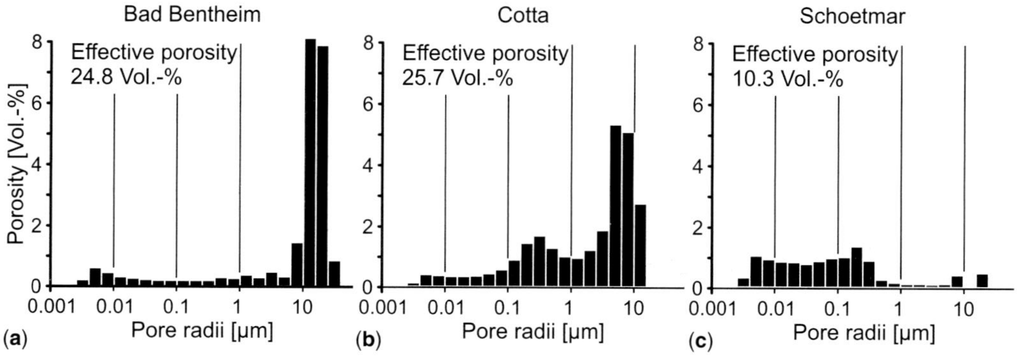


**Fig. 3.** Pore-radii distribution and porosity of the investigated sandstones: (**a**) Bad Bentheim, (**b**) Cotta and (**c**) Schoetmar Sandstone.

Schoetmar exhibits a higher tensile strength varying from 5.31 to 5.35 N $mm^{-2}$. This can be attributed to the relatively low porosity and, thus, better cohesion of the material. A conspicuous anisotropy of the tensile strength can only be determined for the Cotta Sandstone at 15.2%, which is certainly caused by the preferred shape orientation of detrital grains.

The hygric expansion ($\varepsilon$) of the samples varies strongly, which is caused by the different contents of swelling clay minerals. The samples from Bad Bentheim and Cotta show only low expansion values of around 0.20 mm $m^{-1}$ during water absorption. The sandstone from Schoetmar has a distinct hygric expansion between 0.79 mm $m^{-1}$ parallel to and 1.13 mm $m^{-1}$ perpendicular to the bedding. Thus, the hygric dilatation of the Schoetmar sample also shows a strong directional dependence, which is attributed to the preferred orientation of clay minerals. After the initial moisture content is reached, all samples return to their initial length, leaving no residual strain.

## Length change behaviour during salt loading

During salt loading, the length change behaviour varies strongly depending on the sandstone type, its rock fabric and the salt type. The results for the different salt and stone types are discussed below.

### *Sodium sulphate loading*

During the absorption stage in the first loading cycles, the Schoetmar Sandstone shows an obvious hygric expansion in the presence of the sodium sulphate solution (Fig. 4a, cf. Table 3). After drying, a residual strain remains, probably caused by the incorporation of sodium ions in the interlayers of the clay minerals. This length change behaviour occurs up to the fifth loading cycle. Starting with the ninth cycle, the dilatation shows a strong increase during solution uptake. Also, a strong increase of the residual strain is observable, which results in a large material loss. After the 13th cycle, the cylindrical samples of this specimen are more or less completely deteriorated.

The length change of the sandstone from Cotta is characterized by differing behaviour in the two sample directions during sodium sulphate solution loading (Fig. 4b). Up to the eighth cycle the sample that is oriented perpendicular to the bedding plane shows a slight expansion during the solution uptake and a contraction in the drying stage. This results in a continuous slight contraction of the specimens. After the eighth cycle a larger expansion is observable, which results in an obvious residual strain. After the 12th cycle this sample was deteriorated. The sample parallel to the bedding is characterized by a very slight expansion during the absorption stage and a slightly stronger contraction after drying. This behaviour leads to a continuous shortening of the samples.

The Bad Bentheim Sandstone shows increasing contraction during both the solution uptake and the drying stage, although only a slight length change for each cycle is observable (Fig. 4c). At the beginning of the solution uptake, a contraction occurs followed by a small expansion during further cycles. In contrast, a continuous contraction is observable during the drying phase. This behaviour is observed for both sample directions during the first 13 loading cycles, whereas significant decay phenomenon are lacking. Remarkably, a total contraction up to 1 mm $m^{-1}$ was observed. At this time, the cause of the pronounced material shortening remains unknown.

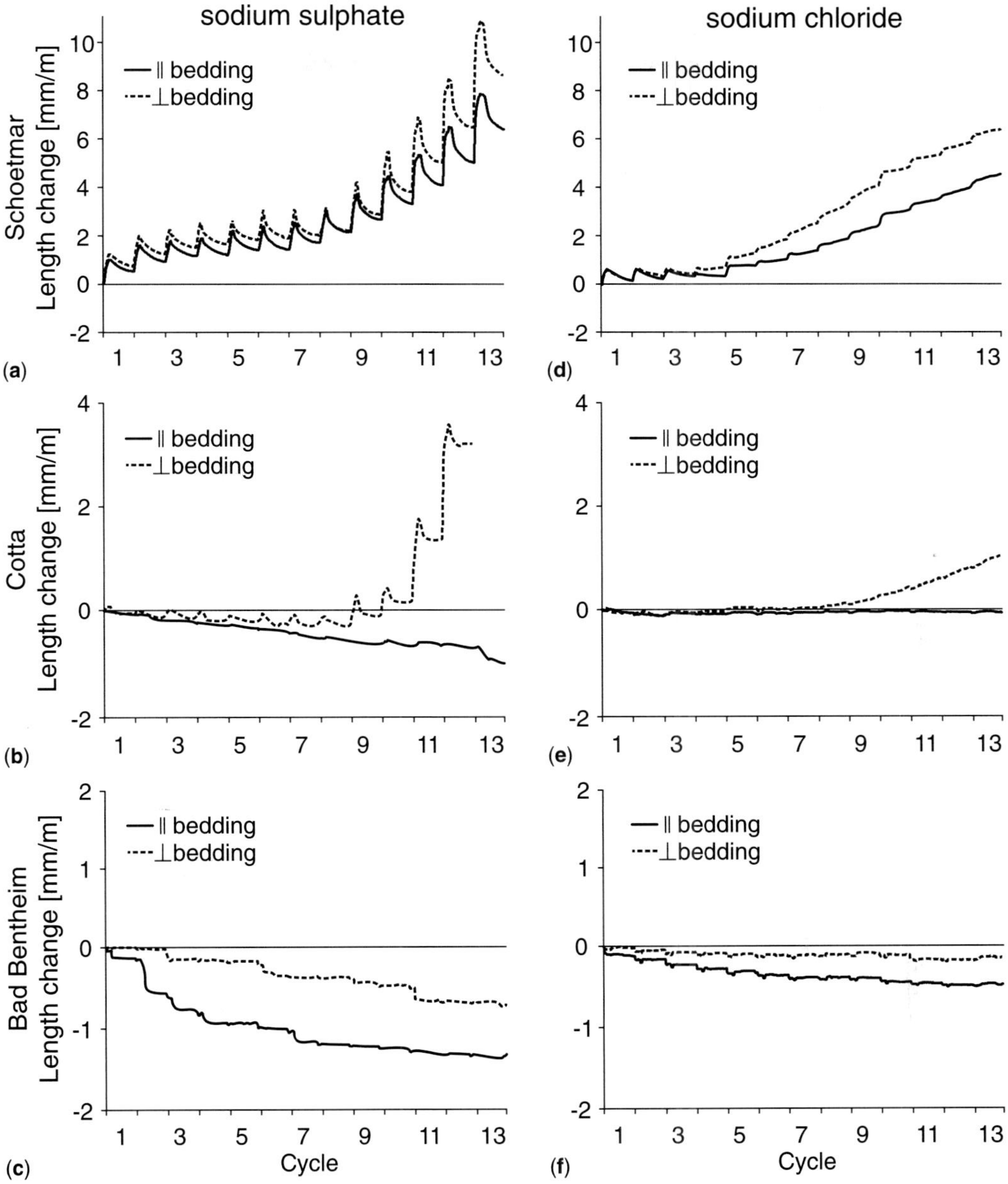


**Fig. 4.** Length change behaviour of sandstone samples parallel and perpendicular to bedding during salt tests with sodium sulphate and sodium chloride for 13 loading cycles (for explanations see the text).

## *Sodium chloride loading*

Up to the third loading cycle with sodium chloride, the Schoetmar Sandstone shows a normal hygric expansion with a slight residual strain after drying (Fig. 4d). The behaviour in the following 10 cycles is characterized by an expansion, both during the solution uptake and drying stage. This is caused a pronounced residual strain.

The sample from Cotta shows a strong directional dependence of dilatation during cyclic loading with sodium chloride (Fig. 4e). Up to the seventh cycle, no significant length change occurs. This behaviour was observed in the sample parallel to the bedding up to the 13th cycle. In contrast, the specimen oriented parallel to the *Z*-direction exhibits a pronounced expansion only in the drying stage. Thus, after the experiments, the rock sample shows an obvious residual strain perpendicular to the bedding.

The Bad Bentheim sample exhibits a slight contraction for all 13 loading cycles with sodium chloride (Fig. 4f). This length change occurs during the wetting as well as during the drying stage and is more or less comparable to the sodium sulphate test.

## Discussion

The pore-space of sandstones is a major determinant for salt weathering, as it represents the hollow space in which crystallization processes take place. The main pore-space properties are the effective porosity, the pore-space distribution, the pore geometries and the pore interconnection. For sandstones, these elements are controlled by the original deposited clastic material and any diagenetic evolution (e.g. compaction and cementation).

High porosity should permit much salt crystallization, resulting in more stresses against the porous solids. This means that a high porosity should be a critical fabric element concerning salt attack. In fact, salt crystallization tests show that very often highly porous materials are more sensitive to salt attack than low porosity sandstones (Ruedrich *et al.* 2005). However, our new data show that the lower porosity Schoetmar Sandstone is more sensitive to salt attack. Consequently, other effects or critical fabric elements must also have a significant influence on the salt weathering.

Several scientific investigations clearly document that a large number of micropores adjacent to capillary pores results in a high damage potential (e.g. Fitzner & Snethlage 1982). For example, Putnis & Mauthe (2001) found that crystal growth preferably occurs in larger pores. According to the thermodynamic model from Wellmann & Wilson (1965), the residual solution in smaller pores represents a solution reservoir for the crystal growth in larger pores. Thus, materials that are characterized by a bimodal pore-size distribution or by a submaximum in the smaller pore ranges are very sensitive to salt weathering. This hypothesis is supported by the present data, as a result the sandstones from Schoetmar and Cotta with bimodal pore-space distributions are more strongly affected by salt loading than the Bad Bentheim Sandstone.

Tensile strength could be used as an expression of the resistivity of a solid material against stresses induced by salt crystallization in its pore space. A correlation between tensile strength and porosity is presented by Ruedrich *et al.* (2005). Low-porosity Sandstones show high tensile strength and vice versa. The investigated samples correspond to the reported correlations. However, the tensile strength of a material does not seem to be the key parameter in predicting the effect of salt weathering, which was clearly shown in the case of the Schoetmar Sandstone.

Hygric dilatation is important for two reasons. First, it affects the length change and, secondly, it may control damage. In the first case, hygric dilatation caused by the swelling of clay minerals, decreases with increasing salt loading. For the first cycles, the residual strain is most probably induced by the incorporation of the salt cations in the intermediate layers of swelling clay minerals. The degree of the hygric expansion depends on the salt types used. The hygric expansion and residual strain is high for sodium sulphate loading and lower for sodium chloride (Fig. 5a, c). For example, Snethlage & Wendler (1997) suggested that hygric expansion in combination with salt loading mainly controls the decay of clay-containing sandstones. Owing to swelling of the clay minerals, the pore spaces increase and could be filled by salts. During drying, the fabric cannot return to its initial position, which results in a residual strain. However, the results of sodium chloride loading give evidence that, while drying, a conspicuous expansion is observable. Thus, a stress development during crystallization takes place.

The length change data for the different rocks show varying responses to salt loading. While the sandstone from Schoetmar exhibits a strong residual strain after the cyclic loading, the Bad Bentheim Sandstone is more or less unaffected. In contrast to the Schoetmar Sandstone, the Bad Bentheim Sandstone even shows a contraction. A final explanation for the pronounced and residual contraction remains unknown. However, a stress-induced fabric collapse resulting from salt crystallization seems questionable, as the samples show only slight granular disintegration at the surface. Therefore, the contraction of the samples indicates that tensile stresses occur within the fabric.

The salt weathering of the Cotta Sandstone shows a strong directional dependence. For the sample oriented perpendicular to the bedding, the dilatation behaviour is comparable to the Schoetmar Sandstone. For the sample oriented parallel to the $X$-direction, the Cotta Sandstone shows a continuous contraction and is therefore comparable with the length change behaviour of the Bad Bentheim Sandstone. This observed anisotropic behaviour can most probably be attributed to the shape preferred orientation of the clastic quartz grains, which also produces a shape preferred orientation of the pore geometry.

The sandstones and also the sample directions show a specific damage potential independent of the salt type. It can be observed that the Schoetmar

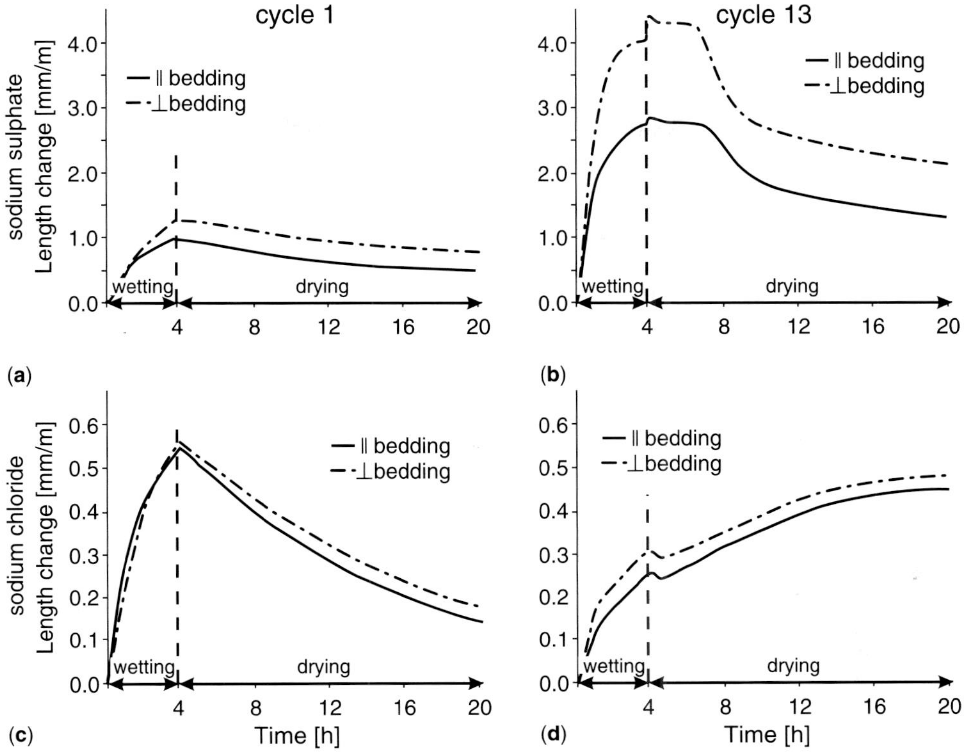


**Fig. 5.** Length change behaviour of the Schoetmar Sandstone during salt tests with sodium sulphate and sodium chloride in the first and 13th loading cycle (for explanations see the text).

Sandstone exhibits progressive deterioration for both sodium sulphate and sodium chloride. The highest residual strain is observed perpendicular to the bedding for both salt types. Consequently, the lowest deterioration occurs parallel to the *X*-direction. The Bad Bentheim Sandstone shows a distinct contraction in each case. This is also observable for the strong anisotropic expansion behaviour of the Cotta Sandstone. For both salt types, a significant residual strain is evident in the *Z*-direction, whereas parallel to the bedding the contraction is less pronounced. However, for both salt types, the amount of residual strain after cyclic loading is different. The sodium sulphate loading generally produces the most significant length changes (expansion and contraction).

The length change behaviour, especially in subsequent loading cycles, indicates that the deterioration process must be different for both salt phases. This is shown for the 13th cycle in Figure 5. The differences are best observed for both sample directions of the Schoetmar Sandstone. For the solution uptake stage of sodium sulphate, a strong dilatation occurs that is more than three times higher than in the initial cycle, and, thus, than the original hygric expansion. The strong expansion is certainly controlled by hydration of the water-free salt phase, thenardite. At the beginning of the drying phase, a further expansion can be observed that is possibly traced back to the crystallization of mirabilite. After this dilatation a pronounced contraction of the samples occurs and indicates that the water-free sodium sulphate phase, thenardite, is developed.

The sodium chloride loaded samples from Schoetmar show a distinctly different length change behaviour, which is also best seen in the 13th cycle (Fig. 5d). While the expansion in the wetting stage is much lower than in the first cycle, a pronounced expansion occurs after a slight contraction at the beginning of the drying stage, resulting in a strong residual strain. This indicates that no hydration was generated during solution uptake, and that dilatation in the drying stage results from a crystal growth of halite.

## Conclusions

The length changes produced by cyclic salt loading with sodium sulphate and sodium chloride for the

Bad Bentheim, the Cotta and the Schoetmar Sandstones allow the following conclusions.

- Different behaviour of the samples upon salt loading is determined by rock properties.
- For some rocks, dilatation is strongly directional, which is further evidence for fabric-controlled decay processes.
- The geometry of pores may significantly influence deterioration.
- The main damage mechanism for sodium sulphate seems to be the development of hydration pressures.
- For the sodium chloride loading, a conspicuous expansion occurs during halite crystallization.
- Although the salt types used show different damage mechanisms, the sandstones exhibit specific sensitivities independent of salt type.
- Both salt types can induce the contraction of a sample.

The investigations show that length change measurements during salt loading are a very helpful tool in understanding weathering processes. However, to obtain more information about the fabric and the salt dependence, further investigations with varying stone types and more salt types must be undertaken.

Thanks go to the reviewers for their comments. Our work was supported by the Deutsche Bundesstiftung Umwelt and the Deutsche Forschungsgemeinschaft (Si 438/17-1).

## References

BRAKEL, J. VAN MODRY, S. & SVATA, M. 1981. Mercury porosimetry: State of the art. *Powder technology*, **29**, 1–12.

CHAROLA, A. E. & WEBER, J. 1992. The hydration/dehydration mechanisms of sodium sulphate. *In*: DELGADO RODRIGUES, J., HENRIQUES, F. & JERIMISAS, F. T. (eds) *Seventh International Congress on the Deterioration and Conservation of Stone, Lisbon*, 581–590.

CHAROLA, A. E. 2000. Salts in the deterioration of porous materials: An overview. *Journal of the American Institute for Conservation*, **39**. World Wide Web Address: http://aic.stanford.edu/jaic/articles/jaic39-03-002_indx.html.

CHATTERJI, S., CHRISTENSEN, P. & OVERGAARD, G. 1979. Mechanisms of breakdown of natural stones caused by sodium salts. *In*: BADAN, B. (ed.) *Third International Congress on the Deterioration and Conservation of Stone, Padova*, 131–134.

CORRENS, C. W. 1949. Growth and dissolution of crystals under linear pressure. *Discussions of the Faraday Society*, **5**, 267–71.

CORRENS, C. W. & STEINBORN, W. 1939. Über die Erklärung der sogenannten Kristallisationskraft. *Zeitschrift für Kristallographie*, **101**, 117–133.

DARWIN, C. R. 1839. *Journal of Researches into the Natural History and Geology of the Countries Visited During the Voyage of HMS Beagle Round the World*. D. Appleton, New York.

DOEHNE, E. 1994. In situ dynamics of sodium sulfate hydration and dehydration in stone pores: Observations at high magnification using the environmental scanning electron microscope. *In*: FASSINA, V., OTT, H. & ZEZZA, F. (eds) *The Conservation of Monuments in the Mediterrane Basin, Venice*, 143–150.

DOEHNE, E. 2002. Salt weathering: a selective review. *In*: SIEGESMUND, S., WEISS, T. & VOLLBRECHT, A. (eds) *Natural Stone, Weathering Phenomena, Conservation Strategies and Case Studies*. Geological Society, London, Special Publications, **205**, 43–56.

DUTTLINGER, W. & KNÖFEL, D. 1993. Salzkristallisation und Salzschadensmechanismen. *In*: *Jahresbericht Steinzerfall - Steinkonservierung 1991*. Ernst & Sohn, Berlin, 197–213.

FITZNER, B. 1969. Die Prüfung der Frostbeständigkeit von Naturbausteinen. *Geologische Mitteilungen*, **10**, 205–296.

FITZNER, B. & SNETHLAGE, R. 1982. Einfluß der Porenradienverteilung auf das Verwitterungsverhalten ausgewählter Sandsteine. *Bautenschutz und Bausanierung*, **3-1982**, 97–103.

KIRCHNER, D. & WORCH, A. 1993. Physikalische Vorgänge bei der Salzkristallisation. *Bautenschutz und Bausanierung*, **16**, 101–103.

MORTENSEN, H., 1933. Die Salzsprengung und ihre Bedeutung für die regionalklimatische Gliederung der Wüsten. *Petermann's Mitteilungen aus Justus Perthes geographischer Anstalt*, **79**, 130–135.

PRICE, C. & BRIMBLECOMBE, P. 1994. Preventing salt damage in porous materials. *In*: ASHOK, R. & SMITH, P. (eds) *Prepr. Contr. Ottawa Congr. Preventive Conservation – Practice, Theory and Research*, **IIC**, 90–93.

PÜHRINGER, J. 1983. *Salt Disintegration: Salt Migration and Degradation by Salt – A Hypothesis*. Swedish Council for Building Research, Stockholm, **D15**.

PUTNIS, A. & MAUTHE, G. 2001. The effect of pore size on cementation in porous rocks. *Geofluids*, **1**, 37–41.

RODRIGUEZ-NAVARRO, C. & DOEHNE, E. 1999. Salt weathering: influence of evaporation rate, supersaturation and crystallization pattern. *Earth Surface Processes and Landforms*, **24**, 191–209.

ROSSI-MANARESI, R. & TUCCI, A. 1991. Pore structure and the disruptive or cementing effect of salt crystallization in various types of stone. *Studies in Conservation*, **36**, 53–58.

RUEDRICH, J., KIRCHNER, D., SEIDEL, M. & SIEGESMUND, S. 2005. Deterioration of natural building stones induced by salt and ice crystallization in the pore space as well as hygric expansion processes. *In*: SIEGESMUND, S., AURAS, M., RUEDRICH, J. & SNETHLAGE, R. (eds) *Geowissenschaften und Denkmalpfleg*. Zeitschrift Deutsche Geologische Gesellschaft, **156**(1), 59–73.

SCHERER, G. W. 1999. Crystallization in pores. *Cement and Concrete Research*, **29**, 1347–1358.

SKINNER, B. J. 1966. Thermal expansion. *In*: CLARK, S. P. (ed.) *Handbook of Physical Constants*. Geological Society of America, **97**, 75–96.

SNETHLAGE, R. 1984. *Steinkonservierung*. Bayerisches Landesamt für Denkmalpflege, Arbeitshefte, **22**.

SNETHLAGE, R. & WENDLER, E. 1997. Moisture cycles and sandstone degradation. *In*: BAER, N. S. & SNETHLAGE, R. (eds) *Saving our Architectural Heritage, The Conservation of Historic Stone Structures*. Wiley, Chichester, 7–24.

SPERLING, C. H. B. & COOKE, R. U. 1980. *Salt Weathering in Arid Environments. I. Theoretical Considerations*. Bedford College Papers in Geography, **9**.

STEIGER, M. 2005. Crystal growth in porous materials – I: The crystallization pressure of large crystals. *Journal of Crystal Growth*, **282**, 455–469.

STEIGER, M., NEUMANN, H.-H., GRODTEN, T., WITTENBURG, C. & DANNECKER, W. 1998. Salze in Natursteinmauerwerk: Probenahme, Messung und Interpretation. *In*: SNETHLAGE, R. (ed.) *Natursteinkonservierung 2*. Fraunhofer IRB, Stuttgart, 61–91.

SUNAGAWA, I. 1981. Characteriztics of crystal growth in nature as seen from the morphology of mineral crystals. *Bulletin Mineralogie*, **104**, 81–87.

TABER, S. 1916. The growth of crystals under external pressure. *American Journal of Science*, **41**, 532–556.

WELLMAN, H. W. & WILSON, A. T. 1965. Salt weathering, neglected geological erosive agent in coastal and arid environments. *Nature*, **205**, 1097–1098.

WELLMAN, H. W. & WILSON, A. T. 1968. Salt weathering or fretting. *In*: FAIRBRIDGE, R. W. (ed.) *The Encyclopedia of Geomorphology*. Reinhold Book Corporation, Stroudsburg, PA.

WINKLER, E. M. & WILHELM, E. J. 1970. Salt burst by hydration pressures in architectural stone in urban atmosphere. *Bulletin of the Geological Society of America*, **81**, 567–572.

WINKLER, E. M. 1975. *Stone: Properties, Durability in Man's Environment*, 2nd edn. Springer, New York.

WINKLER, E. M. 1994. *Stone in Architecture*, 3rd edn. Springer, Berlin.

ZEHNDER, K. & ARNOLD, A. 1989. Crystal growth in salt efflorescence. *Journal of Crystal Growth*, **97**, 513–521.

# Complex weathering effects on durability characteristics of building stone

P. A. WARKE & B. J. SMITH

*School of Geography, Archaeology and Palaeoecology, Queen's University Belfast, Belfast BT7 1NN, Northern Ireland, UK (e-mail: p.warke@qub.ac.uk)*

**Abstract:** Durability characteristics of five stone types are assessed and compared using the standardized sodium sulphate salt crystallization test and a modified laboratory weathering simulation in which a combination of salt weathering ($Na_2SO_4$) and freeze–thaw cycles are used. Data indicate significant differences in durability rankings between the two test methods especially in lower-order durability stone types. Both the standard salt crystallization test and the modified durability test identify Leinster Granite and Stanton Moor B Sandstone as the most durable of the five stone types, with the granite performing well under both sets of conditions. Discrepancy between rankings arises in the lower orders, with Portland Limestone, Stanton Moor A Sandstone and especially Dumfries Sandstone responding differently to the two sets of experimental conditions. In the modified durability test the range of permeability values for each stone type produced the same ranking as that indicated by mean percentage weight change values but mean permeability values for each stone type do not appear to be reliable predictors of weathering response. Differences in durability rankings between the two test regimes are attributed in the first instance to the temperature conditions used, with more extreme and unrealistic heating to 103 °C in the standardized test 'over-weathering' stone while conditions in the modified test allowed the development of stone-specific decay characteristics. Inclusion of salt weathering and freeze–thaw cycles in the modified test introduced complexity into the decay process that more accurately reflects 'real-world' conditions. Data also indicate that relatively minor structural and mineralogical differences between samples of the same stone type can significantly influence weathering behaviour, resulting in distinct rates and patterns of breakdown.

Weathering of building stone involves an often complex progression from 'fresh' to 'failed' stone – a progression that typically proceeds episodically with intervening periods of apparent quiescence. Prediction of stone response to weathering relies, for the most part, on standardized durability tests that confer a fixed assessment of expected durability that, in turn, informs choice of stone for use on particular parts of a building (e.g. Building Research Establishment 1989; Yates & Butlin 1996). Unfortunately, standardized durability tests, such as the sodium sulphate test, only register the two end extremes in the progression from 'fresh' to 'failed' stone with blocks inserted as fresh samples and revisited on disintegration. Such standardized tests tend to assess durability through exposure to a single weathering process, which is an unrealistic representation of the weathering system where rarely, if ever, do processes operate in isolation and where complex interactions and synergistic relationships between processes can enhance overall weathering effectiveness. In addition, because of the comparatively extreme experimental conditions used in, for example, the salt crystallization test, the reliability of resultant data may be in question especially where durability status is not clearly defined. It has been noted that temperature is one of the most significant factors in determining the efficacy of salt weathering, with both the extent and nature of damage being more severe when drying temperatures exceeding 100 °C are used (McGreevy & Smith 1982; Davison 1986).

Whilst recognizing the need for some means of assessing the potential weathering response of stone, it is clear that standard durability tests have their limitations and are not suitable for identifying temporal and spatial subtleties of weathering-related stone decay and, consequently, can fail to accurately predict stone response to weathering under complex 'real-world' conditions where structural and mineralogical properties can change during the exposure lifetime of stone (Smith & Kennedy 1999). Therefore, a more complex testing method is suggested in which samples are exposed to the combined effects of weathering processes operating within environmental parameters that more accurately reflect actual conditions (Warke *et al.* 2006).

However, in proposing a modified durability test, it is not intended to detract in any way from the obvious value of standard tests that are, by necessity, designed to be quick and simple to perform. Instead, it is suggested that in particular instances

*From*: PŘIKRYL, R. & SMITH, B. J. (eds) *Building Stone Decay: From Diagnosis to Conservation.* Geological Society, London, Special Publications, **271**, 211–224.
0305-8719/07/$15.00 

it may be necessary to be able to predict with more accuracy expected long-term weathering changes. This may be especially relevant for replacement stone or stone that is to be used for decorative features or some larger aesthetically significant element of a building façade where differences in weathering response may result in loss of architectural detail and detract from the general appearance of a structure.

This project aims to improve the understanding of stone response to complex weathering, and to demonstrate how the introduction of more realistic and representative testing parameters may improve our ability to more accurately predict the weathering behaviour of stone under 'real-world' conditions. To achieve this, three project objectives were set.

1. Assessment of stone durability under complex weathering conditions in which cycles of high-frequency, low-magnitude salt weathering are combined with less frequent but higher-magnitude freeze–thaw cycles under controlled laboratory conditions.
2. Comparison of durability status of different rock types as defined by the standard salt crystallization test with performance under the modified durability testing procedure outlined in Objective 1.
3. Modelling decay dynamics of different rock types using data from systematic analysis of samples before, during and after exposure to modified and standard durability tests.

## Methodology

### *Materials*

Stone types were selected on the basis of differences in their structural and mineralogical properties (Table 1), their perceived durability characteristics and the fact that they are representative of stone commonly used in construction or as part of conservation programmes. In addition to Dumfries Sandstone, Portland Limestone and Leinster Granite, two types of Stanton Moor Sandstone were identified on the basis of differences in grain-size characteristics within the original bulk sample (Stanton Moor A and Stanton Moor B) – differences that reflect conditions in the original fluvial depositional environment.

### *Modified durability test: experimental procedure*

Sixty-six 75 $mm^3$ blocks from each of the five stone types selected were cut, washed and air-dried. Each 'master' group of 66 blocks was subdivided into 11 subsets, each comprising six blocks with two of these identified as control samples. Because of space restrictions in the environmental chamber, only two stone types at a time could be run through the experimental regime. Each block was placed on a separate tray inside the chamber to collect any debris released. The experimental regime comprised a total of 220 daily weathering cycles, with the weathering 'history' accumulated by each of the 11 sample subsets spanning a range of weathering combinations from subset 1 (with exposure to 220 salt weathering cycles) to subset 11 (which experienced a total of 200 salt weathering and 20 freeze–thaw cycles) (Table 2).

*Detail of salt weathering cycles.* Four blocks from each of the 11 subsets were immersed daily in a 2.5% solution of $Na_2SO_4$ for approximately 20 s. The blocks were immersed on a fine mesh frame that trapped any debris released. Released debris was collected, washed, dried and weighed. Each daily experimental run lasted 20 h and comprised two consecutive weathering cycles, each of 10 h in duration. The first 10 h cycle was a 'wet' cycle because the blocks entered it wet from immersion in the salt solution, while the second 10 h cycle, which followed the first without interruption, was a 'dry' cycle. This combination of 'wet' and 'dry' cycles more closely simulates 'real-world' conditions where stone on buildings often has time to dry between wetting events but still experiences temperature fluctuations.

Each 10 h salt weathering cycle comprised four temperature segments: 1 h during which temperature rose from +10 to +40 °C followed by 4 h at +40 °C then a staged temperature decrease over 1 h to +10 °C followed by a further 4 h at +10 °C. The succeeding second 'dry' cycle had the same temperature parameters. Relative humidity values within the environmental chamber were held at 30% (±5%). A 2.5% $Na_2SO_4$ solution was used for the salt weathering cycles because of the need to avoid conditions so extreme that subtleties of different stages in the decay sequences were lost due to overly accelerated breakdown. A pilot study demonstrated that 5% and 10% solutions of $Na_2SO_4$, especially when applied to sandstone, resulted in complete sample breakdown before a representative number of experimental cycles could be completed. $Na_2SO_4$ was chosen because it has been widely used in other weathering simulation studies (including industry standard durability tests) and therefore its use promotes comparability of data; the crystallization, hydration and thermal characteristics of this salt are well understood, and this salt can occur in the built environment and therefore data are of relevance to better understanding of stone weathering under 'real-world' conditions.

**Table 1.** *General structural and mineralogical properties of stone types used in this study*

| Stone type | Porosity (%) | Ultrasound pulse velocity ($m\ s^{-1}$)* | Permeability (mD)† | General description |
|---|---|---|---|---|
| Leinster Granite | 0–3 | Range: 2594–4023<br>Mean: 3088 | Range: 0.4–4.1<br>Mean: 1.75 | Tertiary feldspar-rich granite with interlocking crystalline structure. Feldspars comprise 42.5% (plagioclase 31.5%), quartz 26.7%, mica 19.7% and clays 11% |
| Portland Limestone | 13–26 | Range: 3914–4481<br>Mean: 4235 | Range: 1–150<br>Mean: 15 | Jurassic oolitic limestone with bimodal-pore size distribution (30–8 and 0.3–0.08 μm). Calcite comprises 57.8%, quartz 2.3% and clays 13.5%. Porosity can be extremely variable especially with regard to larger pores |
| Dumfries Sandstone | 18–25 | Range: 2243–2398<br>Mean: 2333 | Range: 200–1000<br>Mean: 600 | Permian quartz and iron-rich red sandstone with well-defined bedding and a unimodal poresize distribution of approximately 100 μm diameter. Quartz comprises 52%, feldspars 10.5%, clays 18% (smectites) and mica 1%. As well as occurring as isolated deposits, clays commonly form interstitial laminae |
| Stanton Moor A Sandstone (fine–medium grained) | 17 | Range: 2688–3690<br>Mean: 3100 | Range: 7.7–205.7<br>Mean: 61.0 | Carboniferous fine–medium grained (<300 μm) quartz-rich sandstone (Millstone Grit series) with well-developed interlocking quartz and feldspar overgrowths. Quartz comprises 62%, feldspars 14.25%, mica 0.75% and clays (chlorite) 11.25% |
| Stanton Moor B Sandstone (medium–coarse grained) | 13.5 | Range: 2997–3200<br>Mean: 3105 | Range: 4.75–114<br>Mean: 58.19 | Carboniferous medium–coarse grained (>300 μm) quartz-rich sandstone (Millstone Grit series) with well-developed interlocking quartz and feldspar overgrowths. Quartz comprises 62.25%, feldspars 21.5%, mica 2% and clays (chlorite) 7.5% |

* Ultrasound pulse velocity values were recorded for three opposing faces on each block and the average value recorded.
† Three measurements of permeability were taken from each block and the average value in milliDarcys (mD) recorded for each block.

**Table 2.** *Detail of salt (SW) and freeze–thaw (F–T) weathering cycle combinations for each subset sample group used in the modified durability test (Warke et al. 2006)*

| Sample subset | SW | F–T | SW | F–T | SW | F–T | SW | F–T | SW | F–T | SW | F–T | SW | F–T | SW | F–T | SW | F–T | SW | F–T | SW | Summary of weathering cycles programme for each sampling subset |
|---|---|---|---|---|---|---|---|---|---|---|---|---|---|---|---|---|---|---|---|---|---|---|
| 1 | 20 | – | 20 | – | 20 | – | 20 | – | 20 | – | 20 | – | 20 | – | 20 | – | 20 | – | 20 | – | 20 | 220 SW cycles only |
| 2 | 20 | – | 20 | – | 20 | – | 20 | – | 20 | – | 20 | – | 20 | – | 20 | – | 20 | – | 20 | 2 | 18 | 218 SW and 2 F–T cycles |
| 3 | 20 | – | 20 | – | 20 | – | 20 | – | 20 | – | 20 | – | 20 | – | 20 | – | 20 | 2 | 20 | 2 | 16 | 216 SW and 4 F–T cycles |
| 4 | 20 | – | 20 | – | 20 | – | 20 | – | 20 | – | 20 | – | 20 | – | 20 | 2 | 20 | 2 | 20 | 2 | 14 | 214 SW and 6 F–T cycles |
| 5 | 20 | – | 20 | – | 20 | – | 20 | – | 20 | – | 20 | – | 20 | 2 | 20 | 2 | 20 | 2 | 20 | 2 | 12 | 212 SW and 8 F–T cycles |
| 6 | 20 | – | 20 | – | 20 | – | 20 | – | 20 | – | 20 | 2 | 20 | 2 | 20 | 2 | 20 | 2 | 20 | 2 | 10 | 210 SW and one F–T cycles |
| 7 | 20 | – | 20 | – | 20 | – | 20 | – | 20 | 2 | 20 | 2 | 20 | 2 | 20 | 2 | 20 | 2 | 20 | 2 | 8 | 208 SW and one F–T cycles |
| 8 | 20 | – | 20 | – | 20 | – | 20 | 2 | 20 | 2 | 20 | 2 | 20 | 2 | 20 | 2 | 20 | 2 | 20 | 2 | 6 | 206 SW and one F–T cycles |
| 9 | 20 | – | 20 | – | 20 | 2 | 20 | 2 | 20 | 2 | 20 | 2 | 20 | 2 | 20 | 2 | 20 | 2 | 20 | 2 | 4 | 204 SW and one F–T cycles |
| 10 | 20 | – | 20 | 2 | 20 | 2 | 20 | 2 | 20 | 2 | 20 | 2 | 20 | 2 | 20 | 2 | 20 | 2 | 20 | 2 | 2 | 202 SW and one F–T cycles |
| 11 | 20 | 2 | 20 | 2 | 20 | 2 | 20 | 2 | 20 | 2 | 20 | 2 | 20 | 2 | 20 | 2 | 20 | 2 | 20 | 2 | – | 200 SW and two F–T cycles |

Laboratory simulation studies are, by necessity, oversimplifications of 'real-world' systems and, therefore, wherever possible it is important to use parameters that are as close an approximation to those experienced under natural conditions as possible. This consideration guided the choice of temperature conditions for both the salt weathering and freeze–thaw cycles. Reliance on extreme or unrealistic temperature conditions may result in data reflecting the experimental conditions rather than bearing any meaningful similarity to actual stone response in the built environment (McGreevy & Smith 1982; Warke & Smith 1998; McGreevy *et al.* 2000).

*Detail of freeze–thaw weathering cycles.* After every 20 salt weathering cycles, selected sample subsets were exposed to two consecutive freeze–thaw weathering cycles. As Table 2 shows, the first subset in the experimental run to experience freeze–thaw cycles was subset 11 followed by subset 10, 20 salt cycles later and so on until by the end of the experiment only subset 1 was left with no exposure to freeze–thaw cycles. The relevant subset samples were immersed in deionized water for approximately 20 s and then exposed to two consecutive freeze–thaw cycles (one 'wet' and one 'dry'). As with the salt cycles, each experimental run lasted for 20 h with two separate 10 h cycles, each of which comprised four temperature segments: 1 h during which temperature decreased from +20 to +10 °C followed by 4 h at +10 °C and then a staged decrease over 1 h to −10 °C followed by 4 h at −10 °C. The second 'dry' cycle had the same temperature parameters, with the exception of the first hour when temperature underwent a staged increase from −10 to +10 °C. Each of the 11 subsets included two control blocks, one of which remained dry throughout the entire experimental run while the other was wetted each day with deionized water. Both control blocks experienced the same combinations of freeze–thaw and/or salt weathering cycles as their subset counterparts.

## *Standard durability test: experimental procedure*

The standardized sodium sulphate salt crystallization durability test, as outlined in BRE Report 141 (Building Research Establishment 1989) was used to identify the durability status of the stone types used in this project. Six 40 mm cubes of each of the five stone types were cut, washed and oven dried at 103 ± 2 °C until constant weights were achieved. The samples were removed from the oven and allowed to cool in a desiccator until they reached room temperature (*c.* 20 °C) after which they were each weighed ($W_0$). The samples were then labelled with permanent ink and reweighed ($W_1$). Each sample block was immersed in a 14% solution of $Na_2SO_4$ for 2 h at a temperature of 20 °C, after which they were removed from the solution and placed in a preheated humidified oven (103 °C) for 16 h to dry. The samples were removed from the oven, allowed to cool to room temperature and weighed ($W_f$) after which they were immersed in the salt solution again. This sequence of drying and immersion was repeated a total of 15 times. At the end of the 15 cycles each block was weighed for a final time ($W_f$) and the percentage weight loss for each sample calculated (percentage weight loss $= 100[W_f - W_1]/W_0$) along with the mean percentage weight loss value for each of the five stone types analysed. Although the standard testing procedure does not require each block to be weighed after each cycle of wetting and drying, it was decided to introduce this extra step to enable a rudimentary comparison of the rate of sample breakdown between different stone types. Mean percentage weight change data after 5, 10 and 15 cycles are reported.

### *Analysis*

Although the main emphasis in data reported here is on weight loss characteristics, reference is made to salt distribution and evidence of structural deterioration. The analytical techniques used included scanning electron microscopy (SEM), thin-sectioning (TS), atomic absorption spectroscopy (AAS) and ion chromatography (IC). SEM and TS allowed identification of microstructural change, while AAS and IC were used to identify the distribution of salts within the stone fabric. Together, these data derived from analysis of selected samples during and after testing enabled modelling and comparison of the decay dynamics of each stone type.

## Results

### *Modified durability test (combined salt weathering and freeze–thaw cycles)*

Mean percentage weight loss data for each of the 11 subsets of each stone type are shown in Figure 1 and Table 3. Dumfries Sandstone exhibited the greatest amount of breakdown followed by Stanton Moor A, Portland Limestone and Stanton Moor B, with Leinster Granite proving to be the most durable stone.

- Leinster Granite and Stanton Moor B data produced significant correlations between mean cumulative percentage weight loss and the nature and number of weathering cycles, with a positive correlation for the granite and a negative correlation for Stanton Moor B (Fig. 1). For Stanton Moor B a combination of low porosity and permeability, and a limited range of permeability values that reflect the closely interlocked granular structure of this stone with its

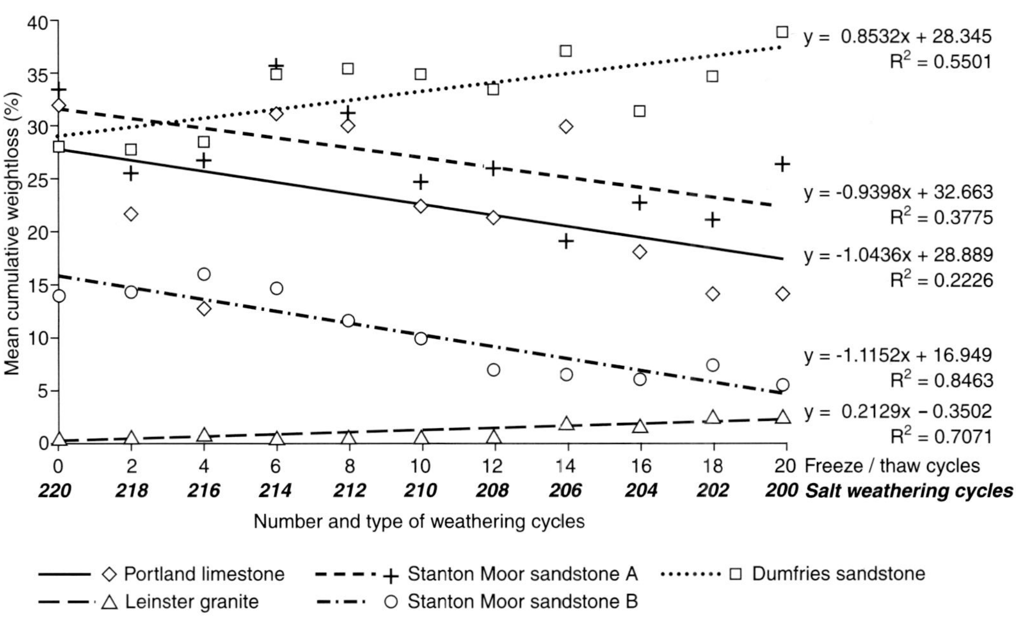


**Fig. 1.** Correlation between mean subset cumulative percentage weight loss and the number and type of weathering cycles to which each sample subset was exposed during the modified durability test.

**Table 3.** *Mean percentage weight loss data for subset sample groups from each stone type*

| No. of weathering cycles | | Mean percentage weight loss for each sample subset | | | | |
|---|---|---|---|---|---|---|
| SW | F–T | Leinster Granite | Stanton Moor Sandstone B | Portland Ilmestone | Stanton Moor Sandstone A | Dumfries Sandstone |
| 220 | 0 | 0.2 | 14.0 | 32.1 | 33.8 | 28.3 |
| 218 | 2 | 0.2 | 14.4 | 21.8 | 26.1 | 27.9 |
| 216 | 4 | 0.9 | 16.1 | 12.8 | 27.1 | 28.7 |
| 214 | 6 | 0.2 | 14.7 | 31.3 | 36.1 | 35.4 |
| 212 | 8 | 0.4 | 11.6 | 30.2 | 31.7 | 35.7 |
| 210 | 10 | 0.3 | 10.0 | 22.5 | 25.1 | 35.1 |
| 208 | 12 | 0.4 | 7.0 | 21.5 | 26.4 | 33.7 |
| 206 | 14 | 1.7 | 6.6 | 30.1 | 19.5 | 37.4 |
| 204 | 16 | 1.4 | 6.1 | 18.2 | 23.2 | 31.7 |
| 202 | 18 | 2.3 | 7.2 | 14.2 | 21.5 | 35.0 |
| 200 | 20 | 2.2 | 5.3 | 14.2 | 26.8 | 39.2 |
| Average | | 0.9 | 10.3 | 22.6 | 27.0 | 33.5 |

SW, salt weathering cycles; F–T, freeze–thaw weathering cycles.

well-developed quartz and feldspar overgrowths, resulted in restricted penetration of salt and moisture into substrate material that facilitated its removal during wetting prior to freeze–thaw cycling. Consequently, the greater the number of freeze–thaw cycles the less the amount of debris lost from Stanton Moor B. Although weight loss in Leinster Granite was comparatively low, data show that the extent of breakdown increased with an increase in the number of freeze–thaw cycles experienced. The difference in response between these two stone types appears to reflect the nature of their grain boundaries and to a lesser extent their respective mineralogies.

- Significant differences in response exist between Stanton Moor A and Stanton Moor B, with the former experiencing more deterioration than the latter. This may be explained by differences in permeability characteristics and a higher percentage clay content in Stanton Moor A, which is associated with increased salt weathering effectiveness because clays can provide points of ingress for moisture and act as foci for salt accumulation (McGreevy & Smith 1984; Warke & Smith 2000; Warke *et al.* 2004). More detailed discussion and data regarding the weathering response of both Stanton Moor A and B are reported in Warke *et al.* (2006).
- The response of Portland Limestone was extremely variable reflecting its heterogeneous structural properties especially with regard to porosity and permeability characteristics – characteristics that influence the extent of salt penetration and the nature of its subsequent accumulation at depth in substrate material.
- Dumfries Sandstone was identified as the least durable stone type in this modified test, with blocks losing on average over one third of their initial weight. Given the combination of high porosity and permeability, and the abundance of clays (smectites) forming interstitial laminae, this response was not unexpected.

In the modified durability test it was interesting to note that the range of permeability values for each stone type produced the same ranking as that indicated by mean percentage weight change values (Table 4). Data indicate that mean permeability values for each stone type were not equally accurate

**Table 4.** *Durability ranking results from modified durability test*

| Stone type | Mean weight change (%) | Permeability (range) (mD) | Mean permeability (mD) | Durability ranking |
|---|---|---|---|---|
| Leinster Granite | −0.93 | 3.7 (0.4–4.1) | 1.75 | Most durable |
| Stanton Sandstone B | −10.26 | 109.3 (4.7–114) | 58 | ↑ |
| Portland Limestone | −22.63 | 149 (1–150) | 15 | |
| Stanton Sandstone A | −27.02 | 198.3 (7.7–206) | 61 | |
| Dumfries Sandstone | −33.46 | 800 (200–1000) | 600 | Least durable |

indicators of weathering response, a point that is exemplified by comparison of Stanton Moor A and B sandstones.

- In Stanton Moor B because of low mean permeability, a restricted range of permeability values and comparatively low-porosity characteristics, salt penetration into substrate material was limited to the surface and near-surface resulting in a very gradual loss of material through granular disintegration – breakdown characteristics that were very similar to those of Leinster Granite.
- In Stanton Moor A, although mean permeability was similar to that of Stanton Moor B, the range of permeability values was much greater reflecting the presence of permeability 'hot spots' on block surfaces that facilitated salt penetration into substrate material where clay minerals provided loci for salt accumulation and the subsequent establishment of more organized subsurface disruption and generalized weakening of intergranular cohesion (Warke *et al.* 2006).

*Initiation of breakdown.* Following a preliminary stage of apparent quiescence each stone type started to break down at different points in the experimental run. The duration of this quiescent stage, when no debris was released, varied considerably, with Dumfries Sandstone blocks being the first to fail after 28–30 cycles followed by Portland Limestone (36–48 cycles), Stanton Moor A (114–126 cycles) and Stanton Moor B (118–130 cycles). Leinster Granite was the only stone type that showed any significant difference between the initiation of debris release in blocks exposed to both salt (SW) and freeze–thaw (F–T) cycles and those exposed to just salt weathering. The former (subsets 10 and 11) started to release debris after 112 (104 SW and 8 F–T) and 116 (106 SW and 10 F–T) cycles, respectively, while the latter (subsets 1 and 2) showed no evidence of breakdown until 148 and 150 salt weathering cycles had elapsed.

SEM examination of selected granite samples showed that blocks exposed to both freeze–thaw and salt cycles exhibited a combination of more open intergranular joints and intragranular microfracturing, which was particularly common in near-surface and surface feldspars with some limited quartz involvement. In contrast, debris release in those blocks exposed to just the salt weathering appeared to be driven primarily by the opening of grain boundaries and a general reduction in intergranular cohesion by the penetration and accumulation of crystallized salt.

*Trigger effect of high-magnitude, low-frequency freeze–thaw cycles.* There was no evidence of a direct link between exposure to freeze–thaw cycles and a contemporaneous increase in the rate of debris released. Exposure to freeze–thaw cycles undoubtedly resulted in an overall increase in debris released from Leinster Granite and Dumfries Sandstone, but any link between specific freeze–thaw cycles and debris release is complex, with a lag-time of variable duration often following a freezing event before significant debris was lost. It is important to note that the incorporation of freeze–thaw cycles into the salt weathering experimental regime affected different stone types in different ways with some releasing more debris while others released less.

### *Standard sodium sulphate salt crystallization durability test*

The standardized sodium sulphate salt crystallization durability test as outlined in BRE Report 141 (Building Research Establishment 1989) identified a durability ranking for the five stone types in which Leinster Granite and Stanton Moor B proved to be the most durable with the granite performing well under both modified and standardized test conditions (Tables 3–5). Discrepancy between rankings arises in the lower orders, with Portland Limestone, Stanton Moor A and especially Dumfries Sandstone responding differently to the two sets of experimental conditions. Mean weight change rates are shown in Figure 2, with all stone types registering an increase in mean weight after five test cycles associated with the accumulation of salt. After 10 cycles the Leinster Granite, Stanton Moor B and Dumfries Sandstones had all continued to gain weight with no significant loss of material, whereas both Portland Limestone and Stanton Moor A Sandstone had started to break down. By the end of the 15 test cycles all but the granite and Stanton Moor B samples showed evidence of major failure and material loss.

In summary, data indicate that the range of permeability values for each stone type provided a reasonably accurate means of predicting stone durability under modified test conditions. It is suggested that the modified durability test provides a more accurate reflection of weathering behaviour of stone because of the use of more than one weathering process, more realistic temperature parameters and a relatively dilute $Na_2SO_4$ solution, which together enable each stone type to resolve resultant weathering stresses in ways that more accurately reflect response under 'real-world' conditions.

### *Modelling decay dynamics*

Summaries of decay dynamics based on data from the modified durability test procedure are presented

**Table 5.** *Durability ranking results from the standard sodium sulphate salt crystallization test*

| Stone type | Mean % weight change after 5 cycles | Mean % weight change after 10 cycles | Mean % weight change after 15 cycles | Durability ranking |
|---|---|---|---|---|
| Leinster Granite | +0.10 (range +0.06–+0.14) | +0.13 (range +0.08–+0.17) | +0.13 (range +0.16–+0.11) | Most durable |
| Stanton Moor B Sandstone (medium–coarse grained) | +0.77 (range +0.73–+0.81) | +0.88 (range +0.44–+1.17) | −6.90 (range −13.0 to −4.7) | ↑ |
| Dumfries Sandstone | +0.88 (range +0.51–+1.39) | +1.17 (range −2.74 to +0.29) | −16.6 (range −43.2 to −6.1) | |
| Portland Limestone | +0.89 (range −2.33 to +0.74) | −15.19 (range −44.30 to −5.94) | −27.10 (range −69.1 to −14.5) | |
| Stanton Moor A Sandstone (fine–medium grained) | +0.84 (range +0.80–+0.88) | −2.24 (range −8.67 to +0.68) | −67.00 (range −82.0 to −38.0) | Least durable |

Positive values indicate weight gain as a percentage of initial weight and negative values weight loss.

in order of durability ranking as determined by the modified test. Cumulative percentage weight loss curves for selected subset samples from each of the five stone types are shown in Figure 3a, b, and comparative conceptual models of breakdown are presented in Figure 4a–e.

*Leinster Granite.* IC and AAS analysis of surface and substrate samples showed that $Na_2SO_4$ penetration was restricted to the upper few millimetres of stone. Granite blocks remained intact for over the first half of the experimental run, after which deterioration proceeded very slowly through the release of individual grains, with SEM identifying preferential exploitation of mica by salt crystallization within cleavage planes which, as the mica was broken down and released, allowed further salt penetration and crystallization to destabilize adjacent grains. Freeze–thaw cycles facilitated the action of salt by opening intergranular boundaries and fracturing individual grains, especially the feldspars.

*Stanton Moor B Sandstone.* Block to block weathering response was consistent and similar to that of the granite with no breakdown during the first half of the experimental run after which disintegration gradually proceeded through release of individual grains. This reflected the restriction of salt and moisture penetration into substrate material because of low porosity and permeability. Even when salt was able to penetrate, its effectiveness was limited because of the closely interlocked structure of this sandstone arising from extensive and well-developed quartz and feldspar overgrowths that created significant intergranular cohesion. Of the two weathering processes, salt cycles appeared to be more effective than the combination of salt and freeze–thaw primarily because of the relative ease with which salts could be washed out during immersion in deionized water.

*Portland Limestone.* Block to block weathering response was extremely variable with regard to the total amount of material lost, reflecting the variable distribution of pore spaces (especially the larger pores) within individual blocks and between blocks. Breakdown was initially gradual with the release of ooliths both individually and in small aggregations. The initial gradual rate of material loss was in some cases followed by a period of more rapid disintegration as salt exploited substrate pores. IC and AAS analysis of substrate material from blocks in the later stages of the experimental run showed the deep penetration of $Na_2SO_4$ filling pore spaces. Ironically, this infilling and effective blocking of deep pores by salt appears to have slowed rates of breakdown in most blocks in the later stages of the experimental run.

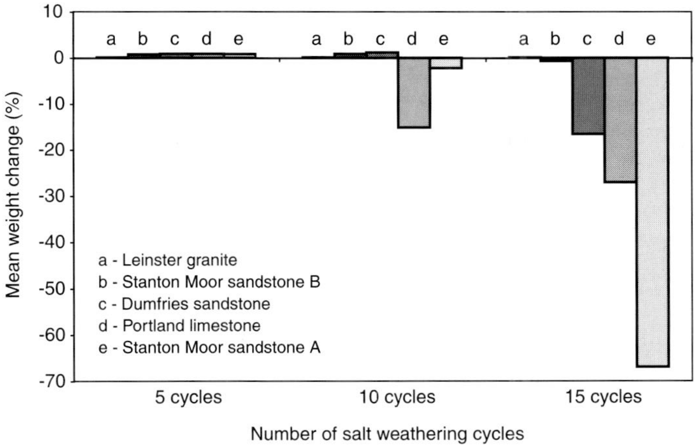


**Fig. 2.** Mean percentage weight change of each stone type sample set used in the standard sodium sulphate salt crystallization test after 5, 10 and 15 test cycles.

*Stanton Moor A Sandstone.* Block to block weathering response was consistent with decay sequences comprising three clearly defined stages. For approximately the first half of the experimental run no debris was released. However, in the few cycles prior to the initiation of breakdown surface conditions changed with a 'bowing' of block surfaces (Warke *et al.* 2006). TS and SEM analyses showed that this surface deformation was related to substrate microfracture development. In the two–four cycles immediately following development of surface 'bowing', blocks broke down rapidly through extensive surface scaling and subsequently through granular disintegration. Weathering response of both grain size varieties of Stanton Moor Sandstone differed significantly, suggesting that differential weathering response could occur on building façades or within larger individual blocks where such relatively small differences in grain size fall well within acceptance limits of natural variability in the choice of stone.

*Dumfries Sandstone.* Weathering response of Dumfries Sandstone was the most extreme with regard to both the early start of deterioration in the experimental run and to the quantity of material released. Despite this, breakdown was normally a gradual process proceeding through extensive granular disintegration. Where blocks contained clay laminations, breakdown could be briefly accelerated due to splitting along these lines of weakness (see Fig. 3b); however, granular release was the

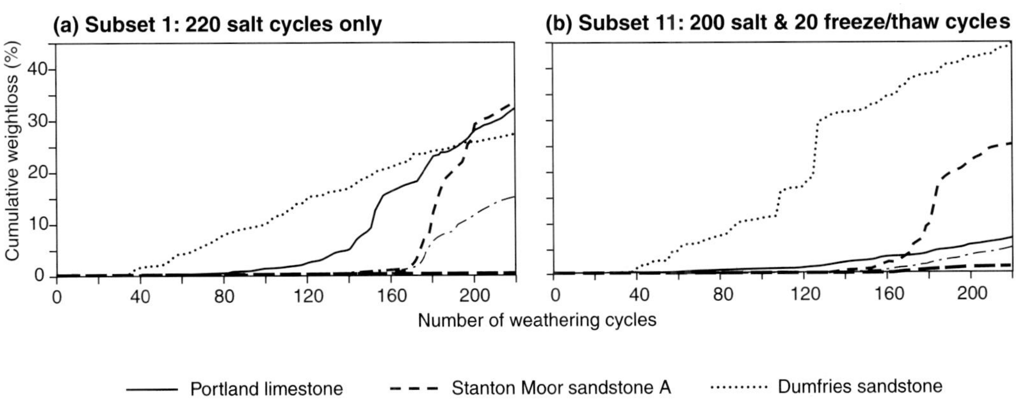


**Fig. 3.** Cumulative mean percentage weight loss curves for sample subsets 1 (220 salt-weathering cycles) and 11 (200 salt and 20 freeze–thaw weathering cycles).

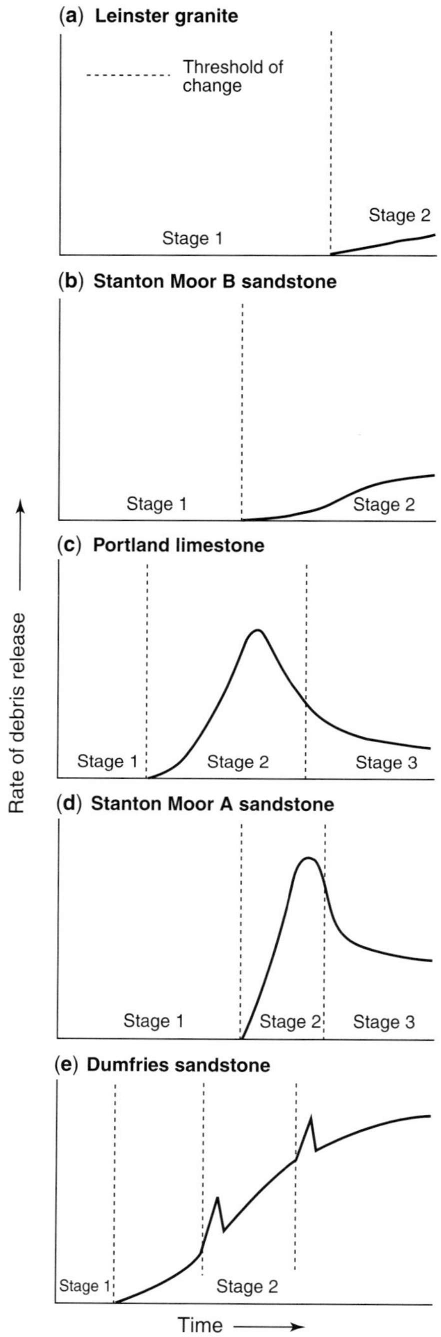


**Fig. 4.** Comparative conceptual models of weathering response during the experimental run of: (**a**) Leinster Granite; (**b**) Stanton Moor B Sandstone; (**c**) Portland Limestone; (**d**) Stanton Moor A Sandstone; and (**e**) Dumfries Sandstone. Stone types are ranked in order of decreasing durability as determined by modified test conditions, with Leinster Granite being the most durable and Dumfries Sandstone the least.

dominant stone response to the experimental conditions.

Each of the five stone types tested followed quite different decay pathways despite exposure to the same experimental conditions (Fig. 4a–e). In this study, data indicate that differential response primarily reflected the influence of intrinsic thresholds, whereby samples were progressively weakened until a point when the 'stress' imposed by repeated weathering could no longer be absorbed and failure occurred. That it was primarily intrinsic thresholds that were controlling breakdown is demonstrated by the fact that when there was a change in external variables through exposure of selected samples to freeze–thaw weathering there was no apparent associated triggering or acceleration in breakdown. This may have reflected either of the following.

- The stone types may not have been sufficiently sensitive to the nature of change imposed and, consequently, the change in external conditions was of insufficient magnitude to trigger failure.
- There may have been a lag between exposure to changed external conditions and system response, with the latter becoming indistinct from response to subsequent salt weathering cycles (Schumm 1991).

The change from a condition of stability to instability is identified as a 'threshold of change' and some of the stone types tested exhibited several thresholds of change, whereby they changed from one state into another. This may equate with the 'characteristic and transient forms' identified by Brunsden & Thornes (1979) with initial failure representing transient form adjustment followed by a period of time when the stone, although releasing material, displays a characteristic form until its sensitivity to external conditions brings it to another intrinsic threshold of change and another transient form. Stanton Moor A Sandstone (Fig. 4d) is a good example of this, whereby through Stage 1 in the decay sequence this stone type exhibited an apparently stable characteristic form until intrinsic structural thresholds were breached resulting in relatively rapid deterioration through extensive surface scaling in Stage 2 (transient form). This was followed by changes in both the nature and rate of breakdown, with granular disintegration replacing scaling and an associated decrease in the rate of material released as the stone assumed its new characteristic form in Stage 3. In comparison to Stanton Moor A Sandstone, the decay dynamics of Dumfries Sandstone (Fig. 4e) were more complex reflecting the different sensitivity of its components. For example, the visible clay laminations proved to be more sensitive than the intervening quartz and feldspar layers, thereby resulting in deterioration characterized by peaks in breakdown and debris release as the clay beds failed.

Inkpen (2005) notes the problem of identifying more sensitive parts of a system before failure occurs. In the case of Dumfries Sandstone this is not difficult because of the obvious nature of the clay laminations and our knowledge of the susceptibility of clays to weathering. It is more of a problem with regard to seemingly homogeneous stone types such as Stanton Moor where relatively minor differences in grain size and permeability characteristics may result in differential weathering response within a single block of stone (Warke *et al.* 2006).

## Discussion

### *Differences in durability rankings*

The standard testing procedure employs relatively extreme conditions with regard to both temperature and salt concentration – conditions that 'force' the rapid breakdown of stone thereby masking subtle differences in disintegration patterns. The most important factor in this 'forcing' effect would appear to be the temperature regime used with repeated heating of samples to 103 °C. High temperatures have long been associated with damage to stone with reports of greatly increased stone deterioration in the presence of salt when temperatures of more than 100 °C were used (Marschner 1978; Price 1978; McGreevy & Smith 1982; Davison 1986). More recently, Logan (2004) demonstrated the effect of repeated cycles of heating to 107 °C on samples of marble with widespread grain-boundary separation and microfracture development being characteristic outcomes. An important point to arise from Logan's work was the observation that loss of material strength was not linear but exponential, with the most significant decline in strength occurring relatively early on within the first 20–30 cycles out of a total experimental run of 200 cycles. Repeated exposure to extreme temperatures increases the likelihood of disruption of the microstructural properties of stone, which in turn facilitates the efficacy of exploitative weathering agents such as salt both from the point of its increased penetration of stone fabric and its increased crystallization pressures (Winkler & Singer 1972; Sperling & Cooke 1980; Goudie & Viles 1997).

The effect of using extreme temperature conditions is compounded by the method of heating (oven-based), whereby all stone samples are 'forced' through the same temperature conditions irrespective of their individual thermal properties. Indirect or oven-based heating results in lithologically indiscriminate cycling of test samples, whereby temperature response is primarily determined by external conditions and not by intrinsic stone properties, and so forcing some stone types, under extreme heating, to reach temperatures that they would never experience under natural conditions (Warke & Smith 1998).

At best, the standardized sodium sulphate salt crystallization test provides a comparatively crude measure of durability, probably most useful as a predictive tool in identifying the most durable of stone types but less so in cases where differences in durability between stone types are less clearly defined.

In comparison, the modified durability test employs less extreme conditions with regard to both temperature and salt concentrations. The difference in temperature conditions is particularly significant because, although an indirect method of heating was used for comparative purposes, the maximum temperature of 40 °C falls within the range of surface temperatures reached by various stone types under natural conditions (Kerr *et al.* 1984; Jenkins & Smith 1990; Goudie 1997) and also under direct heating laboratory experiments (Warke *et al.* 1996; Warke & Smith 1998). McGreevy *et al.* (2000) suggested that excessively high and unrepresentative temperatures can 'over-weather' stone creating microscale damage that may then be exploited by salt and moisture. This 'over-weathering' may have resulted in uncharacteristically poor durability responses for some stone types that may respond very differently to 'real-world' conditions. An example of this disparity between test results and actual response is provided by Smith (1999) who notes that, although Ancaster stone (Jurassic limestone) has been used extensively and successfully in England as a building stone since Roman times in the construction of many major historic buildings that have withstood the test of time, as a stone type it actually fails the standardized salt crystallization test.

Although the same salt ($Na_2SO_4$) was used in both test procedures, the concentration differed with a 14% solution used in the standard test and a 2.5% solution in the modified test. Choice of a 2.5% solution of $Na_2SO_4$ was dictated by the need to avoid conditions so extreme that the subtleties of different stages in the decay sequence of each stone type would be lost in the forcing of rapid breakdown. A pilot study showed that 5 and 10% solutions of $Na_2SO_4$, when applied to sandstone, resulted in complete and very rapid sample breakdown before a representative number of experimental cycles could be completed. This highlights an important issue raised by McGreevy & Smith (1982), and subsequently by Price (1996) and Smith *et al.* (2005), whereby the use of highly concentrated or even saturated salt solutions in simulation studies is unrealistic, contributing to excessive disintegration with resultant data being reflections of experimental design rather than indications of potential response to 'real-world' conditions. There appears to be a very fine line between creating experimental conditions that act as the primary control on stone response and those that enable structural and mineralogical properties of stone to dictate the nature of breakdown. This balance will vary for different stone types, but when designing a comparative stone weathering simulation experiment the conditions employed should aim to identify the weathering response characteristics of the least durable samples. This will probably necessitate longer experimental runs, whereby extreme conditions (high temperatures, higher salt concentrations) are replaced by more realistic environmental parameters and more experimental cycles, i.e. more time.

Under 'real-world' conditions building stone is exposed to the cumulative and sequential effects of different weathering processes, and it is important to at least attempt to include an element of this complexity in testing procedures primarily because of the potential for synergy, whereby the efficacy of one form of weathering is enhanced by the operation of another. In this study salt weathering cycles represented low-magnitude, high-frequency weathering, while freeze–thaw cycles, which occurred less frequently, were intended to represent higher-magnitude weathering conditions.

Freeze–thaw events have been identified as triggers for the release of previously weathered and weakened stone (Camuffo & Sturaro 2001; Hall 2004), and it has been demonstrated experimentally that the extent of frost damage can be greatly increased by the presence of certain salts (Williams & Robinson 1991, 2001). However, within the context of this study, not all stone types responded in the same way to the combined effects of salt and freezing. Data indicate that heterogeneous stone such as Dumfries Sandstone proved to be particularly susceptible to the combined effects of salt and freeze–thaw weathering cycles, which resulted in its being identified as the least durable stone type under modified testing conditions (Table 4). In contrast, under standard testing where samples were exposed to only salt weathering, Dumfries Sandstone proved to be more durable, being ranked above Portland Limestone and Stanton Moor A Sandstone (Table 5).

Under modified test conditions Stanton Moor Sandstone (A and B) and Portland Limestone samples exhibited a decrease in the amount of material lost with exposure to an increased number of freeze–thaw cycles (Fig. 1). This reflects the important role of structural controls in determining stone weathering susceptibility, whereby significant substrate penetration of salt was initially restricted to surface and near-surface layers where it was relatively easily washed off and removed from the system during wetting in deionized water prior to freeze–thaw cycling. Inclusion of both salt weathering and freeze–thaw cycles in one test allowed these lithological differences in weathering response to be demonstrated and appear to have had a significant influence on performance of those stone types with a less clearly defined durability status.

Another potentially influential factor in determining durability status was the size of sample blocks used. Although shape was held constant, in the standard test 40 mm cubes were employed compared to the 75 mm cubes employed in the modified test. That size matters in weathering studies has been demonstrated experimentally by Goudie (1974) and acknowledged in a number of more recent studies and reviews (e.g. Goudie & Viles 1995; Viles 2001; Smith *et al.* 2005), with the comparatively poor performance of smaller samples attributed to a variety of factors including:

- the tendency of 'small samples (to) accentuate edge effects which, during temperature/moisture cycling, influence internal temperature and moisture regimes, salt distribution and, through these, patterns of chemical alteration and internal stress' (Smith 1996, p. 9);
- 'a failure to differentiate the effects of mineralogical and structural variations, such as bedding, that are seen to operate at a larger scale ...' (Smith *et al.* 2005, p. 219).

The smaller size of sample blocks used in the standard test combined with the potentially disruptive effect of repeated heating to over 100 °C appears to have predisposed all but the most durable stone types to early and extensive disintegration. In comparison, the less extreme conditions used in the modified test, combined with larger test samples and the use of both salt and freeze–thaw cycles, have facilitated the development of lithologically distinct decay sequences that data indicate are primarily a reflection of intrinsic structural and mineralogical properties and not of the experimental conditions.

## *Permeability as an indicator of potential durability*

Moisture movement, salt migration and accumulation at depth within substrate material is primarily controlled by permeability characteristics of stone which, in turn, are closely linked to pore properties, particularly the presence and extent of interconnected pore spaces (McGreevy 1996; Goudie

1999; Nicholson 2001). Permeability is a spatially variable property even within relatively homogeneous stone types (McKinley & Warke 2007) and data from this study indicate that the greater the initial range in permeability values, the greater the potential for salt and moisture ingress and retention, and hence disruption of substrate material. In the modified durability test it was interesting to note that the range of permeability values for each stone type produced the same ranking as that indicated by mean percentage weight change values, i.e. durability status. This was particularly notable in the case of Stanton Moor Sandstone samples, where Stanton Moor A proved to be less durable than Stanton Moor B with permeability ranges of 198 (mean 61 mD) and 109 mD (mean 58 mD), respectively, despite having similar mean permeability values (Table 4).

The significance of this apparent relationship between the range of permeability values for a given stone type and its durability status remains somewhat speculative, but may be a potentially fruitful avenue for future research, especially with regard to the use of permeability data as primary indicators of potential weathering response.

## Conclusions

In proposing a modified durability test it is not intended to detract from the value of the established standard salt crystallization test. Although the use of standardized durability testing procedures to evaluate an extremely complex material like stone has aroused considerable debate, it is acknowledged that a standardized approach to testing is essential if there is to be meaningful comparability of results and commonality of terminology within the construction and conservation industries – it is the nature of the testing procedure and not the need for standardization that is at issue. However, while recognizing the value of the sodium sulphate salt crystallization test as a rapid means of assessing durability, it is suggested that an additional more complex secondary testing procedure could be made available, particularly in circumstances where the choice of stone for replacement matching or for use in carved architectural detailing is not a straightforward one. A better understanding of the detailed differences in deterioration pathways between stone types and the factors controlling these should help to better inform decision making regarding the choice of the most suitable stone.

Special thanks must go to G. Alexander in the Cartographic Department of the School of Geography, Archaeology and Palaeoecology for help with preparation of the diagrams, and to laboratory technical staff for assistance during the experimental work. Financial support for this project was provided by an Engineering and Physical Sciences Research Council (EPSRC) grant GR/R54491/01.

## References

Brunsden, D. & Thornes, J. B. 1979. Landscape sensitivity and change. *Transactions of the Institute of British Geographers*, **4**, 463–484.

Building Research Establishment. 1989. *Durability Tests for Building Stone*. BRE Report, **141**. Building Research Establishment, Watford.

Camuffo, D. & Sturaro, G. 2001. The climate of Rome and its action on monument decay. *Climate Research*, **16**(2), 145–155.

Davison, A. P. 1986. An investigation into the relationship between salt weathering debris production and temperature. *Earth Surface Processes and Landforms*, **11**, 335–341.

Goudie, A. S. 1974. Further experimental rock weathering by salt and other mechanical processes. *Zeitschrift für Geomorphologie Supplementband*, **21**, 1–12.

Goudie, A. S. 1997. Weathering processes. *In*: Thomas, D. S. G. (ed.) *Arid Zone Geomorphology: Process, Form and Change in Drylands*. Wiley, Chichester, 25–39.

Goudie, A. S. 1999. Experimental salt weathering of limestones in relation to rock properties. *Earth Surface Processes and Landforms*, **24**, 715–724.

Goudie, A. S. & Viles, H. A. 1995. The nature and pattern of debris liberation by salt weathering: a laboratory study. *Earth Surface Processes and Landforms*, **20**, 437–449.

Goudie, A. S. & Viles, H. A. 1997. *Salt Weathering Hazards*. Wiley, Chichester.

Hall, K. 2004. Evidence for freeze–thaw events and their implications for rock weathering in Northern Canada. *Earth Surface Processes and Landforms*, **29**, 43–57.

Inkpen, R. 2005. *Science, Philosophy and Physical Geography*. Routledge, London.

Jenkins, K. & Smith, B. J. 1990. Daytime rock surface temperature variability and its implications for mechanical rock weathering: Tenerife, Canary Islands. *Catena*, **17**, 449–459.

Kerr, A., Smith, B. J., Whalley, W. B. & McGreevy, J. P. 1984. Rock temperatures from S.E. Morocco and their significance for experimental rock weathering studies. *Geology*, **12**, 306–309.

Logan, J. M. 2004. Laboratory and case studies of thermal cycling and stored strain on the stability of selected marbles. *Environmental Geology*, **46**, 456–467.

Marschner, H. 1978. *Application of Salt Crystallisation Test to Impregnated Stones*. RILEM/UNESCO Symposium, Paris, Report, **3.4**.

McGreevy, J. P. 1996. Pore properties of limestones as controls on salt weathering susceptibility: a case study. *In*: Smith, B. J. & Warke, P. A. (eds) *Processes of Urban Stone Decay*. Donhead, Shaftesbury, 150–167.

McGreevy, J. P. & Smith, B. J. 1982. Salt weathering in hot deserts: observations on the design of simulation experiments. *Geografiska Annaler*, **64A**, 161–170.

McGreevy, J. P. & Smith, B. J. 1984. The possible role of clay minerals in salt weathering. *Catena*, **11**, 169–175.

McGreevy, J. P., Warke, P. A. & Smith, B. J. 2000. Controls on stone temperatures and the benefits of interdisciplinary exchange. *Journal of the American Institute of Conservation*, **39**, 259–274.

McKinley, J. & Warke, P. A. 2007. Controls on permeability: implications for stone weathering. *In*: Přikryl, R. & Smith, B. J. (eds) *Building Stone Decay: From Diagnosis to Conservation*. Geological Society, London, Special Publications, **271**, 225–236.

Nicholson, D. T. 2001. Pore properties as indicators of breakdown mechanisms in experimentally weathered limestones. *Earth Surface Processes and Landforms*, **26**, 819–838.

Price, C. A. 1978. *The Use of the Sodium Sulphate Crystallisation Test for Determining the Weathering Resistance of Untreated Stone*. RILEM/UNESCO Symposium, Paris, Report, **3.6**.

Price, C. A. 1996. *Stone Conservation: An Overview of Current Research*. Getty Conservation Institute, Los Angeles, CA.

Schumm, S. A. 1991. *To Interpret the Earth: Ten Ways to be Wrong*. Cambridge University Press, Cambridge.

Smith, B. J. 1996. Scale problems in the interpretation of urban stone decay. *In*: Smith, B. J. & Warke, P. A. (eds) *Processes of Urban Stone Decay*. Donhead, Shaftsbury, 3–18.

Smith, B. J. & Kennedy, E. 1999. Moisture loss from stone influenced by salt accumulation. *In*: Jones, M. S. & Wakefield, R. D. (eds) *Aspects of Stone Weathering, Decay and Conservation*. Imperial College Press, London, 55–64.

Smith, B. J., Warke, P. A., McGreevy, J. P. & Kane, H. L. 2005. Salt-weathering simulations under hot desert conditions: agents of enlightenment or perpetuators of preconceptions? *Geomorphology*, **67**, 211–227.

Smith, M. R. 1999. *Stone: Building Stone, Rock Fill and Armourstone in Construction*. Geological Society, London, Engineering Geology, Special Publications, **16**.

Sperling, C. H. B. & Cooke, R. U. 1980. *Salt Weathering in Arid Environments. Part 1: Theoretical Considerations*. Bedford College Papers in Geography, **8**.

Viles, H. A. 2001. Scale issues in weathering studies. *Geomorphology*, **41**, 63–72.

Warke, P. A. & Smith, B. J. 1998. Effects of direct and indirect heating on the validity of rock weathering simulation studies and durability tests. *Geomorphology*, **22**, 347–357.

Warke, P. A. & Smith, B. J. 2000. Salt distribution in clay-rich weathered sandstone. *Earth Surface Processes and Landforms*, **25**, 1333–1342.

Warke, P. A., McKinley, J. & Smith, B. J. 2006. Variable weathering response in sandstone: factors controlling decay pathways. *Earth Surface Processes and Landforms*, **31**, 715–735.

Warke, P. A., Smith, B. J. & Magee, R. 1996. Thermal response characteristics of stone: implications for weathering of soiled surfaces in urban environments. *Earth Surface Processes and Landforms*, **21**, 295–306.

Warke, P. A., Smith, B. J. & McKinley, J. 2004. Complex weathering effects on the durability of building sandstone. *In*: Přikry, R. (ed.) *Dimension Stone 2004*. Taylor & Francis, London, 229–235.

Williams, R. B. G. & Robinson, D. A. 1991. Frost weathering of rocks in the presence of salts: a review. *Permafrost and Periglacial Processes*, **2**, 347–353.

Williams, R. B. G. & Robinson, D. A. 2001. Experimental frost weathering of sandstones by various combinations of salts. *Earth Surface Processes and Landforms*, **26**, 811–818.

Winkler, E. M. & Singer, P. C. 1972. Crystallization pressure of salts in stone and concrete. *Geological Society of America Bulletin*, **83**, 3509–3514.

Yates, T. & Butlin, R. 1996. Predicting the weathering of Portland limestone buildings. *In*: Smith, B. J. & Warke, P. A. (eds) *Processes of Urban Stone Decay*. Donhead, Shaftesbury, 194–204.

# Controls on permeability: implications for stone weathering

J. M. McKINLEY & P. A. WARKE

*School of Geography, Archaeology and Palaeoecology, Queen's University Belfast, Belfast BT7 1NN, Northern Ireland, UK (e-mail: j.mckinley@qub.ac.uk)*

**Abstract:** In the light of a well-researched relationship between rock properties and susceptibility of stone to weathering, the role of permeability in weathering is examined. A review of weathering studies indicates the varied use and nature of porosity data, but the paucity of permeability studies in weathering trials. Key factors that control porosity and permeability, depositional characteristics and diagenetic processes are discussed and investigated, with a view to discussing the implications for stone weathering. Results from experimental studies on a range of rock types comprising sandstone, limestone and granite are presented. The relevance of permeability measurement is explored in terms of spatial mapping and quantitative assessment of the deterioration of natural building stone. Increased knowledge and appreciation of the inherited characteristics of a rock is demonstrated to provide valuable insight and a greater understanding of how natural stone heterogeneity is accentuated and exploited by weathering and continued exposure to moisture and salts. Mapping the spatial distribution of permeability provides greater insight into the extent of variability in stone deterioration and presents the possibility of monitoring and predicting the hydraulic properties of stone and how these are modified by weathering processes.

The relationship between rock properties and the susceptibility of stone to weathering has been highlighted in many studies (e.g. Drever 1994; McGreevy 1996; Goudie 1999; Nicholson 2001; Inkpen *et al.* 2004; Smith *et al.* 2005). The role of the rock parameters of porosity (e.g. McGreevy 1996; Nicholson 2001, 2002; Bidner *et al.* 2002; Jornet *et al.* 2002; Burlini 2002; Pera & Burlini 2002; Přikryl & Dudková 2002) and permeability (e.g. Carey & Curran 2000; Russell *et al.* 2002; McKinley *et al.* 2006; Warke *et al.* 2006) has been investigated and discussed in relation to weathering. However, a review of the pertinent literature demonstrates a tendency in weathering studies to concentrate on porosity rather than permeability as the key petrophysical property to monitor during exposure trials and laboratory simulations. Measured porosity parameters used in weathering studies as described in the literature have been varied in nature and terminology, and include total porosity, air void porosity, capillary porosity (e.g. Jornet *et al.* 2002), interconnected porosity, fracture porosity (e.g. Nicholson 2001, 2002), effective porosity (e.g. McGreevy 1996) and microporosity (e.g. Inkpen *et al.* 2004). As the findings from weathering studies strongly indicate the importance of pore properties in influencing rock susceptibility to weathering, a greater knowledge of rock properties becomes essential, underpinned by an increased understanding of the controls on porosity and permeability. The purpose of this paper is to clarify the role of permeability in weathering studies, and to investigate the controls on the rock properties of porosity and permeability with a view to discussing the implications for stone weathering. The results from experimental studies on a range of rock types comprising sandstone, limestone and granite will be discussed. The relevance of permeability measurements is investigated in relation to the spatial mapping and quantitative assessment of the deterioration of building stone.

## Explanation of stone properties

Building stone contains characteristics inherited from its depositional, compaction and cementation or crystallization history. A building stone such as sandstone or limestone contains pores or voids that, when connected in some way, permit the movement of fluids and salts of differing physical and chemical properties. Individual pores or voids vary in size, shape and arrangement, directing the movement of fluids and salts along preferred pathways and at differential rates. Primarily, heterogeneity in pore space is a result of variability in two important aspects of natural stone: porosity and permeability, and the spatial continuity of these rock properties (Cross *et al.* 1993). Tucker (1994) defines porosity as a measure of the pore space and describes two types: absolute porosity and effective porosity. Absolute porosity refers to the total void space within a rock including void space within grains. Effective porosity is used to describe the interconnected pore volume and therefore is more closely related to permeability, which is the ability of a sediment to transmit fluids (Tucker 1994). Permeability will depend on the shape and size of pores or voids and pore connections

*From*: PŘIKRYL, R. & SMITH, B. J. (eds) *Building Stone Decay: From Diagnosis to Conservation.* Geological Society, London, Special Publications, **271**, 225–236.
0305-8719/07/$15.00 

(throats), and also on the properties of the fluids involved (i.e. capillary forces, viscosity and pressure gradient). Calculated from Darcy's law, permeability, in simple terms, is a measure of how easily a fluid of a certain viscosity flows through a rock under a pressure gradient (Allen *et al.* 1988).

However, natural rocks seldom retain their original porosity (Beard & Weyl 1973; Houseknecht 1987). Primary depositional processes produce fabric characteristics that are further modified by compaction and cementation. As a result, two major types of porosity are produced: primary and secondary porosity. Primary porosity is developed as a sediment is deposited and includes inter- and intraparticle/granular porosity (Tucker 1994). Secondary porosity develops during diagenesis by dissolution or removal of soluble material and through tectonic movements producing fracturing. Fractures or vugs may contribute substantially to flow capacity (i.e. permeability properties) but contribute little to absolute porosity (Timmerman 1982). Secondary precipitation, in the form of diagenetic cements, has the potential to seal fractures or vugs.

The key factors that control porosity and permeability in sandstones are, therefore, depositional characteristics (including fabric features) and diagenetic features such as cements (Worden 1998). The predominant cements in sandstones comprise carbonates, clay minerals and quartz cements. Aspects of carbonate, quartz and clay cementation in sandstones have been comprehensively covered in three special publications (Morad 1998; Worden & Morad 2000, 2001). Porosity in limestones tends to be more erratic in type and distribution than for sandstones (Tucker 1994). Porosity types in limestones have been defined (based on Choquette & Pray 1970) as fabric selective, depending on whether pores are defined by the fabric (grains and matrix) of the limestone (e.g. intercrystalline), and non-fabric selective, porosity that cuts across the actual rock fabric (e.g. fracture porosity). Stylolites in limestones can form a type of porosity in terms of acting as conduits for fluid movement or conversely produce a reduction in porosity through the accumulation of clays and insoluble residue (Park & Schot 1968; McGreevy 1996). Porosity in crystalline rocks, including igneous and metamorphic, occurs generally as a result of fracturing, granular decomposition or dissolution, and may be accentuated by mineral alignment or banding.

## Relevance of rock properties in weathering studies, and discussion of previous work on porosity and permeability

Weathering studies (e.g. McGreevy 1996; Goudie 1999; Smith & Kennedy 1999; Nicholson 2001) indicate that the susceptibility of porous stone is related to porosity and pore characteristics in that the presence of interconnected pore spaces, and thus permeability properties of the stone, facilitates the penetration and movement of moisture and salts. However, as mentioned previously, there is a tendency in previous work to focus on porosity parameters to investigate the effect of weathering trials on pore properties, with only a predicted assessment of the movement of moisture and salts. A range of porosity parameters and measurement techniques have been deployed in laboratory simulations comprising mercury porosimetry (e.g. McGreevy 1996; Smith & Kennedy 1999), helium porosity (e.g. Markopoulos & Galetakis 2002), fracture porosity based on ultrasonic velocity (e.g. Nicholson 2002), air void, gel and capillary porosity (e.g. Jornet *et al.* 2002), and petrographic analysis in exposure trails (e.g. Bidner *et al.* 2002; Přikryl & Dudková 2002). The significance of pore connectivity has been stressed by Cooke (1979) and Smith & Kennedy (1999) in the role of accumulated salts at pore-throats, modifying stone response to wetting–drying and heating–cooling simulations. Experimental simulations have become routine in monitoring changes in rock properties, including pore characteristics, induced by weathering using parameters such as fracture porosity to indicate changes in void space (Nicholson 2001). According to Nicholson (2002), fracture porosity represents the aggregate percentage volume of new voids introduced into the rock as a result of induced deterioration in the form of a single fracture or a number of smaller fractures or microcracks. Hence, fracture porosity, Nicholson (2001, 2002) states, provides a greater indication of internal and hidden modification induced by weathering than monitored weight loss. However, moisture movement, salt migration and the distribution of salts at depth may be influenced by pore space but are controlled by permeability characteristics of the stone (McGreevy 1996; Smith *et al.* 2005). Although the relevance of permeability may be inferred in weathering studies, this is an area that is relatively understudied in the literature and requires further investigation. The measurement of permeability in natural building stone has been carried out using several techniques. These include a modified autoclam system that was originally designed for assessing durability of concrete, and which measures air and water permeability (Beggan *et al.* 1996; Russell *et al.* 2002), and a constant-head permeameter that measures water permeability on totally saturated samples (Thomachot & Jeannette 2002). Non-destructive permeability measurements have also been generated using a steady-state gas probe permeameter (see Carey & Curran 2000 for an explanation of the technique) and an unsteady-state portable air probe permeameter (an explanation of the technique is detailed in Jones 1992 and a description of its use in McKinley *et al.* 2006). The history of probe-permeametry development is reviewed in Hurst & Goggin (1995), and recommended practice for the technique is found in Goggin (1993) and Sutherland *et al.* (1993). The highly variable nature of stone deterioration has been acknowledged over the scale of an individual block or slab (Shelford *et al.* 1996) and over the extent of a building façade (Turkington & Smith 2000, 2004). Averaging of permeability measurements was found by Warke *et al.* (2006) to lead to an underestimation of the effect of changes in pore properties on the durability characteristics of building sandstones. These studies highlight the inadequacy of mean porosity and permeability values to investigate the variable nature of natural stone decay and emphasize the need to examine the spatial distribution of rock properties. The advantage of probe permeametry as a technique in the characterization of porous building stone is that it presents the opportunity to produce a high-resolution spatial quantification of permeability variation (Carey & Curran 2000).

## The role of primary depositional controls and diagenetic processes

An interesting study by Weber & Lepper (2002) presented an integrated approach, which combined geological background with properties of two types of siliciclastic dimension stone. A relationship was found between depositional environment, diagenetic overprint and resistance to weathering influences. Quartz-cemented channel-fill deposits were not affected by weathering processes, whereas floodplain deposits were more vulnerable to weathering and experienced a distinct loss of material owing to the presence high amounts of clay matrix and mica content. However, the role of primary depositional controls and diagenetic processes in determining microscale pore characteristics, as they directly influence permeability properties, requires much greater investigation in the context of stone durability. The presence of primary hydraulic structural features such as bedding and laminations in building sandstone was noted in a study by Carey & Curran (2000). High-permeability zones were found to correspond to coarse-grained cross-laminations, whereas finer-grained laminations produced lower permeabilities. The influence of depositional structural controls on permeability has also been recorded by Thomachot & Jeannette (2002), in that permeability was found to be greater parallel to bedding rather than perpendicular to it regardless of petrophysical properties. Grain-size and relatively minor mineralogical differences between samples of the same sandstone were found to affect durability characteristics in salt weathering and freeze–thaw weathering simulations (Warke *et al.* 2004, 2006). Coarser-grained sandstone exhibiting extensive interlocking quartz overgrowths displayed greater durability in both types of experimental trials than the finer-grained sandstone with a slightly higher clay content. In effect, the findings from these studies indicate that weathering such as salt and freeze–thaw accentuates and exploits heterogeneity in natural stone (Thomachot & Jeannette 2002; Warke *et al.* 2004). The vulnerability of clay minerals to salt damage has been explained by their tendency to act as points of moisture ingress and as foci for salt accumulation (Rodriguez-Navarro & Doehne 1999; Warke & Smith 2000). The potential addition of the swelling properties of smectite clays to the disruptive effects of salt crystallization has also been noted (e.g. McGreevy 1996). McGreevy (1996) considered the presence of diagenetic smectite within stylolites to be a contributing control on the susceptibility of chalk to weathering. Low effective porosity in the chalk was counteracted by preferential debris loss around stylolite seams (McGreevy 1996).

Gypsum-related decay in a non-calcareous building sandstone was found to be directly related to the exploitation of an intrinsic source of calcium from igneous-related diagenesis of the original clay-rich arkosic sandstone (McKinley *et al.* 2001). The choice of building stone from 'hardened' sandstone in close contact with an igneous intrusion had a direct bearing on the subsequent deterioration of the building sandstone under salt weathering conditions. Alteration of the authigenic mineralogical make-up of the sandstone, produced directly in response to contact with the igneous source, meant it was particularly vulnerable to exploitation by gypsum salts. Matias & Alves (2002) identified the influence of petrographic factors on the durability of granite stone. Grain-size variation and crystalline heterogeneity were found to produce differential weathering patterns. The indication from these studies is that gaining an understanding of a rock's individual characteristics, inherited from its formation history, provides a better appreciation of the potential for moisture and salt movement and of the disruptive effects of salts.

The porosity and permeability properties of stone in conjunction with its inherited characteristics will change as weathering progresses. Pores may become filled with the accumulation of salts and secondary porosities will be created through mechanical breakdown and the development of microfracture networks (Smith *et al.* 2005). Rodriquez-Navarro & Doehne (1999) investigated the importance of pore size on crystallization and growth patterns of different salts. An important outcome of the work by Rodriquez-Navarro & Doehne (1999) is the emphasis on the importance of the hydraulic properties of the pore system in determining the flow rate and evaporation rate of the saline solution and, thus, resultant crystallization. Porosity provides information on pore structure, but measurement of permeability is essential in the monitoring of the hydraulic properties of pore systems. Research questions still remain as to whether the stone heterogeneity exploited by weathering will persist as stone deterioration continues or whether salt crystallization seals off pore space and effectively homogenizes the pore system. Spatial mapping of permeability variation presents the opportunity to quantitatively assess and monitor the ongoing deterioration of stone.

## The scale of observation

The issue of scale has been identified for a considerable period in geomorphology and weathering studies (e.g. Schumm & Lichty 1965; Smith 1996; Philips 1999; Viles 2001). At the micropore scale the sorting and packing of grains can be markedly variable. Thus, non-uniformity or heterogeneity is inherent, even at the pore scale. However, random variations or heterogeneous elements at the pore scale may be sufficiently small to be considered homogeneous at a larger scale, for example laminae, stratum or microcracks. This uniqueness of a stone decay system has implications for the weathering of stone (Smith 1996). Microscale variations in effective porosity within an individual sandstone building block or on the face of an outcrop have a subsequent impact on permeability characteristics and may contribute to the development of differential weathering and surface retreat (Rodriquez-Navarro & Doehne 1999; Warke *et al.* 2006). However, the impact of small-scale variations can only be fully assessed once their presence is recognized and quantification of their variability achieved (Corbett *et al.* 1992).

Philips (1999) suggests three categories in an attempt to cope with scale linkage. These are:

- hierarchy theory for linking processes at multiple scale-defined hierarchical levels;
- mathematical tools for translating process descriptions or analyses across spatial scales;
- techniques for identifying critical spatial scales.

Considering the issue of scale in relation to the characteristics of building stone, primary hydraulic features such as laminae, stratum or larger bedding features can be classified in terms of a scale-defined hierarchy (Viles 2001). Whereas diagenetic cements may be restricted to pore scale or be restrained by scale-dependent depositional features, fractures tend to cut across scale boundaries. In terms of rock properties, porosity measurements describe the characteristics of the rock at the scale of individual pores. However, as permeability is related to the connection of pores, permeability measurements present the opportunity to explore the linkage of the movement of moisture and salts across scales. Investigation of the spatial distribution of permeability enables the analysis of variability at a pore scale to be integrated with examination of variation at laminae or stratum scale. As Viles (2001) suggests, variogram analysis enables similarity in weathering features such as patterns of relief (e.g. Inkpen *et al.* 2000) to be identified at different scales. Extending the use of geostatistical techniques using parameters from variography for spatial prediction and spatial simulation allows the researcher to predict and simulate processes and resultant features across spatial scales. Investigating the spatial variability of permeability enables zones of high permeability, and thus potential areas of moisture and salt ingress, to be identified. The critical scale of petrophysical features at which porosity and permeability characteristics would affect the durability of the stone can then be identified and assessed.

## Discussion of experimental studies

The results from experimental studies of a range of rock types comprising sandstone, limestone and granite are presented. The aim of the studies is to examine the role of permeability in relation to mapping the spatial variability of rock properties as a quantitative evaluation of the weathering of stone. Cubic blocks (75 × 75 × 75 mm) of fresh cut quarry stone were used for analysis, which were set aside from a set of 66 blocks involved in salt weathering experiments (Warke *et al.* 2006). The rock types comprise a medium- to coarse-grained and a fine- to medium-grained Carboniferous Sandstone (Stanton Moor Sandstone, Millstone Grit Series), a Permian sandstone (Dumfries Sandstone), a Tertiary granite (Leinster Granite) and a Jurassic limestone (Portland Limestone). Permeability measurements were made using an unsteady-state Portable Probe Permeameter (PPP250™, Core Laboratories Instruments, 2001). Unsteady-state permeametry measures pressure decay as a function of time, enabling the computation of gas (air) permeability. Measurements are made by pressing the probe tip fitted with a Neoprene seal against the rock surface. Initial flow pressure declines as gas flows into the rock surface, the decay v. time is recorded and the permeability is calculated as millidarcies (mD) from the pressure decay curve by a DAQ™ card in a laptop

(PPP 250™, Core Laboratories Instruments 2001). In basic terms, the higher the permeability of the sample, the faster the pressure will decay from an arbitrary initial pressure (psig – pounds per square inch gauge) (Jones 1992).

A regular grid scheme was adopted to avoid any bias with regards to bedding or laminae structures. The results shown relate to one block face of each of the rock types. Measurements taken at a sample spacing of 10 mm provided a total of 49 measurements for each block. Permeability distributions and summary statistics for each of the rock types are shown in Figure 1 and Table 1, respectively. Petrographic analysis was performed on all rock types and porosity estimated from optical microscopy (Galehouse 1971).

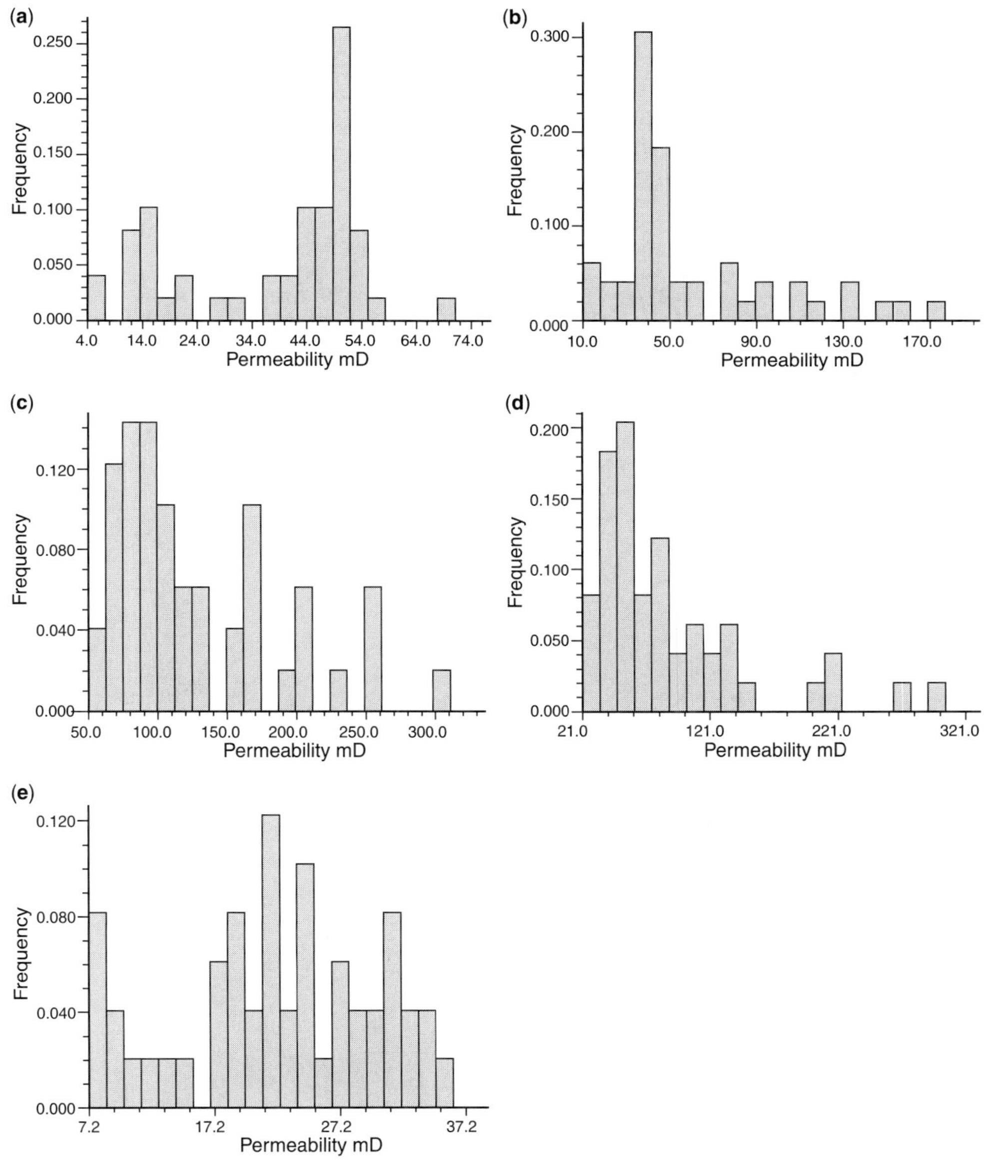


**Fig. 1.** Histograms of permeability distributions for the different rock types: (**a**) medium- to coarse-grained Carboniferous (Stanton) sandstone; (**b**) fine- to medium-grained Carboniferous (Stanton) sandstone; (**c**) Permian (Dumfries) sandstone: (**d**) Jurassic (Portland) limestone; and (**e**) Tertiary (Leinster) granite. Comparability between graphs is best achieved through a comparison of the distribution shape of the histograms. For this reason *y*-axes are not presented on the same scale.

**Table 1.** *Porosity values (vol. %) and summary statistics for permeability data (mD) for the different rock types*

| Statistics | Carboniferous sandstone Medium–coarse grained | Carboniferous sandstone Fine–medium grained | Permian sandstone | Jurassic limestone | Tertiary granite |
|---|---|---|---|---|---|
| Porosity (vol. %) | 13.5 | 17 | 26.5 | 18 | 0 |
| Permeability (mD) | | | | | |
| Mean | 37.84 | 59.74 | 127.25 | 87.47 | 22.13 |
| Maximum | 67.9 | 169 | 298 | 291 | 34.8 |
| Median | 45.4 | 42.5 | 103 | 64.8 | 22 |
| Minimum | 4.01 | 10.3 | 49.7 | 20.5 | 7.19 |
| Range | 63.89 | 158.7 | 248.3 | 270.5 | 27.61 |
| Standard deviation | 16.53 | 38.2 | 49.7 | 60.76 | 7.85 |

Geostatistical analysis was used to characterize the spatial variability of permeability. Parameters from variogram analysis were used for spatial prediction (kriging) and spatial simulation. Sequential Gaussian simulation (SGS), in which simulated values are conditional on the original permeability data and previously simulated values, was used to generate a spatial representation of permeability variation. A single simulated realization is shown in Figure 2 for one block face of each of the rock types. SGS was conducted using algorithms supplied as part of the Geostatistical Software Library (GSLIB; Deutsch & Journel 1998). A full discussion of the geostatistical technique deployed in this study is detailed in Deutsch & Journel (1998), and application of the technique to permeability studies in Lloyd *et al.* (2003), and McKinley *et al.* (2004, 2006).

The distributions displayed in Figure 1 demonstrate a broad range of permeability values for all of rock types. Histograms for the fine- to medium-grained Carboniferous sandstone, the Permian sandstone and the Jurassic limestone exhibit a positive skew and indicate the presence of a high proportion of lower values within a wide range of permeabilities (Table 1). The medium- to coarse-grained Carboniferous sandstone and the Tertiary granite display smaller ranges of permeability values and show histograms tending towards normal distributions. The simulated realizations (Fig. 2) illustrate spatial variability in permeability for all rock types with visible areas of low and high permeability.

## Influence of rock properties on spatial variability of permeability

### *Carboniferous sandstone (Stanton Moor)*

The histograms and simulated realizations highlight differences in the range of values and the spatial variability in permeability between the medium- to coarse-grained and fine- to medium-grained Carboniferous sandstones (Figs 1a, b & 2a, b). Higher mean values for porosity and permeability are recorded for the fine- to medium-grained Carboniferous sandstone, along with a larger range of permeability values than for the medium- to coarse-grained sandstone (Table 1). This difference in rock parameters between the sandstones is reflected in the variation exhibited in grain size and authigenic mineralogy. Although quartz forms the predominant detrital framework mineral in both the sandstones with subordinate plagioclase and K-feldspar, authigenic mineralogy is variable. A tightly interlocking mosaic of quartz cement (Fig. 3a) occurs in both sandstones and would be the most likely cause for reducing porosity and permeability in this sandstone type. However, diagenetic clays in-filling pore spaces (Fig. 3b) and as *in situ* replacement of silicate grains form a significant proportion of the authigenic cement for both sandstones and would potentially provide points of weakness for the ingress of moisture and salts. During weathering simulation experiments, using a combination of frost and salt, the finer-grained Carboniferous sandstone, which contained a higher proportion of diagenetic clays, experienced significantly more deterioration in structural integrity in comparison to the coarse-grained sandstone samples (Warke *et al.* 2006). The indication from the weathering trials was that the greater the range in initial permeability values, the greater the potential for salt and moisture ingress and retention, and hence eventual disruption of the fabric of the stone (Warke *et al.* 2006). Hence, knowledge of permeability variability is more important than generating mean permeabilities in estimating the overall weathering properties of Stanton Moor Sandstone, and understanding the spatial distribution of areas of high and low permeability enables potential points of salt and moisture ingress to be predicted (McKinley *et al.* 2006). The greater range of permeability values may also have influenced the movement of salts and moisture within the stone fabric allowing accumulation at

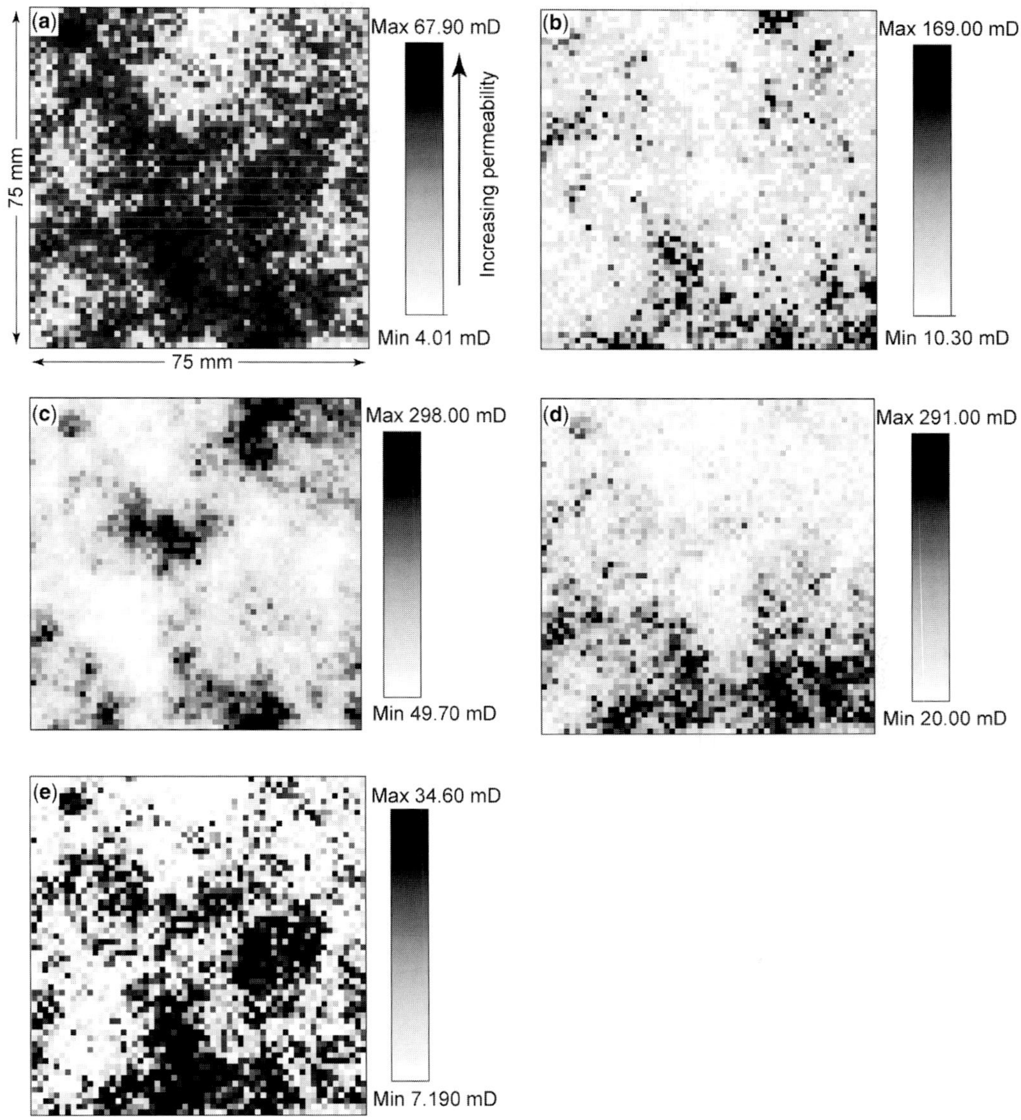


**Fig. 2.** Single SGS realizations for the different rock types: (**a**) medium- to coarse-grained Carboniferous (Stanton) sandstone; (**b**) fine- to medium-grained Carboniferous (Stanton) sandstone; (**c**) Permian (Dumfries) sandstone: (**d**) Jurassic (Portland) limestone; and (**e**) Tertiary (Leinster) granite.

depth. Distinct rates and patterns of breakdown can be related to relatively minor structural and mineralogical differences between blocks of the same stone type and this has been shown to have a significant influence on weathering behaviour.

### *Permian sandstone (Dumfries Sandstone)*

The histogram of the Permian sandstone suggests a positive skewness and indicates the presence of a large proportion of low-permeability values (Fig. 1c). Highest mean permeability and porosity values are recorded for this rock type together with the largest range of permeabilities when compared to the other rock types (Table 1). In terms of detrital mineralogy, quartz forms the predominant framework grain in this red-coloured sandstone. The major feldspar is K-feldspar with subordinate plagioclase. Quartz overgrowths and authigenic feldspar are present in small amounts. Diagenetic

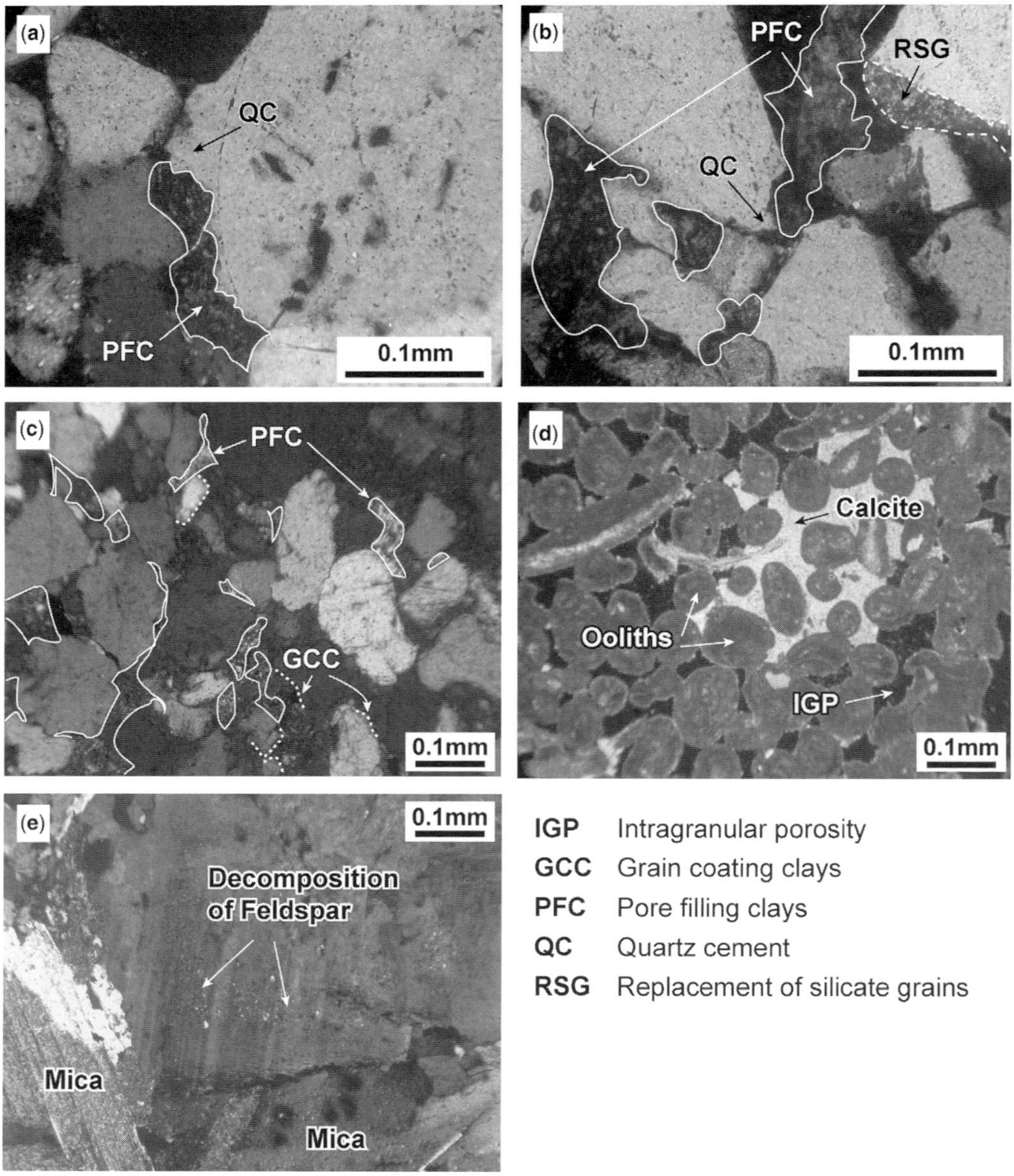


**Fig. 3.** Photomicrographs of the different rock types: (**a**) medium- to coarse-grained Carboniferous (Stanton) sandstone; (**b**) fine- to medium-grained Carboniferous (Stanton) sandstone; (**c**) Permian (Dumfries) sandstone: (**d**) Jurassic (Portland) limestone; and (**e**) Tertiary (Leinster) granite.

clays form the most significant cement in the Permian sandstone and are found exhibiting several habits: *in situ* replacement of silicate grains (mainly feldspars), grain-coating rims and, to a lesser extent, in-filling pores (Fig. 3c). The simulated realization of permeability variation indicates the location of permeable areas in the Permian sandstone (Fig. 2c). High mean porosity and permeability values would suggest reduced durability properties for this building sandstone. However, identification of the mineralogical composition of the rock, including the elevated diagenetic clay content, combined with an increased knowledge of the spatial distribution of high-permeability areas provides an increased understanding and awareness of the location of potential vulnerability of this building sandstone to deterioration.

### *Jurassic limestone (Portland Limestone)*

The main constituent of the Jurassic (Portland) limestone, calcite, is found in various forms. Ooliths, formed of micrite (calcareous mud), are evident throughout the limestone; crystalline calcite is also found as rim coatings on ooliths and as a pore-filling cement (Fig. 3d). Both skeletal fragments and quartz grains are found as the centre of ooliths with concentric accumulation of micritic calcite around the cores. Dark coloured impurities in the ooliths indicate the presence of clays. Intergranular porosity appears volumetrically most significant with less significant secondary porosity. The permeability histogram (Fig. 1d & Table 1) exhibits a broad range of values, but a positive skew indicates the presence of a high proportion of low permeabilities with less abundant higher values. The simulated realization of permeability in Figure 2d clearly demonstrates a spatially distributed zoning of high and low permeabilities. The lower part of the block face records much higher values (>200 mD) than the upper part of the face (<100 mD). High-permeability values that concur with volumetrically significant intergranular porosity indicated from optical microscopy can be related to the absence of recrystallized secondary calcite in this part of the limestone block and the presence of intergranular porosity between ooliths. The inference suggested is that variations in permeability may lead to significant spatial variability in deterioration of the limestone.

### *Tertiary (Leinster) granite*

This feldspar-rich granite displays the lowest mean and smallest range of permeability values compared to the other rock types (Table 1). Large plagioclase feldspars in combination with quartz form the major constituents of the rock. Biotite mica, present in significant amounts, forms large laths set in a surrounding interlocking mosaic of quartz and feldspar (Fig. 3e). Alteration of plagioclase feldspar is a conspicuous feature of the rock and as a result clays form an important constituent of the rock (up to 10%). Porosity is not detected by optical microscopy (Table 1) but may be present as secondary microporosity in diagenetic clays. The spatial distribution of permeability, as represented by the simulated realization in Figure 2e, indicates a variable arrangement of hot spots of higher permeabilities. Evidence from petrographic analysis suggests that alteration and decomposition of plagioclase feldspar (Fig. 3e) has produced microporosity in diagenetic clays, which can be related to the zones of higher permeability. The spatial variability of permeability may also indicate the spatial distribution of microfractures and highlights the role of microcracks as a contributor to overall porosity in the granite block. These would potentially enable the development of conduits or pathways for the migration of moisture and salts into the otherwise low-permeability granite.

Findings from the discussed experimental studies illustrate that the range and spatial distribution of permeabilities provide greater insight into durability properties and predicted weathering behaviour of stone than average porosity or permeability values. When related to fabric and mineralogy (including diagenetic cements), the controls and locations of potential weaknesses of the rock types can be explained and predicted with respect to moisture movement, salt migration and vulnerability to salt damage.

## Conclusions: implications for stone weathering

Permeability has been shown to be an important parameter in the investigation of stone weathering and should be considered in combination with the measurement of porosity. The controls on the porosity and permeability characteristics of stone are inherited from its formation history but may be modified as weathering continues. An increased knowledge of the inherited characteristics of a rock will provide valuable insight and greater understanding of the vulnerability of building stone to deterioration. As stone heterogeneity is accentuated and exploited by weathering, continued exposure to moisture and salts may result in the blocking of pores and homogenization of a pore system. The measurement of permeability provides an effective method of monitoring the hydraulic properties of stone and how these are modified by weathering processes. Further work is required through laboratory experimental simulation and exposure trials to fully assess the implications of this. Mapping the spatial distribution of permeability provides greater insight into the extent of variability in stone durability characteristics than generating mean porosity or permeability values. Geostatistical techniques provide the opportunity to assess and compare the influence of different scales of rock properties and identify the critical scale of variation for stone deterioration. Spatial mapping of permeability characteristics presents the possibility of generating a quantitative assessment of building stone and a prediction of stone durability properties through spatial prediction and simulation.

This work was jointly supported by the Engineering and Physical Sciences Research Council grant GR/R79449/01 and GR/R54491/01. The authors would like to thank the referees for very helpful comments that considerably improved an earlier version of the paper. The authors would also like to thank M. Pringle and G. Alexander in the Cartography Department of the School of Geography for help with the diagrams.

## References


ALLEN, D., COATES, G., AYOUB, J. & CARROLL, J. 1988. Probing for permeability: an introduction to measurements with a Gulf Coast case study. *The Technical Review*, **36**(1), 6–20.

BEARD, D. C. & WEYL, P. K. 1973. Influence of texture on porosity and permeability of unconsolidated sand. *AAPG*, **57**, 349–369.

BEGGAN, J., LONG, A. E. & BASHEER, P. A. M. 1996. The permeability testing of masonry materials. *In*: SMITH, B. J. & WARKE, P. A. (eds) *Processes of Urban Stone Decay*. Donhead, London, 205–211.

BIDNER, T., MIRWALD, P. W., RECHEIS, A. & BRÜGGERHOFF, S. 2002. Stone as sensor material for weathering. *In*: PŘIKRYL, R. & VILES, H. A. (eds) *Understanding and Managing Stone Decay*. Karolinum Press, Prague, 97–111.

CAREY, P. F. & CURRAN, J. M. 2000. High resolution characterization of permeability in arenaceous building stones. *Zeitschrift für Geomorphologie*, **120**, 175–185.

CHOQUETTE, P. W. & PRAY, L. C. 1970. Geologic nomenclature and classification of porosity in sedimentary carbonates. *AAPG Bulletin*, **54**, 207–250.

COOKE, R. U. 1979. Laboratory simulation of salt weathering processes in arid environments. *Earth Surface Processes Landforms*, **4**, 347–359.

CORBETT, P. W. M., RINGROSE, P. S., JENSEN, J. L. & SORBIE, K. S. 1992. Laminated clastic reservoirs: The interplay of capillary pressure and sedimentary architecture. Paper presented at the 67th Annual Technical Conference and Exhibition, Washington, DC, 4–7 October. SPE Paper, **24699**, 365–376.

CORE LABORATORIES INSTRUMENTS. 2001. *PPP-250™ Portable Probe Permeameter, Operators Manual*. Core Laboratories Instruments, Houston, TX.

CROSS, T. A., BAKER, M. R. *ET AL*. 1993. Applications of high-resolution sequence stratigraphy to reservoir analysis. *In*: ESCHARD, R. & DOLIGEZ, B. (eds) *Subsurface Reservoir Characterisation from Outcrop Observations*. Editions Technip, Paris, 11–33.

DEUTSCH, C. V. & JOURNEL, A. G. 1998. *GSLIB: Geostatistical Software Library and User's Guide*, 2nd edn. Oxford University Press, New York.

DREVER, J. I. 1994. Durability of stone: mineralogical and textural perspectives. *In*: KRUMBEIN, W. E., BRIMBLECOMBE, P., COSGROVE, D. E. & STANIFORTH, S. (eds) *Durability and Change: The Science, Responsibility and Cost of Sustaining Cultural Heritage*. Dahlem Workshop Reports, Environmental Sciences Research Report, **15**, 27–37.

GALEHOUSE, J. S. 1971. Point counting. *In*: CARVER, R. E. (ed.) *Procedures in Sedimentary Petrology*. Wiley-Interscience, New York, 385–409.

GOGGIN, D. J. 1993. Probe permeametry: is it worth the effort? *Marine and Petroleum Geology*, **10**, 299–308.

GOUDIE, A. S. 1999. Experimental salt weathering of limestones in relation to rock properties to rock properties. *Earth Surface Processes and Landforms*, **24**, 715–724.

HOUSEKNECHT, D. W. 1987. Assessing the relative importance of compaction processes and cementation to reduction of porosity in sandstones. *AAPG Bulletin*, **71**, 633–642.

HURST, A. & GOGGIN, D. J. 1995. Probe permeametry: an overview and bibliography. *AAPG Bulletin*, **79**, 463–473.

INKPEN, R. J., COLLIER, P. & FONTANA, D. 2000. Close-range photogrammetry of rock surfaces. *Zeitschrift für Geomorphologie Supplement*, **120**, 67–81.

INKPEN, R. J., PETLEY, D. & MURPHY, W. 2004. Durability and rock properties. *In*: SMITH, B. J. & TURKINGTON, A. V. (eds) *Stone Decay: Its Causes and Controls*. Donhead, London, 33–52.

JONES, S. C. 1992. The Profile Permeameter: A new, fast, accurate minipermeameter. Paper presented at the 67th Annual Technical Conference and Exhibition of the Society of Petroleum Engineers, Washington, DC, October. SPE Paper, **24757**, 973–983.

JORNET, A., TERUZZI, T. & RÜCK, P. 2002. Bowing of Carrara marble slabs: comparison between natural and artificial weathering. *In*: PŘIKRYL, R. & VILES, H. A. (eds) *Understanding and Managing Stone Decay*. Karolinum Press, Prague, 161–169.

LLOYD, C. D., MCKINLEY, J. M. & RUFFELL, A. H. 2003. Conditional simulation of sandstone permeability. *In: Proceedings of IAMG 2003, Portsmouth, UK, 7–12 September 2003*. International Association for Mathematical Geology IAMG, Kingston, Canada.

MATIAS, J. M. S. & ALVES, C. A. S. 2002. The influence of petrographic, architectural and environmental factors in decay patterns and durability of granite stones in Braga monuments (NW Portugal). *In*: SIEGESMUND, S., WEISS, T. & VOLLBRECHT, A. (eds) *Natural Stone, Weathering Phenomena, Conservation Strategies and Case Studies*. Geological Society, London, Special Publications, **205**, 273–281.

MARKOPOULOS, T. & GALETAKIS, M. 2002. Investigation of stones used for the restoration/reconstruction of the church of saint Peter Dominican, Heraklion-Crete. *In*: PŘIKRYL, R. & VILES, H. A. (eds) *Understanding and Managing Stone Decay*. Karolinum Press, Prague, 189–196.

MCGREEVY, J. P. 1996. Pore properties of limestones as controls on salt weathering susceptibility: a case study. *In*: SMITH, B. J. & WARKE, P. A. (eds) *Processes of Urban Stone Decay*. Donhead, London, 150–167.

McKinley, J. M., Curran, J. M. & Turkington, A. V. 2001. Gypsum formation in non-calcareous building sandstone: a case study of Scrabo Sandstone. *Earth Surface Processes and Landforms*, **26**, 869–875.

McKinley, J. M., Lloyd, C. D. & Ruffell, A. H. 2004. Use of variography in permeability characterisation of visually homogeneous sandstone reservoirs with examples from outcrop studies. *Mathematical Geology*, **36**, 761–779.

McKinley, J. M., Warke, P., Lloyd, C. D., Ruffell, A. H. & Smith, B. 2006. Geostatistical analysis in weathering studies: case study for Stanton Moor building sandstone. *Earth Surface Processes and Landforms*, **31**, 950–969.

Morad, S. 1998. *Carbonate Cementation in Sandstones*. International Association of Sedimentologists, Special Publications, **26**.

Nicholson, D. T. 2001. Pore properties as indicators of breakdown mechanisms in experimentally weathered limestones. *Earth Surface Processes and Landforms*, **26**, 819–838.

Nicholson, D. T. 2002. Quantification of rock breakdown for experimental weathering studies. *In*: Přikryl, R. & Viles, H. A. (eds) *Understanding and Managing Stone Decay*. Karolinum Press, Prague, 59–74.

Park, W. C. & Schot, E. H. 1968. Stylolites: their nature and origin. *Journal of Sedimentary Petrology*, **38**, 175–191.

Pera, E. & Burlini, L. 2002. Elastic properties of selected Italian marbles. *In*: Přikryl, R. & Viles, H. A. (eds) *Understanding and Managing Stone Decay*. Karolinum Press, Prague, 171–182.

Philips, J. D. 1999. *Earth Surface Systems – Complexity, Order and Scale*. Blackwell, Oxford.

Přikryl, R. & Dudková, I. 2002. Experience with long term experimental exposure of building stones: a 70 years study in the Czech Republic. *In*: Přikryl, R. & Viles, H. A. (eds) *Understanding and Managing Stone Decay*. Karolinum Press, Prague, 135–143.

Rodriguez-Navarro, C. & Doehne, E. 1999. Salt weathering: influence of evaporation rate, supersaturation and crystallisation pattern. *Earth Surface Processes and Landforms*, **24**, 191–209.

Russell, M. I., Basheer, P. A. M., Harmon, N. & Curran, J. M. 2002. Permeation properties of building stone: The Autoclam Permeability System. *In*: Přikryl, R. & Viles, H. A. (eds) *Understanding and Managing Stone Decay*. Karolinum Press, Prague, 75–84.

Schumm, S. A. & Lichty, R. W. 1965. Time, space, and causality in geomorphology. *American Journal of Science*, **263**, 110–119.

Shelford, A., Inkpen, R. J. & Payne, D. 1996. Spatial variability of weathering of Portland stone slabs. *In*: Smith, B. J. & Warke, P. A. (eds) *Processes of Urban Stone Decay*. Donhead, 98–110.

Smith, B. J. 1996. Scale problems in the interpretation of urban stone decay. *In*: Smith, B. J. & Warke, P. A. (eds) *Processes of Urban Stone Decay*. Donhead, London, 3–18.

Smith, B. J. & Kennedy, E. 1999. Moisture loss from stone influenced by salt accumulation. *In*: Jones, M. S. & Wakefield, R. D. (eds) *Aspects of Stone Weathering, Decay and Conservation*. Imperial College Press, London, 55–64.

Smith, B. J., Warke, P. A., McGreevy, J. P. & Kane, H. L. 2005. Salt-weathering simulations under hot desert conditions: agents of enlightenment or perpetuators of preconceptions? *Geomorphology*, **67**, 211–227.

Sutherland, W. J. C., Halvorsen, A., Hurst, A., McPhee, G. & Worthington, P. F. 1993. Recommended practice for probe permeametry. *Marine and Petroleum Geology*, **10**, 309–317.

Thomachot, C. & Jeannette, D. 2002. Evolution of the petrophysical properties of two types of Alsatian sandstone subjected to simulated freeze-thaw conditions. *In*: Siegesmund, S., Weiss, T. & Vollbrecht, A. (eds) *Natural Stone, Weathering Phenomena, Conservation Strategies and Case Studies*. Geological Society, London, Special Publications, **205**, 19–32.

Timmerman, E. H. 1982. *Practical Reservoir Engineering*. Pemwell, Tulsa, OK.

Tucker, M. E. 1994. *Sedimentary Petrology An Introduction*. Blackwell Scientific, Oxford.

Turkington, A. V. & Smith, B. J. 2000. Observations of three-dimensional salt distribution in building sandstone. *Earth Surface Processes and Landforms*, **25**, 1317–1332.

Turkington, A. V. & Smith, B. J. 2004. Interpreting spatial complexity of decay features on a sandstone wall: St Matthew's Church, Belfast. *In*: Smith, B. J. & Turkington, A. V. (eds) *Stone Decay: Its Causes and Controls*. Donhead, 149–166.

Viles, H. A. 2001. Scale issues in weathering studies. *Geomorphology*, **41**, 63–72.

Warke, P. A. & Smith, B. J. 2000. Salt distribution in clay-rich weathered sandstone. *Earth Surface Processes and Landforms*, **25**, 1333–1342.

Warke, P. A., McKinley, J. M. & Smith, B. J. 2006. Variable weathering response in sandstone: factors controlling decay sequences. *Earth Surface Processes and Landforms*, **31**, 715–735.

Warke, P. A., Smith, B. J. & McKinley, J. M. 2004. Complex weathering effects on durability characteristics of building sandstone. *In*: Přikryl, R. (ed.) *Dimension Stone 2004*. Taylor & Francis, London, 229–235.

Weber, J. & Lepper, J. 2002. Depositional environment and diagenesis as controlling factors for petro-physical properties and weathering resistance of siliciclastic dimension stones: integrative case study on the 'Wesersandstein' (northern Germany, Middle Buntsandstein). *In*: Siegesmund, S., Weiss, T. & Vollbrecht, A. (eds) *Natural Stone, Weathering Phenomena, Conservation Strategies and Case Studies*. Geological Society, London, Special Publications, **205**, 103–114.

Worden, R. H. 1998. Dolomite cement distribution in a sandstone from core and wireline data: the

Triassic fluvial Chaunoy Formation, Paris Basin. *In*: HARVEY, P. K. & LOVELL, M. A. (eds) *Core-log Integration*. Geological Society, London, Special Publications, **136**, 197–211.

WORDEN, R. H. & MORAD, S. (eds). 2000. *Quartz Cementation in Sandstones*. International Association of Sedimentologists, Special Publications, **29**.

WORDEN, R. H. & MORAD, S. (eds). 2001. *Clay Cementation in Sandstones*. International Association of Sedimentologists, Special Publications **34**.

# Texture, spatial and orientation dependence of internal strains in marble: a key to understanding the bowing of marble panels?

CH. SCHEFFZÜK[1,2], S. SIEGESMUND[3], D. I. NIKOLAYEV[2] & A. HOFFMANN[3]

[1]*GeoForschungsZentrum Potsdam, Section 5.3, Telegrafenberg, 14473 Potsdam, Germany (e-mail: scheff@gfz-potsdam.de)*

[2]*Frank Laboratory of Neutron Physics, JINR Dubna, 141980 Dubna, Russia*

[3]*Geoscience Centre, University Göttingen, Goldschmidtstrasse 3, 37077 Göttingen, Germany*

**Abstract:** A Carrara marble sample was measured using neutron time-of-flight diffraction on its crystallographic preferred orientation (texture), a strain scan through the sample and strain pole figures to explain the effect of deformation of marble plates. Strong compressional residual strain values of up to $\varepsilon = -1.3 \times 10^{-3}$ measured with residual strain pole figures in a virgin bulk sample have been found. Obviously, the magnitude of residual strain depends on the sample size. Features of the observed texture pole figures and internal strain pole figures are related to each other by their sample orientations. Texture and residual strain measurements were combined with investigations of thermal expansions under dry and wet conditions in different directions to the main stress direction.

Numerous cases of damage to sculptures, architectural heritage or façade marble stones indicate that the deterioration of building stones depends mainly on climate. The clear morphological alteration of marble even within a short exposure time (e.g. Grimm 1999) is well known. Recently, it has also been proposed that physical weathering is the initial stage of deterioration of marble (see Siegesmund *et al.* 2000).

Durability is an important issue to consider when specifying stones as a cladding material for exterior exposure. But the spectacular bowing behaviour of marble slabs has given a negative image to this material. The complete replacement of façade panels of some prestigious buildings like the Amoco building in Chicago (Logan *et al.* 1993), the Finlandia Hall in Helsinki, the Grand Arche de la Defense in Paris or the University Library in Göttingen (see Fig. 1), all made of marble coming from the Carrara area, are often cited as examples of the concerns for the durability of these materials. However, bowing was frequently reported from ancient gravestones (e.g. Grimm 1999). Detailed knowledge of the mechanisms and rates of decay is important in order to protect historical monuments, as well as to reduce loss of marble during reconstruction. Kessler (1919) found that repeated heating may lead to permanent dilatation owing to microfracturing. Based on laboratory testing, Logan *et al.* (1993) explained the bowing of marble slabs on the Amoco building as a result of anomalous expansion–contraction behaviour of calcite together with the release of locked residual stresses. Hypotheses proposed by other researchers have assumed the presence of moisture or gravity variation in the material (e.g. Winkler 1994). The reasons for the observed deformation are still under discussion. It turns out, for example, that natural marble also shows complex anisotropic weathering behaviour. It has been shown (Siegesmund *et al.* 2000; Koch & Siegesmund 2002) that weathering behaviour is influenced by thermal expansion, existing cracks (Siegesmund *et al.* 2000) and preferred crystallographic orientation (here referred to as texture) of calcite and/or dolomite (Ruedrich *et al.* 2001; Zeisig *et al.* 2002).

The present paper focuses on the effects of internal stresses. To verify the theory that locked-in stress – that is, stress generated during the geological history of the rock – could be responsible for the bowing intensity, fresh material from a marble quarry in Carrara (Italy) was used for testing. Neutron diffraction was used as a non-destructive method to investigate the texture, and the spatial and orientation dependence of strain in a bulk marble sample. Based on earlier investigations (Scheffzük *et al.* 2004) of texture and residual strain by neutron time-of-flight (TOF) diffraction, this study will focus on a more detailed investigation of a Carrara marble sample. To do this, the neutron TOF texture diffractometer SKAT (Ullemeyer *et al.* 1998) and the strain–stress diffractometer EPSILON-MDS (Frischbutter *et al.* 2000; Walther *et al.* 2000) at JINR, Dubna

*From*: PŘIKRYL, R. & SMITH, B. J. (eds) *Building Stone Decay: From Diagnosis to Conservation.* Geological Society, London, Special Publications, **271**, 237–249.

**Fig. 1.** Typical bowing behaviour of façade panels at the University Library in Göttingen.

(Russia) have been employed. The measured spectra were processed with a program developed recently to extract detailed Bragg diffraction lines by fitting the peak parameters. Such an approach allows one to demonstrate the crystallographic texture as well as the internal strain. Much attention was paid to accuracy and methodology, and the SKAT texture instrument was used to determine strain pole figures in relation to the residual strain determined by strain scans with higher resolution than the EPSILON-MDS strain diffractometer.

To constrain the bowing to internal strain relationship, the sample was also investigated with respect to the rock fabric, *in situ* stresses and thermal expansion behaviour under dry and wet conditions.

## Rock sample and microfabrics

An oriented block of around 1 $m^3$ was taken from the Carrara area (Fig. 2). The sample was oriented with respect to the foliation plane ([$xy$] plane while [$z$] is perpendicular to the foliation) and the principal stress directions $\sigma_1$, $\sigma_2$ and $\sigma_3$. The stress directions were measured by the over-coring method and revealed to be: $\sigma_1 = (5.13 \pm 1.07)$ MPa, $\sigma_2 = (3.72 \pm 0.61)$ MPa and $\sigma_3 = (1.89 \pm 1.32)$ MPa (Alnaes pers. comm.).

The investigated Carrara marble sample was bright white and contained locally different

**Fig. 2.** Fresh marble block that was used to obtain the sample material for the present investigations. Black lines reflect the foliation. Note that the block is in a situation in which the platform represents the upper surface in nature. For a nature-like orientation, the block must be rotated 180° around the <$a$>-axis.

concentrations of typically grey veins. The veins were folded and showed a streak-like distortion, and usually varied in thickness from 1 to 10 mm. Locally, at a decimetre-scale, a preferred orientation of the grey veins could be observed. The microfabrics were homogeneous and fine grained. The sample showed a nearly perfectly recovered grain fabric with straight grain boundaries and frequent 120° triple junctions (Fig. 3). The grain size varied from 140 to 500 μm, the averaged grain size was about 250 μm. Only a very few grains were twinned. A detailed description of the twinning mechanisms and the application for palaeostress estimation is described by Rowe & Rutter (1990). The macroscopically visible grey veins

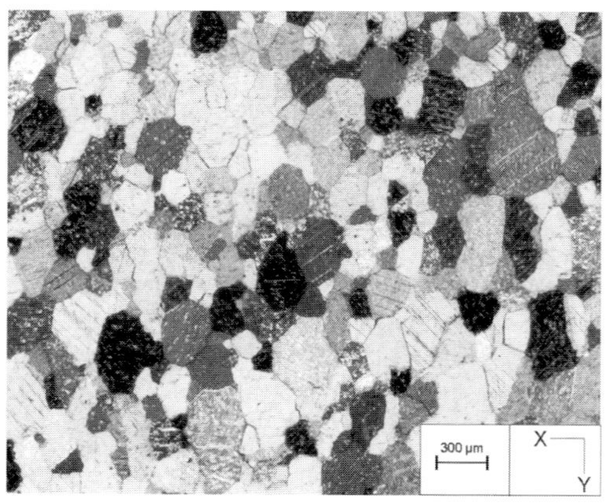


**Fig. 3.** Microscope observations of Carrara marble microstructure: homogeneous nearly perfect recovered microfabrics and fine-grained material, grain size 140–500 μm, grain fabric with straight grain boundaries and frequent 120° triple point junctions.

were more fine grained, but showed very strong undulose extinction. The undulose extinction and the formation of subgrains occurred to such an extent that individual grains were difficult to detect. Grain boundaries were interlobate and a great number of fluid inclusions or graphite are probably responsible for the grey colour of the veins.

## Experiment

In general, crystalline materials can be investigated by diffraction methods, so that by fulfilling the Bragg law

$$2d \sin \Theta = n\lambda \tag{1}$$

(with $d$ the lattice spacing, $\Theta$ the diffraction angle and $\lambda$ the wavelength), Bragg diffraction lines corresponding to their Miller indices are detectable. Based on the high penetration depth of neutrons in matter, neutron diffraction enables one to investigate samples in the centimetre range. In particular, the neutron TOF technique carries out the simultaneous detection of all Bragg reflections in a given range of wavelengths. This technique is well suited to investigating minerals because the peak separation of lines in a line-rich diffraction pattern is possible as a result of the high resolution common for minerals with lower crystal symmetry.

At the pulsed research reactor IBR-2 in Dubna, the diffractometers EPSILON-MDS for strain–stress analysis and SKAT for texture analysis have a neutron flight path of about 101.3 (EPSILON-MDS) and 103.8 m (SKAT) respectively, and makes it possible to investigate Bragg diffraction lines with lattice spacing of up to $d = 5.5$ Å simultaneously.

The spherical sample method of Tobisch & Bunge (1972) is used to place the sample completely into the neutron beam. Consequently, a spherical bulk sample with a gauge volume of up to 100 cm$^3$ can be investigated without geometrical corrections.

Figure 4 demonstrates the sum of all 1368 spectra (for $\varphi = 0°–90°$, $\vartheta = 0°–360°$ with $\Delta\varphi = \Delta\vartheta = 5°$) of the SKAT diffractometer for the calcite powder sample; lattice spacings and their corresponding Bragg reflections are also indicated. The mineral content of other phases, such as graphite, quartz and dolomite, is lower than 5%. The background (Bkg) of the Bragg diffraction peak (see Fig. 5) has been approximated by a linear relationship with the background parameters $b$ and $k$:

$$\text{Bkg} = kt + b. \tag{2}$$

The non-symmetrical Bragg peak was fitted by the bell-shaped asymmetrical Gaussian function of the

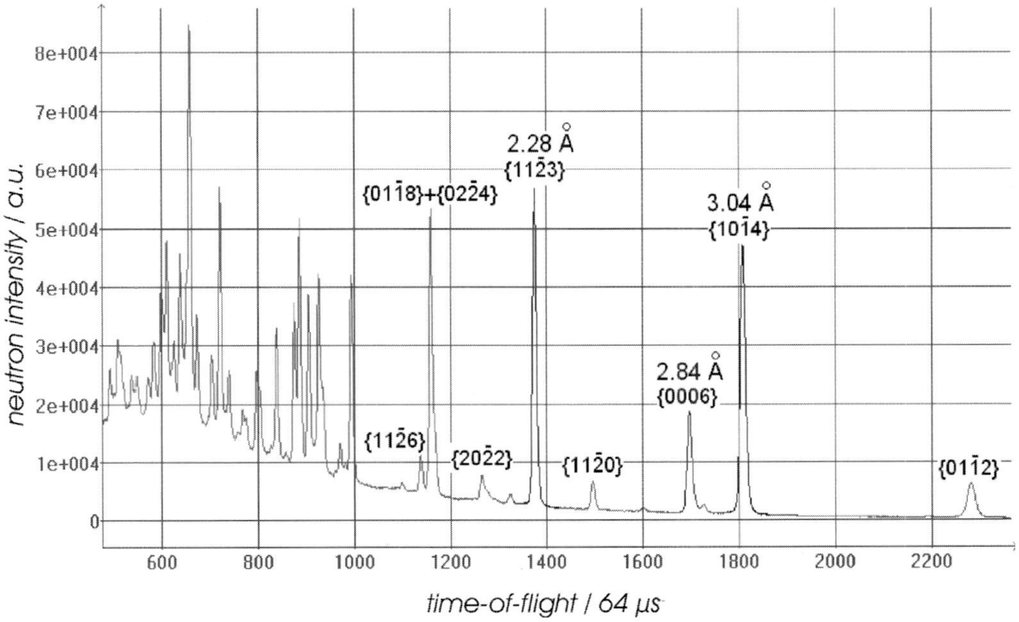


**Fig. 4.** Sum of all the 1368 spectra for calcite powder (SKAT). Bragg reflections indicated by their Miller indices and the corresponding $d$-values are presented.

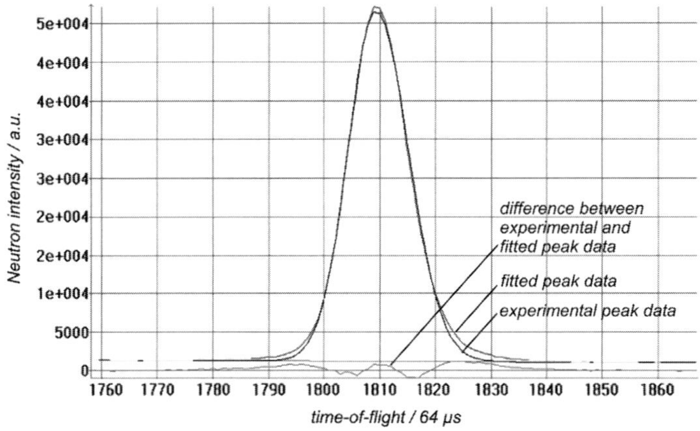


**Fig. 5.** Individual peak $\{10\bar{1}4\}$ approximation: experimentally measured peak, fitted peak, and the difference between measured and fitted curves are plotted.

following form:

$$I(t) = \begin{cases} A_0 \exp\left(-\frac{(t-t_0)^2}{2\sigma_1{}^2}\right) t \le t_0 \\ A_0 \exp\left(-\frac{(t-t_0)^2}{2\sigma_2{}^2}\right) t \le t_0 \end{cases} \quad (3)$$

Here, $A_0$ (φ, ϑ) describes the amplitude, $t_0$ (φ, ϑ) the position and $\sigma_i$ (φ, ϑ) are different half-widths at the half maximum of the Bragg peak. Four parameters with their uncertainties $\Delta A_0$, $\Delta t_0$, $\Delta\sigma_1$, $\Delta\sigma_2$ can be determined for each individual Bragg reflection peak by processing all of the 1368 diffraction pattern. Not only the amount of data has to be processed, but also a large quantity of numerical operations has to be fulfilled. Using the parameters $A_0$ (φ, ϑ), $\sigma_1$ (φ, ϑ), $\sigma_2$ (φ, ϑ), one can produce texture pole figures, the parameter $t_0$ (φ, ϑ) is used to produce the internal strain pole figures.

Figure 5 illustrates the approximated function and the quality of the fit of an individual Bragg peak. One can see that the difference between the experimental and fitted curves is even smaller than the peak background. This is the criteria for high accuracy of the peak fit on one side and of the reasonable choice of the peak shape on the other side.

## Texture

The physical properties of calcite, such as thermal expansion and elastic modulus, are highly anisotropic. Consequently, it is necessary to know whether a sample has a crystallographic preferred orientation (texture) to predict the material behaviour. Although, a strong texture is usually formed during plastic deformation, a weak one does not usually demonstrate primary plastic deformation to any high degree.

A cubic sample (edge length 80 mm) was investigated with the SKAT diffractometer. The powder, prepared from the sample with a grain size of 50 μm, was used as the texture- and strain-free reference of the sample. The Bragg diffraction lines of the trigonal calcite shown in Figure 4 were used to investigate texture and strain. Only those Bragg diffraction lines that do not overlap and have the best peak/background ratio were chosen to analyse the peak parameters. The (0006) Bragg reflection peak was used because the pole figure for this crystallographic index corresponds to the distribution of the *c*-axis, which is the most informative. The accuracy of the peak position estimation obtained for the calcite powder sample is illustrated in Figure 6. Because of the non-spherical sample geometry, the texture data have been corrected by their integral peak–background relation (see Nikolayev *et al.* 2005).

The measured texture of sample g4 shows a low texture strength (cf. Fig. 7). The texture could be characterized as an incomplete fibre texture or the one with very elongated components. The small circle distribution on the $\{10\bar{1}4\}$ pole figure demonstrates evidence for the incomplete fibre texture. The minimum value on all pole figures is larger than 0.69 m.r.d. (multiple of random distribution) and the maximum value is less than 1.25 m.r.d. The smallest minimum value among all pole figures gives an estimate of the volume part of the grains that does not have a preferred orientation. That means that more than 70% of the grain volume fracture is distributed uniformly and only a small volume fracture has a weak preferred orientation. Thus, the measured sample has a very weak texture. The marble block from which the sample was cut possesses an homogeneous texture.

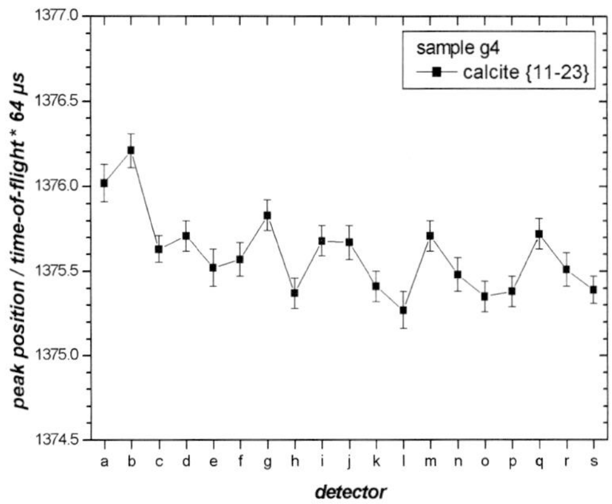


**Fig. 6.** Dependence of the peak position from the detector at the SKAT diffractometer, the error bars demonstrate the quality of peak position reconstruction for the powder sample.

## Residual strain

In mechanics the definition of a strain is derived from a displacement between 'loaded' and 'unloaded' states. In general, the medium is considered contiguous. However, at the scale of the crystallographic lattice the medium could not be considered as a contiguous one. Nevertheless, elastic stress modifies the lattice distances that can be observed by diffraction experiments. So, internal or macroscopic strain can be detected by diffraction methods.

The macroscopic intracrystalline strain is defined as:

$$\varepsilon^{hkl} = \frac{d^{hkl} - d_0^{hkl}}{d_0^{hkl}}. \tag{4}$$

Using the neutron TOF technique the $d$-values correspond with the detected time channels, as seen in Figure 4. Bragg diffraction peaks $(10\bar{1}4)$ and $(11\bar{2}3)$ were analysed using the measured spectra for sample g4. Utilizing powder measurements to estimate the stress-free state, macroscopic internal strain was obtained for this sample. Figure 7 displays the corresponding texture and strain pole figures. Qualitatively, strain pole figures are very similar despite the fact that their corresponding texture pole

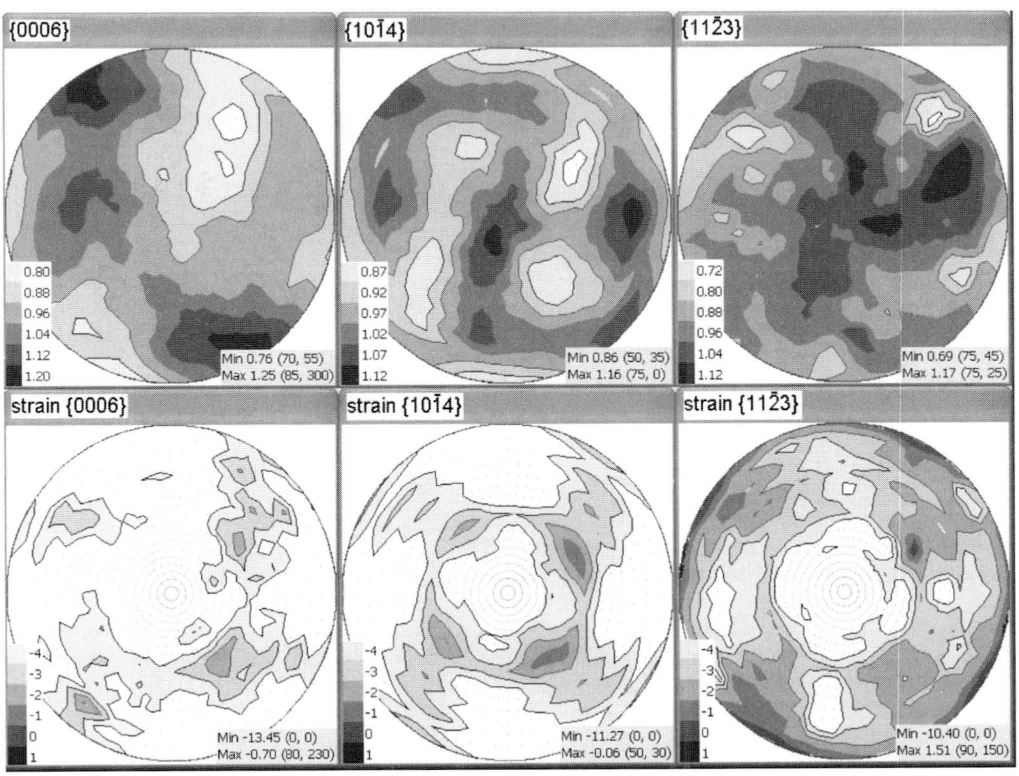


**Fig. 7.** Experimental pole figures (equal area projection; upper row) and internal strain pole figures ($\varepsilon$ in $10^{-4}$, lower row) for sample g4.

figures are different. It means that the microscopic strain, calculated by using different diffraction lines, gives similar magnitudes and features. If the internal strain tensor has constant values then the index surface is ellipsoidal. As one can see from the strain pole figures, they are not ellipsoidal, i.e. the strain field in the studied sample has a much more complicated character. Consequently, the internal strain in the studied sample is not uniform. The strain maxima in Figure 7 are bent over the foliation plane by about 45°.

Next to measurements of sample g4 on SKAT, detailed studies of the sample were carried out on the EPSILON-MDS diffractometer. Figure 8 displays the specific directions for which neutron diffraction patterns were collected by a gauge volume of $5 \times 5 \times 5\ \mathrm{mm}^3$. The sample was translated on the goniometer table by steps of 9 mm to carry out strain scans. Strain values are plotted at various sections. Figures 9 and 10 display the strain values for the Bragg diffraction lines $(10\bar{1}4)$ and $(11\bar{2}3)$. Position points of the sample were measured with an exposition time of 2 h at each position and marked by $\varepsilon_2, \ldots, \varepsilon_8$. Afterwards, the sample was rotated about the $[z]$-axis by 175° counter-clockwise and a further scan was carried out. The corresponding positions are marked by $\varepsilon'_2, \ldots, \varepsilon'_8$. Figure 10 demonstrates in the upper row (Fig. 10a, b) the strain data of scan 1 and in the lower row (Fig. 10c, d) the strain data of another scan (scan 2). The figures reflect the strain inhomogeneities. Furthermore, the upper row shows the strain of two different peaks $(10\bar{1}4)$ and $(11\bar{2}3)$ for the collimators 7 and 8, whereas the lower row shows these reflections $(10\bar{1}4)$ (Fig. 10c) and $(11\bar{2}3)$ (Fig. 10d) for two directions $[x]$ and $[z]$, perpendicular each to other.

The main stress directions determined by mechanical testing with the over-coring method (Alnaes pers. comm.) lies in the foliation plane. They were used as initial data, with the following values in the polar coordinate system:

$$\sigma_1 = (5.13 \pm 1.07)\ \mathrm{MPa} \quad (\varphi = 61.7^\circ, \vartheta = 218^\circ)$$
$$\sigma_2 = (3.72 \pm 0.61)\ \mathrm{MPa} \quad (\varphi = 77.3^\circ, \vartheta = 121.2^\circ)$$
$$\sigma_3 = (1.89 \pm 1.32)\ \mathrm{MPa} \quad (\varphi = 31.5^\circ, \vartheta = 10.9^\circ).$$

Figures 9 and 10 demonstrate that strain values obtained on different crystallographic diffraction

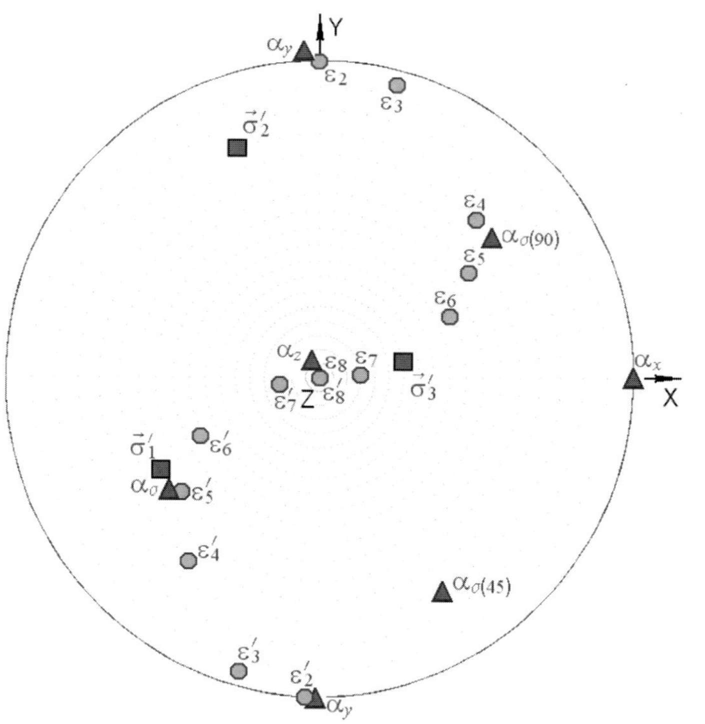


**Fig. 8.** Directions measured with the EPSILON-MDS diffractometer, stress main axes $\sigma_1$, $\sigma_2$ and $\sigma_3$, were obtained by an over-coring technique and orientations of samples used for thermal expansions studies (■, directions determined by mechanical test data; ▲, directions determined by bowing studies; ●, directions determined with neutron diffraction at EPSILON-MDS).

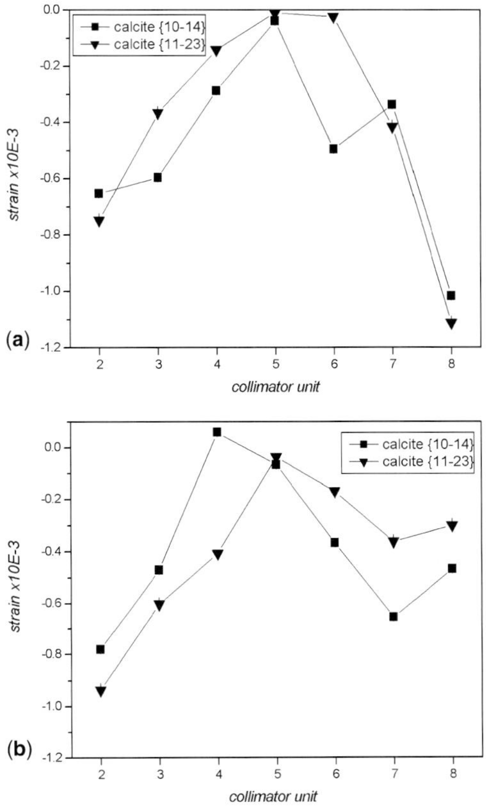


**Fig. 9.** Strain values measured at EPSILON-MDS at neighbouring scan positions in the sample g4 with seven collimator units.

lines are very similar and confirm the reliability of the extracted values. One can also see that the internal strain within the sample has a complicated distribution along the studied strain scan that is far from the uniform values accepted by many models. As shown in equation (5), strain values are dimensionless and express the relative change of lattice spacing. A negative value of strain corresponds to compression and the positive one corresponds to tension. Tables 1 and 2 illustrate the summarized maximum and minimum strain values for directions for which the thermal expansion was studied. The maximum values for different directions have shown qualitatively similar tendencies by measurements on both instruments.

To conclude, internal strain was measured on the investigated sample g4. The internal strains are distributed in a complicated way that does not correspond to a constant value of the strain field. Strain maxima are bent over the foliation plane by about 45°. A careful cross-check of the results has confirmed the reliability of the obtained strain values.

## Thermal expansion and bowing behaviour

Thermal expansion expresses the relative length change of a sample. The relationship with the temperature is non-linear, i.e. the thermal expansion coefficient, $\alpha$, which describes the specific length change ($10^{-6}\,K^{-1}$) depends on the considered temperature interval. In sum, up to now all experimentally determined data of marbles as a function of the heating and subsequent cooling cycle can be classified into four overall categories: (a) isotropic thermal expansion without residual strain; (b) anisotropic thermal expansion without residual strain; (c) isotropic thermal expansion with residual strain; and (d) anisotropic thermal expansion with residual strain (Siegesmund *et al.* 2004). The residual strain may be observed even as a result of very small temperature changes, as shown for the temperature range 20–50 °C by Battaglia *et al.* (1993).

Owing to their anisotropic behaviour, many samples have shown a large difference in residual strain after subsequent cooling down to room temperature. The thermal dilatation as a function of temperature basically had the following characteristics (Figs 11 & 12): during the heating–cooling cycle (20–90–20 °C) the slope increased successively until the destination temperature was reached. All samples generally exhibited a pronounced increase of residual strain after cooling down. In detail, the thermal expansion experiment found a weak directional dependence where the highest residual stresses were observed parallel to the [$x$], [$z$] and $\sigma_1$ directions (around 0.3–0.4 mm m$^{-1}$), while parallel to the [$y$]-direction and 90° to $\sigma_1$ somewhat lower values (0.1–0.2 mm m$^{-1}$) are evident. During the first cycle the residual strain increased significantly for almost all directions. After the second temperature cycle the residual strain did not increase while heating. Koch & Siegesmund (2004) have reported that the increase in the residual strain is limited and may end after the fifth cycle. Under wet conditions the marble showed very different expansion behaviour. They started to expand continuously. This increase in water-enhanced residual strain after five further temperature cycles was relatively high and could reach maximum values of around 0.8 mm m$^{-1}$ (see Figs 11 & 12). The moisture content after heating cycles apparently influences the intensity of marble degradation. Surprisingly, the weak directional dependence under wet conditions was less pronounced and more or less uniform in all directions.

To record the bowing potential of the samples (slabs of 400 × 100 × 30 mm), test-specimens were fixed in an apparatus in which they were

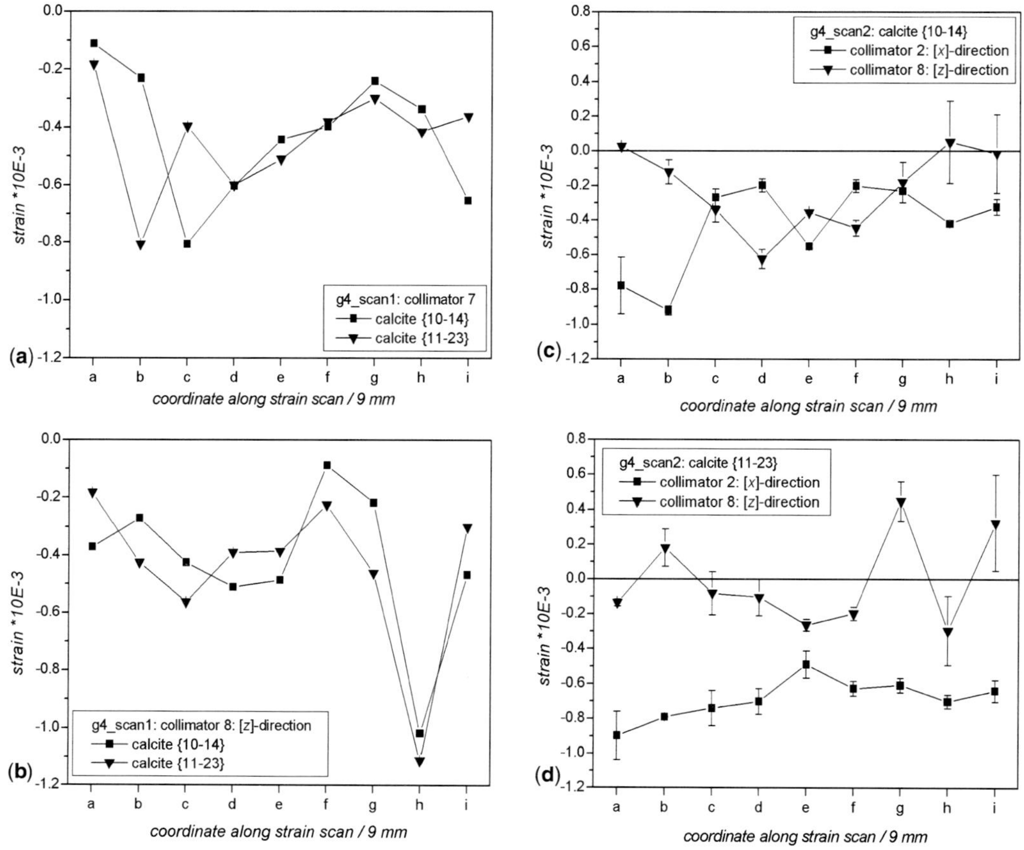


**Fig. 10.** Strain values measured at EPSILON-MDS with one collimator unit for different positions in the g4 sample: (**a**) and (**b**) strain scan 1; (**c**) $\{10\bar{1}4\}$ and (**d**) $\{11\bar{2}3\}$ strain scan 2 reflects two sample directions, $[x]$ and $[z]$, oriented perpendicular to each other.

exposed to varying temperatures on their upper side, and exposed to varying moisture on their underneath side. Both parameters, temperature and moisture, influenced bowing and change within a testing cycle of 24 h. The temperature interval on the upper side of the sample ran from 20 to 80 °C, and maximum values were obtained approximately 1–3 h after the onset of heating. Subsequently, the maximum value is preserved for another 3 h. During all the heating cycle, the test specimens were bedded into coarse-grained sand, which steadily reduced its moisture content in response to the temperature rise. Cooling of the specimens took over 16 h.

**Table 1.** *Residual strain values: the sample main axes and three directions in relation to the main stress direction, $\sigma_1$, measured on the EPSILON-MDS diffractometer*

| Direction | ε($10^{-4}$) max. g4 edge | ε($10^{-4}$) min. | ε($10^{-4}$) max. g4 middle | ε($10^{-4}$) min. | ε($10^{-4}$) max. g4 rotated | ε($10^{-4}$) min. |
|---|---|---|---|---|---|---|
| $[x]$ | | | | | | |
| $[y]$ | 0.24 | −7.77 | 1.86 | −5.57 | −5.17 | −8.45 |
| $[z]$ | −2.64 | −5.58 | −1.17 | −6.56 | −1.56 | −10.65 |
| parallel to $\sigma_1$ | | | | | 5.68 | −1.89 |
| 45° to $\sigma_1$ | | | | | | |
| 90° to $\sigma_1$ | | | | | | |

**Table 2.** *Residual strain values at (0006), (10$\bar{1}$4) and (11$\bar{2}$3) for the sample main axes and three directions in relation to the main stress direction, $\sigma_1$, measured on the SKAT diffractometer*

| Direction | $\varepsilon(10^{-4})$ {0006} | $\varepsilon(10^{-4})$ {10$\bar{1}$4} | $\varepsilon(10^{-4})$ {11$\bar{2}$3} | $\varepsilon_{average}(10^{-4})$ |
|---|---|---|---|---|
| [*x*] | −4.35 | −5.21 | −0.07 | −3.21 |
| [*y*] | −5.76 | −4.44 | −0.17 | −3.46 |
| [*z*] | −13.45 | −11.27 | −10.4 | −11.7 |
| parallel to $\sigma_1$ | −3.22 | −2.71 | −2.18 | −2.7 |
| 45° to $\sigma_1$ | −2.32 | −2.04 | −1.24 | −1.87 |
| 90° to $\sigma_1$ | −4.04 | −3.31 | −1.91 | −3.09 |

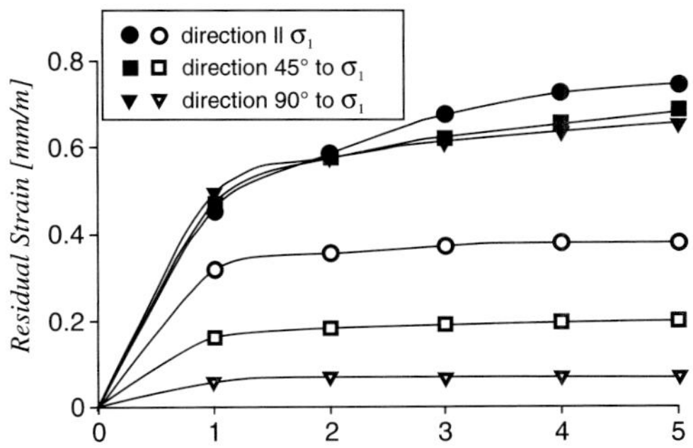


**Fig. 11.** Residual strain represents the permanent length change after heating and cooling cycles (five cycles in total). The thermal expansion was measured in different directions with respect to the *in situ* stress field, i.e. parallel to $\sigma_1$, 90° to $\sigma_1$ and 45° to $\sigma_1$. Filled symbols characterize samples measured under wet conditions, and open symbols under dry conditions.

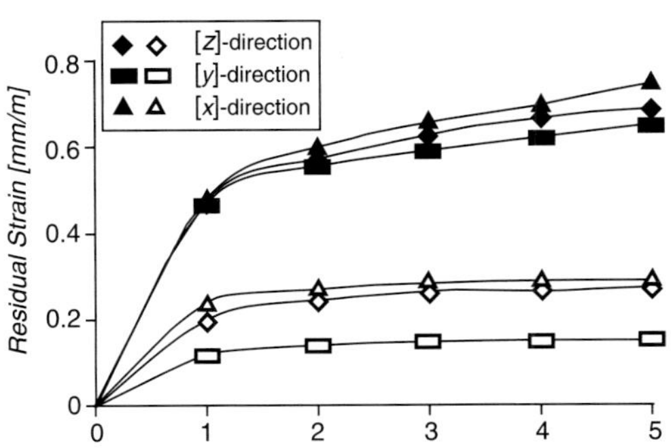


**Fig. 12.** Residual strain represents the permanent length change after heating and cooling cycles (five cycles in total). The thermal expansion was measured in different directions with respect to the rock fabrics (directions [*x*], [*y*] and [*z*]). Filled symbols characterize samples measured under wet conditions, and open symbols under dry conditions.

The results have shown rapid bowing of all specimens during the first five to eight heating cycles (Fig. 13). After 10 cycles, trends can be observed that allow a separation into three groups. Pieces with a relatively low bowing potential at this time (directions [*x*], [*y*] and 90° to $\sigma_1$) demonstrated deformation of approximately 0.75–1.00 mm m$^{-1}$. Other slabs (directions [*z*], parallel to $\sigma_1$ and 45° to $\sigma_1$) have shown bowing rates of 1.2–1.5 mm m$^{-1}$ after 10 heating cycles, whereas at the same time even stronger bowing of approximately 1.7 (direction parallel to $\sigma_1$) and 1.9 mm m$^{-1}$ (direction [*z*]) can be observed. During the following cycles, a decrease of the bowing intensity can be recognized for some plates, especially the one prepared from the [*z*] direction. After 45 cycles, two pieces that were prepared with respect to the general orientation (directions [*x*], [*y*]) showed the lowest values, whereas the direction parallel to $\sigma_1$ as well as the [*z*] direction displays the highest values. In contrast to the thermal expansion the bowing has shown much greater directional dependence, which is significant if the [*z*] and [*y*] direction are compared.

## Discussion and conclusion

Thermal expansion behaviour and the residual strain, i.e. progressive microcracking, may be easily explained by the texture and single crystal coefficients of calcite. The coefficient, $\alpha$, of calcite is extremely anisotropic (Kleber 1990): $\alpha_{11} = 26 \times 10^{-6}\,K^{-1}$ parallel and $\alpha_{22} = -6 \times 10^{-6}\,K^{-1}$ perpendicular to the crystallographic *c*-axis, i.e. calcite contracts normal to the *c*-axis and expands parallel to the *c*-axis while heating. For a marble sample with the texture given in Figure 14 a variation in thermal dilatation of between $4.47 \times 10^{-6}$ and $4.77 \times 10^{-6}\,K^{-1}$ because of the girdle-like and weak *c*-axis distribution was calculated. It is clearly documented that texture-based thermal expansion is more or less isotropic. Thermally induced microcracking

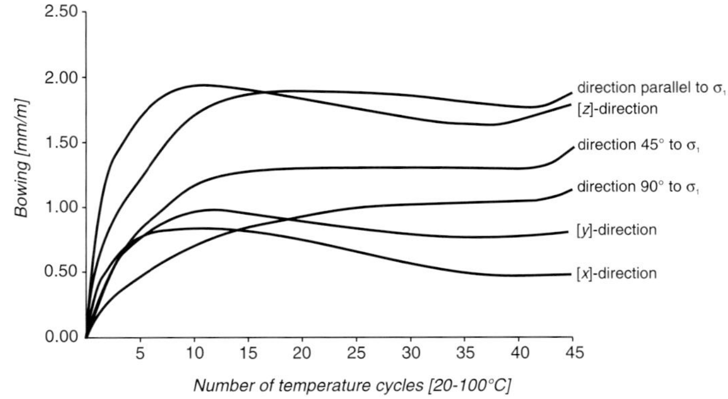


**Fig. 13.** Potential bowing behaviour of samples, which were prepared with respect to $\sigma_1$ and the rock fabrics (directions [*x*], [*y*] and [*z*]).

leads to a residual strain after heating due to tensile, compressive or shear stresses along grain boundaries and thus to deterioration of the rocks' quality. Microstructure-based finite-element simulations reported by Weiss *et al.* (2002, 2003) confirm the experimental results with respect to the magnitude and mechanisms of thermal degradation. Variations in texture may significantly affect the distribution of thermal stresses within marble. There is a strong inverse correlation between thermal stresses and degree of texture, as higher elastic strain energies are associated with weakly textured marbles and vice versa. Moreover, grain-to-grain orientation relationships, frequently called misorientations, and their distributions are also important parameters. Different misorientations lead to variations in the elastic energy density that are of the same order of magnitude to those related to the texture itself (Weiss *et al.* 2003). However, many authors (e.g. Sage 1988 or Koch & Siegesmund 2004) demonstrate that the increase in residual strain stops after a few heating cycles if moisture is absent. Therefore, Bucher (1956) and Winkler (1994) pointed out the

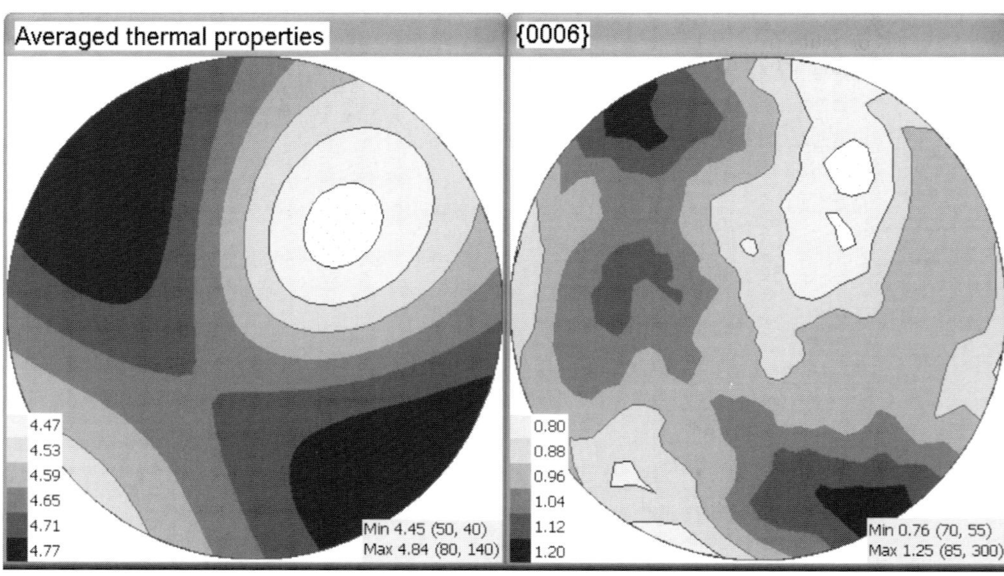


**Fig. 14.** Left: averaged thermal properties: the thermal coefficients for calcite single crystals $\alpha_{\parallel c} = 26 \times 10^{-6}\,\mathrm{K}^{-1}$, $\alpha_{\perp c} = -6 \times 10^{-6}\,\mathrm{K}^{-1}$ were used for calculations: thermal expansion in $[10^{-6}\,\mathrm{K}^{-1}]$. For the isotropic case the thermal coefficient is $\alpha_i = 2/3\alpha_{\perp c} + 1/3\alpha_{\parallel c} = 4.67 \times 10^{-6}\,\mathrm{K}^{-1}$. Right: experimental {0006} pole figure of sample g4.

importance of moisture in the degradation. In the presence of moisture a residual strain increase may be observed causing progressive marble decay. Winkler (1996) explained this behaviour in terms of oriented molecular layers (thickness of 2–3 nm) in capillaries of less than 0.1 μm that could be responsible for swelling by elongation and stone disruption. Poschlod (1990) argued that capillary condensation cannot be responsible for the decay process because the STERN-layer and GOUY–CHAPMANN-layer are not active under normal climatic conditions (below 100 °C). However, although the process of moisture-driven deterioration is not yet understood, the residual strain is much higher (2–3 times higher) if compared with the dry conditions. In addition, the bowing potential of the same marble is unexpectedly high and strongly directional dependent.

Owing to the above-mentioned single crystal thermal properties and the *c*-axis maximum, the maximum thermal degradation is closely linked to this direction that may roughly coincide with the area enclosed by the [*z*] and $\sigma_1$ directions, i.e. the maximum concentration of the *c*-axes within the broad girdle-like arrangement. The weak texture is also clearly documented in the residual stresses, although the misorientation of individual grain-to-grain orientation may produce internal stresses leading to microcracking. Owing to the observed texture and thermal expansion data, the bowing behaviour was unusually high parallel to the [*z*] and $\sigma_1$ directions, with maximum values of 2 mm m$^{-1}$. This cannot be explained solely by either rock fabric or by thermal expansion behaviour.

The methods of neutron diffraction for the investigation of crystallographic preferred orientation, intracrystalline strain by strain scans and residual strain pole figures are suitable for the acquisition of high-quality data based on properties and dimensions of the crystallographic lattice. Combined data from crystallographic texture and strain pole figures give unique data sets that provide an improved understanding of marble deformation behaviour.

In general, the investigated sample is characterized by a very weak crystallographic texture, on the one hand, and large bowing anisotropy, on the other. Macroscopic strain also reveals a high degree of anisotropy. This means that the bowing behaviour of the studied sample could be strongly related to internal stresses. In particular, the observed texture of the cubic sample g4 can be characterized as weak, and all pole figures show low intensities. The {0006} basal plane pole figure is characterized by a girdle-like distribution and low intensities under 1.0 m.r.d., maxima were found in the {0006} pole figure with intensities up to 1.25 m.r.d. Furthermore, features of orthorhombic sample symmetry have been observed in the {10$\bar{1}$4} pole figure. The data sets for texture pole figures could be also used for the analysis of strain by generation of strain pole figures. Mainly compression strain states were observed in the strain pole figures. Mean strain values were found for all crystallographic directions in the range of ε from $-2 \times 10^{-4}$ to $-4 \times 10^{-4}$, but minimum compressional strain values are of about $\varepsilon = -1.3 \times 10^{-3}$. This orthorhombic sample symmetry was also found in all three strain pole figures {0006}, {10$\bar{1}$4} and {11$\bar{2}$3}. So, the relationship between the crystallographic preferred orientation and the strain distribution in the pole figures is obvious because the texture pole figures are also characterized by orthorhombic sample symmetry. The measured compressional residual strain values in the large sample with dimensions of 80 × 80 × 80 mm were up to $-1.3 \times 10^{-3}$ higher than in marble plates with dimensions of 100 × 100 × 30 mm, similar to previous investigations by Scheffzük *et al.* (2004) where compressional residual strains were found to be $-7 \times 10^{-4}$. Residual strain values, measured on quartz (Frischbutter *et al.* 2000), were also in the range of $\pm 4 \times 10^{-4}$. So, it is assumed that the magnitude of the compressional residual strain depends on the dimension of the sample.

Logan (2004) carried out cyclical marble experiments and reported that stored residual strain contributes to a loss of strength within the marble, but, more importantly, it is a critical factor in the development of the bow. This elastic strain may be the result of the geological history of the material, or may be developed as a result of the differential thermal expansion of calcite, because he found that thermal cycles at 36 °C after 30 days produced an increase in the residual strain of up to $10^{-4}$. In contrast, this study has shown that the residual strain in a bulk natural sample has a magnitude of $10^{-3}$, which can be seen to be a result of geological history, taking into account the stress redistribution and/or stress relaxation by sample extraction. In addition, Logan (2004) reported that the loss of tensile strength by accumulation of damage, regardless of the size of thermal cycles, is one factor leading to bowing. This means that strongly bowed panels are characterized by a loss of their mechanical properties and that intracystalline residual stress could be the driving force.

Based on the observed homogeneous microstructure and its texture, the marble sample can be seen as nearly isotropic. Although the strain–stress behaviour is influenced by grain boundaries and pores, a linear relation between strain and stress, described by Hooke's law ($\varepsilon_{ij} = S_{ijkl}\sigma_{kl}$, with $\varepsilon_{ij}$ the local elastic strain, $S_{ijkl}$ the elastic compliances and $\sigma_{kl}$ the local stresses), can be applied. Using the elastic constants for the trigonal calcite $E_{\alpha\text{-axis}} = 87.8$ GPa and $E_{c\text{-axis}} = 57.6$ Gpa, one can estimate the maximum stress value. The observed minimum strain value of $\varepsilon = -1.34 \times 10^{-3}$ in the {0006} strain pole figure corresponds with a residual stress magnitude of about $\sigma_{c\text{-axis}} = 77.5$ MPa. At the present time, no work has been reported in which strain pole figures have been converted to stress pole figures because the local stress $\sigma_{kl}$ and the local strain $\varepsilon_{ij}$ vary from one crystallite to another. Moreover, they vary within the individual crystallites, even if submicroscopic defects, like point defects or dislocations, are ignored. Thus, different lattice orientations in each crystallite in polycrystalline materials imply that the compliances $S_{ijkl}$ are in principle different in each crystallite. Thus, heterogeneous strain and stress distribution occur not only from one crystallite to another, but even inside each crystallite. The transition from strain to stress has been described in several respects by Brakman (1987*a*).

Two directions [*y*] and [*z*], each perpendicular to the other, have been investigated by strain scans across the sample cross-section with a higher resolution. It is obvious that mainly contraction was observed. In general, and in comparison to the [*z*] direction, the [*y*] direction shows strain values with a stronger compression for $(10\bar{1}4)$ as well as for $(11\bar{2}3)$.

The influence of crystallographic texture on diffraction measurements of residual stress provides several approaches for a better understanding of material behaviour in the future. Up to now, the relationship has been discussed mainly for cubic materials (Brakman 1987*b*; van Houtte & de Buyser 1993). The interaction between strain and texture for materials with lower symmetrical crystals is more difficult to establish. Nevertheless, further investigations of structural geological features, rock fabric, weathering, thermal effects, crystallographic preferred orientation, and residual stresses and their interactions are necessary to improve our understanding of the bowing effects observed in marble panels. Results from thermal expansion, together with the observed different intensities of bowing in comparison with neutron diffraction data of residual strain and texture, underline the suggestion that so-called locked-in stress should have an important impact on the bowing of marbles. To detect this locked-in stress, the neutron TOF diffraction measurements can be regarded as a suitable method, but ongoing investigation is necessary to establish this approach as a measure of the bowing behaviour.

Support of the German Bundesminister für Bildung und Forschung (BMBF) through grants 03-DU03X4 (Ch. Scheffzük), 03-DU03G1 (A. Hoffmann) and of the Deutsche Forschungsgemeinschaft (DFG) through grant No. GZ: 436 RUS (D. I. Nikolayev) are gratefully acknowledged. We are very grateful to G. Molli (Pisa) for his help with the samples. We thank one of the anonymous reviewers for helpful comments to improve the paper.

## References

BATTAGLIA, S., FRANZINI, M. & MANGO, F. 1993. High sensitivity apparatus for measuring linear thermal expansion: preliminary results on the response of marbles. *Il Nuovo Cimento*, **16**, 453–461.

BRAKMAN, C. M. 1987*a*. Surface spherical harmonics and intensity and strain pole figures of cubic textured materials. *Acta Crystallographica*, **A43**, 270–283.

BRAKMAN, C. M. 1987*b*. Residual stress analysis using overlapping diffraction peaks. Case of textured cubic materials. *Journal of Applied Crystallography*, **20**, 479–487.

BUCHER, W. H. 1956. Role of gravity in orogenesis. *Bulletin of the Geological Society of America*, **67**, 1295–1318.

FRISCHBUTTER, A., NEOV, D., SCHEFFZÜK, CH., VRÁNA, M. & WALTHER, K. 2000. Lattice strain measurements on sandstones under load using neutron diffraction. *Journal of Structural Geology*, **22**, 1587–1600.

GRIMM, W. D. 1999. Beobachtungen und Überlegungen zur Verformung von Marmorobjekten durch Gefügeauflockerung. *Zeitschrift der Deutschen Geologischen Gesellschaft*, **150**, 195–236.

KESSLER, D. W. 1919. *Physical and Chemical Tests of the Commercial Marbles of the United States*. Technologic Tests of the Bureau of Standards, **123**. Government Printing Office, Washington, DC.

KLEBER, W. 1990. *Einführung in die Kristallographie*. Technik, Berlin.

KOCH, A. & SIEGESMUND, S. 2002. Bowing of marble panels: On-site damage analysis from the Oeconomicum Building at Goettingen (Germany). *In*: SIEGESMUND, S., WEISS, T. & VOLLBRECHT, A. (eds) *Natural Stone, Weathering Phenomena, Conservation Strategies and Case Studies*. Geological Society, London, Special Publications, **205**, 299–314.

KOCH, A. & SIEGESMUND, S. 2004. The combined effect of moisture and temperature on the anomalous expansion behaviour of marble. *Environmental Geology*, **46**, 350–363.

LOGAN, J. M. 2004. Laboratory and case studies of thermal cycling and stored strain on the stability of selected marble. *Environmental Geology*, **46**, 456–467.

LOGAN, J. M., HADEDT, M., LEHNERT, D. & DENTON, M. 1993. A case study of the properties of marble as building veneer. *International Journal of Rock Mechanics, Mining Sciences & Geomechanics Abstracts*, **30**, 1531–1537.

NIKOLAYEV, D. I., LYCHAGINA, T. A., NIKISHIN, A. V. & YUDIN, V. V. 2005. Investigation of measured pole figures errors. *In: Proceedings of ICOTOM 14. Materials Science Forum*, **495–497**, 307–312.

POSCHLOD, K. 1990. Das Wasser im Porenraum kristalliner Naturwerksteine und sein Einfluß auf die Verwitterung. *Münchner Geowissenschaftliche Abhandlungen*, **7**, 1–62.

ROWE, K. J. & RUTTER, E. H. 1990. Paleostress estimation using calcite twinning: Experimental calibration and application to nature. *Journal of Structural Geology*, **12**, 1–17.

RUEDRICH, J., WEISS, T. & SIEGESMUND, S. 2001. Deterioration characteristics of marbles from the Marmorpalais Potsdam (Germany): a compilation. *Zeitschrift der Deutschen Geologischen Gesellschaft*, **152**, 637–663.

SAGE, J. D. 1988. Thermal microfracturing of marble. *In*: MARINOS, P. G. & KOUKIS, G. C. (eds) *Engineering Geology of Ancient Works, Monuments and Historical Sites*. Balkema, Rotterdam, 1013–1018.

SCHEFFZÜK, C., SIEGESMUND, S. & KOCH, A. 2004. Strain investigations on calcite marbles using neutron time-of-flight diffraction. *Environmental Geology*, **46**, 468–476.

SIEGESMUND, S., RUEDRICH, J. & WEISS, T. 2004. Marble deterioration. *In*: PŘIKRYL, R. (ed.) *Dimension Stone 2004*. Taylor & Francis, London, 211–217.

SIEGESMUND, S., ULLEMEYER, K., WEISS, T. & TSCHEGG, E. K. 2000. Physical weathering of marbles caused by anisotropic thermal expansion. *International Journal of Earth Sciences*, **89**, 170–182.

TOBISCH, J. & BUNGE, H. J. 1972. The spherical sample method in neutron diffraction texture determination. *Texture*, **1**, 125–127.

ULLEMEYER, K., SPALTHOFF, P., HEINITZ, J., ISAKOV, N. N., NIKITIN, A. N. & WEBER, K. 1998. The SKAT texture diffractometer at the pulsed reactor IBR-2 at Dubna: Experimental layout and first measurements. *Nuclear Instruments and Methods in Physics Research Section A: Accelerators, Spectrometers, Detectors and Associated Equipment* **A412**, 80–88.

VAN HOUTTE, P. & DE BUYSER, L. 1993. The influence of crystallographic texture on diffraction measurements of residual stress. *Acta Metallurgica et Materialia*, **41**, 323–336.

WALTHER, K., SCHEFFZÜK, C. & FRISCHBUTTER, A. 2000. Neutron time-of-flight diffractometer Epsilon for strain measurements: layout and first results. *Physica B, Condensed Matter*, **276–278**, 130.

WEISS, T., SIEGESMUND, S. & FULLER, E. R., JR. 2002. Thermal stresses and microcracking in calcite and dolomite marbles via finite element modelling. *In*: SIEGESMUND, S., WEISS, T. & VOLLBRECHT, A. (eds) *Natural Stone, Weathering Phenomena, Conservation Strategies and Case Studies*. Geological Society, London, Special Publications, **205**, 89–102.

WEISS, T., SIEGESMUND, S. & FULLER, E. R., JR. 2003. Thermal degradation of marbles: Indications from finite element modelling. *Building and Environment*, **38**, 1251–1260.

WINKLER, E. M. 1994. *Stone in Architecture*. Springer, New York.

WINKLER, E. M. 1996. Technical note: properties of marble as building veneer. *International Journal of Rock Mechanics, Mining Sciences & Geomechanics Abstracts*, **33**, 215–218.

ZEISIG, A., SIEGESMUND, S. & WEISS, T. 2002. Thermal expansion and its control on the durability of marbles. *In*: SIEGESMUND, S., WEISS, T. & VOLLBRECHT, A. (eds) *Natural Stone, Weathering Phenomena, Conservation Strategies and Case Studies*. Geological Society, London, Special Publications, **205**, 65–80.

# The influence of lithology and pore-size distribution on the durability of acid volcanic tuffs, Hungary

Á. TÖRÖK[1], L. Z. FORGÓ[1], T. VOGT[2], S. LÖBENS[2], S. SIEGESMUND[2] & T. WEISS[2]

[1]*Budapest University of Technology and Economics, Department of Construction Materials and Engineering Geology, H-1111 Budapest, Stoczek u. 2, Hungary (e-mail: torokakos@mail.bme.hu)*

[2]*Geoscience Centre of the University of Göttingen, Goldschmidtstrasse 3, 37077 Göttingen, Germany*

**Abstract:** Eight different types of acid tuffs of the Eger Castle (Hungary) and two tuffs from nearby quarries have been studied in detail. Mapping of wall sections reveals that tuffs show weathering forms that are similar to common sedimentary rocks, such as limestones or sandstones. Different lithologies display various weathering features. On pumice-rich tuff ashlars relief due to selective weathering, weathering crusts, multiple flakes and scales occur, while crumbling is common on layered flow tuffs. Conversely, cemented tuff types do not show deep weathering. Pore-size distribution rather than effective porosity controls the weathering susceptibility of tuffs. Frequent larger micropores are the main causes of freeze–thaw-related weathering. Besides clays, newly formed gypsum and calcite are the weathering-related index minerals. Schmidt hammer rebound values mark the weathering process when quarry stones and ashlars of historic walls are compared.

Volcanic tuffs are best known from monuments of Anatolia (Caner-Saltik *et al.* 1994), the Easter Islands (Wendler *et al.* 1996) and even the Philippines (Paterno & Charola 2000). Great varieties of tuffs are used in Europe, for example in Italy (Langella *et al.* 2000), in Germany (Fitzner & Lehners 1990; Egloffstein 1998; Auras & Steindlberger 2005) and in the Netherlands (Nijland *et al.* 2003; van Hees *et al.* 2004). The monuments made of tuff often display various forms of deterioration (Fitzner 1994) and thus the preservation of such sites need special care. For the selection of proper techniques detailed petrographic and physical analyses are required, as has been shown for the basaltic tuffs of Easter Islands (Wendler *et al.* 1996), for volcanic tuffs of Germany (Egloffstein 1998; Steindelberger 2004), and for rhyolite tuffs of Anatolia (Topal & Sözmen 2003) and eastern Hungary (Török *et al.* 2004).

The present study focuses on the acidic volcanic tuffs of NE Hungary. The tuff monuments that often date back to medieval periods are now showing intense deterioration (Török *et al.* 2005). This paper outlines some of the important petrographic and physical properties that influence the susceptibility of tuff to weathering. To achieve this goal, physical parameters, such as Schmidt hammer rebound values, moisture content and water absorption, were measured *in situ*, while pore-size distributions were determined under laboratory conditions. Quarry stones and ashlars from monuments were compared to assess susceptibility to deterioration. Tuffs were grouped according to textural, mineralogical and petrological properties based on microscopic and X-ray diffractometry (XRD) analyses. Lithotypes and weathering forms were correlated on maps of walls, and the weathering sensitivity of tuff types were particularized. This study also aims to provide important data for the restoration of tuff monuments by describing the key properties of these water-sensitive porous stones.

## Material and methods

Intense volcanic activity characterized the Miocene period of north Hungary. The volcanic activity resulted in the formation of lavas and a thick pyroclastic sequence (Hámor 2001). The prevailing lithology that has been used in monuments is rhyolite tuff, although dacitic and andesitic volcanites and pyroclastics are also known from the area. The studied rocks are found in the area of Eger, which is a small historic town in NE Hungary. The utilization of stones as construction materials only began in the 12th century (Kleb 1978). Much of the medieval architecture is now ruined, although remnants of a Basilica and stone ashlars of a later period Gothic palace are still visible. The most prominent building that has been almost entirely built from tuff is the Castle of Eger. The present form of the castle exhibits several phases of reconstruction, beginning as early as the 13th century,

*From*: PŘIKRYL, R. & SMITH, B. J. (eds) *Building Stone Decay: From Diagnosis to Conservation.* Geological Society, London, Special Publications, **271**, 251–260.
0305-8719/07/$15.00 

including at least six different phases of rebuilding. To understand the mechanism of weathering and to assess the importance of stone properties in durability, *in situ* tests were performed on historic walls (16th and 18th century) and newly restored ones in Eger Castle. Lithological characteriztics were documented and samples were taken from each lithotype. Eight tuff types have been identified in the walls including rhyolite and dacite tuffs. Two additional tuffs have been collected and analysed from nearby quarries of Eger-Tihamér and Eger-Demjén, as these quarries provided some of the ashlars of the historic walls. The stone from these quarries might also be used as replacement stones in future restoration. Three of the studied tuffs were exclusively used as replacement stones during the restoration works of the 20th century. The textures of weathered tuffs and fresh samples were analysed by petrographic microscopy. Mineralogical composition of samples from weathered tuffs and quarry stones was compared using XRD (Phillips PW 1800). The weathering forms were described using the nomenclature of Smith *et al.* (1992) and Fitzner *et al.* (1995). The distribution of lithologies and major weathering forms were documented on maps, and the percentage of each weathering form was calculated for the selected walls. Original wall sections and replaced counterparts were compared. The physical properties of ashlars were tested *in situ* using a Schmidt hammer (type Digi Schmidt 2000). The application of this non-destructive surface hardness test method for assessing stone quality of monuments has been described in detail (Török 2003). Water content properties were also measured using MOIST 200 moisture content measuring equipment both with a volumetric detecting head and a surface moisture detecting head. For comparison, water absorption tests using the Karsten tube method were also performed. Other physical properties of tuff types were measured under laboratory conditions. Water absorption and pore-size distribution (mercury method) were measured on small samples from ashlars; while blocks from the quarries were used for testing ultrasonic sound velocity, density and strength.

## Lithotypes and mineralogical composition

The most important properties and components of the tuffs are listed in Table 1. In most tuffs volcanic glass is the predominant component, with variable degrees of crysallization. Light grey, pumice-rich rhyolite tuff (Type I) is interpreted as a lapilli tuff. It contains up to 40% of pumice with a maximum diameter of 10 cm. Millimetre- to centimetre-scale dark grey volcanic clasts and rounded quartz pebbles are also common in this type. Very small quartz crystals and biotites are rare in the glassy groundmass (Fig. 1). XRD analyses have shown that crystobalite and smectite are the accessory minerals.

Medium grey–pink dacite tuff (Type II) has only a few per cent of pumice (maximum 3 cm in diameter) and very small biotite crystals (on the mm scale). Lithic clasts and quartz crystals of mm scale are sparse, but albite content is notable (Fig. 1). Grey–reddish biotite-rich rhyodacite tuff (Type III) contains a significant amount of pumice clasts (20%): 2–3 mm-size biotites (2–3%) and mm-size quartz (1%) phenocrysts are visible in the dark-coloured groundmass (Fig. 1). Black lithic clasts (1–2%) and rounded quartz pebbles (less than 1%) were also identified. Illite is the main accessory mineral (Table 1). Brownish creamy rhyolite tuff (Type IV) is often layered and microporous. It is characterized by common black lithic clasts of maximum 3 cm diameter and rounded pumice clasts. Phenocrysts are rare and are represented by mm-size crystobalite and biotite crystals. It is comparatively rich in secondary clay minerals such as illite. Both pumice and lithic clasts show signs of transporation and the texture reflects deposition. The groundmass is transparent vitreous (Fig. 1). White–creamy soft rhyolite tuff (Type V) is interpreted as an ash flow deposit. It is rich in phenocrysts including quartz (mm size) and biotites (2–3 mm). Small pumice (1 cm) and very small black lithic clasts (mm size) are also present in the flow-bounded groundmass (Fig. 1).

Creamy–brownish mottled rhyodacite tuff (Type VI), a replacement stone, contains very small pumice and black–brownish grey lithic clasts of 0.5–3 cm in diameter. Small biotite crystals are common, while quartz is sometimes found as phenocrysts in the tough, strongly silicified groundmass. Brownish grey, pumice-rich dacite tuff (Type VII) was only used as replacement stone. It contains unsorted pumice lithic clasts of mm size, but some larger ones of cm size are also present. Black, often rounded lithic clasts are sparse, while biotite phenocrysts of 3–4 mm also occur.

Pinkish white biotite-rich rhyolite tuff (Type VIII) is the other common replacement stone type that is found in walls where restoration was completed in 2002. Pumice clasts are sparse, but small biotite crystals are common in the whitish glassy groundmass.

The two rhyolite tuffs from the quarries have somewhat distinctive mineral composition. The creamy white, pumice-rich rhyolite tuff from Eger-Tihamér quarry (Quarry I) is quartz-rich, and contains more biotite and volcanic glass than the greyish white durable rhyolite tuff from the Eger-Demjén quarry (Quarry II). The latter one is

**Table 1.** *Lithology, mineralogy and selected physical properties of tuffs (types I–VI) and replaced tuffs ashlars (types VI–VIII) from Eger Castle and two quarries. The mineral phases are listed in decreasing order of frequency (++ abundant, +common)*

| | Description of lithologies | Main components (XRD) | Schmidt hammer rebound | Water absorption (kg m$^{-2}$ h$^{-0.5}$) | Porosity (vol. %) |
|---|---|---|---|---|---|
| Type I | Light grey, pumice-rich rhyolite tuff | Glass (++), quartz, albite, biotite, smectite, crystoballite | 14 | 56 | 21.2 |
| Type II | Medium grey–pink dacite tuff | Glass (+), albite-anortite, biotite | 21 | 21 | 22.2 |
| Type III | Grey–reddish biotite-rich rhyodacite tuff | Glass (++), biotite, albite, quartz, illite | 23 | 9 | 24.9 |
| Type IV | Brownish creamy layered rhyolite tuff with lithic clasts | Glass (++), plagioclase (++), K-feldspar, biotite, crystobalite, illite | 8 | 60 | 33.0 |
| Type V | White–creamy rhyolite tuff | Glass (+), quartz (++), biotite (+), K-feldspar (+), albite (+), heulandite | 16 | 25 | 31.1 |
| Type VI (replaced) | Creamy–brownish mottled rhyodacite tuff | not measured | 33 | 15 | not measured |
| Type VII (replaced) | Brownish grey, pumice-rich dacite tuff | not measured | 21 | 18 | not measured |
| Type VIII (replaced) | Pinkish white biotite-rich rhyolite tuff | not measured | 20 | 14 | not measured |
| Quarry I (Tihamér) | Creamy white pumice-rich rhyolite tuff | Glass (++), albite (+) biotite (+), quartz, | 38 | 15 | 36.0 |
| Quarry II (Demjén) | Greyish white rhyolite tuff | Glass (+), albite-anortite, quartz, | 51 | 6 | 34.8 |

denser and characterized by grey irregular pumice clasts and unaltered biotites (Fig. 1). The tuff from Eger-Tihamér (Quarry I) is less dense and displays white pumice of various sizes (mm to cm scale) in the vitreous vascicular texture (Fig. 1).

## Surface strength and porosity

The Schmidt hammer rebound values of tuffs used in historic non-replaced walls (Type I–Type V) are between 8 and 23 on average (Table 1). For replacement, stones with Schmidt hammer rebound values between 20 and 33 were used. The quarry stones were found to have the maximum rebound values, particularly the one from the Demjén quarry (Quarry II), as its surface strength was more than double that of the 'strongest' ancient tuff (Type III) of the walls. There are other differences in physical parameters of quarry stones. The tuff from Eger-Tihamér quarry (Quarry I) had a bulk density of 1.454 kg m$^{-3}$, while the air-dry stone from Demjén quarry (Quarry II) measured 1.585 kg m$^{-3}$ on average. The ultrasonic sound velocities show the same pattern; 1.656 and 2.325 km s$^{-1}$ values were recorded on samples from Quarry I and Quarry II, respectively. The compressive strength and tensile strength of these tuffs display similar trends, namely the creamy white pumice-rich rhyolite tuff (Quarry I) has a lower strength than that of the greyish white rhyolite tuff from Eger-Demjén (Quarry II). A fourfold difference between the air-dry and water-saturated indirect tensile strengths of the Eger-Tihamér tuff was recorded under laboratory conditions (Forgó & Török 2004), while the decrease in indirect tensile strength of Eger-Demjén tuff by water absorption was less significant (60%).

Porosity and especially water absorption capacity of ashlars of various tuffs varies significantly. When water absorption is considered, the two end members are Type III and Type IV tuffs, with absorption capacities of 9 and 60 kg m$^{-2}$ h$^{-0.5}$, respectively. All tested tuffs are very porous. The porosity values are between 21.2 and 33.0 for samples taken from ashlars, while for quarry stones the values are 24.3 and 38.6 (vol. %) (Table 1). A significant difference between the porosity of Type I and intensively weathered Type I tuff was found (Fig. 2).

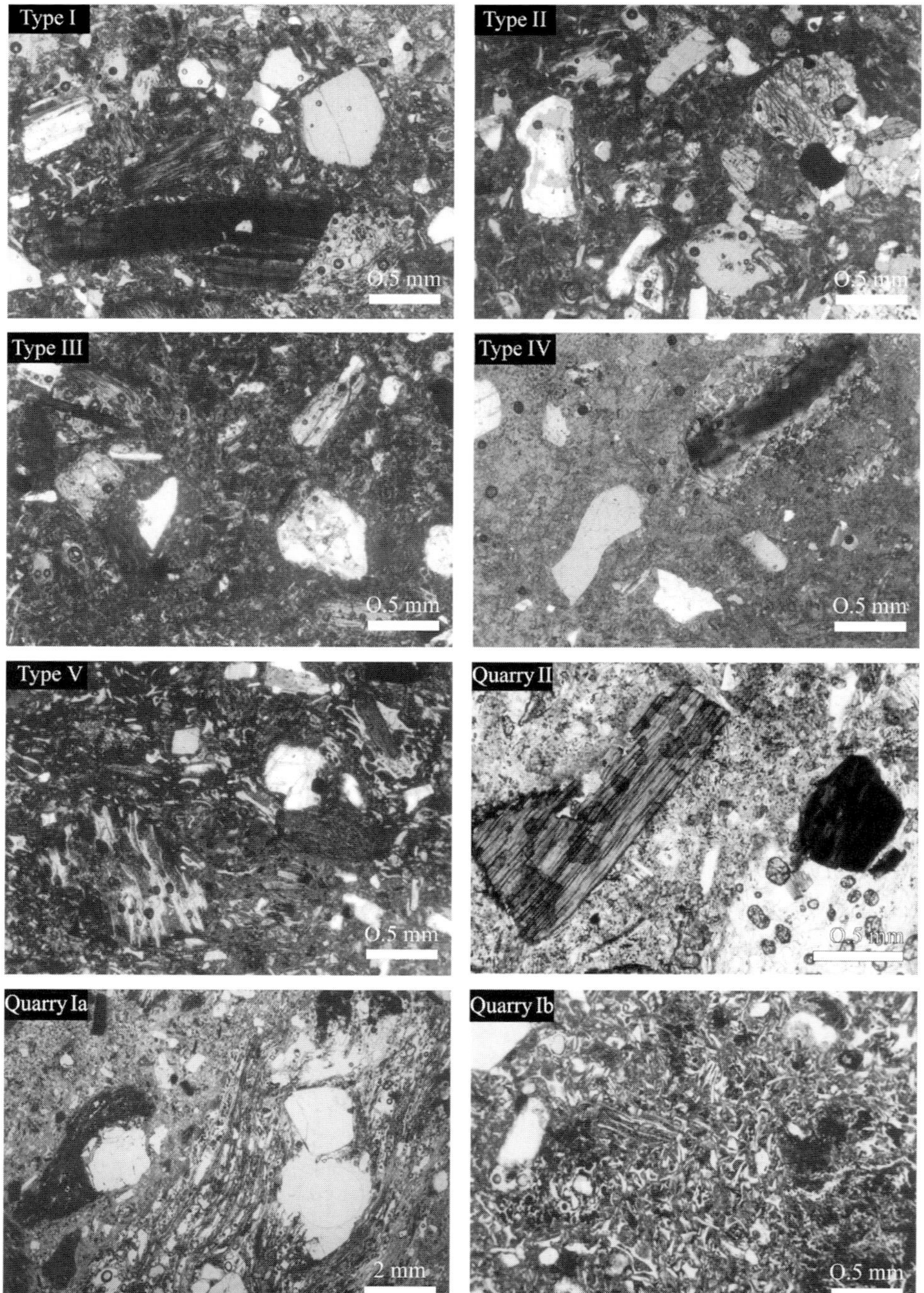


**Fig. 1.** Examples of microfabrics of various tuffs: Type I – angular quartz and biotite phenocrysts in glassy groundmass that shows fluidal texture; Type II – albite crystals, quartz and biotite in glassy groundmass; Type III – angular–acicular quartz, idiomorphic biotite and quartzite lithic clast in dark coloured groundmass; Type IV – altered (chloritic) biotite and quartz crystals in transparent glassy groundmass; Type V – small but frequent biotite and quartz crystals in flow-banded groundmass. Quarry II – unaltered biotite and small angular quartz crystals in vitreous matrix; Quarry Ia – a large pumice clast encloses quartz crystals (middle left); Quarry Ib – vitreous texture with vascicular pores and a few small quartz phenocrystals.

The pore-size distribution of the tuffs used in historic walls and those of the quarry fresh tuffs are very different (Fig. 2). In the tuffs from the walls micropores predominate, while the contribution of macropores to the porosity is generally less than 1%. The only exception is the 'Type III' tuff where small macropores (less than 10 μm) also occur. Bimodal pore-size distribution characterizes Type I, Type II and Type III tuffs. The tuffs from the two quarries are like chalk and cheese, as the

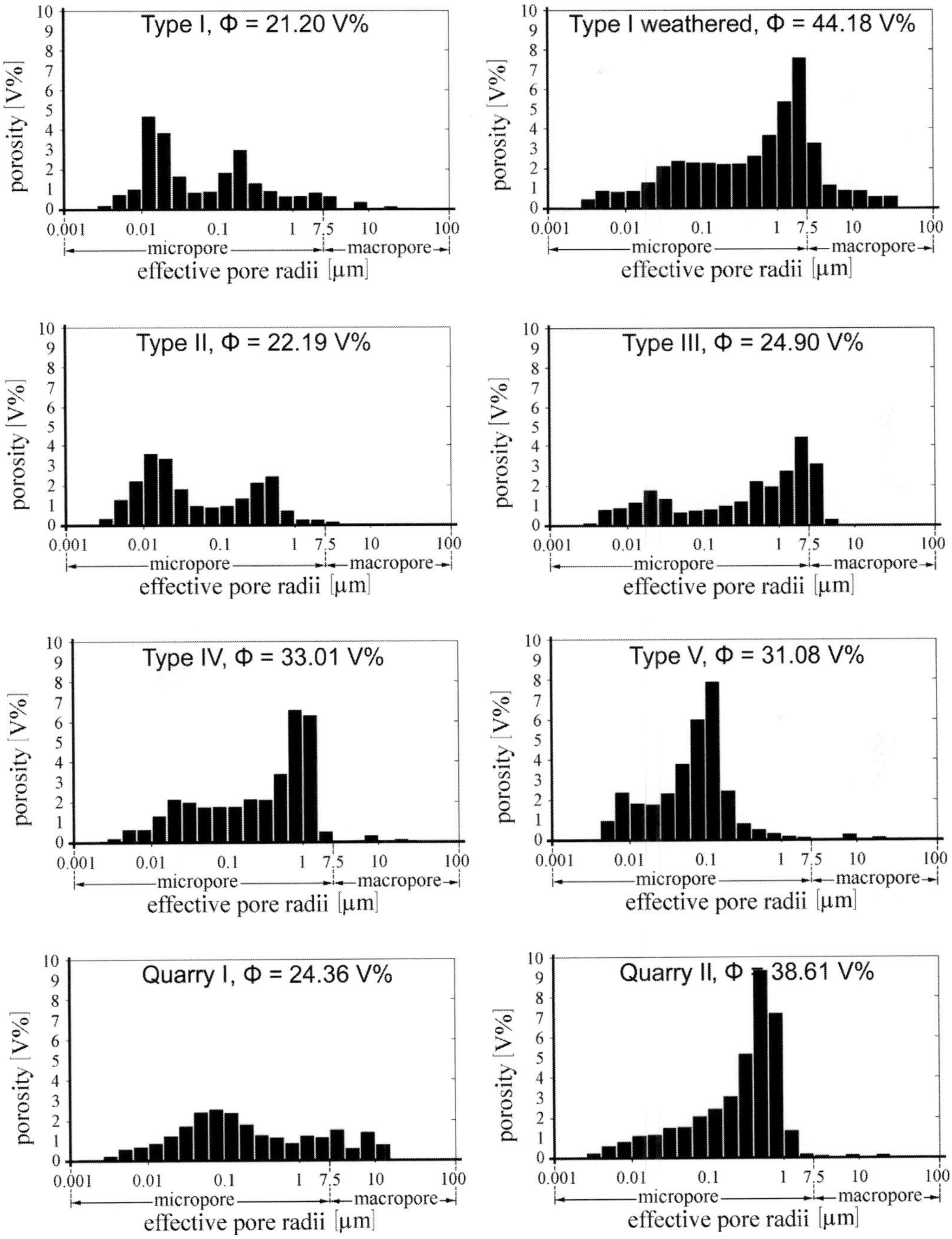


**Fig. 2.** Pore-size distribution of tuff types of the castle wall and of quarries.

one from Demjén is characterized by the dominance of micropores of 0.4–1 μm while that from Egertihamér has both macro- and micropores and a wide size range of effective pore radii.

## Weathering features

Both the tuffs of the original walls and the replaced ashlars show various types of weathering features. The weathering forms in the historic walls, in decreasing order of frequency, are relief due to selective weathering (*sensu* Fitzner & Heinrichs 2002), crumbling, crust formation, and, following detachment, flaking, multiple flaking and scaling. Relief formation is related to the different durability of groundmass and lithic components, i.e. selective and accelerated weathering of sensitive groundmass. The best examples are found on ashlars of Type I, Type IV and Type V tuffs, where pumice or other lithic clasts have a positive relief while the glassy groundmass is stepped back by weathering (Fig. 3). Although crumbling has typically been described for sandstones and marbles, these volcanic tuffs often show very similar features as small parts from the groundmass tend to disintegrate and crumble. This weathering form is very common on Type IV, and also observed on types I and V. Crusts (Fig. 3) and the detachment of scales are also common, but single scales are mostly found on Type II tuff. Frost-related scaling leading to flaking occurs on all types except Type VI (Fig. 3). Weathering is also indicated by XRD results that show the formation of secondary minerals such as calcite and gypsum in weathered tuff samples. Enrichment in clay minerals such as montmorillonite and the appearance of chlorite also marks stone alteration. Salt efflorescence is also relatively common. Salts are generally found near joints and possibly form by the leaching of mortar and mobilization towards the stone surface. Alternatively, they can also be the reaction products of mobilized alkali ions from the tuff (cf. Auras & Steindlberger 2005).

The percentage surface cover of each weathering phenomenon has been calculated for selected walls. For comparison a wall that consists of replaced

**Fig. 3.** Common weathering features of tuffs: (**a**) multiple scaling and flaking related to frost action (Type V); (**b**) relief formation due to selective weathering of groundmass and various lithic clasts exposing pumice clasts (Type I); (**c**) single crust with initial scaling (Type II); and (**d**) crumbling of layered tuff (Type IV) (the scale bar is 20 cm for all photographs).

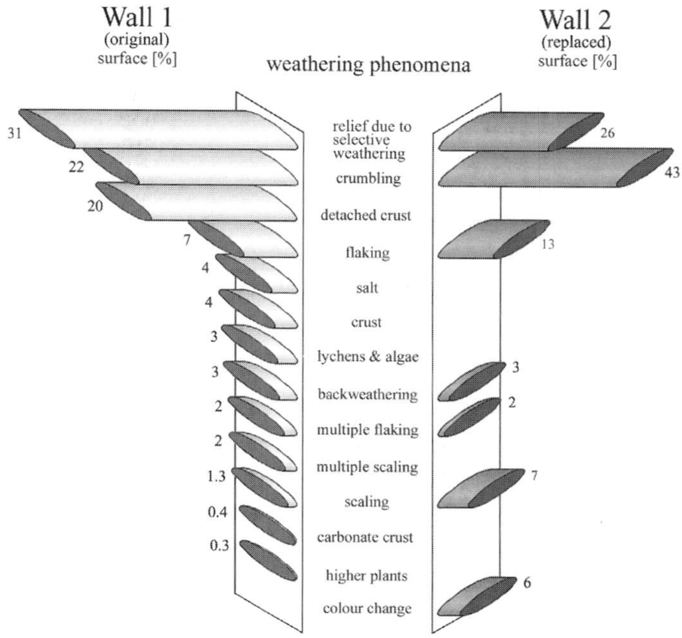


**Fig. 4.** Percentage surface cover of weathering features on the original wall (Wall 1) and on the replaced wall (Wall 2).

ashlars is also shown in Figure 4. Relief formation owing to selective weathering is the most frequent form on historic walls, while replaced ashlars mostly display crumbling. Crust detachment is limited to original ashlars, while flaking occurs on both wall types. Biological activity, lichen and algal colonization, and associated weathering features are only found on original walls (Fig. 4). It is important to note that in the original wall shown in Figure 4 the frequency of lithotypes is: Type I, 46%; Type II, 25%; Type 3, 21%; Type IV, 6%; and Type V, 2%, while in the replaced wall section the refaced part represents only 7% of the wall (Type I, 4%; Type V, 3%) and the rest of the surface is made up of newly introduced stones of Type VI (9%), Type VII (32%) and mostly by Type VIII (52%).

## Discussion

Pore-size distribution is one of the key factors determining stone durability (Punuru *et al.* 1990; Fitzner & Basten 1994; Winkler 1997; Benavente *et al.* 2004). By studying salt crystallization in pores it has been documented that crystallization pressure is lower in larger pores (Gauri *et al.* 1988; Scherer 1999), thus in general smaller capillary pores are more damaged by salt crystallization (Kozlowski *et al.* 1990; Goudie 1999) or by the combination of salt and frost action (Williams & Robinson 2001). By analysing the frequency of weathering features of original walls and the preservation of the studied tuffs; Type I, Type IV and Type V tuffs are the ones that are found to show severe decay. When the mercury porosity of these types is compared to other tuff types it can be noted that Type I has only 21.2% porosity, which is 1% less than that of the durable Type II (Table 1), thus effective porosity of tuff does not necessarily reflect durability, as is also the case for limestones (Benavente *et al.* 2004). Water absorption values are better indicators of durability than porosity, since weathering-sensitive tuff types such as types I and IV have water absorption values that are at least double those of the more durable types II and III tuffs (Table 1).

The pore-size distributions of these acid tuffs show significant variations. It is not unique, as, by studying 11 types of German tuffs, Auras & Steindlberger (2005) also documented diverse ratios of 'gel-pores', 'capillary-pores' and 'air-pores'. When the pore-size distribution and durability of the Hungarian acid tuffs are compared it can be inferred that micropores of a few nanometres in diameter give nearly half of the porosity of the Type I tuff, while these pores are less abundant in the more durable Type II tuff. Thus, it seems that apart from mineralogy and petrology, the frequency of micropores has an important role in controlling the durability of the tuffs. It is in accordance with Fitzner (1994), who also stated that the susceptibility of German tuffs to weathering partly depends on the pore-radii distribution. Steindlberger (2004) also noted that a high level of gel pores (less than 30 nm) leads to water adsorption in the inner parts of German tuffs, while abundant capillary pores (30 nm–10 μm) are responsible for rapid water transport into the tuffs and result in frost damage.

Weathering causes changes in porosity and pore-size distribution by opening up pores (Winkler 1997). These changes are well documented as the porosity of weathered Type I tuff (44%) is more than double than that of Type I (21%). The most significant change is the formation of new and relatively large micropores and the opening of new sets of macropores (Fig. 2). The decreasing amount of small micropores by weathering is not a commonly observed phenomenon for most sedimentary rocks, but it is has been reported for weathering crusts of porous limestone (see Török *et al.* 2007). For volcanic tuffs, well-documented examples are known. Egloffstein (1998) by studying 10 various tuff types has also found that in acid volcanic tuffs weathering often produced a decrease in small micropores (approximately 0.001–0.05 μm) and generated larger micropores and macropores (approximately 0.05 μm and above). One possible explanation for this feature is that weathering of acid volcanic glass produces secondary clay minerals leading to the occlusion of micropores. Auras & Steindlberger (2005) have shown that the closing of micropores can also be attributed to swelling clay minerals as a result of hygric swelling. Thus, it seems that weathering can cause a shift in pore-size distribution toward the larger pores for Type I tuff. An opposite trend, the clogging of pores by weathering, was observed for other tuffs, which was related to the formation of clay minerals (Forgó & Török 2004). The presence of micropores themselves does not directly denote the tuff to be sensitive to weathering, since micropores are also common in durable tuffs such as types II and III. The various grades of cementation and differences in glassy groundmass also contribute to these variations in durability. It is also possible that larger micropores could be contribute to the frost-related weathering of tuffs.

Previous papers (Fitzner 1994; Topal & Sözmen 2003; van Hees *et al.* 2004; Török *et al.* 2005) have stated that the mineralogical composition of tuffs plays an important role in weathering susceptibility. The example of types I and II tuffs is in accordance with these findings, as the first one is more susceptible to weathering and contains a minor amount of smectite, while in the latter this swelling clay was not recorded. In general, montmorillonite and kaolinite are interpreted as forming during the weathering of volcanic glass and feldspars. Likewise, the volcanic glass content of types I and II are also different (Table 1). In addition, Type II tuff is more cemented (see Schmidt hammer rebound values in Table 1), which can be attributed to the different ratio of amorphous silica to devitrified silica found in these tuffs (Fig. 1). Swelling clay minerals are also responsible for decay, and for scaling of Hessian (German) tuffs (Steindlberger 2004) and, very often, of various sandstones (e.g. Smith *et al.* 2002). Limited amounts of swelling clay minerals (smectite) in the Type I tuff might contribute to its weathering susceptibility. Nevertheless, for other types, and even for Type I, weathering is very probably also related to a disadvantageous pore structure with frequent micropores; mostly caused by freeze–thaw cycles. Alternatively, weathering susceptibility can be simply ascribed to a difference in pore-size distribution (Fig. 2) or hygric swelling (Auras & Steindlberger 2005).

Other changes in mineralogy also mark the weathering of the tuffs. Chlorite is considered to be the weathering product of biotite and might be responsible for the greenish tint of some deeply weathered tuffs. Calcite and gypsum are only found in weathered tuffs and thus indicate their ongoing weathering.

The durability of tuffs can also be also predicted by measuring other physical properties (Fitzner 1994; Topal & Sözmen 2003; Auras & Steindlberger 2005). In this study the Schmidt hammer rebound tests have shown that surface strength values often reflect the durability of tuff, as durable tuffs have higher rebound values. It has been noted previously that ryholite tuff is sensitive to water and when saturated its strength decreases dramatically, causing the collapse of cellars and slope stability problems (Kleb 1990). Nevertheless, the rebound values themselves cannot be used as an absolute index of durability, when different tuff types are compared (Topal & Sözmen 2003). As with limestones (Török 2003), the comparison of rebound values for ashlars and quarry stones provides information on the present degree of weathering. It suggests that weathering is leading to a decrease in surface strength. The different surface strengths of replaced and original ashlars indicate that the recently introduced stones are only moderately weathered (Table 1).

## Conclusions

All the studied tuffs can be classified as acid pyroclastic rocks, but similarities and differences in texture and pore-size distribution cause differences in durability. A wide range of weathering forms is observed on the tuffs including relief due to selective weathering, crumbling, crust, detachment, flaking and scaling. These forms are similar to weathering features found on more common lithologies used for construction, such as limestones or sandstones and their occurrence is controlled principally by the fabrics of the tuffs. The percentage of clay minerals, the ratio of phenocrysts v. groundmass and the rate of crystallization of glassy groundmass are the main lithological factors governing the durability. Relief owing to selective weathering is the commonest weathering form on pumice-rich and matrix-rich lapilli tuff, while layered tuff often shows crumbling. Durable dacite tuffs are strongly cemented and have lower water absorption than tuffs prone to weathering. The presence of calcite and gypsum in acid volcanic tuffs are good indicators of weathering.

Effective porosity does not necessarily reflect the durability of the tuffs, and water absorption is a better indicator. However, the sensitivity of the tuff to weathering can be approximately estimated by analysing pore-size distribution; focusing on the presence of small capillary pores or micropores.

Weathering might cause the opening up of pores and a shift in pore-size distribution from micropores towards macropores, alternatively, when clay minerals are formed during weathering, these can also clog pores.

Schmidt hammer data provide information on the durability of the tuffs and clearly indicate that replaced ashlars have greater rebound values than weathered ones.

The funding by the DAAD–MÖB (Hungarian Scholarship Board) is acknowledged. The support of Bolyai János Research Grant (BO/233/04, Á. Török.), the Hungarian Science Fund (OTKA, K 63399, Á. Török.) and Gallus Rehm Fund (L. Z. Forgó.) are also appreciated. The reviews of M. Auras and an anonymous reviewer have improved the quality of the paper. Directorate of the Museum of István Dobó (Eger) provided permissions to work in the castle area where E. Deák guided us. Quarry samples of Eger-Demjén were provided by Mr Rácz and Mr Szilágyi (Korona Ltd, Kerecsend).

## References

AURAS, M. & STEINDLBERGER, E. 2005. Verwitterung und Festigung vulkanischer Tuffe. *Zeitschrift der Deutschen Gesellschaft für Geowissenschaften*, **156**, 167–175.

BENAVENTE, D., GARCÍA DEL CURA, M. A., FORT, R. & ORDÓNEZ, S. 2004. Durability estimation of porous building stones from pore structure and strength. *Engineering Geology*, **74**, 113–127.

CANER-SALTIK, E. N., DEMIRCI, S. ET AL. 1994. Examination of surface deterioration of Göreme Tuffs for the purpose of conservation. *In*: *The Safeguard of the Rock-hewn Churches of the Göreme Valley*. ICCROM, Rome, 85–94.

EGLOFFSTEIN, P. 1998. *Vulkanische Tuffsteine als Werksteine an historischen Bauwerken in Ungarn und Deutschland*. Sven von Loga, Köln.

FITZNER, B. 1994. Volcanic tuffs: The description and quantitative recording of their weathered state. *In*: CHAROLA, A. E., KOESTER, R. J. & LOMBARDI, G. (eds) *Lavas and Volcanic Tuffs. Proceedings of the International Meeting, Easter Island, Chile, 1990*. ICCROM, Rome, 33–51.

FITZNER, B. & BASTEN, D. 1994. Gesteinporosität – Klassifizierung, messtechnische Erfassung und Bewetung ihrer Vervitterungsrelevanz. *In*: SNETHLAGE, R. (ed.) *Jahresberichte Steinzerfall – Steinkonzervierung 1992*. Ernst & Sohn, Berlin, 19–32.

FITZNER, B. & LEHNERS, L. 1990. Rhenish tuff – A widespread, weathering susceptible natural stone. *In*: PRICE D. G. (ed.) *Proceedings of the 6th International Congress of the International Association of Engineering Geology*. Balkema, Rotterdam, 3181–3188.

FITZNER, B., HEINRICHS, K. & KOWNATZKI, R. 1995. Weathering forms – classification and mapping. *In*: SNETHLAGE, R. (ed.) *Denkmalpflege und Naturwissenschaft, Natursteinkonservierung I*. Ernst & Sohn, Berlin, 41–88.

FITZNER, B. & HEINRICHS, K. 2002. Damage diagnosis on stone monuments – weathering forms, damage categories and damage indices. *In*: PŘIKRYL, R. & VILES, H. A. (eds) *Understanding and Managing Stone Decay*. Karolinum Press, Prague, 11–56.

FORGÓ, L. Z. & TÖRÖK, Á. 2004. Influence of petrophysical and petrographical properties on the behaviour of rhyolite tuff, example from Eger, Hungary. *In*: WALRAVEN, J., BLAAUWENDRAAD, J., SCARPAS, T. & SNIJDERS, B. (eds) *Proceedings of the 5th International Symposium in Civil Engineering, 16–19 June 2004, Delft, Netherlands*, Volume 1. Taylor & Francis, London, 589–598.

GAURI, K. L., CHOWDHURY, A. N., KULSHRESHTHA, N. P. & PUNURU, A. R. 1988. Geologic features and durability of limestones at the Sphinx. *In*: MARINOS, P. (ed.) *Engineering Geology of Ancient Works, Monuments and Historical Sites*. Balkema, Rotterdam, 723–729.

GOUDIE, A. S. 1999. Experimental salt weathering of limestone in relation to rock properties. *Earth Surface Processes and Landforms*, **24**, 715–724.

HÁMOR, G. 2001. Genesis and evolution of the Pannonian Basin. *In*: HAAS, J. (ed.) *Geology of Hungary*. Eötvös Kiadó, Budapest, 193–242.

KLEB, B. 1978. *The Past in the Present Life of Eger*. KözDok, Budapest (in Hungarian with English summary).

KLEB, B. 1990. Engineering geological mapping of settlements underbolstered with cellars cut into rocks. *In*: PRICE, D. G. (ed.) *Proceedings of the 6th International Congress of the International*

*Association of Engineering Geology*. Balkema, Rotterdam, 2655–2660.

KOZLOWSKI, R., MAGIERA, J., WEBER, J. & HABER, J. 1990. Decay and conservation of Pinczów limestone I. Lithology and weathering. *Studies in Conservation*, **35**, 205–221.

LANGELLA, A., CALCATERRA, D., CAPPELLETTI, P., COLELLA, A., DE GENARRO, M. & DE GENNARO, R. 2000. Preliminary contribution on durability of some macroporous monumental stones used in historical towns of Campania Region, Southern Italy. *In*: FASSINA, V. (ed.) *Proceedings of the 9th International Congress on Deterioration and Conservation of Stone*, ICOMOS, Venice, 59–67.

NIJLAND, T. G., BRENDLE, S., VAN HEES, R. J. P. & DE HAAS, G. J. L. M. 2003. Decay of Rhenish Tuff in Dutch monuments. Part 1. *Heron*, **48**(3), 149–165.

PATERNO, M. C. & CHAROLA, A. E. 2000. Preliminary studies for the consolidation of Guadalupe tuff from the Philippines. *In*: FASSINA, V. (ed.) *Proceedings of the 9th International Congress on Deterioration and Conservation of Stone*, ICOMOS, Venice, 155–163.

PUNURU, A. R., CHOWDHURY, A. N., KULSHRESHTHA, N. P. & GAURI, K. L. 1990. Control of porosity on durability of limestone at the Great Sphinx, Egypt. *Environmental Geology and Water Science*, **15**(3), 225–232.

SCHERER, G. W. 1999. Crystallization in pores. *Cement and Concrete Research*, **29**, 1347–1358.

SMITH, B. J., WHALLEY, W. B. & MAGEE, R. 1992. Assessment of building stone decay: a geomorphological approach. *In*: WEBSTER, R. G. M. (ed.) *Stone Cleaning and the Nature and Decay Mechanism of Stone. Proceedings of the International Conference, Edinburgh, UK*. Donhead, London, 249–257.

SMITH, B. J., TURKINGTON, A. V., WARKE, P. A., BASHER, P. A. M., MCALISTER, J. J., MENEELY, J. & CURRAN, J. M. 2002. Modelling of rapid retreat of building sandstones: a case study from polluted maritime environment. *In*: SIEGESMUND, S., WEISS, T. & VOLLBRECHT, A. (eds) *Natural Stones, Weathering Phenomena, Conservation Strategies and Case Studies*. Geological Society, London, Special Publications, **205**, 347–362.

STEINDLBERGER, E. 2004. Volcanic tuffs from Hesse (Germany) and their weathering behaviour. *Environmental Geology*, **46**, 378–390.

TOPAL, T. & SÖZMEN, B. 2003. Deterioration mechanisms of tuffs in Midas monument. *Engineering Geology*, **68**, 201–223.

TÖRÖK, Á. 2003. Surface strength and mineralogy of weathering crusts on limestone buildings in Budapest. *Building and Environment*, **38**, 1185–1192.

TÖRÖK, Á., GÁLOS, M. & KOCSÁNYI-KOPECSKÓ, K. 2004. Experimental weathering of rhyolite tuff building stones and effect of an organic polymer conserving agent. *In*: SMITH, B. J. & TURKINGTON, A. V. (eds) *Stone Decay, Its Causes and Controls*. Donhead, London, 109–127.

TÖRÖK, Á., VOGT, T., LÖBENS, S., FORGÓ, L. Z., SIEGESMUND, S. & WEISS, T. 2005. Weathering forms of rhyolite tuffs. *Zeitschrift der Deutschen Gesellschaft für Geowissenschaften*, **156**, 177–187.

TÖRÖK, Á., MÜLLER, C., HÜPERS, A., HOPPERT, M., SIEGESMUND, S. & WEISS, T. 2007. Differences in texture, physical properties and microbiology of weathering crust and host rock: a case study of the porous limestone of Budapest (Hungary). *In*: PŘIKRYL, R. & SMITH, B. J. (eds) *Building Stones Decay: From Diagnosis to Conservation*. Geological Society, London, Special Publications, **271**, 261–276.

VAN HEES, R. P. J., BRENDLE, S., NIJLAND, T. G., DE HAAS, G. L. M. & TOLBOOM, H. J. 2004. Decay of Rehnish tuff in Dutch monuments. *In*: KWIATKOWSKI, D. & LÖFVENDAHL, R. (eds) *Proceedings of the 10th International Congress on Deterioration and Conservation of Stone*, Volume 1. ICOMOS Sweden, Stockholm, 91–98.

WENDLER, E., CHAROLA, A. E. & FITZNER, B. 1996. Easter Island Tuff: Laboratory studies for its consolidation. *In*: RIEDERER, J. (ed.) *Proceedings of the 8th International Congress on Deterioration and Conservation of Stone*, 30 September–4 October, Berlin, ICOMOS, Berlin, 1159–1170.

WILLIAMS, R. B. G. & ROBINSON, D. 2001. Experimental frost weathering of sandstone by various combination of salts. *Earth Surface Processes and Landforms*, **26**, 811–818.

WINKLER, E. M. 1997. *Stone in Architecture: Properties, Durability*. Springer, Berlin.

# Differences in texture, physical properties and microbiology of weathering crust and host rock: a case study of the porous limestone of Budapest (Hungary)

Á. TÖRÖK[1], S. SIEGESMUND[2], C. MÜLLER[2], A. HÜPERS[2], M. HOPPERT[3], & T. WEISS[2]

[1]*Department of Construction Materials, Budapest University of Technology and Economics, Sztoczek u. 2, H-1111 Budapest, Hungary (e-mail: torokakos@mail.bme.hu)*

[2]*Geoscience Centre, University of Göttingen, Goldschmidtstr. 3, D-37077 Göttingen, Germany*

[3]*Institute of Microbiology and Genetics, University of Göttingen, Grisebachstrasse 8, D-37077 Göttingen, Germany*

**Abstract:** Ashlars of the Parliament building and Citadella fortress made of three porous Miocene limestones, a fine-grained limestone, a medium-grained oolitic limestone and a coarse-grained bioclastic limestone, were studied and compared with quarry blocks of the same lithologies. The commonest weathering forms are white (thin and thick) and black (laminar and framboidal) crusts. To assess the processes of crust formation and detachment, descriptions of lithologies and associated weathering features were combined with micro-drilling, pore-size distribution and ultrasonic pulse velocity tests. Microbiological and textural analyses were also performed. The micro-drilling resistance measurements and ultrasonic pulse velocities clearly document the presence of crusts and the degradation of underlying fine- and medium-grained limestones. A textural change, with calcite recrystallization, is also marked by pore occlusion and reduction of microporosity in the crust zone. Crust detachment is initiated by the opening up of microfissures that develop below the cemented crust zones. Fine-grained limestone appears to be less durable than the coarse-grained variety and more prone to rapid crust formation and detachment. Ashlars from where the crusts were removed have lower micro-drilling resistance compared to quarry stones. Microbiological activity appears to play an insignificant role in crust formation and removal. Indeed, the combined effect of air pollution and related gypsum crystallization and more probably freeze–thaw weathering activity lead to crust detachment with rates strongly controlled by the texture and porosity of the limestone substrate.

Weathering crusts found on limestone exposed to air pollution are probably some of the most thoroughly studied weathering phenomena (Kieslinger 1949; Amoroso & Fassina 1983). Previous work has described dark coloured and white weathering crusts, which are further divided according to their morphology and thickness (Smith *et al.* 1992; Camuffo 1995; Fitzner *et al.* 1995; Antill & Viles 1999; Maravelaki-Kalaitzaki & Biscontin 1999). Most of these crusts are enriched in gypsum. The influence of environmental conditions on gypsum crust formation has been thoroughly studied in the field (Amoroso & Fassina 1983; Zappia *et al.* 1998; Fassina *et al.* 2002; Smith *et al.* 2003; Bonazza *et al.* 2004) and under laboratory conditions (Rodriguez-Navarro & Sebastian 1996; Ausset *et al.* 1999; Primerano *et al.* 2000; Cultrone *et al.* 2004). Differences in weathering of various limestones exposed to the same pollution regime have been studied using small test blocks over periods of time (Smith 1996). However, less information is available regarding the physical changes that are triggered by pollution fluxes (Winkler 1966, 1970) or weathering (Bell 1993*a*), and even less information is available on the physical properties of crusts and host rocks. The sparse examples report weathering-related changes in physical properties of various limestones (Christaras 1991; Török 2002*a*, 2003; Török *et al.* 2004) and marbles (Christaras 1996). The loss in strength caused by weathering was also reported for granites (Irfan & Dearman 1978; Kahraman 2001) and for rhyolite tuffs (Topal & Sözmen 2003; Török *et al.* 2005). Research generally focuses on the description of processes and decay products for one type of stone, but rarely describes the variations in physical properties and crust formation on various limestone types (Török 2004).

The aim of this paper is to analyse weathering crust formation on porous limestone and to describe the differences in physical properties of crusts

*From*: PŘIKRYL, R. & SMITH, B. J. (eds) *Building Stone Decay: From Diagnosis to Conservation*. Geological Society, London, Special Publications, **271**, 261–276.
0305-8719/07/$15.00 

developed on fine-, medium- and coarse-grained limestones from one stratigraphic level. The approach taken was to compare ashlars that have experienced several tens or even hundreds of years of exposure on building façades with similar or the same type of porous stone that was collected fresh from a quarry. Two monuments were studied in Budapest; the Parliament building and the Citadella fortress. Both are located in the city centre and thus experience high pollution fluxes (Török 2002*a*). After the description of their lithologies and identification of crust types, microscopic textural analyses, micro-drilling resistance tests and, on core samples, porosimetric and ultrasonic pulse velocity analyses were performed to identify the nature of physical and textural changes with depth below the surface. Microbiological analyses of samples were also performed to assess the role of organisms in weathering crust formation and crust detachment. Previous work (Warscheid & Braams 2000; Krumbein 2003) indicates that organisms may be involved in crust formation, patination or roughening of stone surfaces, as well as inducing material loss by sanding, chipping or scaling. However, other work has shown that organisms may also accompany weathering processes without producing deteriorative effects (Hoppert *et al.* 2005), or form a protective surface on stone preventing or slowing down other weathering processes (Kurtz & Netoff 2001).

The combination of these techniques was used to understand the mechanism of crust formation and to identify circumstances that contribute to crust detachment. In addition, differences in texture and physical properties of both crusts and their host rocks were investigated with specific emphasis on the role of microscale porosity differences in crust removal.

## Methods and environment

### *Methodology*

*In situ* tests were carried out on the walls of the Parliament and Citadella (Fig. 1), which were constructed at the end of 19th century and in the middle 19th century, respectively. The construction history of these monuments is described by Török (2002*a*, *b*); Smith *et al.* (2003) and Török *et al.* (2004). Although limestone was used in several ornamental architectural forms only ashlar blocks were examined in this study. Selected façades were documented, beginning with the characterization of lithological features by visual inspection, followed by a description of decay forms using the nomenclature of Smith *et al.* (1992) and Fitzner *et al.* (1995). The decay forms were mapped and

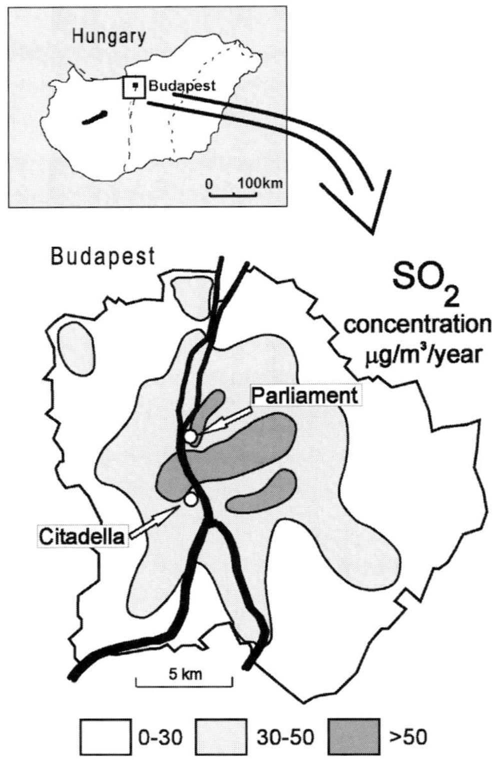


**Fig. 1.** The location of Parliament building and the Citadella fortress in Budapest and the $SO_2$ plume over Budapest (modified after Török 2002).

are reported in Török *et al.* (2004) and Hüpers *et al.* (2005). Mechanical properties of the limestone were determined by means of micro-drilling (for the methodology see Wendler & Sattler 1996). The micro-drilling resistance provides indirect information on the rate of cementation at depth without significant destruction (drilling bit is $\phi = 3$ mm) of stone, allowing the construction of strength v. depth profiles. Ongoing restoration work at the Parliament building enabled the collection of drilled cores (4 cm in diameter) from the main façade. Sampling points are shown in Figure 2. Small samples were also collected from the walls of the Citadella, but no drill cores were obtained as destructive tests are not permitted on the historic walls of this fortress. Ultrasonic velocity measurements were performed on the cores before further sample preparation, using the pulse transmission technique with piezoceramic transducers with a resonant frequency of 350 kHz (UKS 12, Geotron). Following this, the cores were cut into two halves and 19 thin sections were prepared perpendicular to the stone surface using resin impregnation. Porosity data were obtained on small cores (1 cm in diameter) drilled from half-core samples parallel to

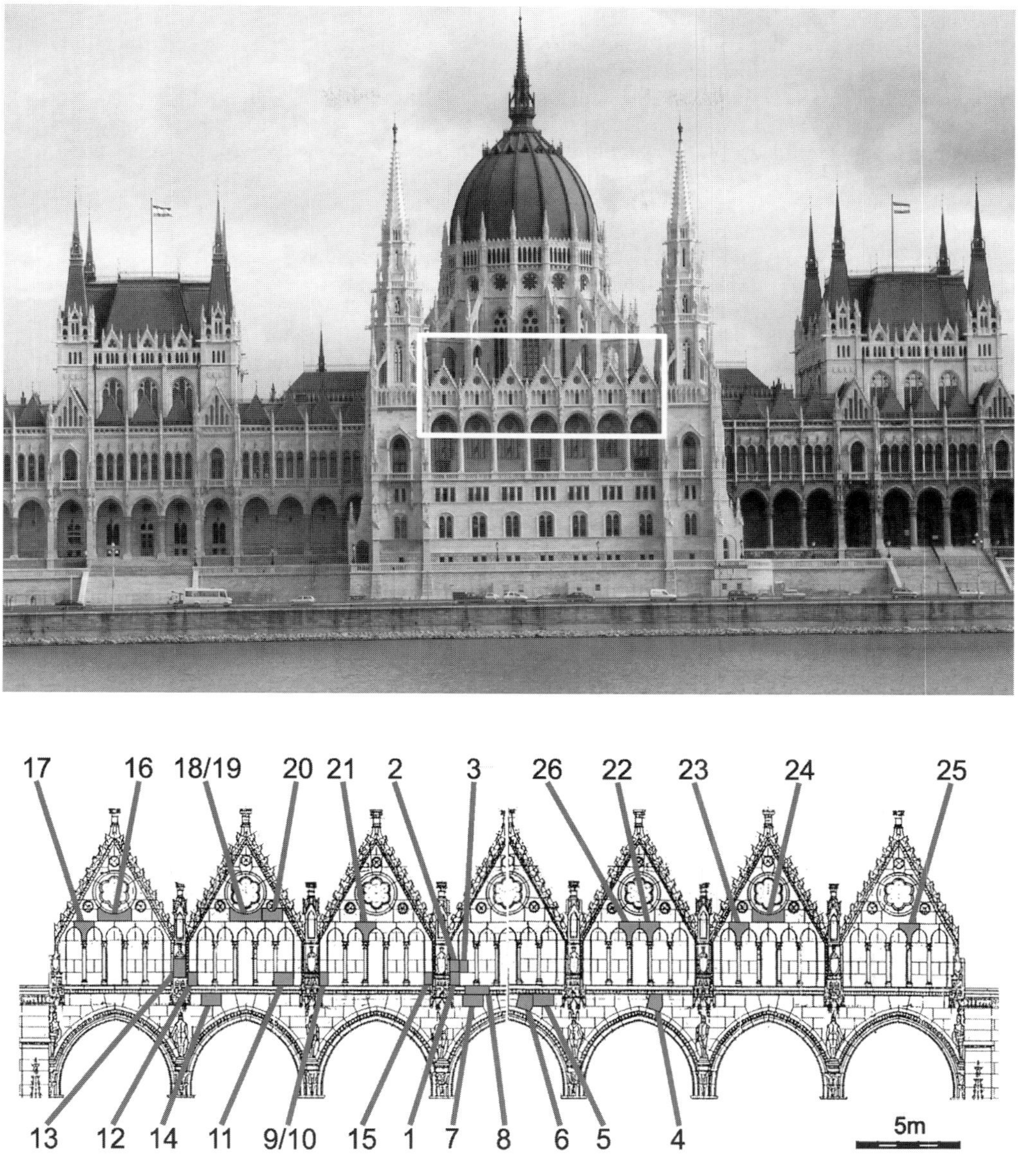


**Fig. 2.** The main façade of the Parliament building (now covered by replacement stones), and the testing and sampling sites (squared area).

the stone face at various depths. By using mercury porosimetry it was possible to determine pore-size distributions within and near the crust and in the underlying host rock. Small samples were also gathered from both sites for testing the mineralogical and microbiological composition of crusts and host rock. X-ray diffraction (XRD, Phillips PW1800) was used to identify the mineralogical composition of 21 selected samples from the walls of the Citadella and Parliament building and six samples from quarry stones. Other samples were prepared and stained for light and transmission electron microscopy (TEM) as described in Hoppert & Holzenburg (1998), Hoppert *et al.* (2002) and Hoppert (2003) to determine colonization of crusts and host rock by micro-organisms.

## *Environment*

Budapest still suffers from air pollution despite the introduction of clean air legislations. In comparison to other European cities the $SO_2$ pollution levels in Budapest are still double (16–18 μg $m^{-3}$, with a winter average concentration of 30 μg $m^{-3}$) those of London and Paris. Airborne dust particles make a significant contribution to the air pollution load showing very high annual concentrations (246 μg $m^{-3}$), which are doubled in autumn and in winter. The average $NO_x$ concentration is 40 μg $m^{-3}$, but more than twice this value can be recorded in the city centre. Further details on air pollution within Budapest compared to other cities are given by Török (2003). Although both buildings are located in the city centre where pollution levels are high (Smith *et al.* 2003), there are differences in their locations and exposure to pollution fluxes and wind/rain. The Parliament building faces the River Danube and is adjacent to a busy road network (Fig. 1). In contrast, the Citadella fortress is located on a small elevated hill approximately 100 m above the Danube on the riverside where constant wind and higher altitude prevent the formation of a pollution plume, and exposes walls to rainwash (Török 2002*a*).

## Lithologies

Soft and porous limestone of Miocene age has been used as a dimension and ornamental stone for centuries in Budapest and surrounding settlements. It has a yellowish-white colour when it is freshly quarried. Some quarries still exist, but most of them were operational during the second half of the 19th century when construction activity was intense and rapid development of the city was marked by construction of public buildings. This porous limestone was used as ashlars, facing stone, slabs, ornaments or even for load-bearing structures. The only active quarry now is found approximately 30 km to the west of Budapest in Sóskút village. Three main limestone types have been identified on the Parliament and Citadella buildings and blocks of similar types were obtained from this quarry, although other lithotypes were also found. The three main lithologies are a fine-grained limestone, a medium-grained oolitic limestone and a bioclastic macroporous limestone. No significant difference in terms of mineralogical composition was found in the quarry samples. The main mineral is calcite (92–97%) in all lithotypes, with minor amounts of quartz. Microscopic analyses showed lithic clasts as well as sand-sized quartz. The largest amount of non-carbonate components was found in medium-grained oolitic limestone with maximum values of 8%, comprising mostly quartz and some feldspar. The most important properties of the lithotypes are listed in Table 1.

The fine-grained type has small visible pores of 0.05–0.5 mm on cut surfaces (Fig. 3a) and its major cement is micrite (Fig. 3b). Small, often microscopic, miliolid foraminifers are the main bioclasts. The major fabric constituents of the second lithotype, the medium-grained oolitic limestone, comprise well- to moderately-rounded calcitic ooids and visible, but evenly scattered, small pores of 0.1–1 mm (Table 1, Fig. 3c). Cross-bedding is occasionally visible on building blocks and even on ashlars. Microscopic analyses revealed that the carbonate constituents are not, from a sedimentological point of view, ooids but micrite-coated grains – called micro-oncoids. Nevertheless, as this limestone type and the other two types are commonly described as oolitic limestone, for the purpose of this study this second type is also referred to as a medium-grained oolitic limestone. The nuclei of these micro-oncoids are often formed by angular–subangular quartz grains or rarely by feldspars (Fig. 3d). This is in agreement with XRD analyses, as the greatest amount of non-carbonate minerals was found in this lithotype. Typical pore sizes observed by microscope are of the order of 0.01–0.2 mm. On pore

**Table 1.** *Properties of limestones*

| Lithologies | Carbonate constituents (grain size in mm) | Microfacies (fabric) | Porosity types | Micro-drilling resistance (km $s^{-1}$) |
|---|---|---|---|---|
| Fine-grained | ++Peloids (0.05–0.2) +Foraminifers (0.1–0.4) | Peloidal Wackestone–packstone | ++Intergranular Mouldic | 0.7 |
| Medium-grained oolitic | ++'Oids' = micro-oncoids Foraminifers (0.1–0.5) | Micro-oncoid Packstone–grainstone | ++Intergranular Intragranular | 1.7 |
| Coarse-grained bioclastic | ++Gastropods (5–15) +Bivalves (1–20) +Micro-oncoids (0.2–5) Intraclasts (0.5–80) Foraminifers (0.2–1) | Bioclastic grainstone | ++Intergranular ++Intragranular +Shelter | 2.1 |

+, common; ++, abundant.

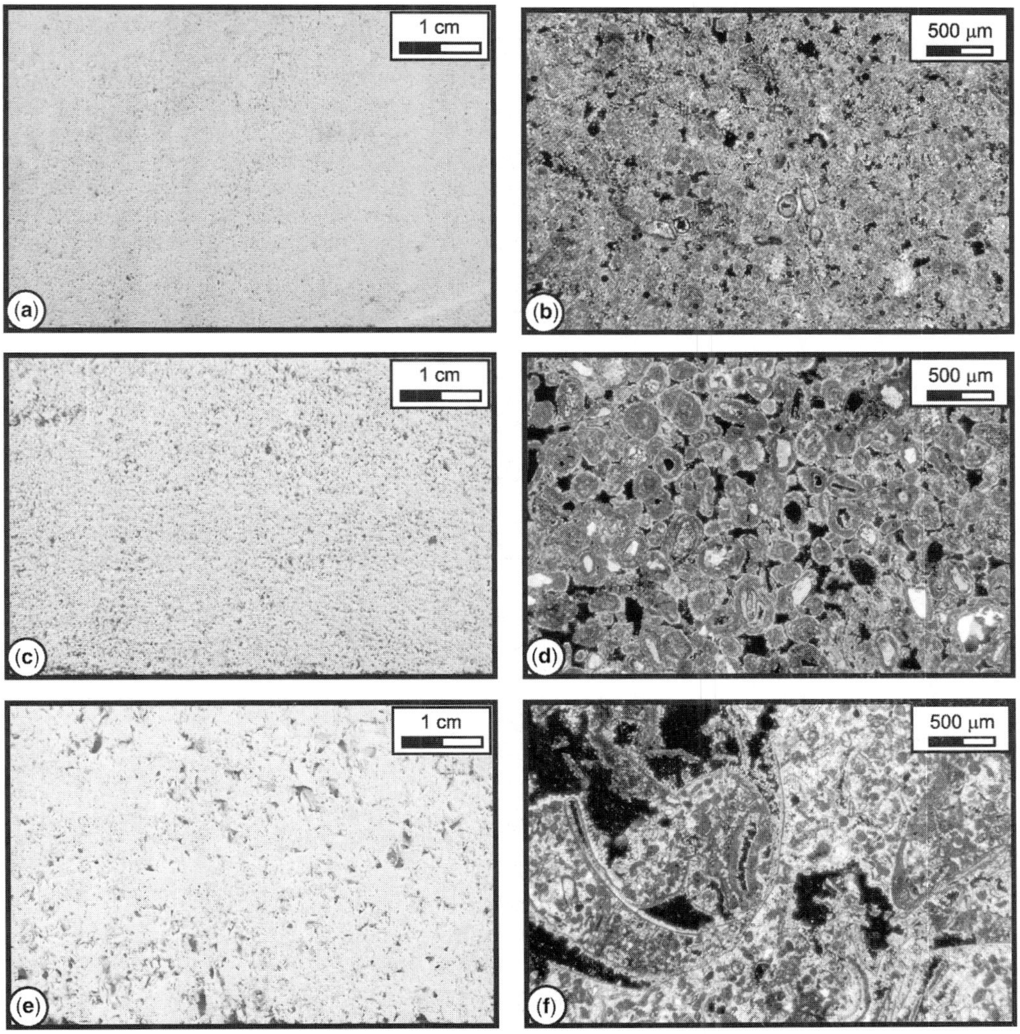


**Fig. 3.** Three lithotypes of porous Miocene limestone: (**a**) & (**b**) fine-grained; (**c**) & (**d**) medium-grained; and (**e**) & (**f**) coarse-grained bioclastic limestone. The fine-grained type is characterized by very small pores (a) and a peloidal-micritic packstone texture (b); while the medium-grained type has larger pores (c) and its major carbonate components are ooids (micro-oncoids) forming a grainstone texture. Angular quartz and quartzite particles are also common, some of these are found in the nuclei of micro-oncoids (d). The coarse bioclastic limestone has variable pores including larger ones (e) and it shell fragments form bioclastic packstone texture (f).

walls thin isopachus microsparitic cement fringes are common, but these calcite crystals do not occlude the pores.

The macroporous bioclastic lithotype is characterized by a bioclastic grainstone fabric with gastropods (Cerithium type), bivalve shell fragments, larger micro-oncoids and other carbonate constituents including intraclasts (Table 1). The pores are large (up to 1 cm) and are irregularly scattered within the stone (Fig. 3e). In this lithotype shelter pores also occur (Table 1, Fig. 3f).

## Weathering crusts

### *Texture and micro-organisms*

The most frequent weathering forms observed on the building façades are crusts. Field survey confirmed data from previous studies (Török 2002*a*, 2005; Török & Rozgonyi 2004) that identified two types of light coloured crust on these porous limestones. Thick (a few millimetres to 1 cm) white crusts develop on surfaces exposed to wind/rain.

The surface of the crust tends to be moderately smooth while the underside is often irregular. Microscopic analyses of these white crusts have shown that the crust zone is characterized by intense dissolution and recrystallization of calcite. The primary fabric of the stone is hardly detectable, and micritic and microsparitic cement replaces the primary components. This can lead to occlusion of pores and to the formation of a cemented zone up to 1 cm in thickness (Fig. 4). A transitional zone exists between the crust and the host rock, and at depth the substrate shows minor cementation. It has also been found that when the substrate is quartz rich the crust also tends to be enriched in quartz. In such cases the quartz particles are incorporated within the micrite. Case-hardened crusts very often show contour scaling, with crust removal exposing an irregular surface prone to either granular disintegration (Török 2003) or the development of a secondary crust (Smith *et al.* 2003). Thin white crusts are generally formed on the fine-grained lithotype, but can also occur on medium-grained oolitic limestones. This 1 mm-thick and fragile crust has a relatively smooth outer surface and if the surface is partly sheltered the crust can be greyish in colour. Fabric analyses of the crust have shown that it consists of micron-sized calcite. Blistering and flaking are the typical weathering features associated with such crusts, with microfissures also developing beneath the crust. At the initial stage, microfissures are only micrometres wide and found up to 1 cm below the crust (Fig. 5). When crust detachment has reached a more advanced stage the fissures open up and microcracks parallel with the surface develop beneath the crust (Fig. 5).

Dark coloured crusts tend to develop in areas protected from wind and rain (Török 2002*a*), but also at exposed sites where high pollution fluxes allow the rapid deposition of particles that are incorporated into the growing crust (Smith *et al.* 2003; Török & Rozgonyi 2004). The framboidal black crust grows on protected parts of walls, generally in hollows, near joints, below cornices or ornaments. This cauliflower-like form consists of small globules or spherules of millimetres to 2 cm in size. It is dominated by gypsum, but calcite and organic carbon can also be found in these crusts. Scanning electron microscopy (SEM) analyses revealed that the crust surface is characterized by rosette-like gypsum crystals that display crystal aggregates containing a mixture of gypsum and calcite crystals (Fig. 6d). Thin black crusts often trace the stone surface and such crusts are also visible on partly exposed ashlars. Microscopic analyses indicates that these dark grey–black crusts consist of micrite with silt-sized quartz particles, opaque and dark organic carbon-rich components also present (Fig. 6c). On the surface of the crust small acicular gypsum crystals occur (Fig. 6b). A microsparitic zone might develop below the crust or the crust can settle directly onto the limestone substrate. It seems that the bond between the thin black crust and the stone surface is strong with microfissures generated below the crust within the host rock (Fig. 6a).

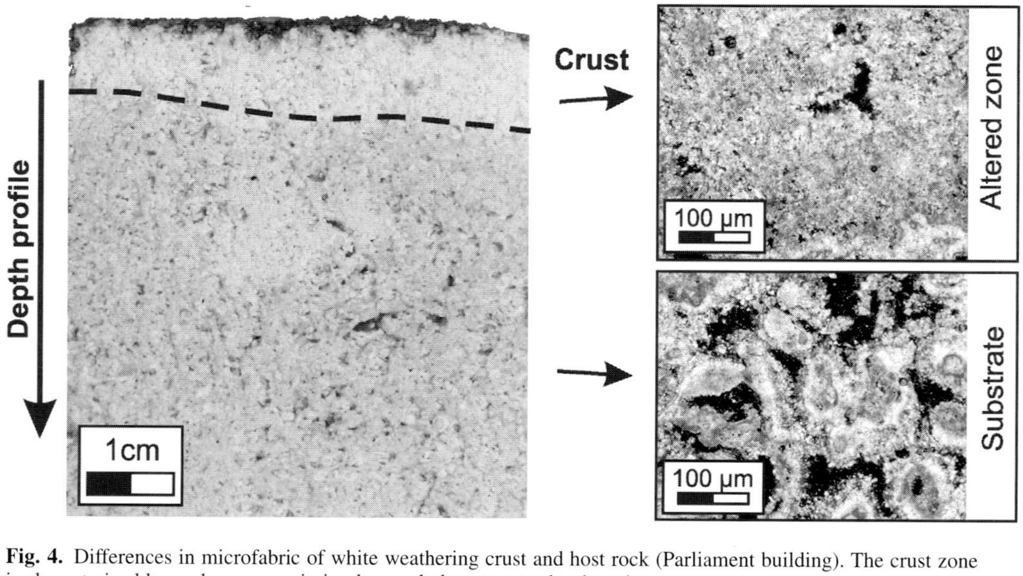


**Fig. 4.** Differences in microfabric of white weathering crust and host rock (Parliament building). The crust zone is characterized by replacement micrite that occludes pores under the microscope and masks the original fabric of the limestone (top right), while the host rock below displays open pores and rounded particles (bottom right).

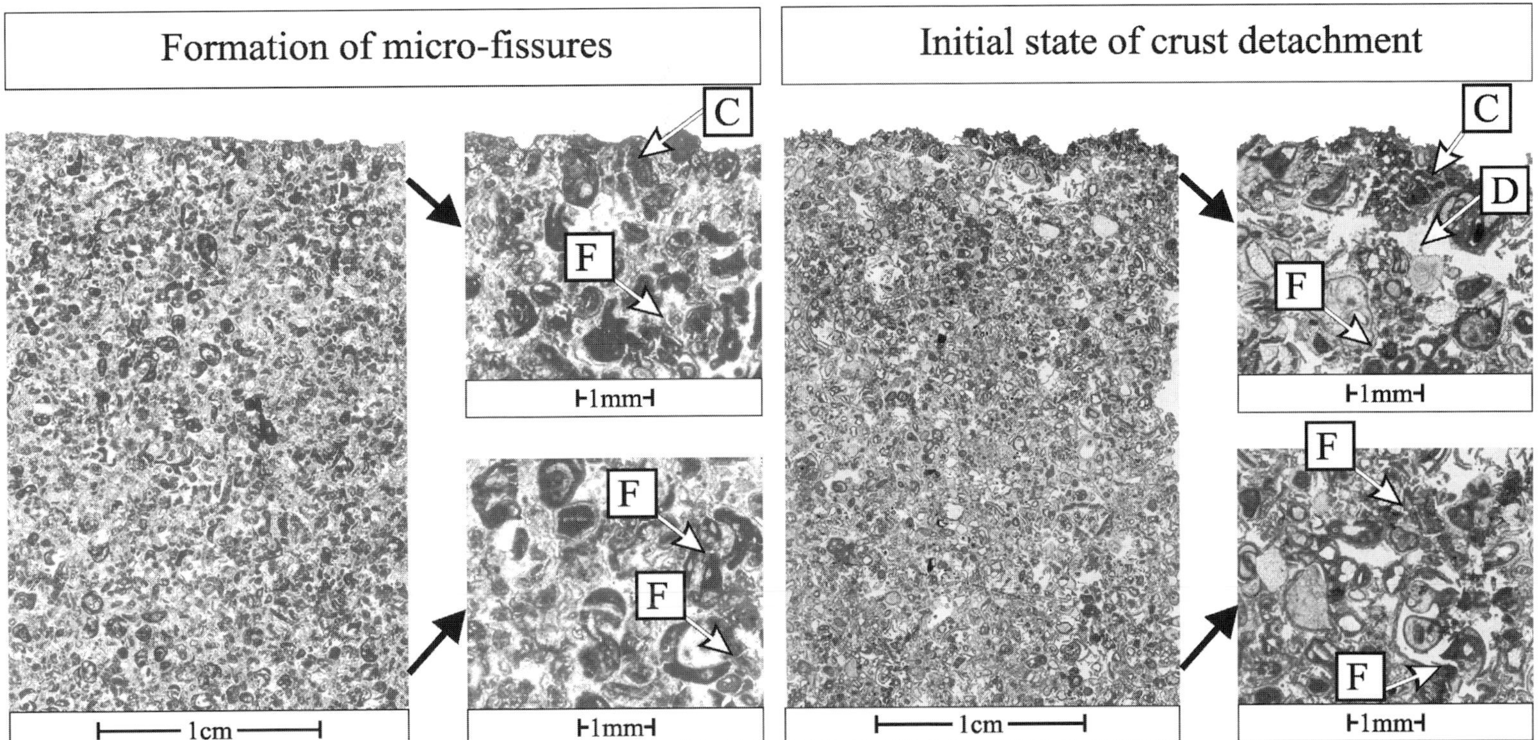


**Fig. 5.** Different phases of crust detachment as observed in thin section (samples are from the Parliament). At a very early stage (left) μm-size microfissures (marked by F) are developed below the crust (C). When crust detachment begins (right) fissures (F) open and a surface parallel detachment zone (D) is formed below the crust (C).

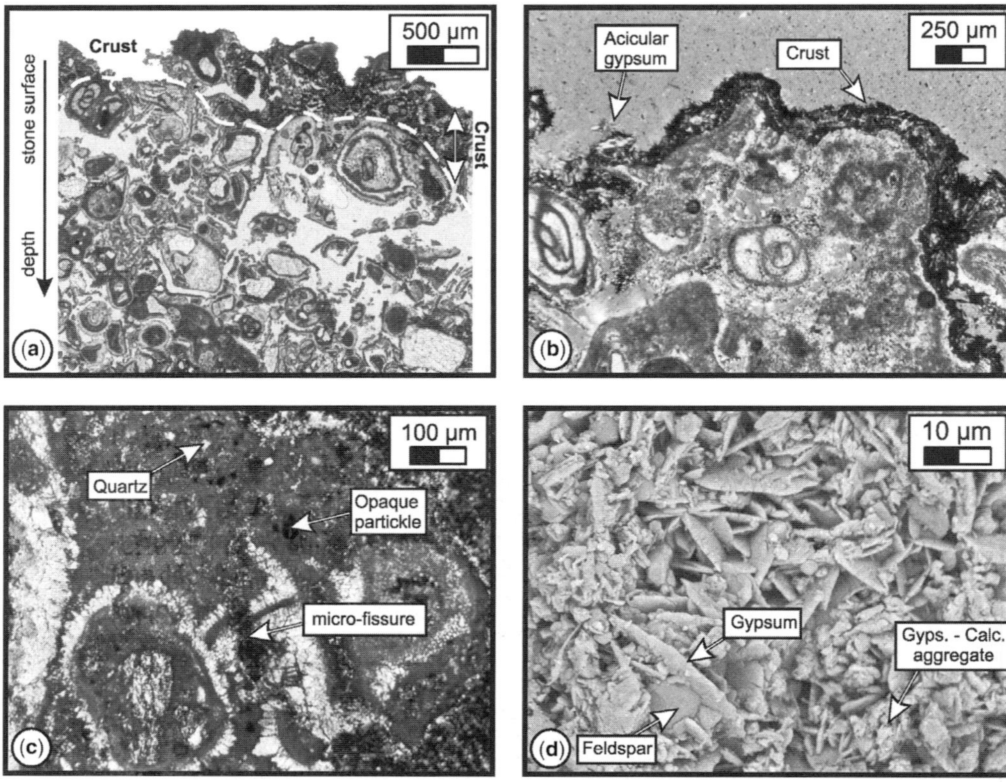


**Fig. 6.** (**a**)–(**c**) Microscopic images and (**d**) SEM micrograph of dark coloured weathering crusts. Dark coloured crusts strongly adhere to the surface while microcracks are generated below this zone (a, Parliament). Thin black crust tracing the surface and also penetrating into the rock (on the right-hand side of the image), while acicular gypsum crystals are mostly found on the crust surface (b, Citadella). Thin black crust with framboidal morphology contains silt-sized quartz particles, opaque components and displays only a few microcracks (c, Citadella). Rosette-like gypsum crystals forming the surface of black crust and also hosting crystal aggregates of gypsum and calcite (SEM image, d, Citadella).

Microbiological studies identified micro-organisms on the crust surface as well as inside crusts and scales taken form the Parliament building as well as the Citadella. They also frequently penetrate ooids in deteriorated surfaces, upon crust removal. Among these organisms, green algae of the genera *Stichococcus* and *Apatococcus* are dominant, accompanied by cyanobacteria of the Genus *Aphanothece*. The cyanobacterial/green algal communities were mostly located on the underside of flaking scales and on the freshly exposed stone surface (Fig. 7a). The exposed surface after crust removal was also frequently penetrated by organisms along small cracks and fissures. The organisms were tightly cemented with the mineral particles (Fig. 7b). Along with green algae, epilithic and chasmoendolithic dematiaceous fungi (genus *Aureobasidium*) colonize crusts and scales. However, these unstable substrata, subjected to flaking or scaling, were never colonized by lichens. On stable crust surface samples from the Citadella, especially under the influence of pollution containing nitrogen compounds (e.g. bird droppings), the nitrophilic lichens *Caloplaca* or *Candelariella* could be detected. All lichen thalli were 0.7 mm in diameter or smaller, indicating that the thalli were not older than approximately 5 years (see Clark *et al.* 2000).

### *Porosity, micro-drilling resistance and ultrasonic velocity*

The crust on fine-grained porous limestone clearly shows a significant change in porosity and pore-size distribution with depth. The crust zone has a porosity of 34.6%, which is nearly 6% more than that of the porosity measured at a depth of 4 cm. The pore-size distribution is also different as the

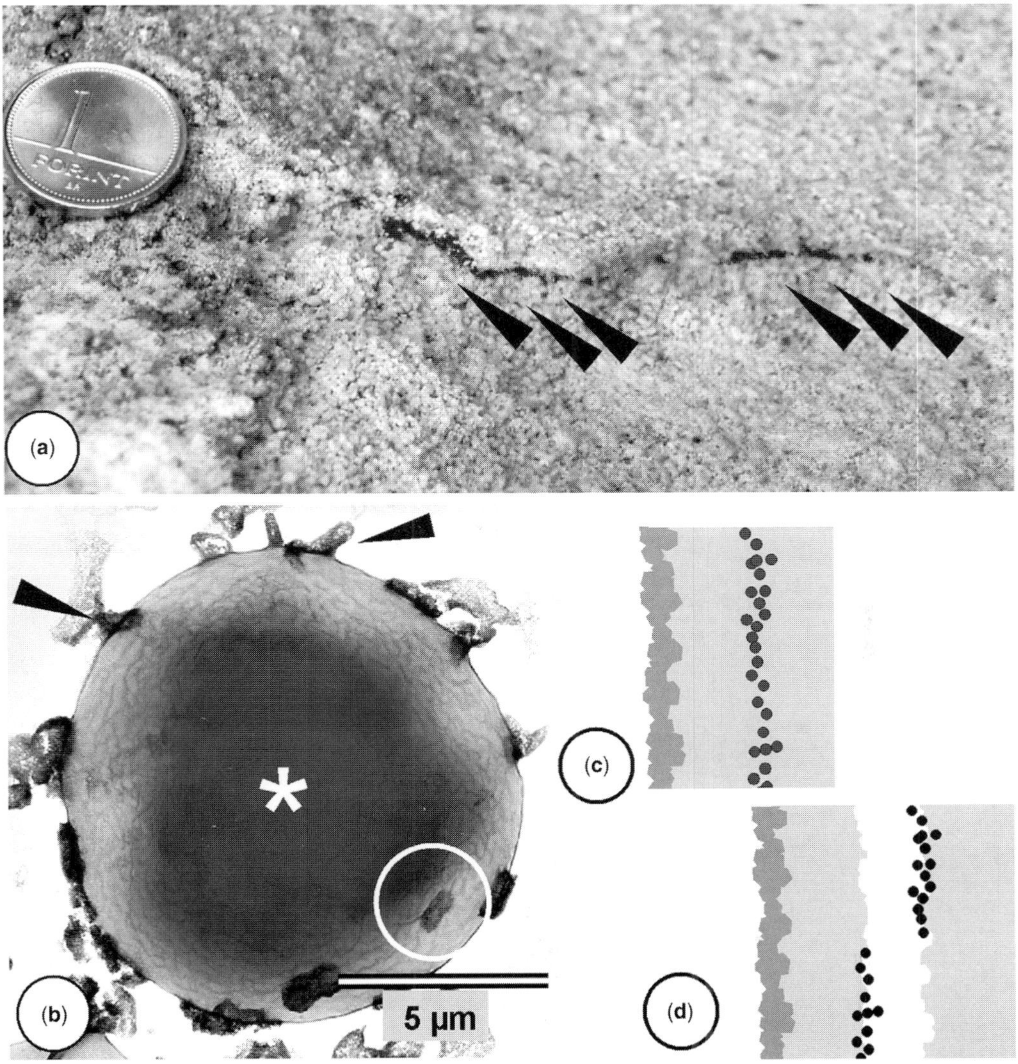


**Fig. 7.** (**a**) & (**b**) Endolithic algal biofilms and (**c**) & (**d**) their location within stones. (a) Endolithic film, marked by arrowheads, visible as dark line 1–2 mm beneath the surface of a scale (the coin is 1.5 cm). (b) Electron micrograph of an algal cell (asterisk) after intensive rinsing in water and negative staining for electron microscopy: small mineral fragments (arrowheads, circle) are directly attached to the cell wall. (c) & (d) Schematic drawing of an algal biofilm in side view with an algal layer (dark dots) (c) before and (d) after scaling. The algae mark the detachment zone.

crust zone displays a bimodal porosity with pore sizes of 1 and larger than 10 $\mu$m, while with increasing depth an increase in pore size is documented (Fig. 8). Micro-drilling data confirm the presence of a few millimetres-thick cemented crust zone, with the micro-drilling resistance of 1.5 s mm$^{-1}$ decreasing to 0.5 s mm$^{-1}$ beneath the crust. Ultrasonic pulse velocity data show that the cemented crust zone has a thickness of approximately 0.5 cm (Fig. 8). The porosity and pore-size distribution of crust-covered medium-grained porous limestone is similar to that of its fine-grained counterparts. Namely, the porosity from the crust zone (33.0%) increases with depth, reaching 38.2% some 4 cm below the stone surface. It is also associated with a shift in pore size from micropores towards macropores from crust to host rock (Fig. 9). The ultrasonic pulse velocity values gradually decrease from the stone surface downward, but micro-drilling resistance does not show

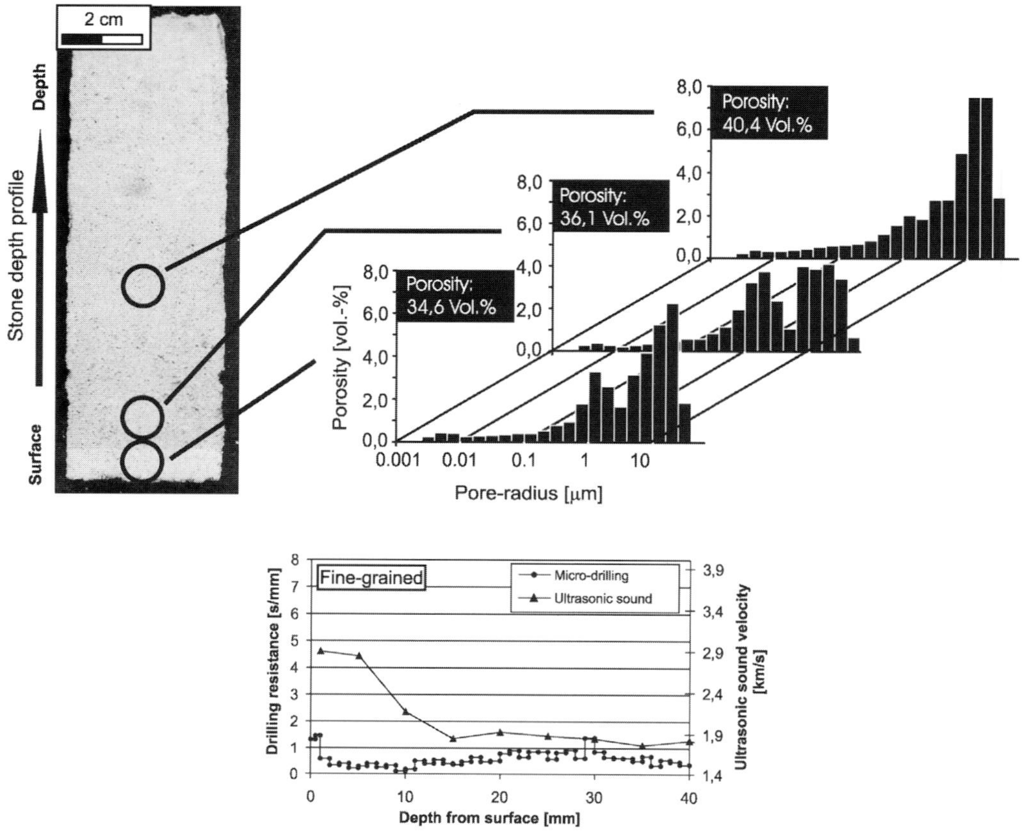


**Fig. 8.** Pore-size distribution (top) and micro-drilling resistance and ultrasonic pulse velocity as a function of depth (bottom) of a weathering crust developed on fine-grained limestone (see the text for a detailed description).

signs of intense cementation in the near-surface zone (Fig. 9). The coarse-grained bioclastic limestone does not have a significant change in porosity from the stone surface downwards and parallel to this no observable trend in pore-size distribution was found (Fig. 10). The only sign of surface cementation is given by ultrasonic pulse velocity and less clearly by micro-drilling resistance (Fig. 10). Owing to the scatter in values no clear interpretation can be made, especially when values of 4–5 s $mm^{-1}$ were obtained at a depth of 3.5–4 cm (Fig. 10). Micro-drilling tests were also performed on ashlars where no crust was present. On these stones the crust had been removed by flaking or scaling. The micro-drilling resistance values of these ashlars are smaller than that of the quarry stones (Table 1) indicating the effects of weathering. Minor differences occur between the values derived from weathered ashlars and unaltered quarry stones of fine-grained lithotype (Fig. 11), while a decrease in drilling resistance was detected in medium- and coarse-grained ashlars in comparison with quarry stones. It was also possible to divide the drilling profiles according to the grade of weathering, especially for medium-grained ashlars (Fig. 11). Moderately weathered testing points are ones where no sign of granular disintegration is recognized. At medium weathered points initial stages of crumbling is observed, while on strongly weathered points intense crumbling and granular disintegration is visible, mostly due to crust detachment and exposure of a pre-weakened stone surface (Fig. 11).

## Discussion

Weathering crusts may temporarily stabilize the stone surface and form a protective layer on ashlars (Török 2003). After crust removal there are two possible scenarios for these porous limestones: (i) rapid loss of stone surface by granular disintegration (Török 2002*a*) or (ii) formation of a new crust that temporarily slows down surface retreat

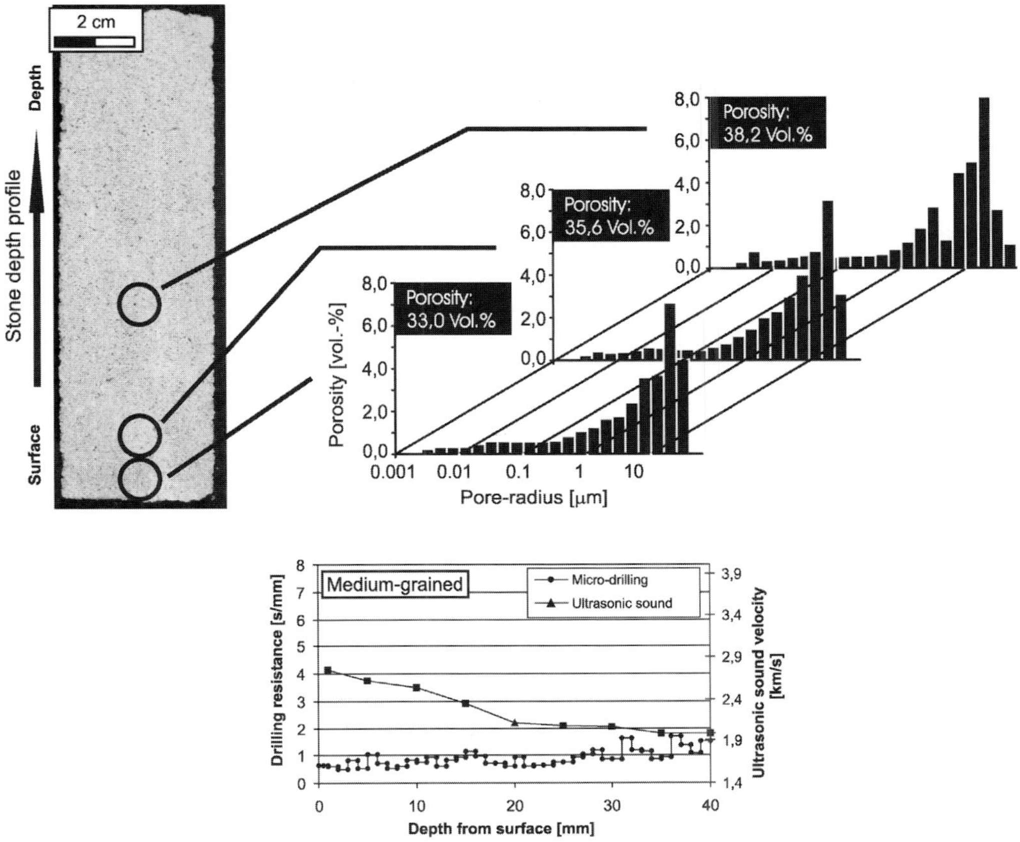


**Fig. 9.** Pore-size distribution (top) and micro-drilling resistance and ultrasonic pulse velocity as a function of depth (bottom) of a weathering crust developed on medium-grained oolitic limestone (see the text for a detailed description).

(Smith *et al.* 2003). Whether crust removal is followed by rapid surface retreat or the stone surface is stabilized by a secondary crust depends on numerous factors including the depth of the weathered zone below the crust and its exposure to wetting–drying (Smith *et al.* 2002) or freeze–thaw cycles. Fabric analyses have shown that crust detachment is initiated by the formation of microcracks within the stone. These microcracks are not limited to the surface zone as they can also be found several centimetres below crusts (Fig. 4). It has also been observed that thin white crusts are characterized by replacement micrites, which partly occlude pores (Fig. 4). Consequently, the crust forms a relatively impermeable layer on the stone surface. This is shown by water absorption tests (Török 2003; Török *et al.* 2004). The difference in porosity between the crust and host rock was detectable by mercury porosimetry of fine- and medium-grained limestone (Figs 8 & 9). Micro-drilling resistance values and ultrasonic pulse velocities show the same trend (Figs 8 & 9). It is well known that for most rocks strength decreases with increasing average grain size. In contrast, increasing micro-drilling resistance (Table 1) and ultrasonic pulse velocities with increasing grain size was measured on these quarry stones. It suggests that the main controlling factor of strength is not grain size but cementation. In this context, the coarser carbonates are more cemented and compact than the fine-grained ones owing to early diagensis and marine cementation. The role of carbonate cements in the durability of limestones has been previously emphasized (Benavente *et al.* 2004). Consequently, the data presented in this paper illustrate the importance of diagentic processes in the cementation and their influence on rock strength. Comparing the values measured on quarry stones (Table 1) and the ones detected on ashlars with crusts (Figs 8–10), it is clear that weathering can be also considered as a diagenetic change that can cause cementation and enable a crust to form on the porous stone surface.

The mechanism of crust detachment is related to the opening up of microfissures and the generation

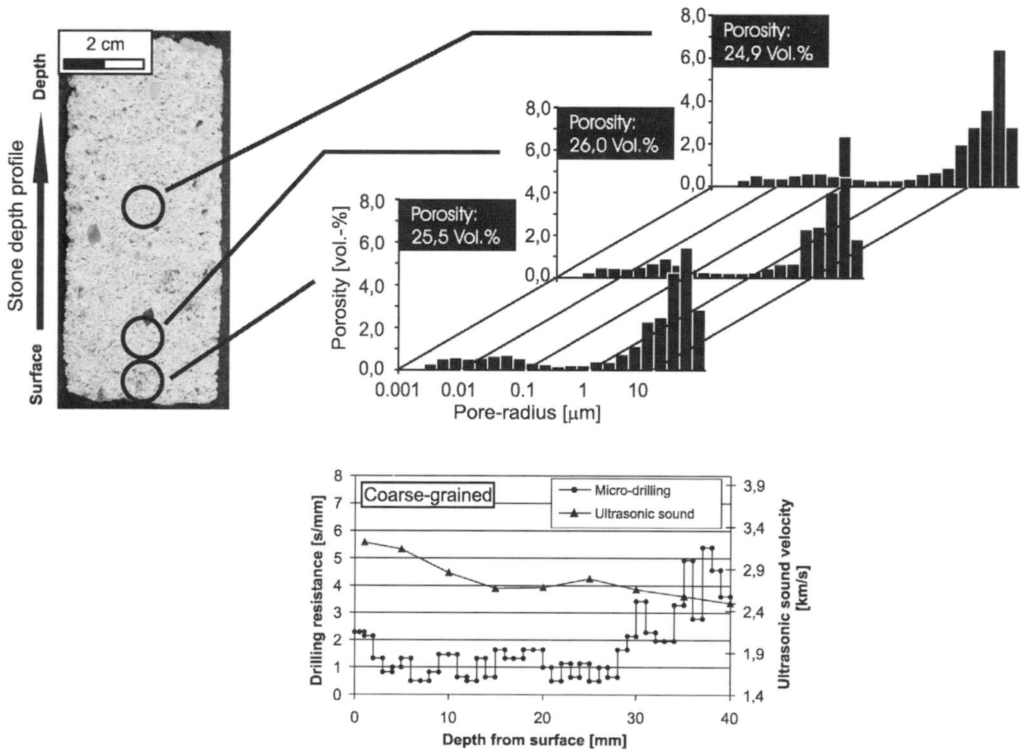


**Fig. 10.** Pore-size distribution (top) and micro-drilling resistance and ultrasonic pulse velocity as a function of depth (bottom) of a weathering crust developed on coarse-grained bioclastic limestone (see the text for a detailed description).

of wider cracks that run parallel or subparallel to the stone surface and are located beneath the cemented crust (Fig. 5). The mineralogy of black crusts is different from white crusts as they consist of silt-sized quartz grains and opaque particles (Fig. 6c). The contribution of flyash (Primerano *et al.* 2000) and the increased organic carbon contents of black framboidal crusts of porous limestone have already been documented for several buildings in Budapest (Török 2002*a*;

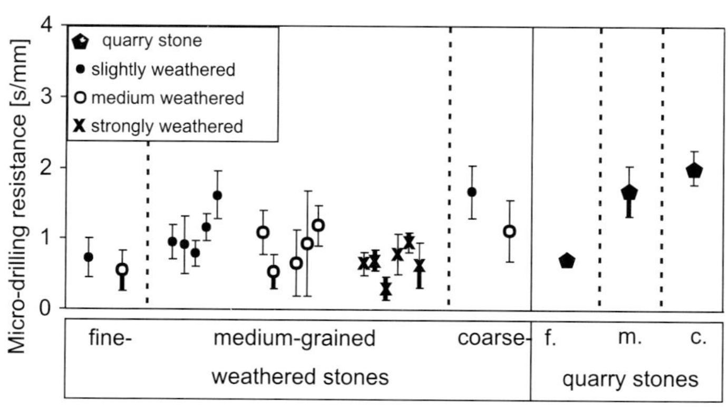


**Fig. 11.** Average micro-drilling resistance values of fine-, medium- and coarse-grained ashlars in comparison to quarry stones. The tests were performed on ashlars from which crusts had already been removed. Error bars are standard deviations of several sets of measurements.

Smith *et al.* 2003; Török & Rozgonyi 2004) and in other cities (Rodriguez-Navarro & Sebastian 1996; Fassina *et al.* 2002; Bonazza *et al.* 2004). It seems that on thin black crusts idiomorphic gypsum crystals are mostly found on the crust surface, while the crust itself is a mixture of calcitic micrite, wind-deposited particles and small gypsum crystals. Surfaces protected from wind and rain, where these crust accumulate, allow the initial settling of dust particles (Primerano *et al.* 2000; Lefévre & Ausset 2002) within stone pores and thus enable the black crust to penetrate into the stone (Fig. 6b). Microcracks are more commonly found in black crusts than in white ones, suggesting that the porosity and permeability of black crusts is somewhat higher than that of the white crusts. This assumption was proven since black crusts tend to have a higher water absorption capacity (Török 2002*a*). It was not possible to find very thick black crusts on the Parliament building where core drilling was allowed, thus pore-size distribution and numerical porosity values are not available for thick black crusts. The crust detachment mechanism of thin black crust (Viles 1993) is similar to white ones, but it is assumed that, owing to a difference in porosity, fabric and cementation, black crusts preferentially form laterally limited flakes or blisters rather than large scaling surfaces. The blackening of larger scaling crusts that have been documented by Smith *et al.* (2003) is mostly related to the fact that secondary and tertiary crusts form in depressions where dust can accumulate. The threshold for crust detachment is apparently controlled by both environmental factors and the rock fabric of the stone. It has been shown that cementation and the adherence of the crust to the surface depends on its carbonate components (Rodriguez-Navarro & Sebastian 1996; Lefévre & Ausset 2002) and the degree of cementation between these components. Fine-grained limestone has a lower micro-drilling resistance and ultrasonic pulse velocity than the coarse-grained one (Fig. 11), and thus when a crust develops it forms a cemented layer on the stone surface increasing these values more abruptly (Fig. 8). In contrast, the weathering crust formed on coarse-grained limestone does not significantly change the physical properties (Fig. 10) and only after crust removal can changes in micro-drilling resistance on exposed coarse limestone surfaces be observed (Fig. 11). This supports field observations where fine-grained limestone ashlars are the ones that show rapid surface retreat and repeated crust formation and detachment, i.e. are more susceptible to weathering. Indeed, a scaling crust is less commonly found on coarse-grained limestone (Török *et al.* 2004) and this lithotype is more resistant and durable in polluted environments and less prone to gypsum-rich crust formation. However, freezing conditions very probably play an important role in triggering crust detachment. The large size of the open cracks below the crusts and the annual 73–87 frosty days in Budapest suggests that freeze–thaw cycles also have a significant role in crust detachment. In this context, the interaction between gypsum-related crust detachment and freeze–thaw-related crust removal cannot be excluded. Indeed, Smith *et al.* (2005) suggested that the accelerated breakdown of rock already pre-weakened by salt weathering is more likely under freezing conditions.

There is no indication that the formation of crusts and scaling was initiated by micro-organisms. However, intensive colonization by algae and cyanobacteria was typically associated with crusts and scales. When exposed to air after flaking, the biofilm became desiccated and lost the typical blue–green colour. Similar biofilms have been frequently observed in natural habitats (see, for example, Bell 1993*b*; Clark *et al.* 2000; Friedmann 2000) and on building stone (for review see Warscheid & Braams 2000). In a thin layer 1–5 mm beneath the surface, the organisms are protected from environmental stress and are exposed to optimum or at least sufficient light intensity for photoautotrophic growth. The green layer did frequently (but not exclusively) mark the predetermined cracking zone of the scale (Fig. 7c, 7d). Also uncolonized thin crusts flake off in an identical way. Thus, it is likely that the algae colonize a predetermined breaking zone, but do not determine this zone by active dissolution processes. This colonization pattern has been frequently observed at the lower parts of the façade of the Citadella, where ground humidity promotes algal growth. As well as the homogeneous biofilm layer, patterns of irregular scattered microcolonies are detectable. Here, the organisms first colonize the surface and penetrate crusts and scales along existing cracks (chasmoendolithic growth). These patterns are exclusively formed by green algae (without cyanobacteria) and fungal hyphae. As uncolonized crusts and ooids exhibit cracks of identical size, in this case a deteriorative activity of the organisms also appears to be unlikely. Micro-organisms accompany crust formation but do not generate weathering forms by their own metabolic activities. There is no indication that the observed algae accelerate the weathering process. Indeed, the role of microbiological activity in crust formation and crust detachment is very doubtful under these environmental and meteorological conditions where pollution and exposure prevents extensive microbiological colonization and where freeze–thaw cycles frequently affect the porous stone.

## Conclusions

Crust formation and detachment among others factors is strongly controlled by the fabric of the limestone substrate. White crusts are composed of replacement calcite micrites and microsparites or when the substrate is quartz rich they contain insoluble quartz grains surviving replacement and calcite recrystallization. White crusts are less porous and contain smaller pores than the host fine- or medium-grained limestone. Microscopic analyses, micro-drilling and ultrasonic pulse velocity tests also confirm that these crusts are better cemented than the underlying host rock. Crust removal is attributed to the opening of micro-fissures and the generation of surface parallel or subparallel cracks below the white crust. The black crusts are actually more porous and rich in silt-sized quartz particles, organic carbon and opaque components. The acicular gypsum crystals are mostly found on the crust surface indicating a rapid crystal growth. It has been also documented that the porosity alone is not necessarily a good indicator of stone durability, meanwhile pore-size distribution and porosity changes with depth provide more information on the severity of weathering. The limitations of the micro-drilling test in detecting the thickness of crusts on coarse-grained limestones (average grain size is of the order of few millimetres) are illustrated by the great scatter of drilling resistance values. Indeed, the 3 mm diameter of drilling bit is too small to make a clear difference between the depositional fabric and weathering-related textural changes. Thin black crusts partly consist of calcite micrite and partly gypsum, with common wind-deposited particles of quartz and carbon-rich components. These crusts are less cemented than their white counterparts and thus smaller flakes and blisters are commonly observed. The crust detachment mechanism for the thin black crust is similar to that for white ones and is controlled by the presence microcracks. The active role of micro-organisms in crust formation is inferior to the non-biogenic processes. Instead, crust formation and detachment is controlled by a combination of air pollution, and most probably meteorological factors such as freeze–thaw cycles.

The support of Bolyai János Research Grant (BO/233/04) (Á. Török) and the Hungarian Science Foundation (OTKA, K63399) (Á. Török) are acknowledged. The funding by the DAAD (German Science Foundation)–MÖB (Hungarian Scholarship Board, project no. 30) is also appreciated. The reviews of P. Warke and R. Přikryl have improved the quality of the paper. We are also very grateful to J. Lukács, B. Andrássy, J. Herkules and to Reneszánsz Co. who provided access to the construction site and restoration works at the Parliament building.

## References

AMOROSO, G. G. & FASSINA, V. 1983. *Stone Decay and Conservation*. Elsevier, Amsterdam.

ANTILL, S. J & VILES, H. A., 1999. Deciphering the impacts of traffic on stone decay in Oxford: some preliminary observations from old limestone walls. *In*: JONES, M. S. & WAKEFIELD, R. D. (eds) *Aspects of Stone Weathering, Decay and Conservation*. Imperial College Press, London, 28–42.

AUSSET, P., DEL MONTE, M. & LÈFEVRE, R. A. 1999. Embryonic sulphated black crusts on carbonate rocks in atmospheric simulation chamber and in the field: role of carbonaceous fly-ash. *Atmospheric Environment*, **33**, 1525–1534.

BELL, F. G. 1993*a*. Durability of carbonate rock as a building stone with comments on its preservation. *Environmental Geology*, **21**, 187–200.

BELL, R. A. 1993*b*. Cryptoendolithic algae of hot semiarid lands and deserts. *Journal of Phycology*, **29**, 133–139.

BENAVENTE, D., GARCÍA DEL CURA, M. A., FORT, R. & ORDÓNEZ, S. 2004. Durability estimation of porous building stones from pore structure and strength. *Engineering Geology*, **74**, 113–127.

BONAZZA, A., SABBIONI, C., GHEDINI, N., FAVONI, O. & ZAPPIA, G. 2004. Carbon data in black crusts on European monuments. *In*: SAIZ-JIMENEZ, C. (ed.) *Air Pollution and Cultural Heritage*. Taylor & Francis, London, 39–47.

CAMUFFO, D. 1995. Physical weathering of stone. *The Science of the Total Environment*, **167**, 1–14.

CHRISTARAS, B. 1991. Durability of building stones and weathering of antiquities in Creta/Greece. *Bulletin of the International Association of Engineering Geology*, **44**, 17–25.

CHRISTARAS, B. 1996. Non destructive methods for investigation of some mechanical properties of natural stones in the protection of monuments. *Bulletin of the International Association of Engineering Geology*, **54**, 59–63.

CLARK, B. M., MANGELSON, N. F., ST. CHAIR, L. L., REES, L. B., BENCH, G. S. & SOUTHON, J. R. 2000. Measurement of age and growth rate in the crustose saxicolous lichen *Caloplaca trachyphylla* using $^{14}C$ accelerator mass spectroscopy. *Lichenologist*, **32**, 399–403.

CULTRONE, G., RODRIGUEZ-NAVARRO, C. & SEBASTIAN, E. 2004. Limestone and brick decay in simulated polluted atmosphere: the role of particulate matter. *In*: SAIZ-JIMENEZ, C. (ed.) *Air Pollution and Cultural Heritage*. Taylor & Francis, London, 141–145.

FASSINA, V., FAVARO, M. & NACCARI, A. 2002. Principal decay patterns on Venetian Monuments. *In*: SIEGESMUND, S., WEISS, T. & VOLLBRECHT, A. (eds) *Natural Stones, Weathering Phenomena, Conservation Strategies and Case Studies*. Geological Society, London, Special Publications, **205**, 381–391.

FITZNER, B., HEINRICHS, K. & KOWNATZKI, R. 1995. Weathering forms-classification and mapping. *In*: SNETHLAGE, R. (ed.) *Denkmalpflege und Naturwissenschaft, Natursteinkonservierung I*. Ernst and Sohn, Berlin, 41–88.

FRIEDMANN, E. I. 2000. Endolithic micro-organisms in the Antarctic cold desert. *Science*, **215**, 1045–1053.

HOPPERT, M. 2003. *Microscopic Techniques in Biotechnology*. Wiley-VCH, Weinheim.

HOPPERT, M. & HOLZENBURG, A. 1998. *Electron Microscopy in Microbiology*. Bios, Oxford.

HOPPERT, M., BERKER, R., FLIES, C., KÄMPER, M., POHL, W., SCHNEIDER, J. & STRÖBEL, S. 2002. Biofilms and their extracellular environment on geomaterials: methods for investigation down to nanometre scale. *In*: SIEGESMUND, S., WEISS, T. & VOLLBRECHT, A. (eds) *Natural Stones, Weathering Phenomena, Conservation Strategies and Case Studies*. Geological Society, London, Special Publications, **205**, 207–215.

HOPPERT, M., KÖNIG, S. & HEGERMANN, J. 2005. Microalgae on building stone. *Zeitschrift der deutschen Gesellschaft für Geowissenschaften*, **156**, 93–101.

HÜPERS, A., MÜLLER, C., SIEGESMUND, S., HOPPERT, M., WEISS, T. & TÖRÖK, Á. 2005. Kalksteinverwitterung – die Zitadella und das Parlaments-Gebäude in Budapest. *In*: SIEGESMUND, S., AURAS, M. & SNETHLAGE, R. (eds) *Stein Zerfall und Konservierung*. Edition Leipzig, Leipzig, 201–209.

IRFAN, T. Y. & DEARMAN, W. R. 1978. Engineering classification and index properties of a weathered granite. *Bulletin of the International Association of Engineering Geology*, **17**, 79–90.

KAHRAMAN, S. 2001. Evaluation of simple methods for assessing the uniaxial compressive strength of rock. *International Journal of Rock Mechanics and Mining Sciences*, **38**, 981–994.

KIESLINGER, A. 1949. *Die Steine von Sankt Stephan*. Verlag Herold, Wien.

KRUMBEIN, W. E. 2003. Patina and cultural heritage – a geomicrobiologist's perspective. *In*: KOZLOWSKI, R. (ed.) *Proceedings of the 5th European Commission Conference 'Cultural Heritage Research: a Pan European Challenge'*. Polska Akademia Nauk, Crakow, 39–47.

KURTZ, H. D. & NETOFF, D. I. 2001. Stabilization of friable sandstone surfaces in a desiccating, wind-abraded environment of south-central Utah by rock surface micro-organisms. *Journal of Arid Environments*, **48**, 89–100.

LEFÈVRE, R. A. & AUSSET, P. 2002. Atmospheric pollution and building materials: stone and glass. *In*: SIEGESMUND, S., WEISS, T. & VOLLBRECHT, A. (eds) *Natural Stones, Weathering Phenomena, Conservation Strategies and Case Studies*. Geological Society, London, Special Publications, **205**, 329–345.

MARAVELAKI-KALAITZAKI, P. & BISCONTIN, G. 1999. Origin, characteristics and morphology of weathering crusts on Istria stone in Venice. *Atmospheric Environment*, **33**, 1699–1709.

PRIMERANO, P., MARINO, G., DI PASQUALE, S., MAVILIA, L. & CORIGLIANO, F. 2000. Possible alteration of monuments caused by particles emitted into the atmosphere carrying strong primary acidity. *Atmospheric Environment*, **34**, 3889–3896.

RODRIGUEZ-NAVARRO, C. & SEBASTIAN, E. 1996. Role of particulate matter from vehicle exhaust on porous building stones (limestone) sulfation. *The Science of the Total Environment*, **187**, 79–91.

SMITH, B. J. 1996. Scale problems in interpretation of urban stone decay. *In*: SMITH, B. J. & WARKE, P. A. (eds) *Processes of Urban Stone Decay*. Donhead, London, 3–18.

SMITH, B. J., TÖRÖK, Á., MCALISTER, J. J. & MEGARRY, J. 2003. Observations on the factors influencing stability of building stones following contour scaling: a case study of the oolitic limestones from Budapest, Hungary. *Building and Environment*, **38**, 1173–1183.

SMITH, B. J., TURKINGTON, A. V., WARKE, P. A., BASHEER, P. A. M., MCALISTER, J. J., MEENLY, J. & CURRAN, J. M. 2002. Modelling the rapid retreat of building sandstones. A case study from polluted maritime environment. *In*: SIEGESMUND, S., WEISS, T. & VOLLBRECHT, A. (eds) *Natural Stones, Weathering Phenomena, Conservation Strategies and Case Studies*. Geological Society, London, Special Publications, **205**, 339–354.

SMITH, B. J., WARKE, P. A., MCGREEVY, J. P. & KANE, H. L. 2005. Salt-weathering simulations under hot desert conditions: agents of enlightenment or perpetuators of preconceptions. *Geomorphology*, **67**, 211–227.

SMITH, B. J., WHALLEY, W. B. & MAGEE, R. 1992. Assessment of building stone decay: a geomorphological approach. *In*: WEBSTER, R. G. M. (ed.) *Stone Cleaning and the Nature and Decay Mechanism of Stone. Proceedings of the International Conference, Edinburgh*. Donhead, London, 249–257.

TOPAL, T. & SÖZMEN, B. 2003. Deterioration mechanisms of tuffs in Midas monument. *Engineering Geology*, **68**, 201–223.

TÖRÖK, Á. 2002*a*. Oolitic limestone in polluted atmospheric environment in Budapest: weathering phenomena and alterations in physical properties. *In*: SIEGESMUND, S., WEISS, T. & VOLLBRECHT, A. (eds) *Natural Stones, Weathering Phenomena, Conservation Strategies and Case Studies*. Geological Society, London, Special Publications, **205**, 363–379.

TÖRÖK, Á. 2002*b*. The influence of wall orientation and lithology on the weathering of ooidal limestone in Budapest, Hungary. *In*: PŘIKRYL, R. & VILES, H. A. (eds) *Understanding and Managing Stone Decay*. Karolinum Press, Prague, 229–240.

TÖRÖK Á. 2003. Surface strength and mineralogy of weathering crusts on limestone buildings in Budapest. *Building and Environment*, **38**, 1185–1192.

TÖRÖK, Á. 2004. Comparison of the processes of decay of two limestones in a polluted urban environment. *In*: MITCHELL, D. J. & SEARLE, D. E. (eds) *Stone*

*Deterioration in Polluted Urban Environments*. Science Publishers, Enfield, 73–92.

TÖRÖK, Á. 2005. Gypsum-induced decay on the limestone buildings in the urban environment of Budapest. *International Journal for Restoration of Buildings and Monuments*, **11**(2), 71–78.

TÖRÖK, Á. & ROZGONYI, N. 2004. Mineralogy and morphology of salt crusts on porous limestone in urban environment. *Environmental Geology*, **46**, 323–339.

TÖRÖK, Á., VOGT, T., LÖBENS, S., FORGÓ, L. Z., SIEGESMUND, S. & WEISS, T. 2005. Weathering forms of rhyolite tuffs. *Zeischrift der Deutshen Gesellschaft für Geowissenschaften*, **156**(1), 177–187.

TÖRÖK, Á., WEISS, T., HÜPERS, A., MÜLLER, C. & SIEGESMUND, S. 2004. The decay of oolitic limestones controlled by atmospheric pollution: a case study from the Parliament and Citadella in Budapest, Hungary. *In*: KWIATKOWSKI, D. & LÖFVENDAL, R. (eds) *Proceedings of the 10th International Congress on Deterioration and Conservation of Stone*. ICOMOS Sweden, Stockholm, Volume II, 947–954.

VILES, H. A. 1993. The environmental sensitivity of blistering of limestones walls in Oxford, England: a preliminary study. *In*: THOMAS, D. S. G. & ALLISON, R. J. (eds) *Landscape Sensitivity*. Wiley, Chichester, 309–326.

WARSCHEID, T. & BRAAMS, J. 2000. Biodeterioration of stone: a review. *International Biodeterioration and Biodegradation*, **46**, 343–368.

WENDLER, E. & SATTLER, L. 1996. Bohrwiderstandsmessungen als zerstörungsarmes Prüfverfahren. *Werkstoffwissenschaften und Bausanierung*, **1**, 145–159.

WINKLER, E. M. 1966. Important agents of weathering for building and monument stone. *Engineering Geology*, **1**, 381–400.

WINKLER, E. M. 1970. The importance of air pollution in the corrosion of stone and metals. *Engineering Geology*, **4**, 327–334.

ZAPPIA, G., SABBIONI, C., RIONTINO, C., GOBBI, G. & FAVONI, O. 1998. Exposure tests of building materials in urban atmosphere. *The Science of the Total Environment*, **224**, 235–244.

# A comparative and critical study of X-ray CT and neutron CT as non-destructive material evaluation techniques

J. VLASSENBROECK[1], V. CNUDDE[2], B. MASSCHAELE[1], M. DIERICK[1], L. VAN HOOREBEKE[1] & P. JACOBS[2]

[1]*Department of Subatomic and Radiation Physics, Ghent University, Proeftuinstraat 86, B-9000 Ghent, Belgium (e-mail: jelle.vlassenbroeck@ugent.be)*

[2]*Department of Geology and Soil Science, Ghent University, Krijgslaan 281/S8, B-9000, Ghent, Belgium*

**Abstract:** X-ray computerized tomography (CT) has traditionally been used as a medical diagnostic tool. This non-destructive technique has developed as an important research tool for a wide variety of scientific subjects. For material research 'medical' CT, microCT and, very recently, nano- or submicroCT have been used as non-destructive material evaluation techniques for engineering and geological purposes. The fact that X-ray CT visualizes the internal structure of natural building stones and yields information on porosity values and pore-size distributions is a major advantage for the study of their conservation. The penetration of fluids like water, consolidants or water repellents inside porous materials is important when dealing with conservation and restoration research. Recently, high-speed neutron tomography has been introduced as a visualization technique for fluids inside porous materials. High-speed neutron tomography can be used as a complementary technique to X-ray tomography as elements like hydrogen, which have a weak attenuation for X-rays, are easy to detect using neutrons. In this paper the basic principles of computerized tomography and more specifically X-ray and neutron tomography are discussed. In addition, application possibilities, advantages and limitations of medical CT, X-ray microCT and high-speed thermal neutron CT are outlined.

Many destructive research techniques are available to analyse stone samples. However, there is a need for non-destructive techniques that provide three-dimensional (3D) visualization of internal structures, as destructive techniques can change these structures. These non-destructive techniques should allow the monitoring of certain phenomena as a function of time, including fluid flow, conservation and restoration actions, and artificial weathering.

A well-known 3D visualization technique, originally used in medicine, is X-ray computerized tomography (CT). Owing to the potential of this technique, it has been used in various geological studies, including palaeontology, sedimentology, petrology, soil science and fluid-flow research. By the middle of the 1990s X-ray microCT started to be used in rock analysis and related research fields. Although X-ray microCT offered a much higher spatial resolution than the original medical X-ray CT systems, X-ray nanoCT (or better described by submicroCT) is sometimes used in material research. Besides X-ray CT, neutron tomography has been used to monitor fluid flow inside natural building stones (Masschaele *et al.* 2004; Dierick *et al.* 2005). X-ray CT and neutron tomography are complementary to each other, as they can provide different information about the same sample. Although the physics of X-ray and neutron interactions are quite different, the basic theories behind X-ray CT and neutron tomography are similar. In this paper both these differences and similarities are discussed.

## Basic principles

Computerized tomography is a tool designed to visualize the internal structure of a sample. A number of projections – taken at different angles by means of a type of penetrating radiation – can be used to reconstruct the 3D distribution of the different elements inside the sample. The two most important radiation probes are X-rays and neutrons.

In a projection image (recorded by a detector) every pixel value corresponds to the amount of attenuation of the radiation along a straight line through the sample between the source and the pixel position. This is mathematically represented by the line-integral of the attenuation coefficient (2D information) along this path. The relation between the original intensity of the source and the intensity at the detector position is known as the Lambert–Beer law:

$$I = I_0 \cdot e^{-\int \mu(s)\cdot\rho(s)\cdot \mathrm{d}s} \tag{1}$$

*From*: PŘIKRYL, R. & SMITH, B. J. (eds) *Building Stone Decay: From Diagnosis to Conservation.* Geological Society, London, Special Publications, **271**, 277–285.
0305-8719/07/$15.00 

Here $\mu(s)$ is the mass attenuation value per unit mass (in $cm^2\ g^{-1}$); $\rho(s)$ is the gravimetric density (in $g\ cm^{-3}$), $\mu(s) \cdot \rho(s)$ is the linear attenuation coefficient (in $cm^{-1}$) and $\int ds$ is the length of the material along the radiation path (in cm).

From this set of projections the linear attenuation coefficient at each point of the sample can be derived (3D information). As attenuation coefficients are dependent on local density and chemical composition, the internal structure of a scanned material can be reconstructed.

It has to be taken into account that the Lambert–Beer law makes the following assumptions, which are in most cases not valid and therefore complicate the measurements.

- The attenuation coefficients should be independent of the path length inside the sample. As attenuation coefficients are generally energy-dependent this implies that the radiation should be monochromatic. Otherwise the spectral distribution of the radiation changes according to the penetration depth (beam hardening), and so do the attenuation coefficients.
- The intensity at a detector point should only be dependent on the absorption along a straight line between the source and the point.

Typically, these assumptions are not met. X-rays produced by tube sources are not monochromatic. In some cases, a beam hardener is used to remove the X-rays with the lowest energies. These X-rays are attenuated very easily and cannot penetrate the whole sample, resulting in artefacts in the reconstruction. Only synchrotrons can produce monochromatic beams with a high flux, but they are less accessible than microfocus X-ray tubes. Beam hardening is less of a problem when using neutron beams.

The intensity at a detector point can be influenced by scattering. The amount of scattering depends on the type of radiation (neutrons or X-rays), the energy of the radiation and the composition of the sample. Positioning the sample close to the detector reduces the influence of scattering, as the scattered particles hit the detector close to the point where they would be detected without scattering. Positioning the sample close to the source results in an almost homogeneous background because the scattering occurs at a large distance from the detector. Positions between detector and source can give corrupt projections and result in a degradation of image quality and loss of resolution. These considerations are only relevant when high-energy X-rays or neutrons are used as, otherwise, scattering is not an issue.

Next to the complications regarding the possible violation of the Lambert–Beer law, the tomographic reconstruction also requires some assumptions. Theoretically, every detector row should only contain the information of one cross-section through the sample, perpendicular to the axis of rotation and the detector. This would imply the use of only a parallel or a fan beam.

Parallel beam geometry is typically used at neutron beam lines and X-ray synchrotron facilities. The radiation emerges from a relatively large opening and the rays are (approximately) parallel to each other. A 2D detector is used to take projections. For each detector row, the data at different projections angles can be arranged in a 2D array, called a sinogram. A sinogram contains all data necessary for the reconstruction of one cross-section.

Fan beam geometry is used in high-power, high-energy X-ray tubes and in some medical scanners. A point source is used and projections are registered by a detector consisting of one row. The use of a point source allows for a magnification of the sample. As only a line detector is needed, more sensitive and advanced means of detection are possible. In addition, collimation in front of the detector is possible (typically by using a pair of lead slabs). This reduces the contribution of scattered X-rays because a large portion of them cannot hit the detector. The main disadvantage of this geometry is the scanning speed, as different scans should be taken for every cross-section that has to be visualized. This can be partially compensated for by moving the sample while scanning, a technique often used in modern medical scanners (helical or spiral CT) (Crawford & King 1990). The combination of high-power tubes and sensitive detector elements also reduces the exposure time, resulting in an acceptable scanning time.

In reality a third scanning geometry can be used, namely cone beam, where a point source is used and projections are taken with a 2D detector. By means of the so-called FDK algorithm (Feldkamp *et al.* 1984), corrections necessary owing to the use of the cone beam can be made to the reconstruction process. The reconstruction will no longer be mathematically exact, but for small opening angles of the cone beam the artefacts are small. The cone beam geometry combines the advantage of magnification with the speed of a 2D scan. This allows quick scans compared to fan beam geometry and is especially beneficial at radiation sources with a low flux. Cone beam geometry is therefore typically used at X-ray microfocus tubes and can also be applied at X-ray synchrotron facilities.

## X-ray CT

### *In general*

Different methods are available to produce an X-ray probe for radiography and tomography. Most methods are based on one of two physical processes: bremsstrahlung and synchrotron radiation. In microCT and medical scanner tubes bremsstrahlung is responsible for the creation of X-rays. This paper will focus on these scanners as they are the most commonly used and most easily available. We refer to Materna *et al.* (1999) for an example of the application of synchrotron-based tomography to the field of geology. It should also be noted that high-energy gamma rays from radioactive sources (typically californium) can be used to visualize large samples, because of their large penetration depth.

In X-ray tubes a mono-energetic electron beam impinges on a solid target that results in the production of X-rays with energies between 0 keV and the energy of the electrons. Superposed on this continuous spectrum are the so-called edges (peak features), which generally contribute little to the total X-ray flux.

A slab of a material with a relatively high atomic number can be placed in front of the beam to attenuate the low-energy X-rays (beam hardener). The material and thickness can be chosen depending on the tube voltage.

The resolution $R$ of the CT system is a very important parameter for material research. It is defined by the resolution of the X-ray source and the detection system and expressed by equation (2) (Mouze 1996):

$$R = \frac{d}{M} + \left(1 - \frac{1}{M}\right) d_s. \tag{2}$$

Here $d_s$ is the spot size of the X-ray source, $d$ is the resolution of the detector and $M$ is the magnification, which is determined by the position of the source–object distance $D_{so}$ and the source–detector distance $D_{sd}$:

$$M = \frac{D_{sd}}{D_{so}}. \tag{3}$$

Consequently, instrumental resolution plays an important role in the resolution that can be obtained in the object visualization. As it is easier to minimize the size of the X-ray focal spot (Van Geet 2001), the spot size will often determine the optimal achievable resolution of the instrument. This optimal resolution is only relevant when the object can be placed very close to the X-ray source (large magnification) and is therefore limited to small objects. The focal spot size is mainly dependent on two parameters: the size of the impinging electron beam and the amount of scattering of the electron beam in the bremsstrahlung target. Larger spot sizes limit the resolution but allow higher electron beam currents because the heat load can be spread over a larger surface. Higher electron (and by consequence higher X-ray) energies and thicker targets also worsen the resolution but provide a more efficient conversion from electrons to X-rays.

### *Interaction processes*

X-rays primarily interact with materials through the photo-electric effect and Compton scattering. In both processes the incoming photon (X-ray) interacts with the electron cloud of the target atoms. The photo-electric effect is dominant at low energies (below 100–200 keV depending on the material). Here the X-rays are absorbed and the attenuation coefficients are strongly dependent on the atomic number. Compton scattering is dominant at higher energies (100–200 keV and above). Not only does it have a low dependency on the atomic number (the attenuation is essentially proportional to the density of the penetrated materials), but as it is not an absorption process it violates the Lambert–Beer law.

To differentiate minerals with very similar mass density but dissimilar compositions, low-energy X-rays are preferentially used. It has to be taken into account that these low energies will limit the maximum object size that can be penetrated. For high absorbing materials and larger objects high energies are required. The disadvantage of working with high energies is that the transmission values will lower in sensitivity for different atomic compositions and scattering increases.

### *Medical CT*

In a medical CT scanner, the patient (or sample) is stationary, while the source and detector rotates at a high speed in a large frame called a gantry. Typically, a high-power X-ray tube with an electron beam energy lower than 150 keV and a focal spot size of the order of 1 mm is used. The target rotates at a high speed to spread the heat load over a larger area, which allows for a high electron beam current, together with a relatively large spot size. The detector (typically consisting of one or more (up to 32) curved linear arrays next to each other) has a very high sensitivity and takes projections with a large signal-to-noise ratio in short exposure times. The linear detector arrays can be composed of two different types of elements: Xenon-filled gas ionization chambers or photodiodes covered with a scintillating material. A collimator in front of the detector reduces the contribution of scattered X-rays.

Medical scanners are optimized for obtaining images with a large contrast while keeping the total dose absorbed by the patient low. The need for a large focal spot (to allow large currents) results in a limitation of the resolution to about 250 μm in modern scanners. Because of the high power of the tube and the sensitivity of the detector, the scanning speed per slice is high (1 s or less) and the total scanning time for a large number of slices is still acceptable. By moving the patient or sample along the axis of rotation during the scan – which results in a helical scanning path – the waste of time due to the movement of the patient/sample between different scans (fan beam CT) can be eliminated. Special reconstruction algorithms are developed for this helical or spiral CT. If more than one row is used, the partial cone beam geometry further complicates the reconstruction algorithm (cone-beam helical or spiral CT). The use of these multislice detectors, however, further increases the scanning speed.

### *MicroCT*

In microCT (and nanoCT) the sample is rotated instead of the source–detector system. The electron beam in the X-ray tube is focused to a small spot by using coils. The target is typically very thin (≈10 μm) to minimize the amount of scattering of the electron beam and, by consequence, the X-ray spot size. Because the electron beam power is focused to a small spot, the current is much lower than in medical tubes, otherwise heat dissipation poses a problem. The combination of a low electron beam current and a thin target results in a low X-ray flux. This means that generally only 2D detectors are used to save scanning time and every projection requires a relatively large exposure time (up to 30 s).

The typical detectors used are fibre optically coupled (intensified) CCD (charge-coupled device) sensors and flat panel detectors. CCD based detectors need a scintillator material to convert the X-ray energy to visible light, which can then be detected and intensified if necessary. The most commonly used scintillator screens are gadolinium oxysulphide (GOS) and thallium-doped caesium iodide (CsI:Tl) screens. Flat panel detectors can be composed of amorphous selenium, which results in direct detection of the X-rays, or amorphous silicon, which requires a scintillator screen. Another kind of flat panel uses CMOS (complementary metal oxide semiconductor) technology to convert the scintillation light to an electronic signal. CMOS technology has the disadvantage of being sensitive to radiation damage because of the crystalline structures used.

The resolution of microCT systems is an order of magnitude better than the resolution of medical scanners. Apart from the focal spot size of the tube, the pixels of the detector should be small enough to prevent very large magnifications. If the detector pixels are smaller than the spot size, the resolution is determined by the detector (and not the X-ray tube) as described by equation (2), but only if the sample is positioned close to the detector ($M \approx 1$). This is, however, never the case in microCT systems, which explains why the spot size of the tube – and not the pixel size – is more important for the optimal achievable resolution.

The actual resolution is often larger than stated above because of the detection mechanism. An X-ray 'hit' on the detector results in a so-called point-spread function, representing the distribution of the corresponding signal over the detector surface. This is normally not limited to one pixel, so the detector resolution is larger than a pixel. The modulation transfer function (MTF), equal to the Fourier transform of the point-spread function, can be used to quantitatively characterize the behaviour of the detector.

Special care has to be paid to the rotation of the sample. Two parameters are very important for the rotation motor: the wobble (rocking motion) of the rotational axis and the radial runout (lateral displacement of the axis during the rotation). These should be very small so they do not distort the reconstruction.

The same considerations can be applied to nanoCT, where the focal spot size of the tubes can be as small as 0.2 μm. Accurate motor control and a good detector choice (due to the lower X-ray flux) become even more crucial.

### *Applications in geology*

In the 1970s CT was primarily used for palaeontological research, but soon soil researchers, petroleum engineers, sedimentologists, petrologists and many others discovered the wide range of possibilities when using X-ray CT. In the beginning only medical CT was being used, while in the 1990s microCT was introduced as a non-destructive evaluation technique.

X-ray CT is basically and most importantly a non-destructive visualization technique, providing 2D and 3D images of the internal structure (porosity, fractures, etc.) of natural building stones. When the X-ray attenuation difference for different materials is large, clear and high contrasting images can be made, such as the location of pyrite inside natural slates (Fig. 1).

As it is possible to calculate the theoretical attenuation of a certain mineral, it is possible to make some predictions concerning their visibility and detectability inside reconstructed CT images. Figure 2 illustrates the theoretical attenuation curves for the minerals calcite, quartz and gypsum. Gypsum has a theoretical attenuation lying in between the attenuation of quartz and calcite, meaning that its resulting grey value will also be

**Fig. 1.** Three-dimensional reconstruction of pyrite in natural slates (approximately 9 mm in height) after scanning with microCT (samples from R. De Taeye).

situated between the grey values of quartz and calcite.

Quartz and calcite, with a mass density of 2.65 and 2.71 g $cm^{-3}$ respectively, have different attenuation coefficients owing to the difference in atomic number of the composed minerals. Only in a small area at low energy will the attenuation coefficient of quartz be higher than the one of calcite (Fig. 2). With increasing X-ray energy, their attenuation coefficients change over, making calcite more attenuating than quartz; their attenuation coefficients converge at approximately 130 keV. Owing to their difference in attenuation, it should be possible to distinguish the two minerals by microCT imaging.

As well as providing information on the internal structure of natural building stones, X-ray CT can be used for the visualization of conservation products, the monitoring of weathering phenomena and many more.

Based on the Lambert–Beer law, products can be made visible by increasing their X-ray attenuation. Higher attenuation can be accomplished when the original products are doped with a material, containing elements with a high atomic number. Products containing iodide, with atomic number 53, or bromine, atomic number 35, will result in a higher attenuation than minerals like quartz and calcite. By mixing conservation products with a higher attenuating product, their visualization inside natural building stones is possible. Cnudde & Jacobs (2004) and Cnudde *et al.* (2004) already demonstrated that by mixing ethylorthosilicated-based consolidants and siloxane-based water repellents with a higher attenuating material like 3-bromopropyltrimethoxysilane, these products are detectable inside natural building stones.

Figure 3 demonstrates the theoretical attenuation of an oligomer siloxane (10 vol.% in white spirit) water repellent (Hydro 10), and its theoretical higher attenuation due to the mixing with different concentration 3-bromopropyltrimethoxysilane. On this curve it is immediately clear that to obtain a visual contrast between the water repellent and

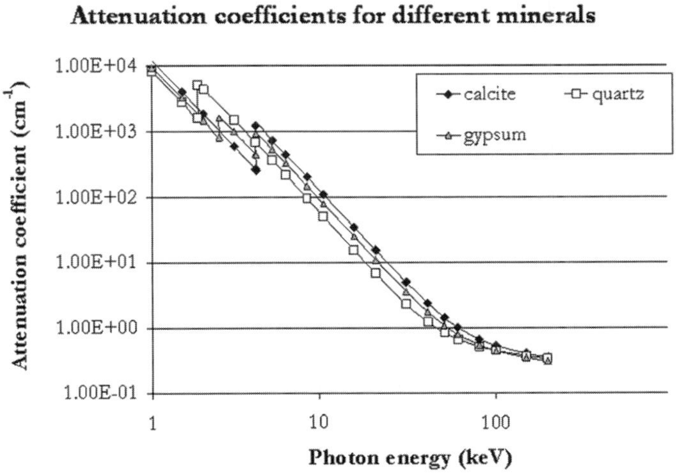


**Fig. 2.** Attenuation coefficients: calcite, quartz and gypsum.

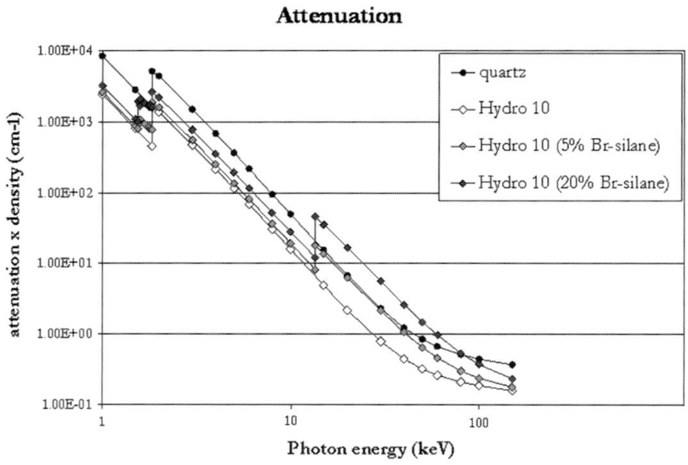


**Fig. 3.** Attenuation spectrum of Hydro 10 with 0, 5 and 20% concentration of 3-bromopropyltrimethoxysilane after polymerization in combination with the attenuation of quartz.

the quartz, mixing with 3-bromopropyltrimethoxysilane is necessary.

Figure 4 demonstrates the localization of the water-repellent product mixed with 20% 3-bromopropyltrimethoxysilane inside the natural building stone, after scanning with X-ray microCT.

Salts, which often manifest themselves by efflorescence or subflorescence, are some of the most damaging agents to stone. The internal pressure accumulated due to salt crystallization often generates spalling and flaking phenomena in building stones. From a previous study (Cnudde & Jacobs 2004) the accumulation of thenardite ($Na_2SO_4$) inside stone material was visible as a result of porosity reduction. Based on the attenuation coefficients of quartz, calcite and thenardite, theoretically no contrast between quartz and thenardite will be present, while this should be the case for calcite and thenardite. On the reconstructed images of salt containing sandstones, salt accumulation can be detected by local porosity reductions.

The monitoring by microCT of biological weathering by bacteria on natural building stones and

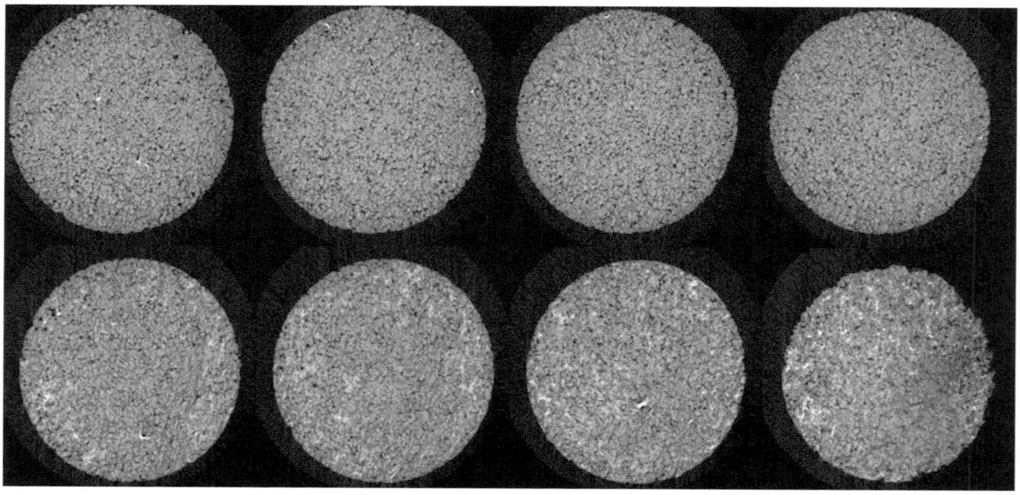

**Fig. 4.** Montage of reconstructed cross-sections taken before (upper row) and after treatment (lower row) with water repellent Hydro 10 mixed with 20% 3-bromopropyltrimethoxysilane (white areas correspond to the location of bromopropyltrimethoxysilane).

concrete, described by De Graef *et al.* (2005), indicates that, although microCT is not able to detect the 0.5–1.5 μm large bacteria, their influence on the superficial surfaces is noticeable.

## Neutron tomography

### *In general*

Neutron beams can be produced by a number of methods. For radiographic and especially tomographic purposes, two kinds of sources are the most common. First, there are nuclear reactors, where uranium or plutonium fission is used to produce a controlled chain reaction that frees energy and produces neutrons. Secondly, in spallation sources, high-energy particles hit a solid or liquid target and knock neutrons out of the nuclei of the target.

In both cases, the neutrons are slowed down by passing through a 'moderator', consisting of cells of water at room temperature (thermal neutrons) or containers of hydrogen (or deuteron) cooled to 20–30 K (cold neutrons) to produce a thermal or cold neutron beam, respectively. Once moderated, the neutron beam is led to the tomography set-up (sample). Cold neutrons are led through neutron guides, where the neutrons are reflected at the inner surface of the guide and emerge from an opening near the set-up. Thermal neutrons pass through a collimator window with a certain width or length (noted as $D$) and propagate to the set-up at a large distance $L$.

The degree to which the thermal neutron beam is parallel is characterized by the $L/D$ ratio, which is an indication of the divergence of the beam. This ratio is mostly between 50 and 800. The neutron flux is lower for larger $L/D$ ratios. The $L/D$ ratio defines what resolution can be obtained by the system if the detector is not the limiting factor. For cold neutrons, the divergence (in degrees) of the beam exiting from the neutron guide defines the resolution under these circumstances. As in X-ray CT, positioning the sample close to the detector increases the influence of the detector resolution on the image resolution.

The neutron beam has a continuous spectrum, but beam hardening can be corrected more easily during reconstruction than in X-ray CT.

### *Interaction processes*

In contrast to X-rays, neutrons interact primarily with the nuclei of a material. The most important processes for cold and thermal neutrons are nuclear absorption reactions and elastic scattering (meaning without energy loss). High-energy neutrons will not be considered as they are less suited for applications in geology.

Because of the numerous nuclear reactions, the (absorption) attenuation coefficients for neutrons have no simple relation to the atomic number of the material of interaction. Light elements like hydrogen, lithium and boron show high cross-sections, but so do some heavy elements like cadmium and gadolinium. On the other hand, both light elements like aluminium and heavy elements like lead are almost transparent for neutrons. As a consequence, neighbouring elements in the periodic system or even isotopes of the same element can show large differences in neutron attenuation.

Scattering can degrade the image quality as described previously. However, scattering is the primary process of interaction for hydrogen. As the scattering occurs over large angles, the sample should be positioned at a large distance from the detector. This means the majority of the scattered neutrons do not hit the detector and those that do give rise to an homogenous background.

For geological samples, thermal neutrons are more interesting than cold neutrons. The latter have a much lower energy that results in very large absorption cross-sections. This means only small samples (typically smaller than 1 cm) can be scanned.

### *Neutron detectors*

The limited resolution of neutron tomography is not only as a result of the divergence of the neutron beam, but also the detection process. Because neutrons are electrically neutral, they cannot interact directly with a detection material. Therefore neutron detectors are often doped with elements with a large cross-section for a specific fission, capture or collision reaction. These reactions produce secondary charged particles which can then be detected. This indirect mechanism of detection gives rise to a rather large point-spread function (several pixels wide).

The most commonly used detector for neutron tomography is the combination of a scintillating screen with a cooled CCD camera. In this configuration the neutrons hit a scintillator and the light emitted from the back of the scintillator is reflected to a CCD camera by a mirror (typically over 90°). The detector resolution is determined by the thickness and composition of the scintillator. Scintillator screens are composed of the same materials as in X-ray detection, but are sometimes doped with lithium to increase the light production. The main disadvantage of this detection system is the small throughput of light from the scintillator to the CCD sensor. However, it is very flexible and

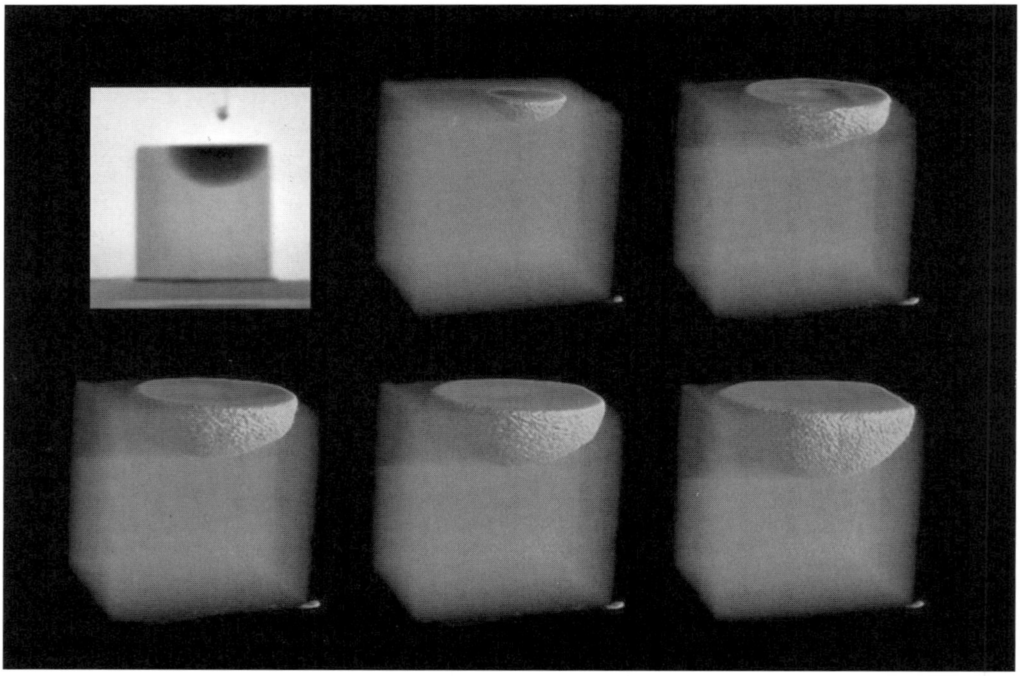

**Fig. 5.** Visualization of water infiltration inside a Maastricht limestone, treated with Hydro 8.

prevents high-energy gamma rays from the source hitting the CCD sensor (because of the mirrored path of the light).

Flat panels can also be used for neutron detection. The use of amorphous silicon or selenium is preferred to CMOS flat panels as radiation damage can pose a major problem. Again, the scintillators can be doped to be able to detect the particles better.

### *Applications in geology*

The way that neutrons interact with materials makes them complementary to X-rays. Whereas materials such as water only result in low absorption and low contrast when using X-rays, they can easily be visualized with neutrons. Dense materials such as metals cause X-ray attenuation but are almost transparent to neutrons. Also, the high neutron fluxes make it possible to follow processes such as water and petrol penetration in time and in three dimensions. As demonstrated by Masschaele *et al.* (2004) and Dierick *et al.* (2005), water, petrol and conservation products can be easily located and monitored inside natural building stones. Figure 5 shows a $4 \times 4 \times 4$ cm$^3$ cube of Maastricht limestone. A small amount of petrol was released on the top and monitored inside the stone as it penetrated the rock. A radiograph is shown (top left), together with a sequence of tomographies within a time frame of 3 min.

When water is visualized and the water content of the sample under investigation is too large, it is possible to use so-called *heavy water*, where the hydrogen atoms are replaced by deuterium atoms, which have a lower thermal neutron cross-section.

## Conclusions

Non-destructive techniques, such as X-ray tomography, and neutron tomography, have been reviewed. Both techniques are complementary and provide important 3D information on the internal structure of materials. Depending on the X-ray CT system utilized, different image resolutions can be achieved for samples of different size. While medical CT is used for large samples, microCT and nanoCT are more suitable for small samples. X-ray CT is a very powerful visualization technique for the internal structures of natural building stones and contributes to the identification of minerals, or at least gives an indication of possible elements. As it can locate doped water repellents and consolidants inside stone samples, a wide range of experiments can be performed in order to obtain more detailed information on the behaviour of these materials. An important link between the internal structure of a natural building stone and the localization of conservation products is possible. In addition, microCT images the localization of salts and the effects of bacterial weathering. As microCT is a non-destructive technique, samples can be monitored during natural and artificial weathering experiments, providing data on changes in porosity and the internal structure of the samples during experiments. Neutron tomography is less widely available, but can provide a tool to image water, water repellents and consolidants inside larger samples (without any doping). For both X-ray and neutron tomography a promising future lays ahead as material research techniques that can act in a complementary fashion to more traditional destructive research techniques.

This study is supported by the Institute for the Promotion of Innovation by Science and Technology in Flanders, Belgium through a PhD grant for V. Cnudde.

## References

CNUDDE, V. & JACOBS, P. 2004. Monitoring of weathering and conservation of building materials through non-destructive X-ray computed microtomography. *Environmental Geology*, **46**, 477–485.

CNUDDE, V., CNUDDE, J. P., DUPUIS, C. & JACOBS, P. J. S. 2004. X-ray micro tomography used for the localisation of water repellents & consolidants inside natural building stones. *Material Characterization*, **53**, 259–271.

CRAWFORD, C. & KING, K. F. 1990. Computed tomography scanning with simultaneous patient translation. *Medical Physics*, **17**, 967–982.

DE GRAEF, B., CNUDDE, V., DICK, J., DE BELIE, N., JACOBS, P. & VERSTRAETE, W. 2005. A sensitivity study for the visualisation of bacterial weathering of concrete and stone with computerized X-ray microtomography (CT). *The Science of the Total Environment*, **341**, 173–183.

DIERICK, M., VLASSENBROECK, J., MASSCHAELE, B., CNUDDE, V., VAN HOOREBEKE, L. & HILLENBACH, A. 2005. High-speed neutron tomography of dynamic processes. *Nuclear Instruments and Methods in Physics Research A: Accelerators, Spectrometers, Detectors and Associated Equipment*, **542**, 296–301.

FELDKAMP, L. A., DAVIS, L. & KRESS, J. 1984. Practical cone beam algorithm. *Journal of the Optical Society of America*, **1**, 612–619.

MASSCHAELE, B., DIERICK, M. *ET AL*. 2004. High-speed thermal neutron tomography for the visualization of water repellents, consolidants and water uptake in sand and lime stones. *Radiation Physics and Chemistry*, **71**, 807–808.

MATERNA, T., JOLIE, J., MONDELAERS, W., MASSCHAELE, B., HONKIMAKI, T., KOCH, A. & TSCHENTSCHER, T. 1999. Uranium-sensitive tomography with synchrotron radiation. *Journal of Synchrotron Radiation*, **6**, 1059–1064.

MOUZE, D. 1996. X-ray microradiography. *In*: AMELINCKX, S., VAN DYCK, D., VAN LANDUYT, J., VAN TENDELOO, G. (eds) *Handbook of Microscopy*, Volume I. VCH, Weinheim, 130–147.

VAN GEET, M. 2001. *Optimisation of microfocus X-ray computer tomography for geological research with special emphasis on coal components (macerals) and fractures (cleats) characterization*. PhD thesis, KUL.

# Rock petrophysics v. performance of protective and consolidation treatments: the case of Mt Arzolo Sandstone

F. CARÒ[1] & A. DI GIULIO[2]

[1]*Dipartimento di Scienze della Terra, Università degli Studi di Pavia, Strada Ferrata, 27100 Pavia, Italy (e-mail: federico.caro@manhattan.unipv.it)*

[2]*Dipartimento di Scienze della Terra, Università degli Studi di Pavia, Strada Ferrata, 27100 Pavia, Italy*

**Abstract:** The petrophysical characteristics of a building stone used in the city of Pavia, northern Italy, are analysed in the light of stone conservation through the application of water repellent and consolidant products. The research focuses on the modification of petrophysical properties as a function of the applied products, and on the critical assessment of the performances of fluorinated and siloxane resins as a function of the variable nature of the same lithotype. The studied material is a calcareous sandstone (Mt Arzolo Sandstone), Late Miocene–Early Pliocene in age, extensively employed as a building material during the 11th and 12th centuries. Experimental investigations on samples from historic quarries were performed before and after application of treatments: i.e. petrographical and fabric analyses; ultrasonic measurements; mercury porosimetry; abrasion resistance measurements; contact angle measurements; capillary and low-pressure water absorptions; water vapour permeability; and colour measurements. Two main lithotypes of Mt Arzolo Sandstone were recognized according to their petrophysical features: the open porosity being similar, differences exist concerning their fabric. These differences influence the physical–mechanical properties of the material and consequently the performances of the applied products. In particular, the difficulties in penetration of products when sandstone of smaller pore size is treated can lead to a significant reduction of the overall performances, which is more evident for products containing organic solvent with high molecular weight.

The petrophysical characteristics of building stones directly influence their whole history, from quarries to monuments, as they affect the material exploitation, workability, use, durability, weathering and, finally, conservation techniques (Carrol 1970).

The local sandstone (Mt Arzolo Sandstone) used in the city of Pavia is a typical example of how petrophysical features can influence the history of a dimension stone and of the related architectural heritage. The intrinsic petrographical and physical properties of this soft, easily workable Upper Miocene–Lower Pliocene calcareous sandstone allowed craftsmen to gain a unique plastic and figurative effect (Fig. 1a). Intensive weathering of this soft stone has, however, accelerated over the last decades as a result, in part, of increased atmospheric pollution within the urban environment (Fig. 1b) (Aguzzi *et al.* 1973).

Sulphation of calcitic cement, hydration of clay minerals, rapid and frequent freeze–thaw cycles and, subordinately, wind action have all been identified as important causes of the weathering affecting Mt Arzolo Sandstone (Aguzzi *et al.* 1973; Riganti *et al.* 1981; Braga *et al.* 1985; Veniale & Zezza 1988). All these factors are emphasized by the humid climate of the area. The resulting effects consist of massive granular disintegration, flaking and, frequently, scaling.

Various authors (Aguzzi *et al.* 1973; Scagni & Vercesi 1987; Braga *et al.* 2000) have pointed out the extreme variability of this lithotype with respect to its petrophysical features, even in a single quarry. This variability is also present on buildings and is of paramount importance to the planning of conservation intervention because of the way in which pre-consolidation and cleaning techniques, and the possible application of protective and/or consolidant treatments, are influenced by substrate characteristics (Young *et al.* 1999).

## Materials

### *The Mt Arzolo Sandstone*

The building materials of the S. Michele Basilica come from the uppermost part of the Mt Arzolo Unit from the Cassano Spinola Conglomerate Formation (Upper Messinian) (Aguzzi *et al.* 1973).

The depositional environment is interpreted as a fan-delta system (Cassano Spinola Conglomerates) progressively evolving towards a fluvial environment (Mt Arzolo Sandstone) (Braga *et al.* 1985). Changing depositional processes deposited clastic sediments characterized by rapid grain-size changes from matrix-rich sands to sandy microconglomerates.

*From*: PŘIKRYL, R. & SMITH, B. J. (eds) *Building Stone Decay: From Diagnosis to Conservation.* Geological Society, London, Special Publications, **271**, 287–294.
0305-8719/07/$15.00 

**Fig. 1.** (**a**) Façade of the S. Michele Maggiore Basilica and (**b**) detail of weathering.

Well-bedded sandstones form lens-shaped bodies, occasionally showing cross-bedded structures (Scagni & Vercesi 1987).

Exploited outcrops are located between the Versa and Staffora Valley (northern Appennines), in the hilly reliefs facing the alluvial Po Plain (Fig. 2) (Aguzzi *et al.* 1973; Braga *et al.* 2000). The exploitation of these quarries is not precisely documented. Nevertheless, during the 12th century they provided Pavia and its province with a large amount of building material. Nowadays, these quarries are abandoned and almost completely hidden by natural vegetation and vineyards.

## *The sandstone in the architectural heritage of Pavia*

The Mt Arzolo Sandstone is the typical and most popular building material of the Romanesque architecture of Pavia. It can be seen in religious and public buildings in the form of decorative elements in portals and windows and in chromatic counterpoint with red bricks.

The stone is used in large amounts in one of the most beautiful examples of Romanesque architecture in Italy, the S. Michele Maggiore Basilica (Fig. 1). Its presence is also particularly outstanding in other religious buildings such as San Pietro in Ciel d'Oro, S. Maria in Bethlem, San Teodoro and San Primo e Feliciano. This sandstone was also used as a building material in buildings that have since been destroyed, such as the double cathedral of S. Stefano and S. Maria del Popolo, on whose ruins the Duomo Cathedral was erected. The Mt Arzolo Sandstone is also found in various elements of houses, where it is limited either to single blocks or to massive wall portions.

## *Treatments*

Three fluorinated polymers and two siliconic resins have been tested. Two fluorinated polymers are elastomers with surface-consolidant and water-repellent properties. These are: (a) a 3% solution of fluoroelastomer copolymer in 60% acetone and 37% *N*-butyl acetate (FECP); and (b) an aqueous emulsion of 5% fluoroelastomer terpolymer (FETP). In addition, a perfluoropolyether with water and oil-repellent properties, i.e. an aqueous emulsion of functionalized perfluoropolyether derivative in 33% isopropyl alcohol (PFPE), has been tested. This compound bears phosphate functional end groups that interact with the calcium ions of the substrate to form high-energy bonds, improving adhesion to the stone surface (Moggi *et al.* 2004).

The tested siliconic resins are siloxane compounds that are commonly used to impart water-repellency to natural and artificial building stones. These are: (a) a 7.5% oligomeric alkyl-alkoxysiloxane in isopropyl alcohol (AAS); and (b) a 7% dimethylpolysiloxane in White Spirit (DMPS). Table 1 reports some chemical and physical properties of the tested compounds.

Products were applied by brush onto the surface of samples previously dried at $60 \pm 5\ ^{\circ}C$, according to UNI 10921 (UNI 2001). Dried samples were weighed before and after the treatment in order to verify the exact quantities of applied

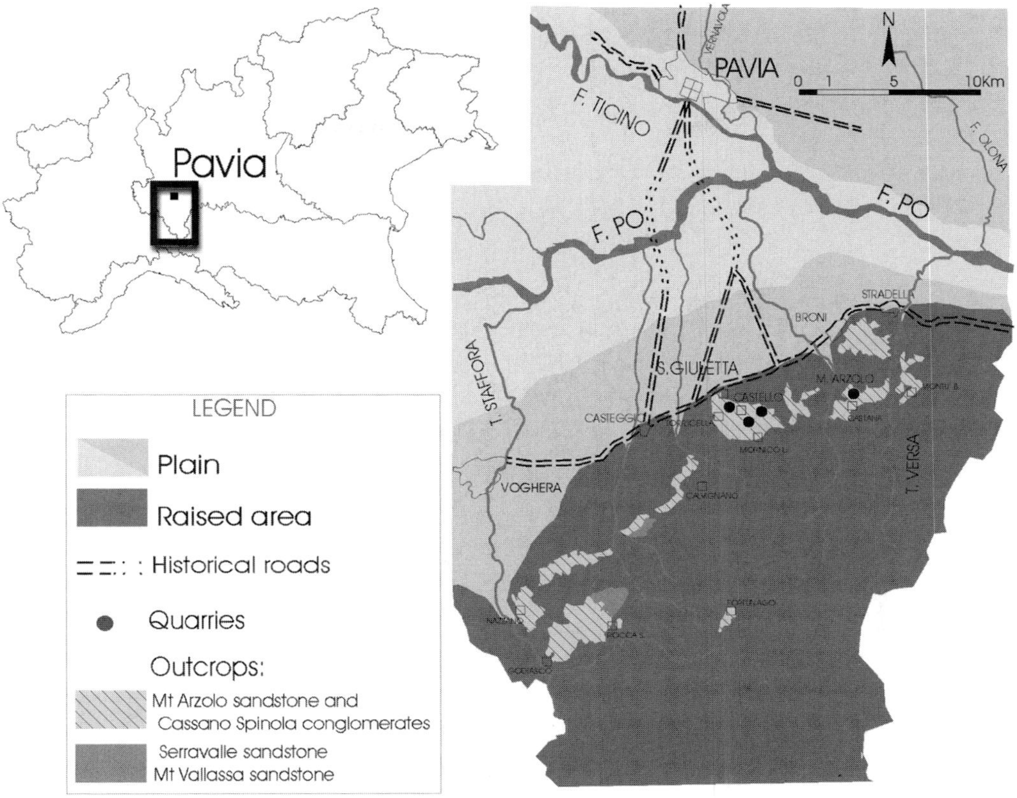


**Fig. 2.** Schematic geological map of the area between Versa Valley and Staffora Valley, and the location of historical quarries of Mt Arzolo Sandstone. Modified after Braga *et al.* (1985).

products, which were different according to the sandstone lithotype and based upon previous studies (Guidetti *et al.* 2000; Moggi *et al.* 2004) (Table 1).

## Methods

Experimental investigations were performed on sandstone blocks taken from an historic and

**Table 1.** *Details of stone treatments*

| Treatment | Nature | Solvent | Dry content (wt%) | Density ($g\ m^{-1}$) | Number-average molecular weight (Da) | Applied quantity ($g\ m^{-2}$) |
|---|---|---|---|---|---|---|
| FECP | Fluoroelastomer | Acetone + *N*-butyl acetate | 3 | 0.84 | $2\times10^5$ | MA 30<br>MB 20 |
| FETP | Fluoroelastomer | Water | 5 | 1.25 | $1\times10^5$ | MA 30<br>MB 30 |
| PFPE | Perfluoropolyether | Water + isopropyl alcohol | 5 | 1.00 | $2\times10^3$ | MA 50<br>MB 30 |
| AAS | Alkyl-alkoxysiloxane | Isopropyl alcohol | 7.5 | 1.05 | – | MA 40<br>MB 30 |
| DMPS | Dimethylpolysiloxane | White Spirit | 7 | 0.80 | – | MA 40<br>MB 40 |

documented quarry abandoned since 1960. The sampled quarry is located SE of the village of Castello, about 20 km south of Pavia (Fig. 2). The Mt Arzolo Sandstone crops out in well-bedded layers ranging in thickness between 0.2 and 1 m, separated by thin pelitic beds.

Samples were collected from the lower part of the section, where unaltered beds show evidence of quarrying. In accordance with UNI 10921 (UNI 2001), 5 × 5 × 2 and 5 × 5 × 1 cm specimens were sawn from the innermost parts of the blocks.

The following measurements were performed on 56 samples, perpendicular to their bedding surfaces, both before and after treatments. The adopted procedures and methods have been developed by the Italian NORMAL Commission, now UNI, an institute for the standardization of methodologies for the study and conservation of stone materials.

According to NORMAL 33/89 (NORMAL 1989), the static contact angle was determined by means of a micrometric eyepiece placed on an aligned optical system that allows the calculation of the contact angle of a drop of bi-distilled water dispensed by a pipette syringe on the plane stone surface; the higher the measured angle, the greater the hydrophobicity of the treated surface. Capillary water absorption was measured according to UNI 10859 (UNI 2000). Low-pressure water absorption and water vapour permeability were determined according to NORMAL 44/93 (NORMAL 1993*a*) and NORMAL 21/85 (NORMAL 1985), respectively, while colour measurements were performed by means of a spectrocolorimeter according to NORMAL 43/93 (NORMAL 1993*b*).

In addition, mercury porosimetry (according to NORMAL 4/80, NORMAL 1980) and non-standardized abrasion resistance measurements were performed before and after treatment. Abrasion resistance was calculated by weighting the material abraded from the stone surface by a rotating disc (Ø = 25 mm) of silicon carbide P320 paper through a Satra Finish Rub Fastness Tester STM 102 (Guidetti *et al.* 1995). Results are expressed in g $m^{-2}$, while surface consolidation, expressed in %, was calculated by comparing the quantity of abraded material before and after the application of treatments.

Petrographical and fabric analyses were performed by point count of 14 thin sections. Indices of anisotropy were determined by means of ultrasonic sound velocity measurements on 10 × 10 × 10 cm specimens according to NORMAL 22/86 (NORMAL 1986).

## Results

### *Sandstone petrophysics*

The petrographical, fabric, physical and mechanical analyses of samples coming from historical quarries allowed the identification of two main sandstone types (hereafter referred to as MA and MB) with different as petrophysical characteristics (Table 2).

According to the classification of Pettijohn *et al.* (1987), the analysed samples fall within the lithic arenite field (average composition: $Q_{27}$ $F_{12}$ $Rf_{61}$). They predominantly contain sedimentary rock fragments (both clastic and carbonate) with minor quartz, feldspars and metamorphic grains. Authigenic components are represented by an abundant fine–medium crystalline carbonate cement.

The magnitude of the anisotropy indices is similar for all the analysed samples and is typically moderate. The mean difference of the P-wave velocities measured along *XYZ* directions ($\Delta MV_p$) is 8.4 ± 2.5%.

Although the two main lithotypes have the same composition they can be distinguished according to their petrophysical characteristics. The MA is a moderately sorted medium sandstone with a mean pore size of 0.3 ± 0.07 μm and around 11% of macropores. The mean open porosity is equal to 13.3 ± 0.7% (Table 2). The MB is a moderately to poorly sorted fine sandstone with a different pore-size distribution: the mean pore size is equal to 0.15 ± 0.05 μm, while macropores are almost absent (<2%) (Table 2). The mean open porosity is equal to 13.7 ± 2.3%. Therefore, both lithotypes have the same pore volume but different pore structure (Fig. 3). This condition directly affects the mean water vapour permeability of samples, measured perpendicular to the bedding surface, which is higher for the MA sandstone (122.4 ± 11.7 g $m^{-2}$ · 24 h), and lower for the MB sandstone (68.1 ± 12.4 g $m^{-2}$ · 24 h). Samples from the MB series also show higher abrasion resistances (Table 2).

### *Efficiency of consolidation*

The efficiency of consolidant and water-repellent treatments was assessed by comparison of different physical properties according to UNI 10921 (UNI 2001). Some properties were similar for both studied lithotypes. However, differences emerged when the compound characteristics directly affected its distribution in the sandstone. The water repellency imparted by the treatments is similar for all the tested compounds and comparable for the two lithotypes. Although the samples show good water repellency, the measure of contact angles reports a very superficial property and does not permit evaluation of how the compounds interact with the substrate.

Similarly, colour changes imparted to the stone after treatments are comparable and always lower than 5. DMPS in White Spirit is the compound that imparted the lowest colour change. It has to be noticed that even $\Delta E^* < 5$ $[\Delta E^* = (\Delta L^{*2} +$

**Table 2.** *Physical characteristics of the two stone types studied*

| Lithotype | Mean grain size (mm) | Sorting ($S_\phi$) | Real density (g cm$^{-2}$) | Open porosity (%) | Total pore volume (cm$^3$ g$^{-1}$) | Mean pore radius (μm) | Pore radius mode (μm) | W.V. perm$_{(20\,°C)}$* (g m$^{-2}$ · 24 h) | Abrasion Index (g m$^{-2}$) |
|---|---|---|---|---|---|---|---|---|---|
| MA | 0.30 ± 0.02 | 0.72 ± 0.03 | 2.56 ± 0.01 | 13.3 ± 0.7 | 0.06 ± 0.01 | 0.28 ± 0.07 | 1.34 ± 0.17 | 122.4 ± 11.7 | 36.8 ± 1.2 |
| MB | 0.20 ± 0.04 | 0.76 ± 0.12 | 2.55 ± 0.01 | 13.8 ± 2.3 | 0.07 ± 0.01 | 0.15 ± 0.05 | 0.29 ± 0.16 | 68.1 ± 12.4 | 23.5 ± 3.4 |

*Water vapour permeability at 20 °C.

$\Delta a^{*2} + \Delta b^{*2})^{1/2}$, where L*, a* and b* are the CIE-Lab chromatic co-ordinates] can be perceived by the human eye, when the major changing chromatic co-ordinate, according to the CIE-Lab colour space, is the lightness ($L^*$).

Water absorption and surface consolidation are the parameters that best illustrate the differences among treatments and between the sandstone lithotypes.

As expected, the elastomers are the compounds with the best surface consolidant properties. The surface consolidation is clearer for the MA sandstone, which is less compact than the MB. Conversely, products with no strengthening properties (PFPE, AAS and DMPS) seemed to impart a slight consolidant effect to the MB sandstone samples, i.e. to the material characterized by a lower abrasion index (Table 2). This phenomenon is more pronounced for the functionalized PFPE which forms high-energy bonds with the substrate (Moggi *et al.* 2004).

The greater surface consolidation registered for the FECP is linked to a high protection against water sorption (Fig. 4). This organic consolidant product is carried and deposited inside the pores by evaporation of the solvent. However, it was observed that owing to the rapid evaporation of acetone and of *N*-butyl acetate, the viscosity of FECP increased during application by brushing and this hampered further penetration into the stone. As a consequence, it formed a superficial seal that increased mechanical properties, decreased water sorption and reduced the water vapour permeability of the sandstone.

The reduction of water vapour permeability after the application of FECP is more noticeable for the MB sandstone, where penetration proved to be more difficult. In contrast, the higher residual water vapour permeability derived from the application of the FETP is combined with a low degree of protection against water absorption (Fig. 4). In this case, the inhomogeneous distribution of agglomerated particles with a mean size of 0.3 μm, which form the aqueous emulsion (Moggi *et al.* 2004), permitted the transfer of water vapour through the stone but did not guarantee the essential protection against water sorption.

Mecury porosimetry registered an overall decrease in total pore volume after the application of the treatments, ranging from 0.8 to 26.3%.

Some differences can be noticed between the applied products. In particular, products with higher molecular weights (FECP and FETP) only filled pores larger than 1 μm as no tangible variation of the dimension of modal pore size ($\phi = 1.34 \pm 0.17$ μm and $0.28 \pm 0.28$ μm, respectively, for MA and MB sandstones) was registered for either MA or MB sandstones (Fig. 5). The distribution of the PFPE in the pore spaces seems to be more homogeneous than the other fluorinated compounds, resulting in a tangible

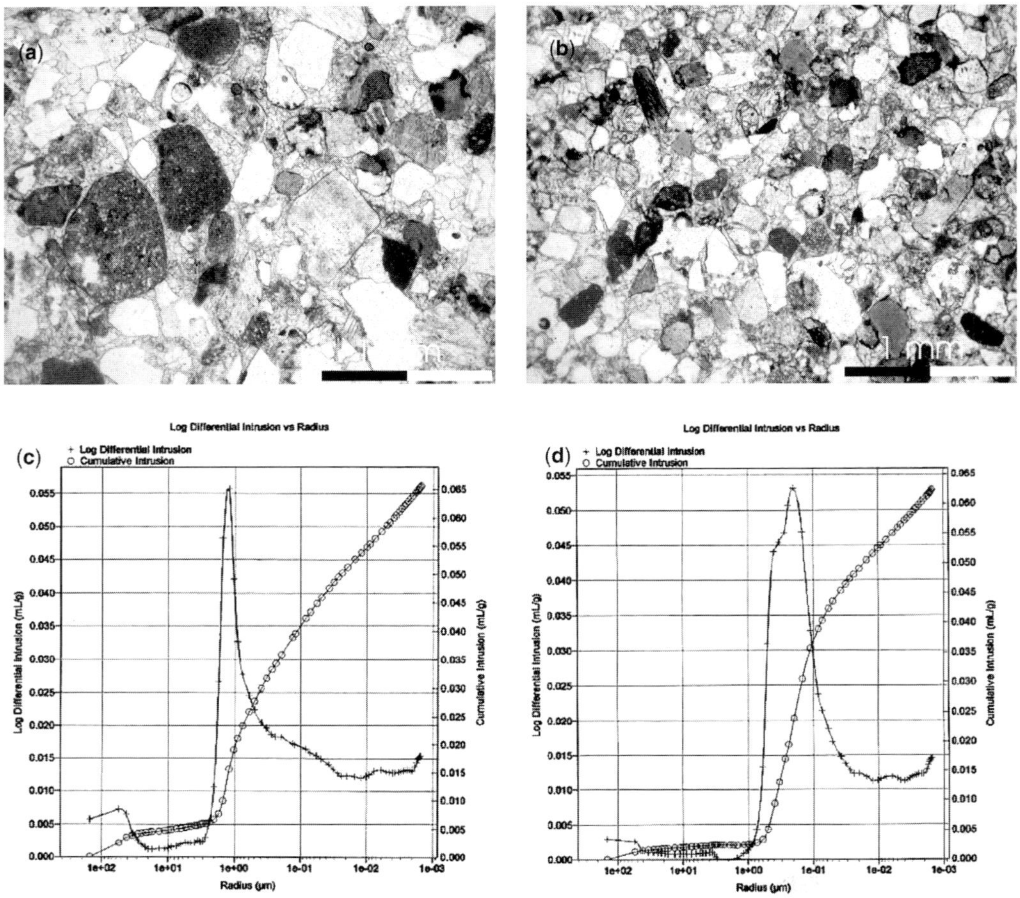


**Fig. 3.** Micrograph (parallel nicol) of (**a**) MA and (**b**) MB sandstones. Example of differential and cumulative pore-size distributions of (**c**) MA and (**d**) MB sandstones.

reduction of the modal pore radius for both MA and MB sandstones (Fig. 5). The variation in pore-size distribution after application of siloxane compounds is comparable to that of the PFPE. Application of AAS and DMPS produced a modification of the pore structure that affected the modal pore radius in the MA and, subordinately, in the MB sandstones (Fig. 5).

## Concluding remarks

Examination of the Mt Arzolo Sandstone confirmed that materials coming from the same quarry can present different petrophysical characteristics. The main difference is in the permeability of the Messinian sandstone, which is directly affected by the mean grain size of the samples for a given sorting class. The mean porosities of the two lithotypes identified are largely similar and essentially independent of grain size, but vary with sorting (Beard & Weyl 1973). Differences in physical and mechanical properties affect the performances of possible conservation treatments. In general, the penetration of such treatments into the MB sandstone is more difficult because of its lower permeability.

Treatment products with higher molecular weights (fluoroelastomers) tend to remain confined to the superficial layers of the samples, forming a surface seal that can occlude smaller pores (Vicini *et al.* 2001). This phenomenon is more obvious in the MB sandstone and strongly affects residual water vapour permeability. The distribution of the compounds in the sandstones was also affected by the nature of the solvent. Product concentration

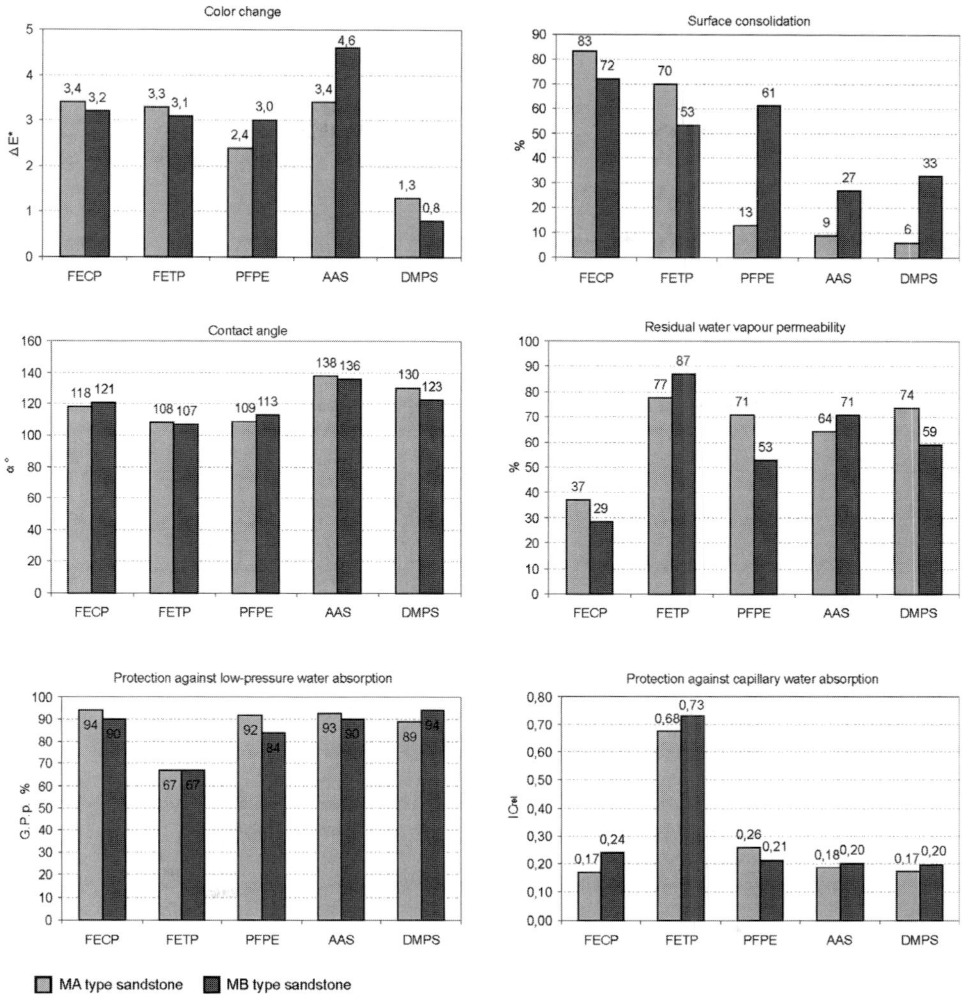


**Fig. 4.** Results of experiments on the two lithotypes of Mt Arzolo Sandstone after the application of protective treatments.

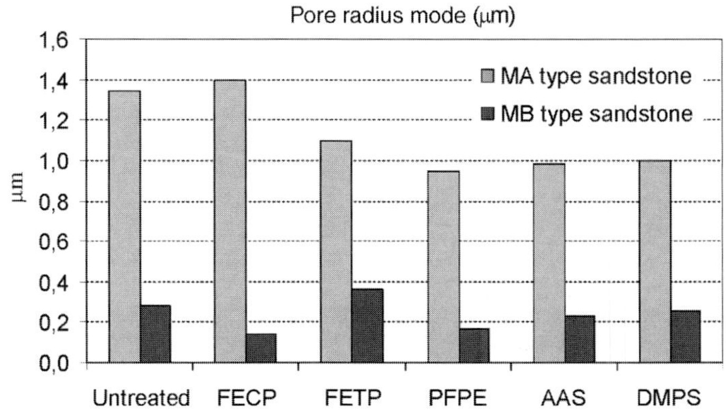


**Fig. 5.** Mid points of the maximum frequency class of pore diameter of untreated and treated samples of Mt Arzolo Sandstone.

close to the surface can influence the superficial resistance of the treated material and increase protection against water. However, these results alone do not necessarily indicate good performances by the products.

According to the experimental results, the preferred type and quantity of protective and consolidation treatments that should be used depend on the petrophysical characteristics of the sandstone to be treated.

The authors are grateful to Dr G. Moggi for his valuable help in the characterization of protective treatments and discussion of results. They would also like here to acknowledge the valuable commentaries provided by the two reviewers, Dr M. Young, The Robert Gordon University, Aberdeen, and Dr D. Nicholson, Manchester Metropolitan University, and by Dr R. Přikryl, Charles University, Prague.

## References

AGUZZI, F., FIUMARA, A. *ET AL*. 1973. L'arenaria della basilica di S. Michele in Pavia: ricerche sull'alterazione e sugli effetti dei trattamenti conservativi. *Atti della Società Italiana di Scienze Naturali e del Museo Civico di Storia naturale di Milano*, **114**, 403–463.

BEARD, D. C. & WEYL, P. K. 1973. Influence of texture on porosity and permeability of unconsolidated sand. *AAPG Bulletin*, **57**, 349–369.

BRAGA, G., CAZZANIGA, C., DI GIULIO, A. & ZEZZA, U. 2000. Stones from Oltrepo pavese in the architecture of the province of Pavia. *In*: CALVI, G. & ZEZZA, U. (eds) *Proceedings of the International Congress Quarry–Laboratory–Monument, 26–30 September, 2000, Pavia, Italy*, La Goliardica Pavese, Pavia, Volume 1, 155–160.

BRAGA, G., VENIALE, F. & ZEZZA, U. 1985. La pietra del San Michele in Pavia. *In*: *Atti del Convegno La Pietra del San Michele: restauro e conservazione, Società per la conservazione dei monumenti dell'arte cristiana in Pavia*, Tipografia Ponzio, Pavia, 83–94.

CARROL, D. 1970. *Rock Weathering*. Plenum Press, New York.

GUIDETTI, V., MASSA, V., RIZZO, F. & ZEZZA, U. 2000. Considerations on protective treatments applied to sandstones from Oltrepo Pavese. *In*: CALVI, G. & ZEZZA, U. (eds) *Proceedings of the International Congress Quarry–Laboratory–Monument, 26–30 September 2000, Pavia, Italy*, La Goliardica Pavese, Pavia, Volume 1, 339–344.

GUIDETTI, V., MATULLO, M. & PIZZIGONI, G. 1995. Methodologies for the study of the efficiency of stone reaggregant products on artificial samples. *In*: *Proceedings of LCP Congress 'Conservation and Restoration of Cultural Heritage', Montreux*, Ecole Polytechnique Fédérale de Lausanne, Lausanne, 237–247.

MOGGI, G., GUIDETTI, V., PASETTI, A. & VICINI, S. 2004. Sistemi innovativi a base di polimeri fluorurati per la conservazione di materiali lapidei artificiali. *In*: *Architett ura e Materiali del Novecento – Conservazione, Restauro, Manutenzione: Atti del Convegno di Studi Bressanone, 13–16 luglio 2004, Arcadia Ricerche*, Arcadia Richerche, Bressanone, 1249–1259.

NORMAL. 1980. *Distribuzione del volume dei pori in funzione del loro diametro*. CNR-ICR, Roma.

NORMAL. 1985. *21/85 Permeabilità al vapor d'acqua*. CNR-ICR, Roma.

NORMAL. 1986. *22/86 Misura della Velocità di Propagazione del Suono*. CNR-ICR, Roma.

NORMAL. 1989. *33/89 Misura dell'angolo di contatto*. CNR-ICR, Roma.

NORMAL. 1993*a*. *44/93 Assorbimento d'acqua a bassa pressione*. CNR-ICR, Roma.

NORMAL. 1993*b*. *43/93 Misure colorimetriche di superfici opache*. CNR-ICR, Roma.

PETTIJOHN, F. J., POTTER, P. E. & SIEVER, R. 1987. *Sands and Sandstones*, 2nd edn, Springer, New York.

RIGANTI, V., PEROTTI, A., FIUMARA, A., VENIALE, F. & ZEZZA, U. 1981. Applicazione di tecniche strumentali al controllo del degrado delle pietre nei monumenti: il caso della Basilica di S. Michele in Pavia. *Atti della Società Italiana di Scienze Naturali e del Museo Civico di Storia naturale di Milano*, **122**, 109–138.

SCAGNI, G. & VERCESI, P. L. 1987. The Messinian between Versa and Staffora Valleys (Pavese–Vogherese Appennine): paleogeography pattern. *Atti Ticinesi di Scienze della Terra*, **31**, 1–20.

UNI. 2000. *10859 Natural and Artificial Stones. Determination of Water Absorption by Capillarità*. Ente Nazionale di Unificazione, Milan.

UNI. 2001. *10921 Natural and Artificial Stones. Water Repellents – Application on Samples and Determination of their Properties in the Laboratory*. Ente Nazionale di Unificazione, Milan.

VENIALE, F. & ZEZZA, U. 1988. New research on sandstone of San Michele Cathedral in Pavia (Italy). *Atti Ticinesi di Scienze della Terra*, **31**, 253–268.

VICINI, S., MARGUTTI, S., MOGGI, G. & PEDEMONTE, E. 2001. In situ copolymerisation of ethylmethacrylate and methylacrylate for the restoration of stone artefacts. *Journal of Cultural Heritage*, **2**, 143–147.

YOUNG, M. E., MURRAY, M. & CORDINER, P. 1999. *Stone Consolidants and Chemical Treatments in Scotland*. Report to Historic Scotland, Edinburgh.

# Overview of recent knowledge of patinas on stone monuments: the Spanish experience

C. VAZQUEZ-CALVO, M. ALVAREZ DE BUERGO & R. FORT

*Instituto de Geología Económica, CSIC-UCM, Facultad de Ciencias Geológicas, Universidad Complutense de Madrid, C/José Antonio Nováis 2, 28040 Madrid, Spain (e-mail: carmenvazquez@geo.ucm.es)*

**Abstract:** The historic treatment of stonework has often been linked to the artificial application of patinas, mainly for aesthetic and protective reasons. Increasingly, however, researchers have identified a possible combined origin for patinas that has linked natural, biological processes to those associated with an artificial, man-made origin. This suggests that, although coatings may have been initially applied on purpose, they transform over time with the aid of micro-organisms and other chemical interactions. The original mixture applied to create a patina could include lime and/or gypsum, water, natural pigments and organic additives. However, their present-day mineralogy is varied and includes a wide range of minerals from calcium carbonates to calcium sulphates, calcium oxalates, calcium phosphates, silicates (quartz, feldspar, clay minerals) and iron oxides/hydroxides. Patinas have been studied in detail in Greece and Italy, but rarely in Spain. In this paper, existing knowledge on Spanish patinas is co-ordinated and previous and current research summarized. Emphasis is placed on artificial patinas initially applied to protect stone. These both appear to effectively protect the stone substrates on which they were applied and provide an insight into historical techniques of stone conservation. Because of this their preservation should be a strong consideration in restoration projects. Ongoing research focuses on the challenges of reproducing patinas, based on historical references.

The history of the stone as a building material is closely linked to the application of many kinds of coating. In most cases of painted stones, the primary reason was probably aesthetic. According to Grissom *et al.* (2002), variations in custom regarding the painting of stonework throughout history and in different parts of the world depended on the quality of available stone and economic issues. Torraca (1988) indicates that the choice of coatings for the protection of stone must be associated with the establishment of a maintenance routine. Wright (1998) complains of the overestimation of 'naked stone' and the development of a new trend in the restoration of monuments that he has named *petrofilia* (stonephilia), claiming that there is a 'colour language' and a literature of finished surfaces present in most of monuments.

The use of natural organic additives in surface films or coverings – patinas – of mineral origin (lime or gypsum) is a traditional technique applied for many centuries to stone materials for their conservation and protection. As such, it was considered as a customary practice, the successful effects of which have lasted in many cases until the present day. Maxová (2000) points out that because works of art were protected (surface coated) from early times, it was inevitable that easily procurable natural substances were used (e.g. linseed oils, beeswax, lime water). Because of their historical significance, patinas have also generated considerable research. For example, the patina known as 'scialbatura' has been extensively studied on several monuments and stone sculptures from Italy (Franzini *et al.* 1984; Lazzarini & Salvadori 1989; Realini & Toniolo 1996) and Greece (Knoll 1968; Korres & Bouras 1983; Kouzeli *et al.* 1988; Maravelaki-Kalaitzaki 2005). Two international symposia regarding oxalate films were also held in 1989 and 1996 (Centro CNR 'Gino Bozza' Politecnico di Milano 1989; Realini & Toniolo 1996). In contrast to these studies elsewhere around the Mediterranean, research on patinas in Spain is scarce. What literature exists is summarized in the following sections. However, based on the recommendation of pioneering studies such as that by Franzini *et al.* (1984), this paper will also examine in depth the characteriztics of the patinas on single monuments and laboratory production of patinas. In doing so it aims to establish the 'state of knowledge' of anthropogenic patinas with specific reference to the Spanish experience.

## Patinas: origin, role, terminology, composition

### Origin

Some patinas may have a biological origin (natural origin, related mainly to lichens: Liebig 1853;

*From*: PŘIKRYL, R. & SMITH, B. J. (eds) *Building Stone Decay: From Diagnosis to Conservation.* Geological Society, London, Special Publications, **271**, 295–307.

Del Monte & Sabbioni 1987), whilst others are artificially created on purpose (Knoll 1968; Lazzarini & Salvadori 1989). The criteria followed by some authors in attributing an artificial origin to studied patinas instead of a natural one are mainly: a sharp contact between patina and substrate (although Lazzarini & Salvadori 1989 say that when the treatment is applied to deteriorated marbles it may penetrate along fissures in the marble), the absence of micro-organisms, remains or related structures that can be observed by scanning electron microscope (SEM), and the absence of similar patinas on either adjacent buildings of the same stone and period and/or on stone in the original quarry (Franzini *et al.* 1984; Appolonia *et al.* 1996). There are, however, an increasing number of authors who endorse a combined, rather than a unique origin for coverings observed on historic stonework (Franzini *et al.* 1984; Martín-Gil *et al.* 1999; Pavía Santamaría & Caro Calatayud 2000). Even Lazzarini & Salvadori (1989), defenders of the artificial origin of the patinas, do not exclude the possible participation, at a later stage, of micro-organisms or lichens in the chemical transformation of the treatments; for instance, the ability of some micro-organisms and lichens to secrete oxalic acid and its later transformation to calcium oxalate (Monte 2003) may increase the presence of this mineral.

## *Ancient recipes*

Historically, many different treatments for protecting stone have been used. These are summarized in Table 1, together with the various additives used. The act of compiling this table highlighted two factors: the importance of historical documentation of the techniques used; and the way in which many patinas remained hidden below surface soiling and were only identified during sensitive cleaning of stonework.

## *Role*

Creating a thin surface film or patina serves two main purposes: one has to do with aesthetics and the other with protection. In terms of aesthetics, patinas may be designed to produce a brighter and/or warmer shade than the original colour of a stone façade. They can also be used to homogenize colour differences related to stone drawn from different strata within a quarry, to imitate other stone, to reduce the new appearance of artwork or to artificially 'age' replaced stone. For example, Danesi & Gambardella (2005), in their study of 17th century restoration practices for stone sculptures, mention how important was to 'render the ancient colour' (*dare il colore antico*) in order to make newly sculpted and white marble elements look like the old ones. In terms of protection, the purpose is obvious, in that the coating acts as a protective mineral skin that is further strengthened by the addition of various organic materials. Unfortunately, there are not many studies that compare patinated with equivalent un-patinated surfaces; however, Cezar (1998) did determine some properties of stone treated and untreated with calcium oxalate. They found that treated samples showed a higher resistance to acid and alkali, a lower porosity and a greater hardness than those that were not treated. The loss of patina is also known to increase the deterioration rate of the substrate (Fig. 1). When the surface of any stone begins to scale (with or without a patina or any kind of layer or film on top), it is well known that decay will accelerate due to an increase in the exposed surface. This could imply that it was only the reduction in surface area consequent upon the creation of a patina that afforded some limited protection to the substrate. However, there are other aspects of patinas, such as the presence of calcium oxalates, with solubilities much lower than those of calcium carbonate, that imbue them with, in relative terms, beneficial, acid-resistant properties (Matteini *et al.* 1996). To fully understand the deterioration mechanisms of natural and artificial patinas, and in turn to formulate some basic guidelines for their preservation, it is clearly important that more detailed, systematic study be undertaken.

## *Terminology*

Patina is sometimes an imprecise, vague and ambiguous term. It comes from the Latin word *patĭna* – or dish – from the varnish with which ancient dishes were covered. The term was used by Filippo Baldinucci to refer to the time-dependent darkening of frescos and oil paintings in 1681 (Krumbein 2002). Several authors have discussed the appropriate meaning of this word in the field of stone conservation (Kouzeli *et al.* 1988; Montanari 1996; Krumbein 2002; Alessandrini 2004). It is also used in another disciplines; for instance, it was used during the 18th century to refer to the colour changes that occur on copper and bronze due to the oxidation of metals (Krumbein 2002).

There are many other terms or expressions to refer to these films: scialbatura (Del Monte & Sabbioni 1987), epidermis or marble skin (Korres & Bouras 1983), *veladura* (Danesi & Gambardella 2005), *pellicole ad ossalato* (Alessandrini 2004). Martín-Gil *et al.* (1999) defined patinas as 'hardening pastes deliberately applied on the recently sculpted stone of ancient historic-artistic monuments'. Lazzarini & Salvadori (1989), gathering and summarizing the observation of many authors,

**Table 1.** *Historic treatments in stone conservation and the organic additives related to them*

| Authors | Historic treatments in stone conservation and organic additives related to them |
|---|---|
| Egyptians, around 150 BC – Sickels (1981) | The use of blood and animal glues as organic additives is reported |
| Vitruvius, 1st century BC (1960) | Reports the use of blood, eggs, albumen, animal glues, fig tree milk, wax, linseed oil and lard as organic additives |
| Pliny the Elder, 1st century AD (1989) | Reports the use of blood and milk, specifically the addition of milk and saffron for plasters preparation |
| Cennini, Gothic painter (1933) | Reports treatment for waterproofing the stone: boiled hot linseed oil, boiling hot pitch or tar and for preparation of the stone: calcined bird bones |
| Borghini, 16th century (1967) | Reports recipes for the marble finish (*velatura*): liquid wash made of soot boiled in urine or vinegar, or urine with cinnamon and clove, or even nut oil with colouring pigments |
| Boselli, 17th century sculptor (1978) | Reports some recipes consisting basically of the mixing of cheese and lime in boiled water and also soot boiled in water |
| Lewin (1966) | Makes a long list of recipes for stone protection, one of them (#76, dated from the 19th century) consists of a mixture of limewater (*latte di calce*), acetic acid and sugar from cane or molasses |
| Sickels (1981) | Reports the use of milk and dairy products (curd, cheese), which improve the properties of the mixture to which they are added in different ways (curing accelerators, adhesives, plasticity modifiers, consolidants and agglutinating agents, thickeners, etc.) |
| Ashurst & Ashurst (1989, p. 12) | Report that in the British Isles mortar additives were limited in ancient times to casein (milk), eggs (whites), linseed oil, fresh blood, beeswax, keratin (from animal hooves and horns), tallow (animal fat), beer, malt and urine. Waxes, fats and oils introduced some water-repellent properties to mortar, sugary materials reduced the water required and retarded carbonation or set, beer and urine acted additionally as air entrainers. Survivors in the 19th century from the old practices were largely linseed oil and tallow |
| Ashurst & Ashurst (1989, p. 44) | Mention, as basic constituent of limewashes, lime, to which pigments may be added for colour, and tallow, linseed oil or casein for a more durable treatment |
| Appolonia *et al.* (1989) | Mention the use of coatings or films consisting of gypsum and calcium caseinate, and gypsum, casein and lime |
| Lazzarini & Salvadori (1989) | Report that patinas on Roman monuments might be made by calcium caseinate and a little quantity of ochre |
| Barahona (1992*b*) | Reports the Spanish use of natural coagulating agents, such as bulls blood, which was used to improve the hardness of the render, and also the use of molasse, milk or egg white to improve or simply control the curing process |
| Rampazzi *et al.* (2004) | Document ancient treatments for the conservation of stone, mortar and stucco: egg, milk, natural resins, oils, gum arabic and molasses in ancient treatments for the conservation of stone |

described patina as 'a homogeneous and extensive layer, often coloured variously yellow, brown, pink or red, over a variety of substrates including stone'.

Kouzeli (undated) defined the term as:

'a thin layer formed on the surface of an object by the passage of time, which is the result of natural processes, human intervention or a combination of the two. It adheres to the material it covers, it is not harmful to it but on the contrary it protects it; it does not change the appearance of the object, in fact it contributes to its aesthetic appearance. The term was initially used to describe the superficial layers on metals, but its use has been extended to include other materials such as wood and stone. Such superficial layers on the stone of monuments are found all over the world. To understand the nature, the structure and the role of the patina on stone monuments and furthermore its origin, is a difficult task to which many scientists have contributed and are still contributing'.

Ordaz & Esbert (1988) defined it as 'a coating or superficial and thin film that forms on the stones for several reasons'. It is also defined as a 'superficial modification of the material that does

**Fig. 1.** Different images showing the protective effect of patinas on the stones they cover. The loss or lack of patinas leads to a rapid decay of the stone surface.

not necessarily imply degradation or deterioration processes'.

Limewashes could be assumed as some kind of patinas, which Ashurst & Ashurst (1989, p. 44) define as 'a traditional surface finishes for many building materials; although once an almost universal treatment, the quality of limewashes varied enormously'.

There are also changes in terminology related to local, regional and even national circumstances (e.g. the Spanish terms of *encalado* and *enjalbegado*: Barahona 1992*a*; Villanueva 1992).

### *Composition*

The mineralogical composition of patinas has been referred to by many different authors (Guidobaldi *et al.* 1982; Alessandrini *et al.* 1988; Kouzeli *et al.* 1996; Previde Massara & Perego 2000; Alvarez de Buergo *et al.* 2002, 2004; Alvarez de Buergo & Fort 2003; Polikreti & Maniatis 2003; Vazquez-Calvo *et al.* 2006). Some of the minerals that may comprise a patina are: calcium carbonates, calcium sulphates, calcium oxalates, calcium phosphates, silicates (quartz, feldspars, etc.), clay minerals and iron oxides and/or hydroxides. This does not mean that all these minerals are present in one single patina. One of the special features of patinas is that their present mineralogy is the result of mineralogical and chemical transformations of an original mixture. From ancient and historical documentation, from the composition of present-day patinas, and from observed chemical and mineralogical transformations it can be said that patinas were generally made of a mixture mainly composed of inorganic and organic compounds, mixed with lime and/or gypsum (binders) in water, natural pigments for colouring and organic additives that would aid the protection of the stone. This mixture was applied as a paint and surface finish on the stone of external building façades and monuments.

## Spanish background and current research on patinas in Spain

There are few references to the presence of protective patinas on the surface of Spanish monuments and buildings. Some of the work carried out in this field in Spain is shown in Table 2. One of the authors referred to in the table is Cabrera-Garrido (1992, 1996). He supports the idea that Spanish stone has traditionally been rendered in the past. He insists on the fact that many of the superficial gypsum layers (on limestone, sandstone or even granite) that are attributed to atmospheric pollution are the result of a gypsum render. It is necessary to stress the appearance of the gypsum layer as it could cause some misunderstanding. This layer of gypsum should not be confused with weathering or pollution crusts. Thin section studies reveal in some cases the existence of minor amounts of gypsum crystals in the pores of the lime-based patinas (Alvarez de Buergo *et al.* 2002), but in most of the cases the patina is formed by a gypsum layer with variable proportions of clay minerals that might have been used to provide consistency to the paste, as well as to obtain a certain colour (Alvarez de Buergo *et al.* 2004). For instance, in the specific case of the Monastery of Uclés (Cuenca), the building is located on the top of a hill, in a rural environment, with no atmospheric pollution as possible contributor to gypsum crust formation (Alvarez de Buergo *et al.* 2004). The term crust usually implies (at least for black crusts) associated decay of the stone on which it develops, which often leads to spalling (at the crust–stone contact). This is not usually observed in the case of patinas.

As well as the authors referred to in Table 2 who have studied specific buildings and monuments, there are others such as Gárate (1990) who have referred to dyes, colourings or tinges as traditional techniques used at diverse times and in diverse cultures. A common technique was to dissolve earth in limewater, sometimes with the addition of organic substances that may constitute the right environment and starting point for the development of micro-organisms. The same author (Gárate 1999) also mentions that is was a common practice in Spain to use fats and waxes on monuments, which reacted with lime or limewater to created a calcium 'soap' that protected the stone.

Some of the patinas on the stones of Spanish monuments have been analysed and found to have a natural origin (Saiz-Jimenez 1995; Garcia-Valles *et al.* 1996; Prieto *et al.* 1997; Valls del Barrio *et al.* 2002). Some of these authors conclude that, although their origin is biological, these patinas act, in many cases, as a protective layer on the stone.

In 2002 the Spanish Historic Heritage Institute (IPHE) held a workshop dealing with the intervention criteria on stone materials (Instituto de Patrimonio Histórico Español 2003). The term patina is mentioned in several chapters. At the diagnosis stage, the need of microstratigraphical studies for the description and comprehension of existing coverings is established. At the cleaning stage, attention is paid to the existence of natural and artificial patinas. The latter are seen as coatings applied intentionally in the past for protective or aesthetic purposes. Their historic nature leads to a recommendation for their preservation. Their removal is only to be considered when their presence puts the cultural good at risk.

Several buildings in the central area of Spain have been selected for studying this kind of protective treatment on the stone. The main objectives of these ongoing studies are the location of patinas around Spain, identification of the types of stones on which patinas were applied, their nature and composition, their degree of conservation, the causes of their decay, and, above all, whether they preserve the underlying stone.

Patinas from six provinces of Spain (Burgos, Cuenca, Guadalajara, Madrid, Salamanca and Valladolid), belonging to three Autonomous

**Table 2.** *Summary patina research carried out on Spanish monuments*

| Reference | Monument/building | Results from the research carried out on patinas |
|---|---|---|
| Cabrera-Garrido (1992, 1996) | Santiago de Compostela Cathedral | Lime-based renderings during the Romanesque, gypsum based renderings for the Gothic period and 17th century |
| Cabrera-Garrido (1992, 1996) | San Gregorio and San Pablo Churches (Valladolid)<br>Infantado Palace (Guadalajara)<br>Torrelaguna Church (Madrid) | Mentions that gypsum mixed with ochre and binded with oil and protein-based substances, like paint, are present in the façades of these buildings. He believes that coatings probably correspond to remains from the 17th century |
| De la Torre & Sebastián (1996) | Comares Tower, 14th century (La Alhambra, Granada) | Thin patina (2–25 μm thick), orange-coloured, mainly composed by calcite and calcium oxalates. EDS analyses: Ca, Si, Al, Cl, S, Mg, K, Na and Fe. Origin attributed to surface conservation treatments applied in the past |
| Pavía Santamaría *et al.* (1996) | Buildings of La Rioja region (12th–18th century) both in urban and rural environments | Patinas (20 μm thick), ranging from orange to yellow colours on different building materials (sandstone, limestone and their mortar joints) calcite, quartz, feldspar, clay minerals, dolomite, hematite, sulphates and calcium oxalates. EDS analyses: P, Fe, Cl and S. Artificial origin assigned: stonemasons applied them following the architectural practice in certain historic periods |
| Navarro-Gascón *et al.* (1996) | San Juan de los Caballeros Church (Segovia)<br>Cuenca Cathedral | Patina applied on dolomitic substrate composed by phosphates, or mixed phosphates and oxalates, and a more recent one on top of this with only oxalates. Reverse sequence has also been detected (phosphates over oxalates) |
| Navarro-Gascón *et al.* (1996) | Silos Monastery (Burgos) | Layers of phosphates or mixed phosphates and oxalates (from bone phosphates–animal glue mixtures), and a more recent one on top of this (attributed to 14th–15th century), with only oxalates (from pigments bound with egg yolk) |
| Alvarez de Buergo (1997) | Madre de Dios Convent, Alcalá de Henares (Madrid) | Patina composed of calcium oxalates, calcite, gypsum and silicates. EDS analyses: Al, Si, S, Ca, Fe and K. Secondary elements P, Mg, Na, Ti and Cl |
| Añorbe (1997) | Monuments of Zamora | Orange patina (up to 400 μm thick) mainly on sandstones and conglomerates composed of whewellite and wedellite, and low contents of calcite, gypsum, iron oxyhydroxides, quartz and feldspars. EDS analyses: Ca, Si, Al, Fe and K, and occasionally S, Cl, P and Mg. Origin attributed to the ageing of past protective treatments |
| Pavía Santamaría & Caro Calatayud (2000) | Monuments of Valladolid | Coverings (EDS analyses: Ca, S, Fe, P, Cl and Zn) on granites and limestone interpreted as the past application of a render on the stone composed by lime/gypsum and a casein-based organic additive such as milk or egg white. They conclude that while the coatings applied on granitic substrates where richer in gypsum, the ones applied on limestone surfaces where richer in lime and organic products |
| Esbert *et al.* (2000) | Gerona Cathedral<br>Burgos Cathedral | Films correspond to gypsum-based coatings applied to the stone in the past for protective and homogenization purposes |
| Alvarez de Buergo & Fort (2003) | Nuevo Baztán Palace (Madrid) | See text for additional details |
| Alvarez de Buergo *et al.* (2004) | Uclés Monastery (Cuenca) | See text for additional details |
| Vazquez-Calvo *et al.* (2006) | Monuments of Lerma | See text for additional details |

Regions (Madrid, Castilla y León and Castilla-La Mancha), have been sampled, resulting in around 90 patina samples (Table 3, Figs 2 & 3). A wide range of analytical techniques can be used for the analysis of this type of film, including microscopic ones extensively described by Vazquez-Calvo *et al.* (2005). The main results obtained of some of the already analysed patinas are summarized below.

### *Palace-Church of Nuevo Baztán, Madrid*

The patina of the Palace-Church of Nuevo Baztán, Madrid (Alvarez de Buergo *et al.* 2002; Alvarez de Buergo & Fort 2003) (Fig. 3a) consists of a multilayered ochraceous film (80–150 μm) on a limestone substrate. Detected minerals are calcite, clay minerals (phyllosilicates), gypsum, quartz, potassium feldspar, calcium oxalates (whewellite and weddellite), calcium phosphates, iron oxides and hydroxides. The patina is defined by the presence of the following elements: Ca, P, Cl and S, and with lesser intensity by: Al, Fe, K and Na. The appearance of the patina is that of a smooth paste that covers the stone homogeneously. The patina is well preserved. There is a fissuring system that does not develop beyond the limestone–patina contact area. It affects the substrate directly beneath the patina running parallel to the surface. A minor fissure system was also detected in some areas of the stone running perpendicular to the surface. This patina has been interpreted as the result of a past treatment, possibly consisting of a mixture of lime, gypsum, animal milk and ochre-earth pigments, where milk casein acted as an agglutinative and adhesive, milk fat had a waterproof function and ochre served as a colouring pigment.

### *Monastery of Uclés, Cuenca*

From the dolostones and limestones of the façades of the Monastery of Uclés, Cuenca (Alvarez de Buergo *et al.* 2004) (Fig. 3b) three 100–500 μm thick patinas have been analysed. Mineralogically, they are calcite, gypsum-calcite and gypsum-based, with varying amounts of other minerals including dolomite, quartz, montmorillonite and calcium oxalates (whewellite and wedellite). The relative use of lime or gypsum corresponds with historical trends. During the 16th century gypsum-

**Table 3.** *Monuments from which patina fragments have been sampled*

| Label* | Monument/building | Town | Province |
|---|---|---|---|
| 1 | Monastery of San Blas | Lerma | Burgos |
| 1 | Collegiate church of San Pedro and San Pablo | Lerma | Burgos |
| 1 | Arch-Prison | Lerma | Burgos |
| 1 | Santa Teresa Square (Arcades with Ascensión Convent) | Lerma | Burgos |
| 2 | Church of Santiago | Cigales | Valladolid |
| 3 | Vázquez's House | Ciudad Rodrigo | Salamanca |
| 3 | Church of San Pedro y San Isidro | Ciudad Rodrigo | Salamanca |
| 3 | Chapel of La Tercera Orden | Ciudad Rodrigo | Salamanca |
| 3 | Hospital of the Pasión | Ciudad Rodrigo | Salamanca |
| 3 | Cathedral of Santa María | Ciudad Rodrigo | Salamanca |
| 3 | Church of San Agustín | Ciudad Rodrigo | Salamanca |
| 3 | Private house | Ciudad Rodrigo | Salamanca |
| 4 | Church of la Natividad de Nuestra Señora | Valdetorres de Jarama | Madrid |
| 5 | Church of San Juan Evangelista | Talamanca de Jarama | Madrid |
| 6 | Palace-Church of Nuevo Baztán | Nuevo Baztán | Madrid |
| 7 | Monastery of Uclés | Uclés | Cuenca |
| 8 | Church El Salvador | Cifuentes | Guadalajara |
| 8 | Convent of Santo Domingo | Cifuentes | Guadalajara |
| 9 | Palace of Infantado | Guadalajara | Guadalajara |
| 10 | Church of Santa María | Aranda de Duero | Burgos |
| 10 | Church of San Juan Bautista | Aranda de Duero | Burgos |
| 11 | Monastery of Santa María de La Vid | La Vid | Burgos |
| 12 | Collegiate church | Peñaranda de Duero | Burgos |
| 12 | House next Collegiate | Peñaranda de Duero | Burgos |
| 13 | Monastery of Santo Domingo de Silos | Silos | Burgos |
| 14 | Monastery of San Pedro Arlanza | Hortigûela | Burgos |
| 15 | Church of Santo Tomás | Covarrubias | Burgos |

*See Figure 2 for a general location of samples.

**Fig. 2.** Map of Spain with locations of the patina sampling.

based patinas were substituted by lime-based ones (Cabrera-Garrido 1994). These films are interpreted as artificial patinas designed to both protect the masonry and to unify the colour of the façades.

### *Lerma, Burgos*

In Lerma, Burgos, the patinas on three buildings have been studied (Vazquez-Calvo *et al.* 2006): the Monastery of San Blas (Fig. 3c, e), Arch-Prison (Fig. 3d) and the Collegiate of San Pedro. The substrate is, in all three cases, a biosparitic limestone (Hontoria Limestone). The films are multilayered. Typical mineralogy is calcite, hydroxyl-apatite and calcium oxalates (mainly weddellite). The most abundant chemical elements in the patinas are Ca, Si, Al, P, K and Fe (and Cl, Mg, Ti, S and Pb as secondary elements). This contrasts with the substrate that is mainly characterized by the presence of Ca. In this specific case, techniques including X-ray fluorescence (XFR) and Laser Induced Breakdown Spectroscopy (LIBS) have proven to be very useful for the chemical characterization of patinas and for distinguishing types of patina. The analysed patinas are the result of the application of films for protective purposes and mainly comprise oxalates and phosphates, and it appears that their presence has contributed to the conservation of the Hontoria Limestone.

## Possible reproduction of patinas

The further aims of this study are the recovery of *patinature* (patination) as a traditional technique, as well as the protection of patinas themselves as protective and aesthetic films and historic features of our built Heritage. One of the challenges that stem from this is the reproduction of patinas. There have been some attempts to reproduce them following ancient recipes (Camaiti *et al.* 1996) and by exposing samples to a natural ageing of 2 years instead of artificial weathering. But the task of deciphering past procedures from present-day relicts is not easy. Some ancient recipes for protecting and colouring stone do still exist and there have been some attempts to reproduce original patinas (Franzini *et al.* 1984) and to improve them through the addition of organic substances. These authors obtained calcium oxalates by mixing calcium oxide and egg white in a glass slide. Camaiti *et al.* (1996) have tested some treatments based on ancient recipes with ingredients, such as water, milk, lime, casein, linseed oil, albumen and sugar, in different proportions and mixtures and applied to marble specimens. The most successful results in terms of calcium oxalate formation were obtained with linseed oil and casein. In the same way, Kouzeli *et al.* (1996) tested several mixtures with ingredients such as casein, calcium phosphate, wheat water, lime, starch, glue and oxalic acid applied to Pentelic marble. The results indicate that only the oxalic acid-based treatment produced calcium oxalates on the surface, but in the form of a white dust that was easily removed.

The Opificio delle Pietre Dure in Florence (Italy) proposed the formation of artificial calcium oxalate for the protection of murals during the 1980s, with such interesting results that in recent years similar treatments have also been applied to calcareous

**Fig. 3.** (**a**) Palace of Nuevo Baztán, Madrid, and appearance of a patina sample. (**b**) Monastery of Uclés, Cuenca and appearance of a patina sample. (**c**) Monastery of San Blas, Lerma, Burgos and appearance of a patina sample. (**d**) Arch-Prison, Lerma, Burgos and appearance of a patina sample. (**e**) Cross-section of a sample of patina under polarizing microscope (parallel nicols).

stone artefacts (Lanterna *et al.* 2000). Cezar (1998), based on this experience researched the effectiveness of the conversion of English limestones by using ammonium oxalate. Cariati *et al.* (2000) carried out some laboratory oxidation tests by UV irradiation of $H_2O_2$ and by UV irradiation of $O_3$ on Candoglia and Carrara marbles, together with organic substances selected from those most frequently used and cited as ingredients of ancient recipes for the conservation treatment of stone materials. These included glucose, gum arabic, molasses, albumen and yolk (hen eggs), full-fat cow milk (not pasteurized) and linseed oil. Calcium oxalates were obtained through an oxidation methodology that reproduced the natural reactions occurring in the troposphere. Finally, in Spain, there have been some attempts to artificially age stone using milky by-products according to Spanish traditional techniques for pottery ageing (Monzó & García del Cura 1999).

With all of this in mind, the authors of this paper are working on the reproduction of ancient patinas with a double purpose: to preserve the material (patina) and the technique (patination). Several recipes have been applied to limestone tablets, as well as on glass slides. In this continuing study, compounds such as lime putty, water, goat and sheep milk, powdered and infant milk, casein, calcium caseinate, oxalic acid, ammonium oxalate, linseed oil, animal bone glue and ochre pigments have been used in different recipes and proportions to obtain covering films on the stone tablets that will be analysed in the near future.

## Summary

The coatings used for the finishing of heritage stone surfaces – known as patinas – date back to very early constructions. The application of patinas to the stone surfaces is thus considered to be a traditional technique, and patinas were applied for aesthetic and/or protective reasons. Although they are generally described as artificial, subsequent modification by micro-organisms and/or lichen activity should not be ruled out. Because of this ongoing chemical evolution, it may be more logical to assign them a combined artificial and natural origin. There is no unanimity for a single term to designate these kinds of film, and terms such as patina, scialbatura, oxalate film, superficial layer, surface finishing, coating, covering, skin, epidermis are in wide use, as well as a variety of local names.

The original composition of patinas can be lime and/or gypsum, water, natural pigments and organic additives, and the ingredients and ratios vary according to local custom and to changing fashions. These original mixtures turn into patinas with varied mineralogy, but mainly consisting of calcium carbonates, calcium sulphates, calcium oxalates, calcium phosphates, silicates, clay minerals and iron oxides/hydroxides.

Although patinas are present on many historic monuments and buildings in Spain – one of the richest such resources in the world – there has been little study of them and few references within scientific literature. A review of the state of knowledge concerning patinas in Spain is, however, presented here, together with new studies and analyses of the artificial patinas on stone buildings and monuments.

Finally, it has been identified that not only is the analysis and characterization of patinas essential, but also their reproduction, based on their present-day composition and recipes extracted from historic accounts. This should utilize laboratory investigations under controlled conditions and with thorough records of the experimental recipes, including those that fail.

The case studies referred to in this paper show that patinas appear to have protected the stone substrates on which they were applied. Although detailed studies of their protective aspects should be central to any restoration programme, consideration should also be given to their preservation as an integral element of the structure's history and as a record of a traditional building technology.

This study has been financed by Project BIA-2003-4073 (Spanish Ministry of Education and Science), and by a Ramón y Cajal Contract (M. Alvarez de Buergo). Thanks are also given to the MATERNAS programme ('Durability and Conservation of Natural Traditional Materials from the Architectural Heritage') financed by Comunidad de Madrid (Regional Government of Madrid) (0505/MAT/94). C. Vazquez-Calvo thanks the CSIC (Spanish Council for Scientific Research) Thematic Network for Historical and Cultural Heritage for a scholarship cofinanced by the European Social Fund. We would like also to express our gratitude to S. Brüggerhoff and the other, anonymous, reviewer whose perceptive comments have helped to improve the quality of the paper. Both editors, B. Smith and R. Přikryl, are deeply acknowledged for helping in improving this paper.

## References

ALVAREZ DE BUERGO, M. & FORT, R. 2003. Protective patinas applied on stony façades of historical buildings in the past. *Construction and Building Materials*, **17**, 83–89.

ALVAREZ DE BUERGO, M., FORT, R. & GÓMEZ-HERAS, M. 2004. The Monastery of Uclés (Cuenca, Spain): characterization and deterioration of building materials. *Materiales de Construccion*, **54**(275), 5–22.

ALVAREZ DE BUERGO, M., FORT, R., LÓPEZ DE AZCONA, M. C. & MINGARRO, F. 2002. Analysis of the ochre patina on the limestone of Palacio de Nuevo Baztán, Madrid, Spain. *In*: GALÁN, E. & ZEZZA, F. (eds) *Protection and Conservation of the Cultural Heritage in the Mediterranean Cities*, Volume 1. Balkema, Lisse, 391–396.

ALESSANDRINI, G. 2004. Patine su materiali lapidei. *In*: TIANO, P. & PARDINI, C. (eds) *Le Patine. Genesi, significato, conservazione*. Nardini Editore, Florence, 15–28.

ALESSANDRINI, G., BUGINI, R. & PERUZZI, R. 1988. I Trattamenti superficiali effetuati nel passato. *In*: *La Certosa di Pavia: passato e presente nella facciata della chiesa*. CNR, Roma, 291–319.

ALVAREZ DE BUERGO, M. 1997. *Caracterización, alteración medioambiental y restauración en paramentos del patrimonio arquitectónico*. Centro de Estudios y Experimentación de Obras Públicas (CEDEX), Ministerio de Fomento Monografía, **M-58**.

AÑORBE, M. 1997. *Valoración del deterioro y conservación en la piedra monumental Caracterización y diagnosis de la piedra monumental de la ciudad de Zamora*. Centro de Estudios y Experimentación de Obras Públicas (CEDEX), Ministerio de Fomento Monografía, **M-56**.

APPOLONIA, L., GRILLINI, G. C. & PINNA, D. 1996. Origin of oxalate films on stone monuments. I. Nature of films on unworked stone. *In*: REALINI, M. & TONIOLO, L. (eds) *II International Symposium: The Oxalate Films in the Conservation of Works of Art, Milano, Italia*. EDITEAM s.a.s.Gruppo Editoriale, Castello d'Argile (BO), 257–268.

APPOLONIA, L., MIGLIORINI, S. & VAUDAN, D. 1989. Lo scialbo degli affreschi di epoca ottoniana della Cattedrale di Aosta. *In*: *Centro CNR 'Gino Bozza' Politecnico di Milano, Le pellicole ad ossalati: origine e significato nella conservazione delle oper d'arte*, Atti del Convegno. Vega, Milano, 245–254.

ASHURST, J. & ASHURST, N. 1989. *Mortars, Plasters & Renders. Practical Building Conservation*, Volume 3. English Heritage Technical Handbook. Gower Technical Press, London.

BARAHONA, C. 1992*a*. Revocos tradicionales. *In*: *Patrimonio Histórico. Castilla-La Mancha (ed) Revestimientos sobre fachadas*. Servicio de Publicaciones de la Junta de Comunidades de Castilla-La Mancha, 29–35.

BARAHONA, C. 1992*b*. *Revestimientos Continuos en la Arquitectura Tradicional Española*. Centro de Publicaciones Secretaría General Técnica, Ministerio de Obras Públicas y Transportes. Madrid OGGI Creaciones Gráficas y Publicitarias, S.A.

BORGHINI, R. 1967. Il Riposo, in cui Della Pintura e della Scultura siFavella de' più illustri Pittori, e Scultori, e dell più famose opere loro si fa menzione; e le cose principali appartenenti a dette arti s'insegnamo. *In*: ROSCI, M. (ed.) *Analytical index and Bibliographic essay*. Labor Riproduzionie Documentazioni, Milano. (1st edn Firenze 1584).

BOSELLI, O. 1978. *Osservazioni sulla scoltura antica*. P. Dent Weil, Firenze (1st edn *c*. 1650).

CABRERA-GARRIDO, J. M. 1992. Contaminación y Patrimonio. Punto de vista del Restaurador. *In*: *Encuentro Europeo sobre Patrimonio Histórico Artístico y Contaminación. Consorcio para la Organización de Madrid Capital Europea de la Cultura 1992, Madrid*, Artes Gráficas, Dincolor, 59–63.

CABRERA-GARRIDO, J. M. 1994. *Los recubrimientos históricos y la protección de la piedra. Las pátinas*. Colegio Oficial de Aparejadores y Arquitectos Técnicos de Madrid (COAATM), 57–63.

CABRERA-GARRIDO, J. M. 1996. Estudios de recubrimientos de fachadas antiguas: la patina de la piedra y el color de la arquitectura. *In*: *XI Congreso de Conservación y Restauración de Bienes Culturales*. Castellón, Diputació de Castelló, Servei de Publicacions, 869–876.

CAMAITI, M., FOMMEI, C., GIAMELLO, M., SABATINI, G. & SCALA, A. 1996. Trattamenti di superfici lapidee secondo antiche ricette: primi resultati sulla formazione di ossalati di calcio. *In*: REALINI, M. & TONIOLO, L. (eds) *II International Symposium: The Oxalate Films in the Conservation of Works of Art, Milano, Italia*. EDITEAM s.a.s.Gruppo Editoriale, Castello d'Argile (BO), 287–298.

CARIATI, F., RAMPAZZI, L., TONIOLO, L. & POZZI, A. 2000. Calcium oxalate film on stone surfaces: experimental assessment of the chemical formation. *Studies in Conservation*, **45**, 180–185.

CENNINI, C. 1933/1437. Il Libro dell' Arte. Translated by THOMPSON, D. V. JR. Dover, New York, 1933, by Yale University Press. World Wide Web Address: http://www.noteaccess.com/Texts/Cennini/.

CENTRO CNR 'GINO BOZZA' POLITECNICO DI MILANO (ed.) 1989. *Le pellicole ad ossalati: origine e significato nella conservazione della oper d'arte*. Atti del Convegno. Milano 25–26 Ottobre, Vega, Milano.

CEZAR, T. M. 1998. Calcium Oxalate: A Surface Treatment for Limestone. *Journal of Conservation & Museum Studies*, **4**. Institute of Archaeology, University College of London, World Wide Web Address: http://www.ucl.ac.uk/archaeology/conservation/jcms/issue4/cezar.html.

DANESI, A. & GAMBARDELLA, S. 2005. Materials and techniques used in the 17th century restoration of sculptures from the antiquities collection of the Palazzo Lancellotti Ai Coronari. *In*: BARBANERA, M. & FRECCERO, A. (eds) *Art, Conservation, Science. The Lancellotti Collection Project*. (The Swedish Institute in Rome. Projects and Seminars, 3:1). World Wide Web Address: http://www.svenska-institutet-rom.org/lancellotti/gambardella.pdf.

DE LA TORRE, M. J. & SEBASTIÁN, E. 1996. A study of oxalate patinas inside historic buildings (The Alhambra, Granada). *In*: REALINI, M. & TONIOLO, L. (eds) *II International Symposium. The Oxalate Films in the Conservation of Works of Art, Milan, 25–27 March 1996*. EDITEAM s.a.s.Gruppo Editoriale, Castello d'Argile (BO), 209–217.

De la Torre, M. J., Sebastián, E. & Rodríguez-Gordillo, J. 1995. Pátinas en La Alhambra: estudio mediante SEM y BSEI. *In*: *XVII Reunión Bienal de la Sociedad Española de Microscopía Electrónica*, Gráficas Oviedo, S. A., Oviedo, 306–308.

Del Monte, M. & Sabbioni, C. 1987. A study of the patina called 'scialbatura' on Imperial Roman marbles. *Studies in Conservation*, **32**, 114–121.

Esbert, R. M., Alonso, F. J., Díaz-Pache, F. & Ordaz, J. 2000. Técnicas microscópicas al estudio de recubrimientos artificiales de la piedra de monumentos. *In*: Galán, E. (ed.) *Protection and Conservation of the Cultural Heritage of the Mediterranean Cities. Proceedings of the 5th International Symposium on the Conservation of Monuments in the Mediterranean Basin, Sevilla, Spain, 5–8 April 2000*. Kronos, S. A., Sevilla, 234–235.

Franzini, M., Gratziu, C. & Wicks, E. 1984. Patine ad ossalato di calcio su monumenti marmorei. *Società Italiana Mineralogia e Petrografia*, **39**, 59–70.

Gárate, I. 1990. *Artes de la cal*. Instituto Español de Arquitectura, Universidad de Alcalá, Madrid.

Gárate, I. 1999. *Artes de los yesos*. Yeserías y estucos. Instituto Español de Arquitectura, Universidad de Alcalá, Madrid.

Garcia-Valles, M. T., Krumbein, W. E., Urzi, C. & Vendrell-Saz, M. 1996. Biological pathways leading to the formation and transformation of oxalate-rich layers on monument surfaces exposed to Mediterranean climate. *In*: Realini, M. & Toniolo, L. (eds) *II International Symposium. The Oxalate Films in the Conservation of Works of Art, Milan, 25–27 March 1996*. EDITEAM s.a.s.Gruppo Editoriale, Castello d'Argile (BO), 317–334.

Grissom, C. A., Charola, A. E. & Henriques, F. M. A. 2002. To paint or not to paint: a difficult decision. *In*: Galán, E. & Zezza, F. (eds) *Protection and Conservation of the Cultural Heritage of the Mediterranean Cities*. A. A. Balkema, Rotterdam, 585–592.

Guidobaldi, F., Laurenzi Tabasso, M. & Meucci, C. 1982. Marble monuments of the Roman Imperial Age: Past surface treatments. *In*: Gauri, K. L. & Gwinn, J. A. (eds) *Proceedings of the Fourth International Congress on Deterioration and Preservation of Stone Objects*, The University of Louisville, Louisville, KY, 175–196.

Instituto de Patrimonio Histórico Español. 2003. Retablos. Anexo Criterios de intervención en materiales pétreos. *Revista del Instituto de Patrimonio Histórico Español*, **2**.

Knoll, H. 1968. *Die Trajanssäule*. Rep. Instituten für Anorganische Chemie, Freie Universität Berlin.

Korres, M. & Bouras, X. 1983. *The Restoration of the Parthenon*. Ministry of Culture, Committee for the Preservation of the Acropolis Monuments, Athens.

Kouzeli, K. Undated. The composition and structure of the patina on the Parthenon and other Greek monuments. World Wide Web Address: (www.thebritishmuseum.ac.uk/gr/kouzelicompositionpaper.pdf).

Kouzeli, K., Beloyannis, N., Tolias, C. & Dogani, Y. 1988. *Ancient and Byzantine Conservational Treatments on the Parthenon. Deterioration and Conservation of Stone*. Institute of Conservation and Restoration of Cultural Property and Nicholas Copernicus University, Torun, 687–694.

Kouzeli, K., Lazari, C., Economopoulos, A. & Pavelis, C. 1996. Phosphatic patinas on Greek monuments (Acroolis of Athens and other ancient and Byzantine monuments): general discussion and further documentation on the presence of oxalates. *In*: Realini, M. & Toniolo, L. (eds) *II International Symposium. The Oxalate Films in the Conservation of Works of Art, Milan, 25–27 March 1996*. EDITEAM s.a.s.Gruppo Editoriale, Castello d'Argile (BO), 83–93.

Krumbein, W. E. 2002. Patina and cultural heritage a geomicrobiologist's perspective. *In: Biodeterioration and its Control – Biotechnologies in Cultural Heritage Protection and Conservation. Proceedings 5th EC Conference, Krakow, 16–18 May 2002*. Polska Akademia Nauk, Krakow, 39–47.

Lanterna, G., Mairani, A., Matteini, M., Rizzi, M., Scuto, S., Vincenzi, F. & Zannini, P. 2000. Mineral inorganic treatments for the conservation of calcareous artefacts. *In*: Fassina, V. (ed.) *Proceedings of the 9th International Congress on Deterioration and Conservation of Stone*, Volume 2. Elsevier, Amsterdam, 387–394.

Lazzarini, L. & Salvadori, O. 1989. A reassessment of the formation of the patina called scialbatura. *Studies in Conservation*, **34**, 20–26.

Lewin, S. Z. 1966. The preservation of natural stone, 1839–1965, an annotated bibliography. *Art and Archaeology Technical Abstracts*, **6**(1), 185–272.

Liebig, J. V. 1853. Ueber den Thierschit. *Liebigs Annalen der Chemie und Pharmazie*, **LXXXVI**, 113–115.

Maravelaki-Kalaitzaki, P. 2005. Black crusts and patinas on Pentelic marble from the parthenon and Erechtheum (Acropolis, Athens): characterization and origin. *Analytica Chimica Acta*, **532**(2), 187–198.

Martín-Gil, J., Ramos-Sanchez, M. C. & Martín-Gil, F. J. 1999. Ancient pastes for stone protection against environmental agents. *Studies in Conservation*, **44**(1), 58–62.

Matteini, M., Moles, A., Lanterna, G. & Nepoti, M. R. 1996. Preliminary monitoring on painted surfaces of a mineral protective treatment based on artificially formed calcium oxalate. *In*: Realini, M. & Toniolo, L. (eds) *II International Symposium. The Oxalate Films in the Conservation of Works of Art, Milan, 25–27 March 1996*. EDITEAM s.a.s.Gruppo Editoriale, Castello d'Argile (BO), 83–93.

Maxová, I. 2000. Changes in properties of stone treated with historical or modern conservation agents. *In*: Fassina, V. (ed.) *Proceedings of the 9th International Congress on Deterioration and Conservation of Stone*, Volume 2. Elsevier, Amsterdam, 395–402.

MONTANARI, V. 1996. Patina: surface deposit, alteration or out-ward appearance? preliminary considerations for the conservation of stone surfaces in architecture. *In*: RIEDERER, J. (ed.) *8th International Congress on Deterioration and Conservation Stone*, Volume 2. Möller Druck, Berlin, 929–933.

MONTE, M. 2003. Oxalate film formation on marble specimens caused by fungus. *Journal of Cultural Heritage*, **4**, 255–258.

MONZÓ, J. C. & GARCÍA DEL CURA, M. A. 1999. Efecto de la bioalteración en la corrección de impactos visuales cromáticos de canteras de rocas carbonáticas. *Ingeopres*, **77**, 76–85.

NAVARRO-GASCÓN, J. V., GÓMEZ GONZÁLEZ, M. L. & CAYO GARCÍA, M. D. 1996. Estudio de la policromía y pátinas de los relieves del claustro del Monasterio de Santo Domingo de Silos (Burgos). *In*: *XI Congreso de Conservación y Restauración de Bienes Culturales, Castellón*. Diputación de Castellón, Servicio de Publicaciones, Castellón, 589–602.

ORDAZ, J. & ESBERT, R. 1988. Glosario de términos relacionados con el deterioro de las piedras de construcción. *Materiales de Construcción*, **38**(209), 39–45.

PAVÍA SANTAMARÍA, S. & CARO CALATAYUD, S. 2000. Contribución al estudio de los recubrimientos sobre piedra monumental. *In*: GALÁN, E. (ed.) *Protection and Conservation of the Cultural Heritage of the Mediterranean Cities. Proceedings of the 5th International Symposium on the Conservation of Monuments in the Mediterranean Basin, Sevilla, Spain, 5–8 April 2000*. Kronos, S. A., Sevilla, 123–124.

PAVÍA SANTAMARÍA, S., COOPER, T. P. & CARO CALATAYUD, S. 1996. A contribution to the study of the patina on monumental stone (patina on calcareous sandstone from La Rioja, Northern Spain). *In*: REALINI, M. & TONIOLO, L. (eds) *2nd International Symposium. The Oxalate Films in the Conservation of Works of Art*, Milan, 25–27 March 1996. EDITEAM s.a.s.Gruppo Editoriale, Castello d'Argile (BO), 221–232.

PLINY THE ELDER. 1989/1st century AD. *Natural History in 10 Volumes*, Volume 10, *Libri XXXVI–XXXVII Pliny*. Harvard University Press, Cambridge, MA.

POLIKRETI, K. & MANIATIS, Y. 2003. Micromorphology, composition and origin of the orange patina on the marble surfaces of Propylaea (Acropolis, Athens). *The Science of the Total Environment*, **308**(1–3), 111–119.

PREVIDE MASSARA, E. & PEREGO, G. 2000. Study of the colourings of the St. Peter's façade (Vatican). *In*: FASSINA, V. (ed.) *Proceedings of the 9th International Congress on Deterioration and Conservation of Stone*, Volume 2. Elsevier, Amsterdam, 425–433.

PRIETO, B., SILVA, B., RIVAS, T., WIERZCHOS, J. & ASCASO, C. 1997. Mineralogical transformation and neoformation in granite caused by the lichens Tephromela atra and Ochrolechia parella. *In*: KOESTLER, R. J. (ed.) *International Symposium on Biodeterioration and Biodegradation*, **40**, Elsevier Science, 191–199.

RAMPAZZI, L., ANDREOTTI, A., BONADUCE, I., COLOMBINI, M. P., COLOMBO, C. & TONIOLO, L. 2004. Analytical investigation of calcium oxalate films on marble monuments. *Talanta*, **63**, 967–977.

REALINI, M. & TONIOLO, L. (eds). 1996. *II International Symposium: The Oxalate Films in the Conservation of Works of Art, Milano, Italia*, 25–27 March 1996. EDITEAM s.a.s.Gruppo Editoriale, Castello d'Argile (BO).

SAIZ-JIMENEZ, C. 1995. Deposition of anthropogenic compounds on monuments and their effect on airborne microorganisms. *Aerobiologia*, **11**, 161–175.

SICKELS, L. 1981. Organics v. synthetics: Their use as additives in mortars. *In: Mortars, Cements and Grouts Used in the Conservation of Historic Buildings*. ICCROM, Rome, 25–52.

TORRACA, G. 1988. General philosophy of stone conservation. *In*: LAZZARINI, L. & PIEPER, R. (eds) *The Deterioration and Conservation of Stone. Notes from the International Venetian Courses on Stone Restoration*. Studies and Documents on the Cultural Heritage, **16**, UNESCO, 243–270.

VALLS DEL BARRIO, S., GARCIA-VALLES, M., PRADELL, T. & VENDRELL-SAZ, M. 2002. The red–orange patina developed on a monumental dolostone. *Engineering Geology*, **63**, 31–38.

VAZQUEZ-CALVO, C., ALVAREZ DE BUERGO, M., VARAS, M. J. & FORT, R. 2005. Study and characterization of patinas by means of microscopic techniques. *In*: HUGHES, J. (ed.) *10th Euroseminar on Microscopy Applied to Building Materials, Paisley (Scotland)*.

VAZQUEZ-CALVO, C., ALVAREZ DE BUERGO, M. & FORT, R. 2006. Patinas in the architectural heritage of Lerma, Burgos (Spain). *In*: FORT, R., ALVAREZ DE BUERGO, M., GOMEZ-HERAS, M. & VAZQUEZ-CALVO, C. (eds) *Heritage, Weathering and Conservation*. Balkema, Rotterdam, **2**, 969–974.

VILLANUEVA, L. 1992. Los revestimientos continuos al exterior. *In: Patrimonio Histórico. Castilla-La Mancha. Revestimientos sobre fachadas*. Servicio de Publicaciones de la Junta de Comunidades de Castilla-La Mancha, 13–25.

VITRUVIUS. 1960/1st century BC. *De architectura, libro VII, cap. IX*. [*The Ten Books on Architecture*]. Dover, New York.

WRIGHT, D. C. 1998. Los acabados de los monumentos novohispanos y la petrofilia al final del siglo XX. *In: La abolición del arte, XXI Coloquio internacional de historia del arte*. *In*: DALL, A. (ed.) *Instituto de Investigaciones Estéticas*, Universidad Nacional Autónoma de México, Mexico, 143–180 (rev. version 4 April 2003). World Wide Web Address: http: //www.prodigyweb.net.mx/dcwright/acabados.htm.

# Green walls?: integrated laboratory and field testing of the effectiveness of soft wall capping in conserving ruins

H. A. VILES[1] & C. WOOD[2]

[1]*Oxford University Centre for the Environment, South Parks Road, Oxford OX1 3QY, UK (e-mail: heather.viles@ouce.ox.ac.uk)*

[2]*Building Conservation and Research Team, English Heritage, 23 Savile Row, London W1S 2ET, UK*

**Abstract:** Soft wall capping, which involves placing a cap of soil and turf (or other vegetation) on the top of ruined walls, is a potentially low cost, easy to maintain, ecologically sensitive and effective method of conserving ruined monuments. An integrated programme of laboratory and field testing has been designed to test the performance of soft capping in comparison with hard capping at a range of sites in England. A sample of ruined walls has been soft capped and monitored using repeat photography, with more detailed wooden dowel monitoring of wall moisture and electronic monitoring of temperatures and moisture levels at the base of soft caps at some sites. Experiments designed to test the thermal blanketing capability of the soft caps have been run in an environmental cabinet on scaled-down versions of soft and hard caps, and similar set-ups have also been monitored outdoors in Oxford. Short-term data from both field trials and laboratory tests illustrate the success of soft wall capping under a wide range of environmental conditions, but longer-term monitoring is needed to evaluate more fully the conservation benefits of soft capping.

Conserving ruined buildings and monuments poses many problems. Ruined walls are vulnerable to attack from rainwater seeping into their cores, from diurnal and seasonal temperature fluctuations, as well as from moisture coming in from the ground. The main conservation strategies used in the past to tackle the decay produced by such factors have been either to let nature take its course or to build hard caps on the top of the walls (often allied with consolidation of the existing wall fabric). Hard capping aims to minimize water flow into the wall from rainwater through the creation of a stone and mortar cap. The objective is to shed water away down the face of the wall as quickly as possible. Various different types of mortar and hard cap design can be used in order to create a visually attractive and effective capping depending on individual circumstances. Hard capping can, however, often be expensive and time-consuming to build and require long-term expensive maintenance to ensure that rainwater is shed quickly away from the walls.

Soft wall capping has been suggested as a possible alternative to hard capping by several authors, and has been used at a range of English sites within the past 15 years (Tolley *et al.* 2000). Soft wall capping involves placing a cap of soil and turf (or other vegetation) on the top of ruined walls. It may have several major advantages over hard capping strategies. First, it is relatively cheap and simple to install. Secondly, it uses natural materials from the local area and may contribute to biodiversity and nature conservation at the site. Thirdly, maintenance costs are likely to be much lower than for hard wall capping. Fourthly, it should reduce the flow of water over the face of the wall as evapotranspiration by the vegetation takes up some water. Fifthly, and perhaps most importantly, soft capping may offer an effective method of slowing decay caused by inappropriate 20th century interventions. Pilot studies in the laboratory have illustrated the potential role of soft wall capping in: (a) providing a thermal blanket for the underlying stone; and (b) acting as a sponge to soak up incident precipitation (rainwater and melting snow) and prevent water ingress (Viles *et al.* 2002). However, there has been much discussion about whether, under real-world conditions, soft wall capping actually works to minimize water ingress into wall tops and to reduce temperature fluctuations within the stonework. Some authors have even suggested that soft wall capping can damage the stone itself, through roots and acids from the soil producing chemical weathering of the stone. For many years in England the generally accepted conservation practice has been to remove vegetation from walls, and so any proposal to introduce a conservation technique that adds vegetation in any form needs careful evaluation.

*From*: PŘIKRYL, R. & SMITH, B. J. (eds) *Building Stone Decay: From Diagnosis to Conservation.* Geological Society, London, Special Publications, **271**, 309–322.
0305-8719/07/$15.00 

## Aims and objectives

Given the debate outlined above, English Heritage have commissioned a 4 year research project with the overall aim of investigating the efficacy of soft capping, comparing this to hard capping under a range of conditions, and establishing best practice in its application and maintenance. No judgement is being made in our research about the philosophical or aesthetic merits of this treatment, although these are key issues that need to be considered in any overall assessment of the value of this conservation technique. The research is being carried out jointly by English Heritage and the University of Oxford, and builds upon an earlier pilot, laboratory-based study (Viles *et al.* 2002). The specific objectives of the research project are: to set up and monitor soft and hard capping field trials on a range of ruined monuments; to develop allied laboratory simulations to provide more detailed testing of the role of various designs of soft capping in providing thermal blanketing and preventing water ingress to different types of stonework; and to produce guidance notes on soft wall capping for conservation practitioners. In terms of thermal blanketing we aim to examine to what extent the soft wall capping reduces the threat of thermal expansion and contraction of blocks of stone. Observations of the state of stonework at several of the monuments studied in this project indicate cracking caused probably by movement as a result of such expansion and contraction. In terms of moisture ingress, we aim to elucidate whether the soft wall caps are effective at soaking up incoming moisture and preventing the wall tops (and wall faces lower down) becoming wet. If soft wall capping can effectively reduce both temperature fluctuations and moisture ingress then they will be of great value in preventing many forms of stone decay as experienced on ruined monuments. The research project is ongoing and this paper reports on our initial findings from an integrated programme of field trials and laboratory experiments.

## Field trials

### *Overall strategy and methodology*

In order to provide a good test of the performance of soft capping under real-world conditions it is important to select sites that represent a wide range of environmental and climatic types, as well as different wall types. It would also be desirable to test a range of soft capping designs. It is also important to find effective ways of monitoring the performance of the capping, and to be able to make realistic comparisons between soft wall capping and other conservation strategies (e.g. hard capping). Furthermore, to provide a robust test of soft wall capping performance monitoring should ideally continue for a decade or more. However, as the research project is temporally and financially limited only two phases of field site establishment have been carried out: phase 1 involving three sites in the Yorkshire Region; and phase 2 involving sites in Gloucestershire, London and two additional sites in Yorkshire. It is hoped that through using the seven study sites we have represented a broad range of environmental conditions, several different material types and a number of architectural styles. Figure 1 shows the location of all the study sites.

The generalized soft capping method used at the phase 1 sites was to use a standard soil (a medium clay loam topsoil to BS 3882), mixed with and without purple slate fragments (5–30 mm in size) in a ratio of 1:3–1:4 slate:soil. This was overlain by turfs cut from the site. The slate has been included to test its capacity to hold moisture during drought conditions. At each of the phase 1 sites a trial was also carried out using hard capping, for which the wall top was stripped of damaged stone and mortar and a proper cap made with lime mortar and stone blocks. For the phase 2 sites a range of different soft capping methods has been used, including seeding and the use of commerical turf (at Howbury).

### *Test sites – first phase*

The first phase of field trials of soft wall capping involved three sites within English Heritage's Yorkshire Region, i.e. Byland Abbey (UK Ordnance Survey (OS) Grid Ref SE 549 789) and Kirkham Priory (OS Grid Ref SE 735 657) both in North Yorkshire, and Thornton Abbey (OS Grid Ref TA 115 190) in North Lincolnshire. These sites were chosen for a number of reasons. They all have ruined walls that are deteriorating, and have some existing soft capping. They also presented us with the opportunity to test soft wall capping on walls of different heights, made of different materials and within different climatic settings. The Yorkshire Region of English Heritage were also keen to be involved in the project, and had implemented a limited programme of soft wall capping in the late 1980s and early 1990s. Thus, they had some experience of the technique, and their sites provided some evidence of the long-term performance (>10 years) of soft capping. All of the ruined sites were constructed using different stones and brick and are subject to different microclimates.

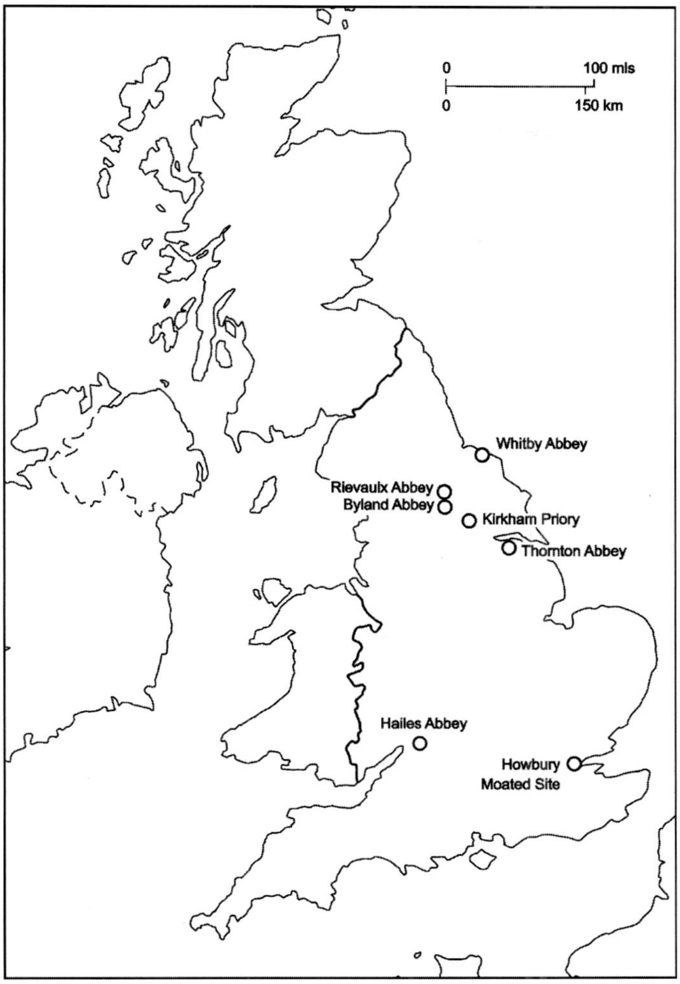


**Fig. 1.** Location for study sites for soft wall capping research project.

Byland Abbey dates from the late 12th and early 13th century, with later additions, and is built of locally quarried sandstone. It is situated in a low-lying area backed by the Hambleton Hills. There is ample evidence today of deterioration of the stonework which seems to have been exacerbated by previous repairs carried out in cementitious mortars and grouts. Most of the walls have been hard capped by the Ministry of Works in the 20th century, whilst some low walls were soft capped in the 1980s. One long (*c.* 30 m) section of wall, around 2 m high, was soft capped in this project, to give a series of short (*c.* 3.5 m-long) sections covered with differing thicknesses of soil (5, 10 and 15 cm) with and without regolith, one 3.5 m section with hard capping, and a control area with no conservation technique. Three other small sections were also soft capped using 5–10 cm-thick soil and turf.

Kirkham Priory was founded in the 1120s with most building carried out during the 12th century. Situated within the Derwent Valley, the Priory was constructed of local limestone. Today it is suffering from minor deterioration. Hard capping has been widely carried out, with some soft capping of low walls and patches on higher walls carried out in the late 1980s–early 1990s (Fig. 2). This previous soft wall capping exercise largely involved placing commercially available turf on low walls (Ogilvy 1996). We installed three test strips of soft wall capping on high walls (>4 m high), using 10 cm-thick soil and turf, with one comparative hard capping area installed adjacent to one of these strips.

**Fig. 2.** Kirkham Priory, showing soft wall capped patch on top of the arch dating from the late 1980s–early 1990s.

Thornton Abbey was founded in 1139 and reconstructed from the 1260s onwards. The site uses a range of building materials, such as Lincolnshire Limestone, chalk, sandstone and brick. Thornton is situated in a very exposed position, near the North Sea coast, and is clearly suffering from extensive deterioration that may be exacerbated by the power stations to the east. Previously, hard capping has been carried out from the 1920s onwards, and in the early 1990s some soft caps on low walls were laid using commercially available turf and seeding (Ogilvy 1996). Figure 3 illustrates the nature of decay at Thornton and the good state of the existing soft wall caps. At Thornton we installed three sections of soft wall capping on low walls (10–75 cm high), with two sections of hard capping nearby on two wall sections around 1.5 m in height.

Overall, at the three Yorkshire ruins in phase 1, we have established 15 soft wall caps and three hard caps. These test caps vary significantly in terms of wall height and width, building material and construction, orientation, degree of exposure to prevailing weather and degree of deterioration. Furthermore, the soft wall capping method used varied in terms of thickness of underlying soil, whether slate fragments were included, the nature of turfs and the method of anchoring turfs in place. At each site the walls consist of two stone or brickwork faces with a central rubble core. Monitoring of the performance of the soft capping at all

(a) (b)

**Fig. 3.** Decay problems and existing soft wall capping at Thornton Abbey. (**a**) Frost damage to low limestone walls on nave; and (**b**) soft wall capping of low cloister wall dating from the early 1990s.

sites is being carried out using simple repeat photography and visual inspection, with additional automated monitoring of temperature and moisture levels under the 30 m-long test wall at Byland from December 2004, and wooden dowel monitoring of the moisture levels in this same wall to provide more detailed data sets. Temperature and moisture are measured with thermistor temperature probes and Watermark sensors (to measure soil moisture contents) attached to a telemetric datalogging system, with measurements taken every 30 min. The wooden dowel survey method is commonly used in investigating moisture in walls, and utilizes thin (*c.* 6 mm) timber dowels put into pre-drilled holes for around 4 weeks at a time (Larson 2004). The dowels absorb moisture from the surrounding stonework, and possess similar water-holding capacities, so provide a reasonable estimate of water contents in the wall itself. This technique is cheap and simple to perform and provides good data of spatial and temporal resolution. However, it only provides a first-order assessment of moisture conditions and should be complemented by other techniques. Forty-one dowels were installed in December 2004 at approximately 15–30 cm below the wall top on both sides of the 30 m-long test wall at Byland Abbey and have been monitored approximately monthly since then.

### *Test sites – second phase*

The second phase of soft capping was carried out from November 2004 to February 2005, and was designed to extend the range of monuments and environmental conditions covered in the study by using four different monuments in very different settings.

Howbury Moated Site in East London near Slade Green was soft capped in November 2004. Howbury provided us with an ideal opportunity to test the performance of soft capping on thin walls (all less than 40 cm wide) within the urban atmosphere of London where air pollution levels are likely to be higher than at the predominantly rural sites that make up the rest of the study. The climate faced by this site in the SE of England is also likely to be very different to that experienced by the sites in Yorkshire, with warmer, drier summers and less harsh winters. Four sections of the monument were soft capped. Two sections between 3.5 and 4 m long on stone and brickwork walls were capped with soil and turf from a garden centre. The turf sections, when completed, had a soft cap of around 7.5–10 cm in thickness. A further two sections, each 0.9–1.4 m long (one stone, one brick), were capped with soil and then seeded with British Seed Houses WFG2 mix at approximately 5 g m$^{-2}$. The seeded caps were about 5 cm deep at maximum.

At Hailes Abbey in Gloucestershire (OS Grid Ref SO 050 300) soft wall capping was carried out in January 2005 along a 17 m-long section of wall, 90–120 cm in width and roughly 3–4 m in height. Hailes Abbey is a late Cistercian monastery, founded in 1246 and built of local Cotswold limestone. Today, the ruins are suffering from high rates of decay in places, probably as a result of freeze–thaw weathering. The wall, which has been soft capped, had been previously covered by roofed scaffolding for well over a year, leading to drying out of the entire wall. Using turf from on site and standard soil, a soft wall capping some 10 cm thick has been established covering all of the previously scaffolded and roofed wall top. The roofing was removed when the soft capping work was completed.

Rievaulx Abbey (OS GR SE 577 849), founded by St Bernard of Clairvaux in the 12th century, became one of the wealthiest monasteries in England. Today, it has many severe decay problems, and much of the ashlar is covered with lichens and mosses. Some early soft wall capping trials had been carried out at Rievaulx on the South Transept. At Rievaulx the trifolium floor above two arcades within the nave of the main abbey church were capped, using turf from on site and standard soil. The bays are approximately 6–7 m above ground level with large arches below, as shown in Figure 4. The stonework

**Fig. 4.** Soft wall capping being installed at Rievaulx Abbey, February 2005.

around the bays is showing extreme flaking and blistering in patches, surrounded by less damaged areas with heavy moss and lichen coverage. Stonework covered with moss and lichens below one of the capped bays was brushed down to remove as much of the surface growths as possible.

Whitby Abbey (OS NZ 904 115) is located in a highly exposed, cliff-top position. Founded in 657, it was destroyed during the Viking invasion and rebuilt around 1220. At Whitby, a section of wall approximately 3–4 m in height and approximately 2 m long over an arched door was soft capped with turf cut on site and standard soil. The exposed setting here provides a very harsh test of the survival and performance of soft wall capping.

The four sites established during phase 2 of the project have been again monitored using repeat photography and visual observations, with wooden dowel measurement at Hailes Abbey (results not reported in this paper).

## Laboratory testing

In order to provide a wider range of testing regimes than can be created by field trials, a programme of laboratory testing has been designed. The basic methodology is an improved version of that trialled in 2001 (Viles *et al.* 2002). Two types of test are currently being carried out. First, using a programmable environmental cabinet within which humidity and temperature can be cycled, the thermal blanketing effect of different types of soft caps is being tested in order to compare this treatment with that of hard capping. The thermal blanket experiments have been run using the basic set-up shown in Figure 5. For each experiment the boxes (*c.* 25 × 25 × 25 cm in dimensions) containing hard and soft caps were placed into a Fisons environmental cabinet and the air temperature cycled over three–five daily cycles of air temperature. The cycle has been designed to simulate extreme conditions, with temperatures moving from 30 °C at the heat of the day down to −1.5 °C in the cool of the night. Secondly, three similar sized boxes, one containing a 5 cm deep soft cap overlying a stone slab, one containing a 10 cm soft cap overlying a stone slab and the final one containing a 5 cm hard cap constructed as at our field sites on a stone slab, have been exposed in Oxford and temperatures monitored to provide a link between the field and laboratory testing. Temperatures have been recorded every 15 min using thermistor temperature probes connected to Gemini Dataloggers TinyTag loggers. For these experiments stone slabs from the monitored ruins were used.

Laboratory testing has some advantages over field experiments, as more external factors can be controlled thus reproducing the amount of unexplained variability within the experimental setup and the impacts can be dramatically speeded up. However, laboratory experiments can also be seen to be less 'realistic' than field trials. For example, the size of soft wall caps in all dimensions but depth has had to be scaled down, and the testing regime used within the environmental cabinet is unrealistically harsh (each 24 h cycle heats the cabinet up to a hot summer's day and cools down to a cold winter's night). Combining laboratory testing and field trials in an integrated programme should provide a balanced overview of the short-term performance of soft capping.

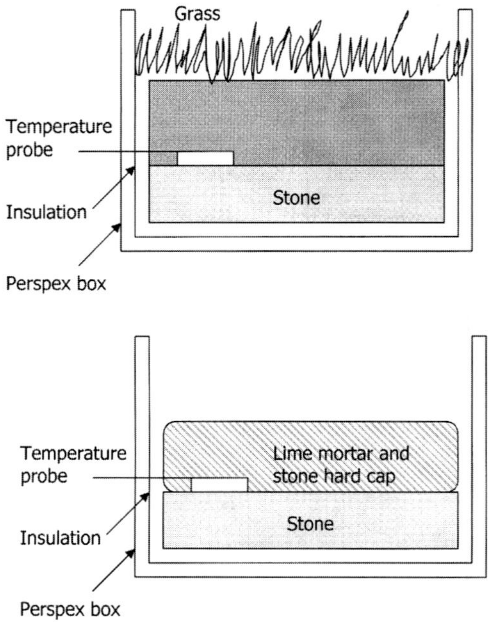


**Fig. 5.** Experimental set-up for the thermal blanket experiments.

## Results

Observations of the long-established (>10 years) soft caps at Byland, Kirkham and Thornton show their resilience and generally good state with healthy turf growth. For our test sites, we have been able to monitor the changing state of the soft wall caps from phase 1 of the project during the year after their establishment using simple photographic resurveys. Figure 6 illustrates the growth of one of the small soft caps at Byland over this period. As can be seen, the caps very quickly took on a natural outline as the turf established, with drying out of the sides of the turf during dry periods and recovery in wetter phases. Almost no

**Fig. 6.** Byland site 3 repeat photography to illustrate the changing state of the soft wall capping. (**a**) May 2004; (**b**) August 2004; (**c**) October 2004; and (**d**) February 2005. Scale is 20 cm high.

significant change in the overall shape of the cap was observed from May 2004 to February 2005. Botanical surveys carried out in July 2004 by John Thompson, Consultant Ecologist, 5 months after establishment of the trials showed that the turf at each site was composed of up to 10 common and widespread grasses. The mixture is typical of that expected at low altitudes on neutral soils in NE England. The principal species were Perennial Ryegrass, Yorkshire Fog, Common Bent and Red Fescue. According to UK Meteorological Office data 2004 was a rather normal rainfall year for the East and NE of England, with total rainfall of 866 mm (115% of the long-term average derived over the period 1961–1990). April, August and October were particularly wet months in comparison with long-term averages, whilst November and December were notably dry.

Soft wall caps at Kirkham Priory, Thornton Abbey and Hailes Abbey showed similar patterns of successful establishment and luxuriant growth. However, at Howbury Moated site, urban foxes damaged some of the soft wall capping early on and the seeding experiment was unsuccessful largely because of human interference, whilst at Rievaulx Abbey the turf has not become well established because the geometry of the building has prevented enough rain reaching the grass. Finally, at Whitby Abbey the soft wall cap, apart from one small area of turf that became partly detached by wind very early on, was seen to be in excellent condition given the extremely harsh weather conditions which the site experiences.

Temperature and soil moisture data from Byland in March 2005 and July 2005 are presented in Figures 7 and 8 as examples of the sort of data being produced. As can be seen, in March 2005 temperatures within the first few centimetres of the hard cap dipped below zero on several occasions, whilst those under soft caps of all thicknesses stayed above zero. Furthermore, the temperatures under soft caps showed much smaller diurnal ranges than those within the hard cap. Soil moisture data for March 2005 shows generally wet conditions within the soft caps, whilst a Watermark probe some 10 cm below the hard cap within

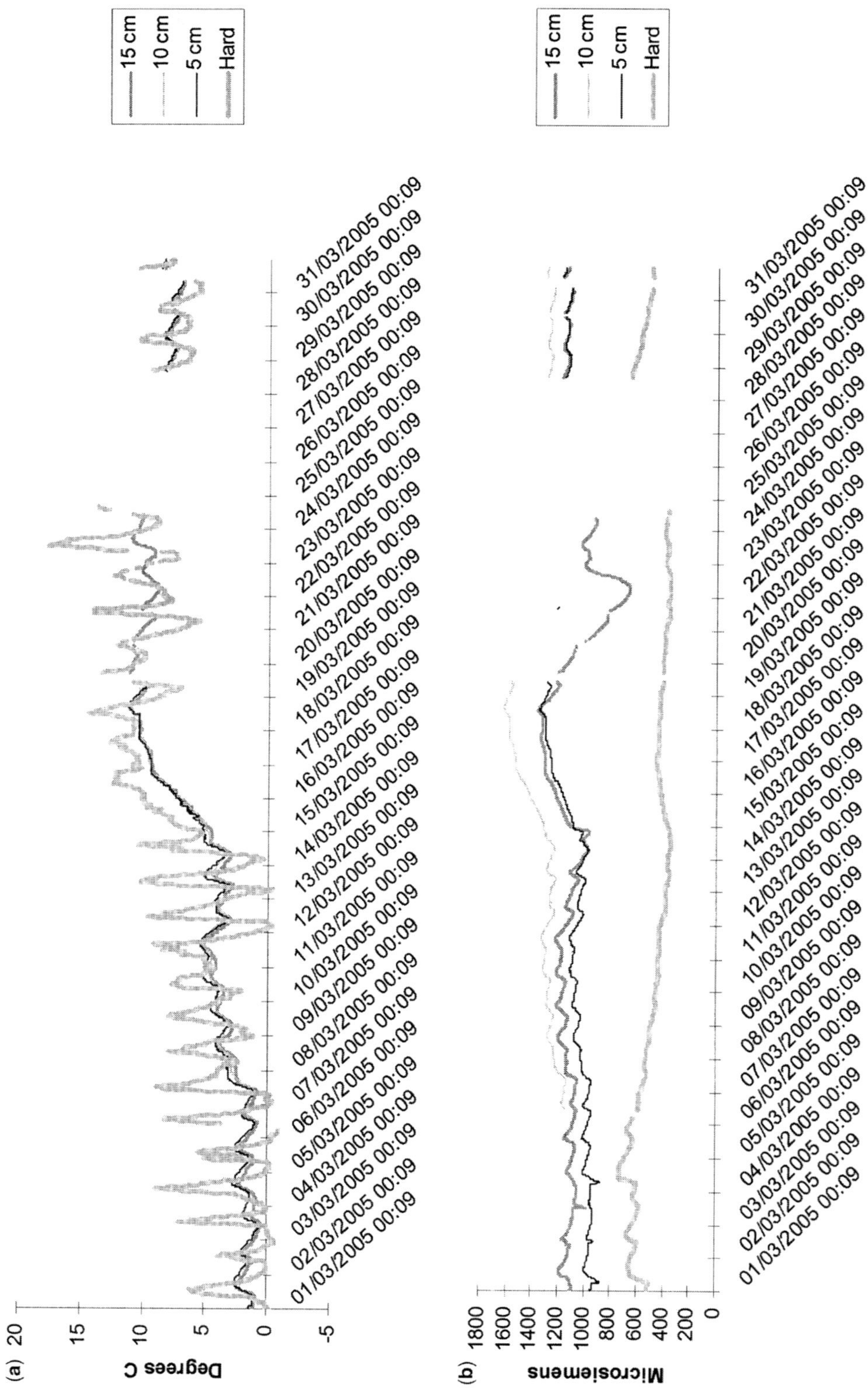


**Fig. 7.** (**a**) Temperature from the base of soft wall capping of different thicknesses and from just below the surface of the hard capping for March 2005; and (**b**) moisture levels from the base of soft wall capping of different thicknesses and from under the hard cap for March 2005.

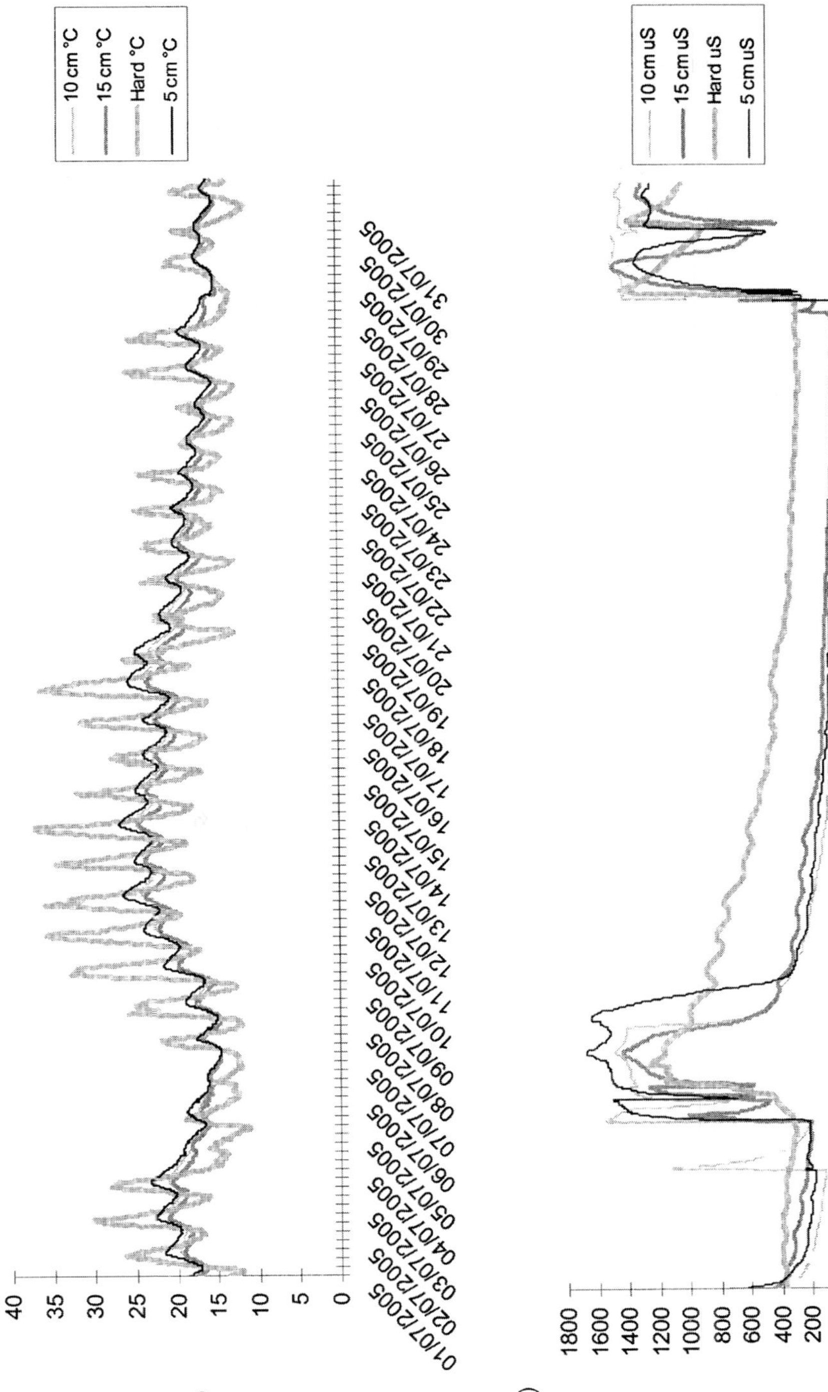


**Fig. 8.** (**a**) Temperature from the base of soft wall capping of different thicknesses and from just below the surface of the hard capping for July 2005; and (**b**) moisture levels from the base of soft wall capping of different thicknesses and from under the hard cap for July 2005.

the core of the wall recorded much drier conditions. During July, as shown in Figure 8a, the soft caps experienced much more muted temperature fluctuations than near the surface of the hard cap. Figure 8b illustrates the much drier soil conditions in the soft caps in comparison with March 2005, with periods of rainfall reflected in sudden rises in moisture levels. Also, moisture levels within the soft caps are now very similar to those in the core of the wall below the hard cap. These results illustrate the effective thermal blanketing provided by the soft wall capping in comparison with the hard cap, and the variable moisture conditions experienced within the soil.

Wooden dowel monitoring data from the same site at Byland for March and July 2005 (shown in Fig. 9) illustrate the variability between late winter and summer conditions in terms of the wetness of the walls and the spatial patchiness of wetness, probably caused by flaws in the mortar within sections of the ruined wall. There is no clear evidence from these 2 months of data that the soft capping is drying out the walls, but outliers in the data set caused by 'wet patches' in the wall may be complicating the situation. The stone used at Byland has a relatively high water absorption capacity of around 18%. Further research using more advanced measurement techniques is needed

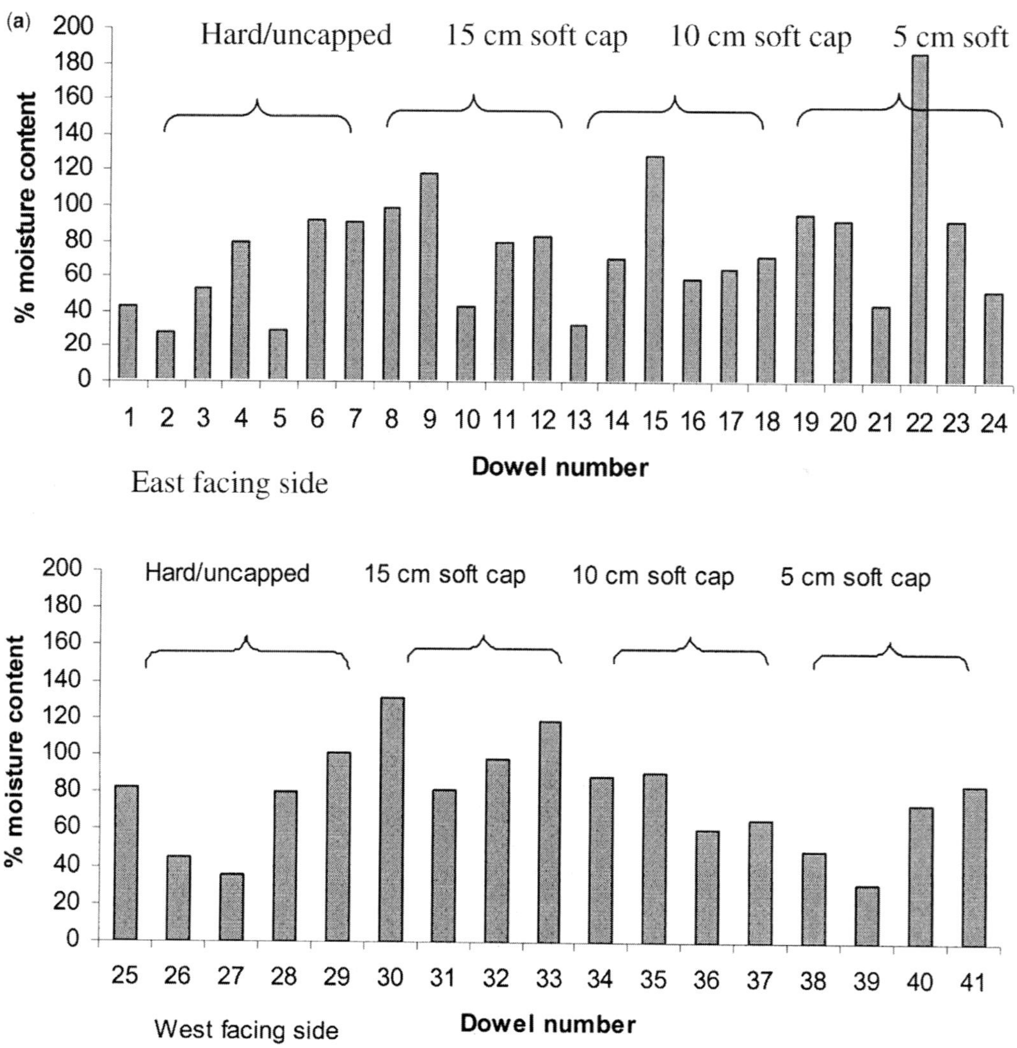


**Fig. 9.** Wooden dowel data from (**a**) March and (**b**) July 2005 for different sides of the wall underlying soft caps of varying thickness and hard capped/uncapped sections. Note that there are no results for dowels 7 and 15 in the July 2005 data set.

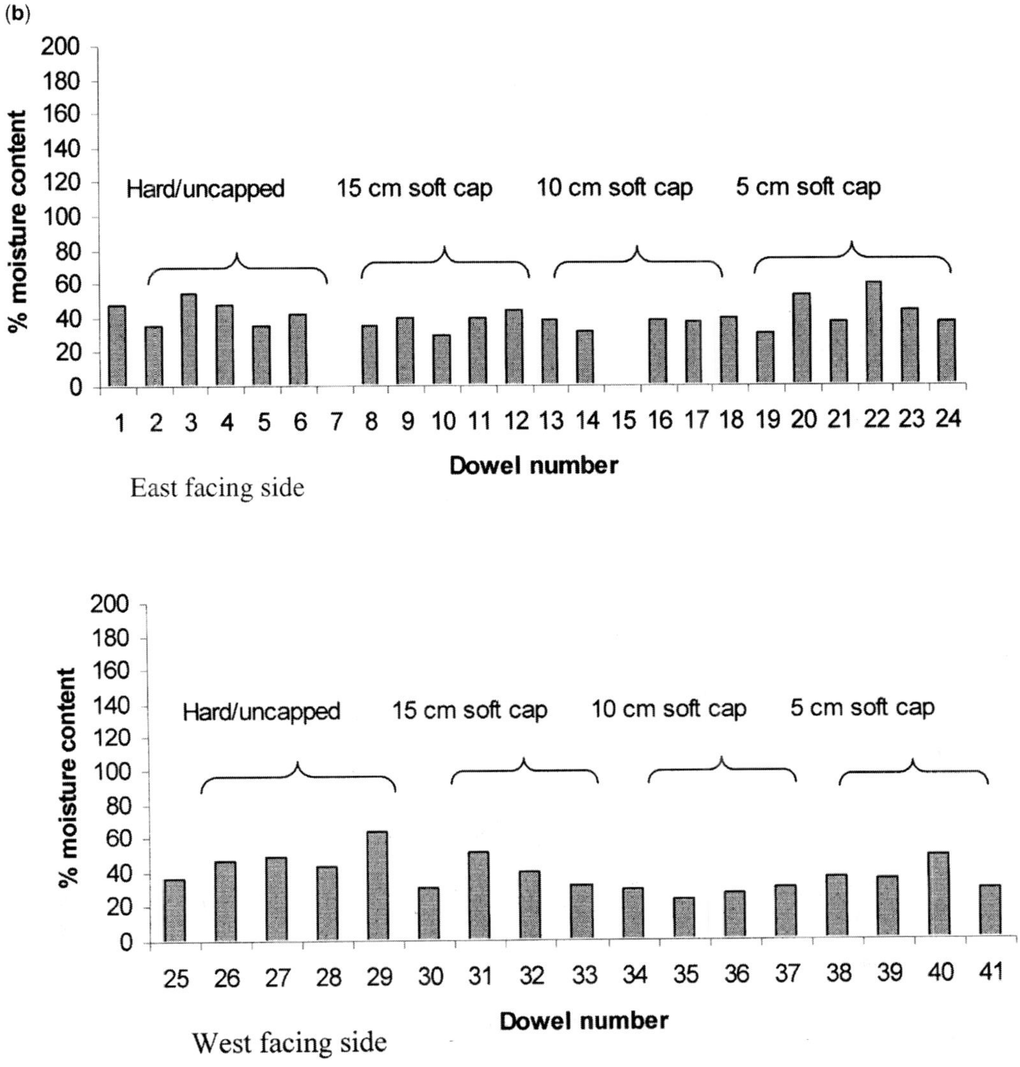


**Fig. 9.** *Continued.*

to clarify the exact role of soft wall capping in influencing moisture levels in the underlying walls, which can be affected by a wide range of variables including stone porosity and mineralogy, wall construction and microclimatic conditions.

Results from one thermal blanketing experiment in the laboratory are shown in Figure 10. These results show the differences in thermal response at the stone surface of a slab of Kirkham stone covered by a hard cap, a 5 cm-thick soft cap and another 5 cm-thick soil cap with slate fragments in the soil mix. As can be seen from Figure 10, the most obvious differences are in the maximum and minimum temperatures experienced by the stone under the hard v. soft caps. When air temperatures plummeted during the experimental cycle to $-1.5\,^{\circ}\mathrm{C}$, the temperatures beneath the hard cap dropped to only about $1\,^{\circ}\mathrm{C}$ and those under the soft caps stayed at around $7.5\,^{\circ}\mathrm{C}$. Similar muting of the temperatures was also observed at the top of the temperature curve, where a maximum air temperature of $30\,^{\circ}\mathrm{C}$ was reduced to $27\,^{\circ}\mathrm{C}$ under the hard cap and $20\,^{\circ}\mathrm{C}$ under the soft caps. Note also that the soft caps show a lagged response, with minimum temperatures experienced at the rock surface about 3–4 h after the minimum air temperatures. The slate fragments did not appear to have any significant effect on the thermal blanketing role of the soft cap.

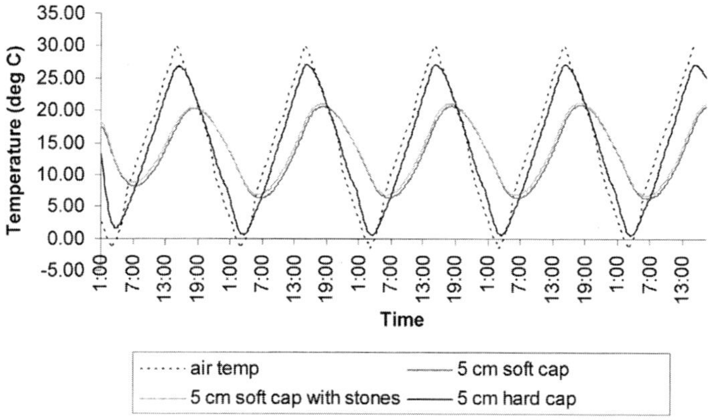


**Fig. 10.** Thermal blanket experiment results comparing a 5 cm-thick hard cap with a 5 cm soft cap and also a 5 cm-thick soft cap with stones included in the soil mix. The *x*-axis denotes time of day in hours.

Data from the boxes exposed to real climatic fluctuations within Oxford can be compared with both the laboratory test findings and the Byland field monitoring data (see Fig. 11 which shows data from March and July 2005). As with the laboratory testing, the data in Figure 11 shows that the soft caps are more effective than the hard cap in reducing temperature variations at the stone surface under both warm and cool conditions. The data in Figure 11a also indicate the occurrence of locally cold conditions (almost down to −5 °C air temperature) that are linked with freezing of the hard cap, but with soft cap temperatures still above zero showing that even the scaled-down soft caps used in the experiments are highly efficient thermal blankets. Figures 7 and 11a present data from Oxford and Byland for March 2005, which allows comparison of performance of the scaled-down boxes with that of real soft caps under similar conditions. Similar trends are visible, but with much less pronounced diurnal variations in the Byland data, indicating that larger areas of soft capping provide a more effective thermal blanket. Importantly, the two data sets show similar trends in terms of the behaviour of hard and soft caps around 0 °C, illustrating the effectiveness of even scaled-down soft caps in reducing the number of freeze–thaw cycles experienced at the stone surface. The July 2005 data sets also show the same general trends, but again with more pronounced and less lagged diurnal fluctuations in the Oxford experimental boxes.

## Discussion and conclusions

The research project has only just begun to yield results so it is too soon to expect to be able to draw definitive conclusions. Furthermore, the very nature of this work means that the next phase and its experimental design will be influenced by the results of the current phase of testing. Nonetheless, we have already found that soft wall capping is easy to establish and, under most conditions as long as there is enough rain and sunlight received by the turf, appears to grow quickly and copes well with periodic drying out. Furthermore, soft wall capping performs well in the short term in terms of thermal blanketing, in both laboratory and field situations. The role of soft wall capping in preventing moisture ingress is, as yet, less clearly established by our research. However, several strands of evidence suggest that the walls underneath soft wall capped sections are generally drier than those under uncapped and hard capped sections. Visual observations during rainfall at Kirkham Priory, for example, reveal that the soft caps are more effective at shedding water away from the wall face and preventing runoff down the face than hard capping. Laboratory experiments are currently being developed to investigate water penetration and water-holding characteristics of the soft wall caps under controlled conditions in order to provide more conclusive results. The experimental and monitoring methods used in this project have proven to be robust and successful, although some problems have been experienced with interpreting the photographic resurveys in detail because of variable conditions of lighting at different times of year. More detailed analyses will be carried out in the later phases of the project on both stone properties and local microclimatic conditions to help explain in more detail the nature and causes of moisture ingress to walls below soft wall capping.

The experiments on site are long term in nature and it may be a few years before any definite conclusions can be drawn from this work. Furthermore, issues over the aesthetic and visual acceptability of soft wall capping on both low and high wall heads will need evaluating to complement our scientific findings. Monitoring is also needed to investigate the long-term management requirements for soft

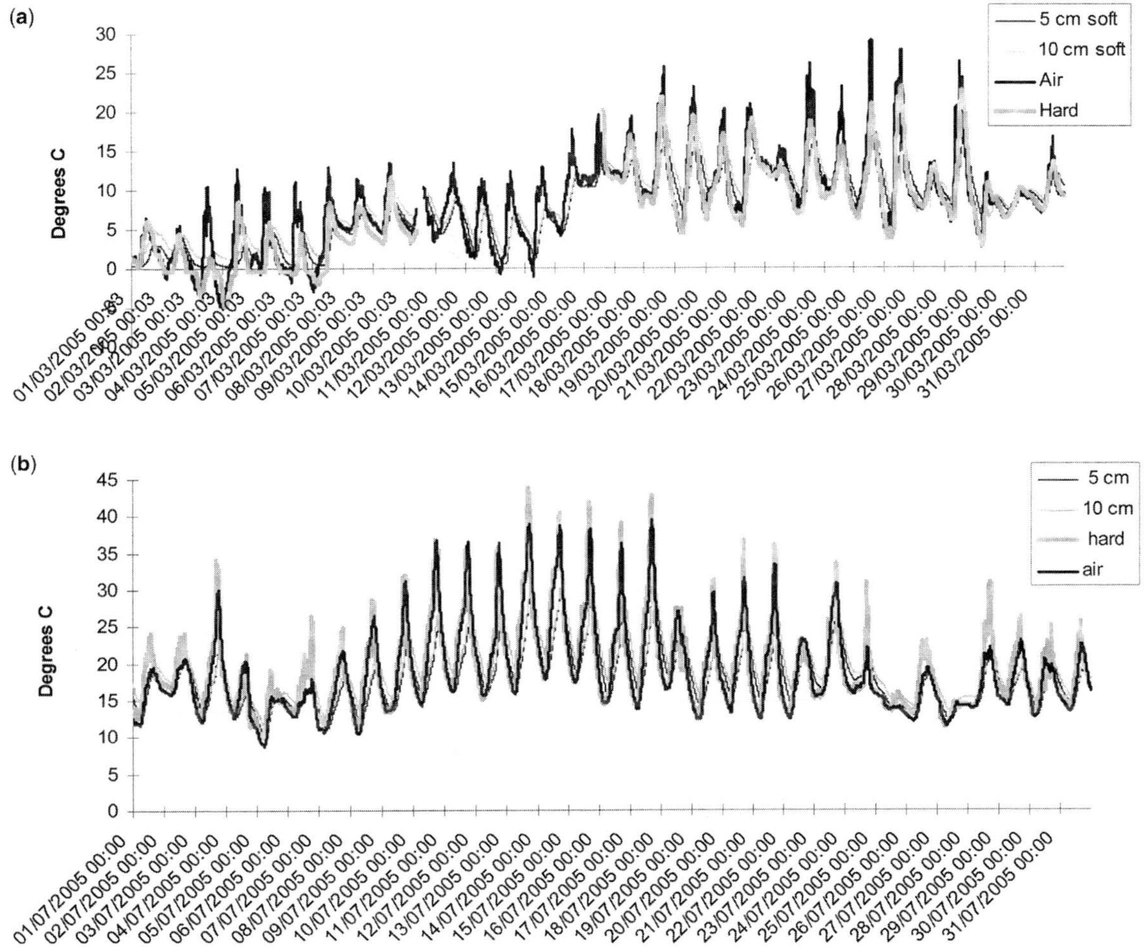


**Fig. 11.** Temperature data from experimental soft caps (5 and 10 cm thick) and experimental hard cap (5 cm thick) exposed outdoors in Oxford. (**a**) March 2005; and (**b**) July 2005.

wall capping, in the light of possible changes in flora and fauna as the capping develops. However, our integrated laboratory and field testing programme has already illustrated that, in the short term, soft wall capping can be an effective and simple technique for conserving ruined walls.

## References

LARSON, P. K. 2004. Moisture measurement in Tirsted Church. *Journal of Architectural Conservation*, **10**, 22–35.

OGILVY, R. I. K. 1996. *Observations on the Practice of Soft Topping of Walls: Historic Properties North Region, 1989–1996*. Unpublished Report to English Heritage.

TOLLEY, R., CHANNER, J, COPPOCK, G., THOMPSON, J. & WESTON, K. 2000. Wigmore Castle, Herefordshire, the repair of a major monument: An alternative approach. *Association for Studies in the Conservation of Historic Buildings Transactions*, **25**, 21–49.

VILES, H. A., WOOD, C. & GROVES, C. C. 2002. Soft wall capping experiments. *English Heritage Research Transactions*, **Stone 2**, 59–73.

# Index

**Note**: Page numbers in *italic* denote figures. Page numbers in **bold** denote tables.